Physics of Collisional Plasmas

Grenoble Sciences

The aim of Grenoble Sciences is twofold:

- to produce works corresponding to a clearly defined project, without the constraints of trends nor curriculum,
- to ensure the utmost scientific and pedagogic quality of the selected works: each project is selected by Grenoble Sciences with the help of anonymous referees. In order to optimize the work, the authors interact for a year (on average) with the members of a reading committee, whose names figure in the front pages of the work, which is then co-published with the most suitable publishing partner.

Contact: Tel.: (33) 476 514 695
E-mail: Grenoble.Sciences@ujf-grenoble.fr
website: http://grenoble-sciences.ujf-grenoble.fr

Scientific Director of Grenoble Sciences
Jean BORNAREL, Emeritus Professor
at the Joseph Fourier University, Grenoble, France

Grenoble Sciences is a department of the Joseph Fourier University, supported by the **French National Ministry for Higher Education and Research** and the **Rhône-Alpes Region**.

Physics of Collisional Plasmas is an improved version of the original book "**Physique des plasmas collisonnels**" by Michel MOISAN and Jacques PELLETIER, EDP Sciences - Grenoble Sciences' collection, 2006, ISBN 978 2 86883 822 3.

The Reading Committee of the French version included the following members:

- **Michel AUBÈS**, Professor at the Paul Sabatier University, Toulouse, France
- **Jacques DEROUARD**, Professor at the Joseph Fourier University, Grenoble, France
- **Anne LACOSTE**, Professor at the Joseph Fourier University, Grenoble, France and
- **Cédric DE VAULX** and **Didier RIEU**

Translation from original French version performed by **Greame LISTER**, Scientist at OSRAM Sylvania Inc.

Typesetted by: **Danielle KEROACK**, Ph.D Physics

Cover illustration: **Alice GIRAUD**
(with extracts from: Plasma light - glow_net, *flickr*)

Michel Moisan • Jacques Pelletier

Physics of Collisional Plasmas

Introduction to High-Frequency Discharges

Translation by Graeme Lister

Springer

Prof. Michel Moisan
Département de Physique
Université de Montréal
Montreal, Québec, Canada

Dr. Jacques Pelletier
Centre de Recherche
Plasmas-Matériaux-Nanostructures
LPSC
Grenoble, France

Translator
Dr. Graeme Lister
OSRAM Sylvania Inc.
Beverly, MA, USA

Original title: "Physique des plasmas collisionnels, application aux décharges haute fréquence", Michel Moisan et Jacques Pelletier, Collection Grenoble Sciences - EDP Sciences, 2006

ISBN 978-94-007-4557-5 ISBN 978-94-007-4558-2 (eBook)
DOI 10.1007/978-94-007-4558-2
Springer Dordrecht Heidelberg New York London

Library of Congress Control Number: 2012941402

Printed on acid-free paper

Springer is part of Springer Science+Business Media (www.springer.com)

Preface

During the 1960s, the public face of plasma physics was almost exclusively represented by plasma confinement, with the goal of developing a reactor to produce electricity by thermonuclear fusion. Such a reactor is still being developed, without any guarantee as to its successful achievement, but since then the applications of plasma physics have increased and diversified: one of the best known, besides lighting, is etching in the fabrication of microelectronic computer chips, for which plasma is indispensable. At present, the use of plasmas continues to expand and, from recent research publications, a seemingly limitless number of applications will eventually see the light of day. In this development, plasmas created by radiofrequency and microwave fields play a particularly important role.

The present text is basically concerned with plasma physics of interest for laboratory research and industrial applications, with emphasis on the understanding of the physical mechanisms involved, rather than on minute details and high-level theoretical analysis. At the introductory level to this discipline, it is very important to assimilate its characteristic physical phenomena, before addressing the ultimate formalism of kinetic theory, with its microscopic, statistical mechanics approach. In this textbook, the physical phenomena have been translated into more tractable equations, using the hydrodynamic model; this treats the plasma as a fluid, in which the macroscopic physical parameters are the statistical averages of the microscopic (individual) parameters. This textbook is intended for students in their early years at the graduate level, and for engineers who are interested in applications. Its level of difficulty lies somewhat below that of JL Delcroix and A Bers (from Université Paris XI, Orsay and Supélec, Gif-sur-Yvette, France, and MIT, Cambridge, MA, USA, respectively), which provides a series of complementary and interesting theoretical treatments.

This book is divided into four chapters.

Chapter 1 is the introductory part of the textbook. It begins with a description of the plasma, an ionised gas, as a collective and electrically neutral gaseous medium, followed, for illustrative purposes, by a few selected scientific and industrial applications. Then, the fundamental concepts of plasma physics are introduced, with progressively increasing detail: the chapter aims to present the basic parameters required to reach a starting knowledge of the plasma medium, such as the Debye length, the electron plasma frequency, the various types of collision between particles and their description through specific cross-sections. The concepts presented in this introduction will be developed as a first approach, i.e. as the first step in an iterative process, to be completed by the detailed and quantitative presentations in the remaining chapters.

Chapter 2 is a thorough examination of the trajectory of a single charged particle (assuming no interaction whatsoever with other particles), subject to an electric field $\boldsymbol{E}$, a magnetic field $\boldsymbol{B}$ or both. In the case of electric fields $\boldsymbol{E}$, special attention will be paid to those at RF and microwave frequencies, designated jointly as high-frequency (HF) fields, in preparation for the modelling of HF discharges developed in Chap. 4. The presence of a magnetic field $\boldsymbol{B}$ results in a cyclotron motion, encountered for example in electron cyclotron-resonance discharges (Chap. 4). The combination of $\boldsymbol{E}$ and $\boldsymbol{B}$ fields in different spatial configurations, and then the inclusion of the spatial inhomogeneity of the $\boldsymbol{B}$ field, reveals the so-called drift velocities, which have to be "tamed" for an efficient operation of Tokomaks, nowadays investigated as possible controlled-fusion reactors.

In contrast to Chap. 2, collisions between particles are taken into account in **Chapter 3**, to establish the hydrodynamic description of the plasma, considered as a fluid. Such a description is obtained from the macroscopic quantities calculated from the distribution function of the (microscopic) velocities of individual particles. The transport equations, i.e. the equations describing the space-time evolution of these quantities, are obtained from integration of the Boltzmann equation over the distribution function of velocities. The concepts of mobility (of charged particles) and diffusion of particles are then introduced, where free mobility and diffusion tensors are deduced from the (momentum transport) Langevin equation. Further, it is shown that, under sufficiently dense plasma conditions, the space-charge electric field makes electrons and ions diffuse together in the so-called ambipolar diffusion regime. Finally, toward the end of the chapter, a first example of a scaling law in plasmas is developed. Then, in the last section, the formation of sheaths located at the interface between the plasma and the walls is described, together with a straightforward and original derivation of the Bohm Criterion, which provides the velocity of the ions as they enter the sheath.

Chapter 4, the last chapter, is dedicated to the mechanisms involved in HF sustained discharges, which are developed based on an entirely new and original approach. The key element is θ_l, the average power lost by an electron through its collision with heavy particles, in this way supplying power to the plasma. It is shown that θ_a, the power taken on average per electron from the HF field, adjusts so that $\theta_l = \theta_a = \theta$, i.e. to compensate for the loss of charged particles. This implies, for instance, that the intensity of the

E field in the plasma is not set by the operator, but by this balance requirement. A further consequence is that, for given operating conditions and HF power density, whatever the means of supplying the HF field to achieve the discharge, the θ value should be the same in all cases. The parameter θ is also instrumental in demonstrating that, contrary to common belief, the **E**-field intensity goes through a minimum at electron cyclotron resonance. The influence of varying the field frequency on the EEDF, and ultimately on plasma properties, is documented both theoretically and experimentally, in the case of low-pressure (< 10 torr) plasmas. The case of high-pressure plasmas (including atmospheric pressure) is centred on the phenomena of discharge contraction and filamentation in rare gases with low thermal conductivity, emphasising the role of molecular ions in these monoatomic gas discharges. Interrupting the kinetic cycle leading to dissociative recombination (of molecular ions) by introducing traces of rare gases with an ionisation potential lower than that of the carrier gas leads to the disappearance of discharge contraction and filamentation.

In addition to the content of the main text, there are a large number of remarks and footnotes, for clarification, or to qualify certain points more precisely. Forty five problems, with detailed solutions, which are an indispensable complement to this book, are distributed at the end of the first three chapters. A set of Appendices provides clarifications of the subjects treated in the main text, together with a number of mathematical developments, and useful mathematical formulae. Finally, an alphabetic index of important terms is supplied, with a page reference to their first appearance in the text given in bold type.

Montréal and Grenoble, *Michel Moisan*
January 2012 *Jacques Pelletier*

Acknowledgements

The authors would first like to congratulate Graeme Lister on his excellent contribution in translating the content of our textbook, originally written in (non-basic) French, into (English) English. Born in Australia, Graeme Lister (Ph.D. on instabilities in plasmas) has been involved since the beginning in modelling plasmas, first for thermonuclear research, in prestigious institutes (Princeton, Garching) and, afterward for RF operated lamps, as a senior scientist in International companies (Thorn Lighting, Osram Sylvania). He is currently working for Ceravision Limited in the development of microwave HID lighting. Needless to say that he has an interest in HF discharges!

Another invaluable contribution to this book has been provided by Danielle Kéroack (Ph.D. in solid state physics) who edited the English manuscript (after having done the same with the French text), while also carrying out her activities in the laboratory and administrative duties in the Plasma-Québec strategic network. Her calculation from basic equations of many figures and her patience with Springer editorial guidelines are particularly noteworthy.

We are happy to acknowledge the contribution made by several people to the content of this book. We would first like to thank Professor Ana Lacoste (Master of plasma physics at Université Joseph-Fourier in Grenoble) at the origin of a number of developments, particularly in Chapter 4, and for providing some problems with their solutions. The content of Chapter 4 was originally developed by young researchers in plasma physics, Kremena Makasheva, Yassine Kabouzi and Eduardo Castaños Martínez as post-doctoral fellows and Ph.D. student in plasma physics respectively, particularly concerning the atmospheric-pressure HF discharges. The new results on this topic appearing in the English edition result from the Ph.D. research of Eduardo Castaños Martínez. The content of this book has also greatly benefited from the many questions and comments from students in the classroom and in the laboratory. Finally, we were lucky to have a group of kind consultants and readers. We need to mention the long-running advice provided by Dr. Antoine Royer, in commenting on the physics of the text and re-writing part of it. Finally, we are grateful to John Michael Pearson and Richard MacKenzie, both professors of physics at Université de Montréal, for answering our (almost daily) queries on the English language.

Contents

Symbols

Vectors are represented by letters in bold characters, $\boldsymbol{A}$. Tensors are also printed in bold: a tensor of order 2 is underlined once, $\underline{\boldsymbol{A}}$; a tensor of order 3, twice, $\underline{\underline{\boldsymbol{A}}}$

$\langle\,\rangle$	average taken over the velocity (or energy) distribution function of particles
a_0	radius of the first allowed electron orbit of the hydrogen atom
$\boldsymbol{B}$	magnetic induction
c	speed of light in vacuum
C_r	flux of particles reflected by a magnetic mirror
$\mathrm{d}\Omega$	elementary solid angle
$\mathrm{d}\boldsymbol{w}$	elementary volume in the velocity space, also noted d^3w: in cartesian coordinates $\mathrm{d}w_x\mathrm{d}w_y\mathrm{d}w_z$
$\boldsymbol{D}$	electric displacement (or induction) field
$\mathcal{D}(\boldsymbol{r},\boldsymbol{w})$	density of probability of presence
D_e, D_i	coefficient of free diffusion of electrons, ions
D_a	ambipolar diffusion coefficient
D_s	effective diffusion coefficient
e	absolute value of the elementary charge
$\hat{\mathbf{e}}_i$	base vector of the i axis of the coordinate frame
$\boldsymbol{E}$	electric field
$\boldsymbol{E}_D$	space-charge electric field
$\boldsymbol{E}_{Da}$	space-charge electric field under pure ambipolar diffusion conditions
$\boldsymbol{E}_{dm}$	electric field related to the magnetic drift
$\mathcal{E}_{\mathrm{kin}}$	kinetic energy
$\bar{\mathcal{E}}_{\mathrm{kin}}$	average kinetic energy
$\mathcal{E}_j$	energy of level j of an atom
f	frequency of a field, of a wave
f_{pe}, f_{pi}	eigen frequency of electrons, ions in a plasma
$f(\boldsymbol{r},\boldsymbol{w},t)$	velocity distribution function of particles
$\boldsymbol{F}$	force
$\boldsymbol{F}_{cd}$	centrifugal force related to magnetic drift curvature
$\boldsymbol{F}_L$	Lorentz's force

$\boldsymbol{F}_{Lm}$	magnetic part of the Lorentz's force
g_j	statistical weight (quantum degeneracy) of the jth level of an atom
h	Planck's constant
$\boldsymbol{H}$	magnetic field
$\underline{\boldsymbol{I}}$	unit tensor of order 2
$\mathcal{J}$	Jacobian (determinant of a coordinate transformation matrix)
$\boldsymbol{J}$	current density
$\boldsymbol{J}_c$	conduction current
$\boldsymbol{J}_p$	polarisation current
$\boldsymbol{k}$	wavenumber vector
k_B	Boltzmann's constant
k_{ij}	reaction coefficient
l_{es}, l_{is}	electron, ion sheath thickness
ℓ_x	free path between two successive collisions
ℓ	mean free path between two successive collisions
m_e	electron mass
m_i	ion mass
M	atom mass
$\underline{\boldsymbol{M}}$	tensor related to the magnetic force
n	plasma density
n_i, n_e	density of ions, electrons
n_g	plasma density at the sheath edge
N	density of molecules (atoms)
N_0	density of atoms at one torr pressure and 0°C
N_d	number of particles scattered elastically by a unique scattering centre
N_D	number of particles in a Debye sphere
N_n	density of nuclei
N	total number of particles in a system
$\mathcal{N}_0$	density of atoms in the ground state
$\mathcal{N}_j$	density of atoms in the jth excited level of an atom
$\boldsymbol{p}$	linear momentum vector
p	gas pressure
p_0	reduced gas-pressure
p_h	pitch of a helix
P_x	macroscopic cross-section for an interaction of type x
P_a	power absorbed by electrons, per unit volume, as averaged over a HF field period
P_t	total absorbed power
$\boldsymbol{\mathcal{P}}_\alpha$	total linear momentum gained or lost by particles of type α
q	charge of a particle
$\underline{\underline{\boldsymbol{Q}}}$	tensor of thermal flux energy
$\boldsymbol{r}$	position vector
r_B	Larmor's radius

r_{Be}, r_{Bi}	Larmor's radius of electrons, ions
$\boldsymbol{R}_g$	instantaneous position of the guiding centre
R	inner radius of the discharge tube
$\mathcal{R}$	mirror ratio
R	reactance
$\boldsymbol{\mathcal{R}}_\alpha$	total kinetic energy gained or lost by particles of type α
s	impact parameter
s_0	average critical impact parameter
$S(f)$	collision operator
t	time
T	temperature of a system under thermodynamic equilibrium
T_e, T_i	temperature of electrons, ions
T_{eV}	temperature in electron-volt
T_g	temperature of the neutral atom gas
$\mathcal{T}$	period of the HF field
$\mathcal{T}_c$	period of the HF field
$\boldsymbol{u}$	speed of a particle relative to the particle average speed
u_k	characteristic energy of electrons
u	energy of a particle
U	potential difference; can also refer to energy
U_{eV}	electron energy in electron-volt
$\boldsymbol{v}$	average speed in the hydrodynamic sense (Sect. 3.3)
v_B	Bohm's velocity
v_g, v_{ph}	group velocity, phase velocity of a wave
v_{th}	most probable velocity of a particle in a Maxwell-Boltzmann distribution
$\boldsymbol{w}_\alpha$	velocity of a particle α
$\boldsymbol{w}_{\alpha\beta}$	relative velocity between two particles, denoted α and β
$\boldsymbol{w}_{de}$	average drift velocity due to the electric field
$\boldsymbol{w}_{dc}$	average drift velocity (over a cyclotron period) due to magnetic field curvature
$\boldsymbol{w}_{dm}$	average drift velocity (over a cyclotron period) due to magnetic field
W	work
Z	positive charge(s) of the ion
α_i	ionisation degree
β	wavenumber
γ	thermodynamic adiabaticity ratio
$\boldsymbol{\Gamma}$	particle flux (number of particles incident on a unit surface, per second)
δ	energy transfer coefficient following an elastic collision
δ_c	characteristic penetration depth of a HF field
ϵ_0	vacuum permittivity
ϵ_p	electric permittivity of the plasma relative to that in vacuum

η	saturation coefficient of a relay state in a multi-step ionisation process
η_v	viscosity coefficient of the fluid
θ	power absorbed per electron; also designates the polar angle
θ_a	average power absorbed per electron
θ_l	average power lost per electron
κ	thermal conductivity of the gas
λ	wavelength
λ_D	Debye's length
$\lambda_{De}, \lambda_{Di}$	Debye's length for electrons, ions
$\varLambda$	characteristic diffusion length
μ	orbital magnetic moment
μ_e, μ_i	mobility of electrons, ions
μ_0	magnetic permeability of vacuum
$\mu_{\alpha\beta}$	reduced mass of particles of type α and β
ν	average collision frequency for momentum transfer
$\nu_i, \bar{\nu}_i$	average ionisation frequency
ν_{id}	single-collision ionisation frequency
ν_{ie}	step-wise ionisation frequency
ν_0	photon frequency
ν_r	average frequency for volume recombination
ν_{ar}	three-body recombination frequency
ν_{mr}	dissociative recombination frequency
ρ	charge density
ρ_c	radius of magnetic curvature
ρ_{ie}	step-wise ionisation coefficient
$\hat{\sigma}$	microscopic differential cross-section
$\hat{\sigma}_t$	all-angle integrated (or total) microscopic cross-section
$\hat{\sigma}_{tc}$	total microscopic cross-section simply accounting for the number of collisions
$\hat{\sigma}_{tm}$	total microscopic cross-section for momentum transfer
σ	electric conductivity
τ	characteristic time
$\varUpsilon(\boldsymbol{r}, \boldsymbol{w}, t)$	any given microscopic parameter
$\phi(r)$	electric potential
ϕ_0	applied potential
ϕ_g	plasma potential at the sheath edge
ϕ_p	plasma potential
φ	azimuthal angle
$\varPhi(r)$	electric potential energy
$\underline{\boldsymbol{\varPsi}}$	kinetic pressure tensor
ω	angular frequency of an alternating electric field
ω_c	cyclotron angular frequency
ω_{ce}, ω_{ci}	cyclotron angular frequency of electrons, ions
ω_{pe}, ω_{pi}	plasma angular frequency of electrons, ions

Acronyms

AC	Alternating current
CM	Centre of mass
DC	DC Direct current (e.g., DC discharges)
ECR	Electron cyclotron resonance
EEDF	Electron energy distribution function
EM	Electromagnetic
HF	High frequencies
ICP	Inductively coupled plasmas
ISM	Industrial, scientific and medical (frequency)
LTE	Local thermodynamic equilibrium
MW	Microwaves
RF	Radiofrequencies
SI	International system
TE	Thermodynamic equilibrium
UV	Ultraviolet

Physical constants

Electron mass	$m_e = 9.10938 \times 10^{-31}\,\mathrm{kg}$
Unsigned charge on a electron	$e = 1.60219 \times 10^{-19}\,\mathrm{C}$
e/m_e ratio	$e/m_e = 1.75882 \times 10^{11}\,\mathrm{C\,kg^{-1}}$
Hydrogen atomic mass	$M_{\mathrm{H}} = 1.67372 \times 10^{-27}\,\mathrm{kg}$
Helium atomic mass	$M_{\mathrm{He}} = 6.64648 \times 10^{-27}\,\mathrm{kg}$
Vacuum electric permittivity	$\epsilon_0 = 8.85419 \times 10^{-12}\,\mathrm{F\,m^{-1}}$
Vacuum magnetic permeability	$\mu_0 = 4\pi \times 10^{-7}\,\mathrm{H\,m^{-1}}$
Avogadro's number	$N_A = 6.02214 \times 10^{26}\,\mathrm{kg^{-1}\,mole^{-1}}$
Lochsmidt's number	$N_L = 2.68678 \times 10^{25}\,\mathrm{m^{-3}}$
Stefan-Boltzmann's constant	$\sigma_{SB} = 0.56704 \times 10^{-7}\,\mathrm{W\,m^{-2}\,K^{-4}}$
Boltzmann's constant	$k_B = 1.38066 \times 10^{-23}\ \mathrm{J\ K^{-1}}$
Planck's constant	$h = 6.62607 \times 10^{-34}\,\mathrm{J\,s}$
	$\hbar = h/2\pi = 1.05457 \times 10^{-34}\,\mathrm{J\,s}$

(Source: NIST, United States)

Other constants

Ion mobility of He^+ in He under standard conditions (760 torr (101.350 kPa), 273 K)	$\mu_i = 10.4 \times 10^{-4}\,\mathrm{m^2\,V^{-1}\,s^{-1}}$
Approximate average frequency for momentum transfer through electron-neutral collisions in helium at the reduced pressure p_0	$\nu/p_0 = 2.4 \times 10^{9}\,\mathrm{s^{-1}}$
Molecular density at 1 torr and 0°C	3.53×10^{22} molecules $\mathrm{m^{-3}}$

Chapter 1
The Plasma State: Definition and Orders of Magnitude of Principal Quantities

1.1 Definition and essential nature of plasma

Plasma is a medium composed of electrons and ions, free to move in all spatial directions; this gaseous medium is distinguished from a classical gas, composed exclusively of electrically neutral particles, by the nature of the interaction between charged particles.

In a classical gas, the interaction between electrically neutral particles is short range and, provided the gas pressure is not greatly in excess of an atmosphere, the interactions generally involve only two particles (binary interactions). In this case, if two particles are travelling toward each other and separated by a distance r, the interaction is at first attractive (proportional to $1/r^7$, referred to as the Van der Waals force) then, immediately before "contact", this force abruptly changes to repulsive (sometimes represented by a force $\sim 1/r^{13}$, Sect. 1.7.9)[1]. To the contrary, the interaction between charged particles (attractive or repulsive, depending on the charges involved) is *long range*, because the Coulomb force between particles is $\sim 1/r^2$ (Sect. 1.7.1), which implies that each charged particle interacts simultaneously with a large number of other charged particles. As a result:

1.1.1 A plasma behaves as a collective medium

Consider, as an example, a plasma in which the particles are, to a first approximation, stationary (very small thermal motion) and suppose that the ions and electrons do not recombine to form neutral atoms: the result is a stationary state where, spatially, the positive and negative charges are distributed

[1] This interaction is often described in a simplified fashion as a collision between "billiard balls", neglecting the (initial) attractive phase of the interaction.

M. Moisan, J. Pelletier, *Physics of Collisional Plasmas*,
DOI 10.1007/978-94-007-4558-2_1,

alternately and almost uniformly; the charge distribution in two dimensions is illustrated very schematically in Fig. 1.1.

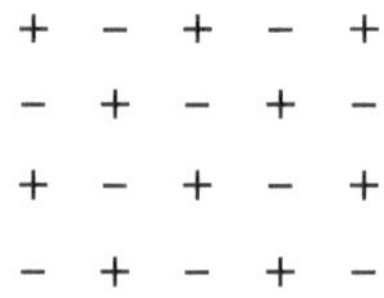

Fig. 1.1 (Highly) idealised spatial distribution of positive and negative charges, in the case where the particles in the plasma are (almost) at rest.

A uniform distribution of charges implies, in particular, that there is no significant local variation in the electric field intensity. However, if a hypothetical perturbation occurs, which displaces only one charge, all the neighbouring charges will move to compensate the local deviation from equilibrium thus created. This demonstrates that a plasma consists of particles which may behave collectively.

1.1.2 A plasma is a macroscopically neutral medium

Consider a given volume of plasma. The charged particles are moving randomly (thermal motion) but, the Coulomb forces they exert on each other, may induce relative displacements, such as to create a significant change in the local charge distribution: the (average) displacement of the particles increases with thermal energy, but decreases with the density of charged particles. From Poisson's equation[2]:

$$\nabla \cdot \boldsymbol{E} = \rho/\epsilon_0 \, , \tag{1.1}$$

where $\boldsymbol{E}$ is the (local) electric field intensity, ρ is the net (local) density of charges (positive or negative), and ϵ_0 is the permittivity of vacuum; the intensity of $\boldsymbol{E}$ increases as ρ increases[3], and consequently the neutrality restoring forces induced by the charge separation become important[4]. For this reason, and provided the volume of the plasma considered is much greater than the typical distance between the particles, the volume will contain, statistically, an equal number of positive and negative charges. The (average) maximum distance of non-neutrality is called the Debye length, designated by λ_D; the dependence of λ_D on charge density and average (thermal) energy will be considered in Sect. 1.6. We can therefore assert that a plasma contained in

---

[2] This is a variation of Maxwell's equation, $\nabla \cdot \boldsymbol{D} = \rho$, where $\boldsymbol{D}$ is the vector displacement (electric induction).

[3] Equation (1.1) in one dimension leads to $E = \rho x/\epsilon_0$.

[4] For a given ρ, the restoring force $|eE|$ increases with x, the distance to the neutrality position ($x = 0$).

a volume V which is much greater than the Debye sphere, $\frac{4}{3}\pi\lambda_D^3$, is macroscopically neutral.

In general, we can say that a plasma is a quasi-neutral medium (that is, neutral in a volume much greater than a Debye sphere) and, in fact, we can write $n = n_e = n_i$, where n is the plasma density, and n_e and n_i represent the electron and ion density respectively, assuming that the ions are singly and positively charged.

1.1.3 First examples of plasmas

Before continuing, consider, as well-known examples, two very different types of plasma:

- the Sun: this is a completely ionised medium, where there are no electrically neutral atoms; at the centre of the sun, the atoms have even lost all their electrons. Astrophysicists have shown that 99.9% of the (visible) material in the Universe is in the form of plasma, which is thus the most common state of matter.
- the light emitting region of a fluorescent lamp: the bulb is filled with a rare gas (typically argon) at about 3 torr ($\approx$ 400 Pa)[5] and a small drop of mercury with a partial vapour pressure of the order of a few mtorr at the operating temperature of about 40°C. An electric field (typically at 50 or 60 Hz AC) of sufficient strength is applied to the gas with the aid of two electrodes mounted as shown in Fig. 1.2, which renders the gas electrically conducting, producing what is called an *electric discharge* in the gas; part of this discharge emits light. In the case of a fluorescent lamp, the principal emission is *UV* radiation from mercury atoms (Hg I 254 nm line), which is converted to visible light by a phosphor deposited on the tube wall. The gas, in this case, is only partially ionised, and "cold" ($\approx$ 400 K), whereas it is "hot" in the case of a star.

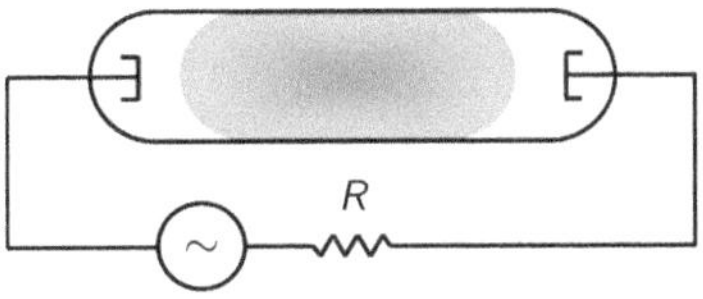

Fig. 1.2 Schematic showing the principle of an electric discharge with an alternating current as, for example, in the tube of a fluorescent lamp. For alternating current, *R* is a reactance (a resistance for continuous current), which is necessary for a stable discharge.

---

[5] The torr is a practical unit of pressure, used in a large number of experimental measurements, while the corresponding unit in the International System (SI) is the pascal (1 torr $\approx$ 133 Pa). The advent of pressure gauges giving the value in pascal should, in future, result in the disappearance of the torr.

General remarks:

1. Terminology: the difference between *ionised gas* and *plasma*. The majority of laboratory discharges are not really plasmas, because they contain not only charged particles, but also electrically neutral atoms and molecules, creating, more correctly, an *ionised gas*. Strictly speaking, the name *plasma* should be reserved for a gas that only contains charged particles, but in practice, the two terms plasma and ionised gas are often confused; the term *cold* plasma necessarily refers to an ionised gas.
The difference between a plasma and an ionised gas may be characterised by defining the *degree of ionisation* α_i of the medium,
$$\alpha_i = \frac{n_i}{n_i + N} , \tag{1.2}$$
where N is the density of the electrically neutral molecules (atoms). For $\alpha_i < 10^{-4}$, one should really refer to an ionised gas, rather than a plasma because, in this case, the majority of interactions are electron-neutral collisions, which are short range. However, even in such a situation, it is possible to propagate an electromagnetic wave (EM) through the (few) charged particles, but the attenuation is, in this case, linked to the electron-neutral collisions, rather than the Coulomb interactions.
2. Plasma is the fourth state of matter. The sequence "solid-liquid-gas-plasma" corresponds to increasing average energy of the constituents, plasma being the state with the highest energy. Thus, if the average energy of the electrons reaches 5 to 10% of the ionisation energy level of the atoms (molecules) (Sect. 1.7.9), an ionised gas is formed, but only partially; when the average energy is close to, or exceeds, the ionisation energy, the gas approches complete ionisation. In the laboratory, this "heating" is obtained by means of an electric field, or by photons.
3. Plasmas, radiating media. A plasma is a *thermodynamic system* which (Sect. 1.4.2), in addition to charged particles (and electrically neutral atoms, in the case of an ionised gas), contains photons which are emitted and absorbed by the particles.
It should be noted, however, that a medium need not be a plasma or an ionised gas to emit photons: it is sufficient for the atoms to be excited without being ionised.
4. Presence of *negative ions*. In addition to positive ions, with charge Ze where e is the absolute value of the elementary charge on an electron, many gas discharges (particularly in those gases referred to as *electro-negative*, for example SF_6) contain negative ions (with a single negative charge, for example H^-, O^-, O_2^-, Cl^-, SF_x^-), which are formed by the capture of an electron by an atom. Nevertheless, one always has quasi-neutrality, such that:
$$-(n_e e + n_{i-} e) + \sum_z n_z Z e = 0 , \tag{1.3}$$

where n_z is the density of positive ions of charge Ze (so called multi-charged ions) and n_{i-} the density of negative ions, with charge $-e$.
It should also be noted that some plasmas, such as those containing nitrogen, mercury and rare gases, do not contain negative ions.

5. Origin of the term "plasma". This term was first introduced into the literature by Tonks and Langmuir in 1929 to describe the "positive column" (Chap. 4) of certain electric gas discharges. Taken from the Greek $\pi\lambda\alpha\sigma\mu\alpha$, this word means "modelled shape" (for example of wax or clay), but can also mean fiction, false appearance! The connection between the etymological sense and the physical phenomenon that it describes is rather slim.

1.2 Areas of research and applications (examples)

Although most research work in plasma physics has been motivated by applications, this discipline, due to the large variety of observable phenomena in plasmas, has made important contributions to a number of different areas of fundamental physics, for example non-linear effects.

Plasma physics is a field that calls for electromagnetism, hydrodynamics, statistical mechanics, and atomic and molecular physics. In order to provide an overview of the vast domain of plasma physics, we will examine some of its branches, with emphasis on applications.

1.2.1 Controlled thermonuclear fusion

In the hope, in the future, of replacing energy currently produced by fossil fuels as well as by nuclear fission reactors, fusion reactions of the type:

$$\mathrm{D} + \mathrm{T} \rightarrow {}^4\mathrm{He} + \mathrm{neutron} + 17.6\,\mathrm{MeV}^6 ,$$
$$\mathrm{D} + \mathrm{D} \rightarrow \mathrm{T} + \mathrm{proton} + 4.0\,\mathrm{MeV} ,$$

have been considered, where deuterium (D) and tritium (T) are isotopes of hydrogen. Theoretically, 1 kg of D-T could provide the same energy as 10^7 litres of fuel oil. These reactions are possible if the nuclei of deuterium and tritium can come sufficiently closely in "contact"; this requires minimum incident energies of 10 keV to overcome the repulsive Coulomb potential between the positive charges. Two methods of heating and confinement are currently being pursued: *magnetic confinement*, which is close to a hypothetical reactor that could be coupled to an electricity network, and *inertial confinement*, which is currently at a fundamental study stage, and which uses

[6] $1\,\mathrm{MeV} = 10^6 \times 1.6 \times 10^{-19}\,\mathrm{J}$ (see Sect. 1.4.2 for details).

a completely different approach. At present, in both cases, it has not been possible to obtain a positive balance from fusion (energy produced greater than the energy required to achieve the reactions), since the loss phenomena have yet to be controlled. Let us briefly examine the two approaches:

- Magnetic confinement devices
  The confinement of charged particles in a magnetic field (Sect. 2.2) is essential to avoid the energy losses to the walls, and their destruction[7]. The type of reactor generally used has a toroidal configuration (forming a system closed on itself), introduced at the Kurchatov Institute in Moscow, under the direction of Academician L.A. Artsimovitch, and is called a tokamak[8]. It consists of a main magnetic field, referred to as toroidal, and a number of other magnetic fields of less intensity (further details will be given at the end of Chap. 2). The plasma is initially heated by induction, using the principle of the transformer, the secondary being the plasma. Extra current and energy are added to the plasma, for example, by high frequency (HF) fields corresponding to the normal modes of the system (for example cyclotron resonance) or to plasma waves. However, the impurities emerging from the walls as a result of their bombardment by particles from the plasma absorb a large fraction of the energy required to overcome the repulsive nuclear potential between the elements, preventing fusion reactions to continue; this problem has yet to be completely resolved. Further, several types of instabilities may occur, which lead to the plasma being "extinguished" or to touch the walls.
  Commencing in the early 1950s by the military, part of the research on fusion was made public in 1958, which led to important civilian financial investments by a number of countries. However, towards the middle of the 1990s, some governments became more critical with regard to this work, and reduced their budgets (the Tokamak in Varennes, Québec, was shut down by the Canadian Government, for example), arguing that it was still too far from a commercial reactor; in fact, at the time of writing (2010), the condition of self-maintenance in such reactors has yet to be achieved. Nevertheless, this research has continued in several installations in Europe, including the Joint European Torus (JET) in Culham, England, and Tore Supra in Cadarache, France. The main purpose of JET is to study transport instabilities, while Tore Supra has utilised super-conducting coils, which allows the machine to operate with increased toroidal magnetic field intensity, while minimizing ohmic losses. These various studies have led to the ITER Project, a large-scale tokamak using super-conducting coils and financed by the international community (Fig. 1.3). This installation should begin operation in Cadarache in 2019.

[7] The idea of magnetic confinement was first suggested by A. D. Sakharov and I. E. Tamm.

[8] Russian acronym for a toroidal chamber and magnetic coils: ТОРОИДАЛЬНАЯ КАМЕРА and МАГНИТНАЯ КАТУШКА.

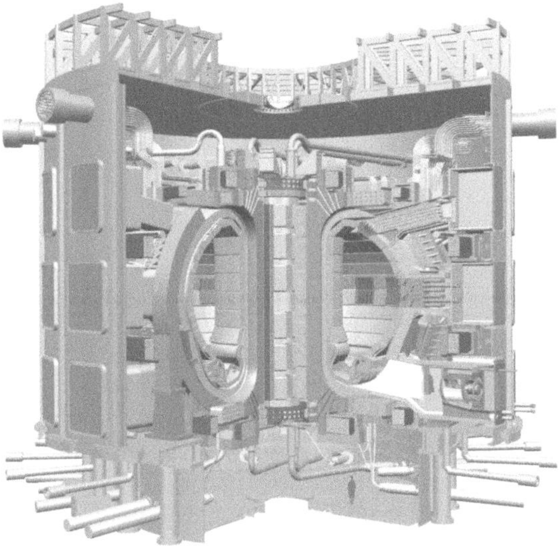

Fig. 1.3 Schematic view, in section, of the ITER reactor. The minor horizontal and vertical radii are 2 m and 3.7 m respectively, whereas on JET they are only 1.25 m and 2.10 m respectively. The major radius of ITER is 6.2 m, compared to 2.96 m for JET. The electric power required for continuous operation is 110 MW (ITER EDA Documentation Series No. 24, published by IAEA, Vienna, 2002).

- Inertial confinement systems
  The principle of inertial confinement is to fire an intense UV laser beam into a deuterium pellet, "peeling" it and inciting the compression of the extracted material towards the centre of the pellet; in order to generate fusion, the transfer of energy to the material must be faster than the subsequent expansion in the reactor chamber, which requires a high power, short pulse laser.

1.2.2 Astrophysics and environmental physics

Stars and the flux of plasma emitted by the sun, called the *solar wind*, are two distinct examples of plasma (in the strict sense), the first being extremely dense, the second, to the contrary, very dilute, and effectively collisionless.

Closer to the surface of the Earth, the *ionospheric layers* are ionised by the solar wind. The charged particles in these layers (the F layer, for example: $n_e \approx 5 \times 10^6\,\mathrm{cm}^{-3}$, $T_{eV} = 50\,\mathrm{eV}$, where T_{eV} is the temperature of the electrons in electron-volt) are confined by the Earth magnetic field, which forces them to oscillate between the Earth's two poles. These ionospheric layers play an important role in the transmission of low frequency waves ($f \leq$ 20–30 MHz). Effectively, they serve as mirrors for these waves, allowing them to be transmitted from one point to another around the world; on the contrary, for higher frequencies, there is no reflection, and the waves "travel" in a straight line, and so it is necessary for the emitting and receiving antennae to be opposite each other to establish communication (for example, Earth-satellite communication). A wave will reflect from the ionospheric layer if its frequency f satisfies $f < f_{pe}$, where f_{pe} is the electron plasma frequency (Sect. 1.5), a characteristic frequency for electrons in the gas. Thus, for the ionospheric F layer, where $n_e \approx 10^5$–$10^6\,\mathrm{cm}^{-3}$, $f_{pe} = 2.8$–$9\,\mathrm{MHz}$.

Still in the field of communication, it is also interesting to consider the effects of a thermonuclear explosion in the upper atmosphere, which would produce a very high density plasma, preventing communication from frequencies of a few MHz up to very high frequencies, notably communication with satellites ($\approx$ 4–12 GHz); such a plasma, because of the electromagnetic (EM) energy generated, could completely destroy communication systems. In the same way, this phenomena of reflection or opacity of waves was the origin of the loss of radio contact with the first space capsule, at the moment it returned to the terrestrial atmosphere: heating of the vehicle, by friction with the ambient air (even though the air density is extremely small at that altitude) resulted in the formation of a plasma surrounding it.

1.2.3 Laser pumping

One of the necessary conditions for obtaining the laser effect is that the density of atoms in the upper energy state of the radiative transition should be greater than that of the lower state, the opposite situation to that occurring in thermodynamic equilibrium (Sect. 1.4.2). In order to achieve this *population inversion*, one can irradiate the atoms with an intense luminous source (optical pumping; for example, by a lamp radiating in the UV), as well as using the properties of the gaseous plasma containing the atomic or molecular radiators (plasma pumping). The He-Ne laser is an example of laser pumping by a plasma: the helium and neon atoms are excited by electron collisions in the He-Ne discharge; this leads to an energy transfer from one excited level of helium to an excited level of neon at almost the same energy (referred to as resonance transfer), the neon level corresponding to the upper level of a radiative transition for laser emission, for example at 632.8 nm. This transfer of energy is particularly efficient because the excited state of helium that feeds the corresponding level in neon is a metastable state, i.e. it has a much longer lifetime than a radiative state, which makes it more susceptible to transfer its internal energy directly to another atom (molecule).

1.2.4 Plasma chemistry

Recall that electrons play a dominant role in the formation or rupture of chemical bonds. In an electric discharge at reduced gas pressure (meaning below atmospheric pressure), one generally finds[9] that $T_e \gg T_i \geq T_g$, where T_e, T_i and T_g are the electron, ion and gas temperatures[10], respectively. Thus, one can give sufficient energy to the electrons, which favour chemical reactions, without the need to heat the ions and atoms from which, in principle, one can obtain an energy efficiency and reaction yield superior to that produced by conventional chemistry in thermal equilibrium (Sect. 1.4.2).

A particularly convincing example of this chemistry in a non-equilibrium plasma is the formation of ozone from O_2, in discharges referred to as corona or dielectric barrier discharges at high-pressure. These discharges have the property of being cold, that is the atoms and molecules are at ambient temperature while T_e is several eV. This is an energetically efficient method, used throughout the world for treatment in waste-water plants, ozone being a strong oxidant and destructive to bacteria.

One can also use an electric discharge to destroy the effluent emitted from industrial processes, atoms and molecules that are toxic for people, dangerous for the ozone layer or contributing to the greenhouse effect. After these particles have been passed through a discharge formed in a different gas (referred to as the carrier gas), or even creating a discharge directly with the molecules to be destroyed, it is possible in some cases to obtain a destruction efficiency or *detoxification* close to 100%: these processes are fast and often less expensive than conventional techniques, such as in very high temperature furnaces which, in addition, contribute to pollution of the environment. These developments have led to the use of microwave plasma systems[11] which can eliminate effluent gases, notably the (per)fluorides (SF_6, CF_6, C_2F_6....) produced by the microelectronics industry. The same non equilibrium technique can be used to purify rare gases such as xenon and krypton, obtained beforehand by cryogenic distillation of air, removing in particular fluoride impurities (e.g. CF_4) and hydrocarbons (e.g. CH_4) originating in the environment and having condensation temperatures close to those of krypton and xenon.

---

[9] The electric field in the discharge principally accelerates the electrons, because of their very small inertia compared to that of ions: thus the energy "enters" the discharge through the electrons (exercise 2.1). Since, in addition, the transfer of energy between electron-neutral and electron-ion during a collision is very small (Sect. 1.7.2), because of the mass ratio (in contrast to ion-neutral and ion-ion collisions) and under the condition that the number of these electron collisions is not very high, one obtains $T_e \gg T_i$.

[10] The concept of temperature to characterise the energy of a group of particles assumes that their energy distribution function is Maxwellian (Sect. 1.4.2 and Appendix I).

[11] These discharges are, however, warmer than those using the corona effect, and accordingly are closer to thermal equilibrium.

1.2.5 Surface treatment

Plasma surface treatment consists of modifying the state of a surface by one of the following generic methods:

- *deposition* of a thin film of a given material (metal; semi-conductor; dielectric; polymer) on the surface;
- *chemical reaction* with the surface itself (oxidation; nitriding) or *physico-chemical* transformation of the surface (modification of adherence, surface energy);
- *erosion* of the surface, either by a chemical reaction, which involves the formation of a volatile molecule, from one or more atoms from the surface and atoms or radicals provided by the plasma, or a physical action, *sputtering* by ion bombardment, such that ions eject atoms from the surface by a mechanical process, or by a chemical reaction assisted (induced) by ion bombardment, which combines chemical and physical actions.

Thus, a plasma produced from the gas CF_4 creates, in the volume, atoms (such as F), radicals (such as CF_x) as well as ions (such as CF_y^+) and more complex species necessary for interactions with the surface that, under suitable operating conditions, can equally well lead to etching of materials (Si, W, SiO_2) as is illustrated in Fig. 1.4, or deposition of teflon-like thin films by plasma-induced polymerisation. In the fabrication of micro-electronic chips, due to the requirements of smaller and smaller miniaturisation, the use of plasma continues to expand its range of applications: *surface cleaning, etching* (production of "patterns" in the substrate by surface erosion), deposition, *ion implantation* (doping by introducing ions deep in the material), *lithography* (impression and "photographic" development of resins allowing the transfer of patterns to define electric circuits), oxidation and thermal treatments.

Fig. 1.4 Example of anisotropic etching of SiO_2 (courtesy of CORIAL, France).

Of the multitude of elementary steps required for the fabrication of integrated circuits, the operations uniquely realised by plasmas represented, at the beginning of the 2000s, close to 50% of the total number of these steps. The introduction of plasma equipment for micro-electronics, and more generally for the micro/nanotechnology industries, constitutes an important outlet and a great impetus for plasma physicists and engineers.

An example of plasma deposition is the fabrication of polycrystalline diamond thin films.

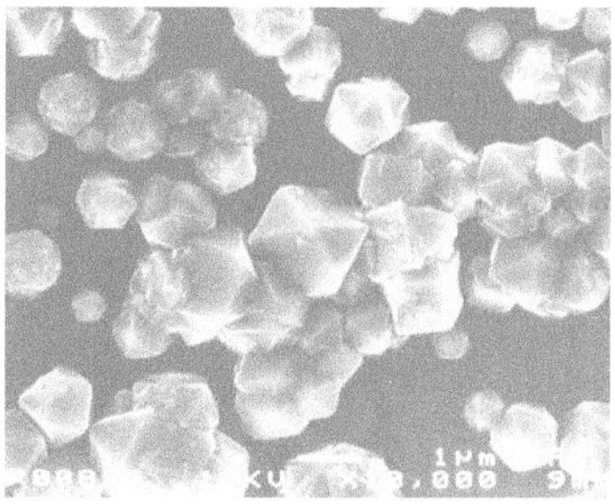

Fig. 1.5 Diamond crystals at the beginning of deposition on a silicon substrate. Once the first layer is completed, the growth rapidly progresses vertically.

Diamond's interesting hardness, heat transport and dielectric properties make it the material of choice in power electronics, as well as for cutting different materials. It is possible, using a plasma, to deposit thin films of polycrystalline diamond, i.e. an assemblage of small crystals of diamond. Their size can vary from 20 nm to a few microns (Fig. 1.5), depending on the operating conditions. The crystals unite, during their growth, through the formation of grain junctions, mostly consisting of amorphous carbon. Such a layer is typically 1 to 5 μm thick. In general, the plasma which is used for this process contains about 1% of a carbon containing material (such as CH_4), the remainder being hydrogen; the operating pressure is between 10 and 100 torr ($\approx$ 1.3–13 kPa) and the deposition takes place on a heated substrate ($\approx$ 500–1000°C). The dissociation of hydrogen molecules in the plasma supplies the atomic hydrogen, necessary to prevent the growth of graphite, an allotropic phase of carbon whose formation would otherwise be thermodynamically favoured with respect to the growth of diamond under these operating conditions.

1.2.6 Sterilisation of medical devices

Deactivation of micro-organisms can be achieved by direct exposure to a discharge in a gaseous mixture, or in a flowing discharge-afterglow[12] of such a gaseous mixture, as shown in Fig. 1.6.

---

[12] A *flowing discharge-afterglow* (*remote plasma*) is obtained from a gas that has been excited and ionised by a discharge and then rapidly drawn into an adjacent vessel, called the afterglow chamber, where there is no longer an electric field. To achieve this, it is necessary for the gas feed to have a sufficiently high flow rate, since the species created in the discharge have a limited lifetime ($\leq$ 1–100 ms).

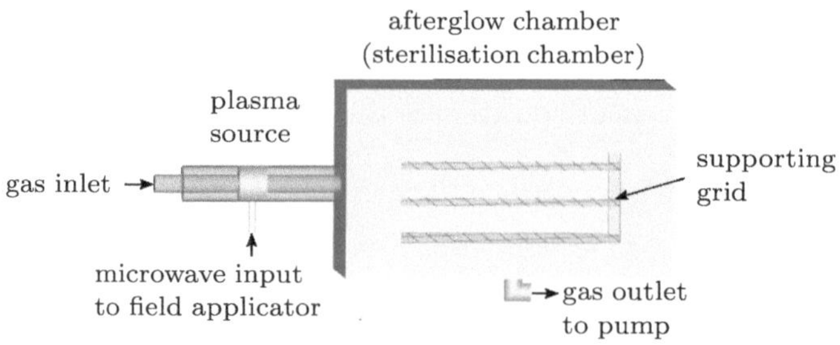

Fig. 1.6 Schematic of the principle of a cold-plasma steriliser of the flowing afterglow type (Université de Montréal).

The biocide species, in the case of the afterglow from a mixture of N_2-O_2 are, on one hand, *UV* photons from excited NO molecules and, on the other hand, atomic oxygen. Excited NO molecules are formed by collisions between nitrogen atoms and oxygen atoms, both originating from the dissociation of N_2 and O_2 molecules by the discharge in such a gas mixture. Under conditions in which the fraction of O_2 in the N_2-O_2 mixture yields a maximum in the emitted *UV* radiation, the exposed micro-organisms (bacterial spores in this case) are completely deactivated as a result of multiple lesions caused by the *UV* photons on their genetic material. Further, atomic oxygen, which is highly reactive, absorbs on the surface of the micro-organisms to form chemically volatile substances, resulting in the removal of material (erosion) of the micro-organism, such that it reduces their size and further facilitates its deactivation by UV photons[13].

1.2.7 Elemental analysis (analytical chemistry)

In order to find the atomic composition of a sample, it must first be atomised: by ion bombardment, in the case of a solid, or by dissociation (fragmentation) of molecules in the case of liquids (previously transformed to an aerosol) and in gases; in the third case, this is achieved with the help of a plasma, usually formed from argon or helium. The atoms present can then be detected, by optical spectroscopy, due to the characteristic radiation of those that have been brought to an excited state, or by mass spectrometry. Their concentration can be obtained by reference to standard samples containing the same atoms, preferably in a molecular matrix (ensemble) that is not very different from the sample to be analysed. This highly sensitive method permits the analysis of doses referred to as ultra-traces (of the order of one nanogram, or even one picogram, per gram of sample). The plasmas used to achieve this are sustained, for example, by high frequency electric fields (microwave and radio frequency).

---

[13] Probably, atomic oxygen can also diffuse into the interior of the micro-organisms and induce lethal lesions.

1.2.8 Lighting

Ionized gases have many applications for lighting, where they generally operate in the arc regime (thermionic cathodes)[14]: these comprise: i) low-pressure mercury vapour lamps (household fluorescent lamps, Fig. 1.2) and sodium vapour lamps (outdoor lighting); ii) high-pressure mercury-vapour based lamps, taking advantage of a high plasma density provided by sodium iodide and metal-halide salts (outdoor lighting).

Lighting is an important market which, however, has not seen spectacular advances in recent years. There have been some attempts to activate certain lamps using high frequency (HF) discharges, with the aim of prolonging life and providing a more energy efficient discharge. The first lamp to be introduced for domestic purposes using a high frequency electric field ($\sim 2.6\,\mathrm{MHz}$, a band authorized for lighting) was by General Electric in 1994; a transistor placed in a cavity inside the lamp provided the high frequency power, which allowed the lamp to be directly substituted for an incandescent (tungsten-halogen) lamp (with a factor 4 improvement in energy efficiency, and 10 times the life, but at a much higher cost to the consumer). Environmental concerns have led lighting companies to investigate ways of eliminating mercury from discharge lamps, and this is currently an active field of research.

1.2.9 Plasma display panels

In *plasma display panels*, the image is obtained as a result of electric discharges created in a number of cells (pixels) of a few hundred microns assembled into large surface panels (more than a million cells for a 42" (1.07 m) diagonal panel). The cells are filled with a mixture of gases, mainly xenon, at a pressure below atmospheric. The UV photons emitted in each micro-discharge excite luminescent phosphors, which re-emit visible photons in one of the three fundamental colours, red, blue and green, depending on the cell. Using this technique, it is possible to construct extremely large screens, with exceptional image quality, good contrast and high brightness. Plasma screens have an important place in the global television market throughout the world.

---

[14] DC and low-frequency AC discharges can be operated in two distinct modes, the *glow regime* (high-voltage cathode fall) and the *arc regime* (low-voltage cathode fall). In the glow regime, the emission of electrons is principally by ion bombardment on the (cold) cathode whereas in the arc regime the electrons are provided by thermoionic emission from the cathode (which can be moderately hot (1000 K–1400 K) when covered with an emissive coating, but much hotter otherwise, easily exceeding 2000 K). By this reckoning, standard fluorescent lamps are arcs, but a cold cathode lamp is a glow. The transition from glow to arc regime is characterised by a substantial decrease of the operating voltage and an increase of discharge current (Cathode fall measurements in fluorescent lamps, [27]).

1.2.10 Ion sources

Positive ion sources are used for a number of applications, including strongly assisted ion surface treatment (e.g. etching by ion machining), microelectronics (doping by ion implantation), nuclear and subatomic physics (mono- and multi-charged ions for accelerators), and spatial (Hall effect sources for ionic propulsion, spaceflight experiments).

Negative ion sources are an efficient method for obtaining high-energy neutral beams. This is the case, for example, for D^- ions which are neutralised into neutral D^0 beams: the interest in D^- negative ions is due to the fact that, after they have been accelerated to energies of MeV, their ion-neutral rate of conversion by charge exchange to neutral D^0 is much higher than that for D^+ positive ions. A beam of very high-energy neutrals D^0 can be used to greatly increase the plasma temperature in a tokomak, since it can be introduced into the vessel without being affected by the confining magnetic field.

1.2.11 Ion propulsion thrusters

These engines obtain their propulsive force from the ejection of heavy, high-speed particles, following the action-reaction principle (conservation of momentum), resulting in a momentum created in the opposite direction to the ejection of the particles. In an ion thruster, the heavy particles are usually xenon atoms which have been ionised once: xenon is used because it is a rare gas, thus only weakly chemically reactive (the engine lifetime is expected to be of order 15 to 20 years), with the highest atomic mass if one excludes radioactive radon (the propulsive force increases with the mass released). The xenon ions are accelerated in an electric field such that they attain a sufficient ejection velocity, but they must be neutralised before leaving the engine in order for the system to remain electrically neutral.

The electrical energy required to ensure the ionisation of heavy atoms (the "fuel") is provided by solar panels, which is extremely economical compared with a classical combustion engine. It is envisaged to increase the ejected ions energy by using a nuclear reactor. A number of telecommunication satellites orbiting the Earth use such ion thrusters. They are utilised to reposition the satellites daily to ensure optimum communication with the Earth, correcting for perturbations in position (altitude, directional position) due to the variable interaction forces exerted on them by the sun and the moon. Moving and stopping them is easy, and they can be positioned very accurately.

Current ion thruster motors, with their weak power, cannot be used for launchings from the Earth, but they are very effective in the spatial environment. They can impart a high speed to a spacecraft, but it takes notably more time than a conventional chemical launcher, which provides a considerable acceleration on ignition. The space probe SMART-I from the European Space Agency (ESA) has travelled more than a million km, and only consumed 60 standard litres (with respect to atmospheric pressure) of xenon.

1.2.12 Further applications

This brief overview of applications and studies of plasmas shows that this field of physics has already obtained some remarkable successes in many areas, including the domestic sphere, and it is equally rich in possibilities for future applications (for example fusion, sterilisation). In order to have an even wider view of the applications of plasmas, the reader is referred, for example, to A. Bogaerts et al (2002).

1.3 Different types of laboratory plasmas

The development and optimization of plasma applications require utilizing and, even, designing adequate plasma sources. This objective can be generally met via one of the three following main generic techniques.

1.3.1 Discharges with continuous current or alternative current at low frequency

In this case, the electrodes which are used to create the discharge are directly in contact with the plasma (Fig. 1.2). The plasma is formed, in a transition stage, by a process of electron multiplication called the *avalanche* (or breakdown) due to the application of a potential difference: the few electrons initially present, accelerated by the electric field, ionise the atoms (molecules) of the gas by collisions, thus augmenting the number of electrons. This growth in the number of electrons ceases after a few hundred micro-seconds, when a stationary state is attained.

In periodic low frequency discharges, the frequency of the maintenance current is assumed to be sufficiently small, such that all the electrical parameters of the plasma are in equilibrium with the applied field. In other words, at each instant in the period of the oscillating field, the plasma can be considered as having attained its stationary state.

1.3.2 High frequency (HF) discharges

Radiofrequencies (RF) and microwaves (MW) are jointly designated as high frequencies (HF). Microwave engineers generally consider that the microwave (MW) frequency domain starts at 300 MHz whereas, when it comes to sustaining a discharge with an electromagnetic field, the lower MW frequency, on practical grounds, is rather 100 MHz. This is because it is possible to make impedance matching circuits based on distributed components (as opposed to lumped components, e.g. constituting LC "matchboxes") at frequencies as low as 100 MHz. Furthermore, discharges can be sustained under electron-cyclotron resonance conditions (Sect. 4.2.3) at frequencies also as low as 100 MHz, although it is commonly thought to require frequencies above 1 GHz. Worldwide ISM authorized MW frequencies are 433.92 MHz, 2.45 GHz and 5.80 GHz. The RF range extends from approximately 1 MHz to 100 MHz, where the worldwide industrial, scientific and medical (ISM) authorized frequencies mostly used are: 13.56, 27.12, and 40.68 MHz.

The electrodes carrying the RF field can be placed inside the vessel (for example, two parallel conducting plates, in which case the discharge is said to be capacitive), or exterior to the vessel (for example, the coils of an inductive discharge (Fig. 4.4)), in which case the vessel must be constructed from a dielectric material transparent to RF radiation. MW plasmas are generally maintained by a *field applicator*[15]. The operating field frequency of the discharge can be chosen so as to optimise the plasma properties for certain applications: examples are given in Chap. 4.

1.3.3 Laser induced discharges

There are two distinct regimes, depending on the incident power of the laser:

- if the photon flux is weak, the wavelength of the laser should be such that it corresponds to the difference in energy between two atomic or molecular levels (this type of transition is referred to as absorption) such that they are raised to an excited state. Following this, a collision between two excited atoms can result in one of them becoming ionised. Direct (single-step) ionization is also possible.
- if the photon flux is strong, the *multi-photonic effect* (where several photons "sum" their energy) becomes important and allows direct ionisation of the gas, without having recourse to collisions.

---

[15] A field applicator designates electrodes or, more specifically, any kind of device that serves to impose the EM field configuration creating the discharge.

1.4 Electron density and temperature of a plasma

These are the two principal characteristics of a plasma, considered from the point of view of the particles.

1.4.1 Range of electron density values in a plasma

These values cover a range which is so large that it is preferable to use a logarithmic scale to classify them. In Tab. 1.1 below, in addition to gaseous plasmas, we have included "dense matter" plasmas, because they have analogous physical properties.

Table 1.1 Different types of plasma, with their corresponding electron densities

Gaseous plasmas	$\log_{10} n_e$ (cm^{-3})
Strongly ionised gases	
Interstellar gas[a]	0
Solar wind[b]	0.5
Ionosphere, F layer (250 km altitude)	5.7
Solar corona	7
Tokamak (fusion experiments)	14
Plasma produced by a laser in a solid target	19–23
Nuclear explosion	20
Weakly ionised gas	
Ionosphere, D layer (70 km altitude)	3
Laboratory discharge, low-pressure	10–12
Laboratory discharge, atmospheric pressure	14–15
Dense matter plasmas	
Electrons in metals	23
Interior of stars	27
Interior of white dwarves	32

[a] Few interactions between the particles (plasmas said to be collisionless), but large influence of external fields.
[b] The solar wind is essentially composed of protons and electrons.

1.4.2 Definition of plasma "temperature" and the concept of thermodynamic equilibrium (TE)

The temperature, T, is a parameter by which the global energy of a medium can be characterised, notably the energy of thermal motion of the particles, since this relates to the average energy (Appendix I (I.11)). It is only possible to speak of a particle temperature if the distribution in energy (velocity) is Maxwellian (short for a Maxwell-Boltzmann distribution); if not, in addition to the average energy of the particles, it is necessary to know the distribution in energy of these particles. We will see that, in a system in thermodynamic equilibrium, a single value T is sufficient to characterise the distribution of photons and particles at the same time. A system in TE is completely characterised by its temperature T and the density N_n of its heavy constituent particles. More exactly, the density N_n includes neutral atoms (molecules) and ions, both in ground state and excited states: it is preferable to refer to N_n as the total density of nuclei, to avoid ambiguity (see problems 1.3 and 1.4).

Consider a system consisting of atoms (neutral and ionised) together with EM radiation (photons), this radiation being linked to the excited states of the atoms and ions as well as to the Coulomb interactions between charged particles (Bremsstrahlung, Sect. 1.7.1). This ensemble is in *complete thermodynamic equilibrium* if there are sufficient interactions between the various components of the system, such that each type of energy exchange process in a given energy direction (for example, increase of energy of the "particle" during the interaction) is statistically rigorously compensated by the same type of process in the inverse direction (loss of energy by the same type of particle in our example): this requirement of compensation (detailed balance) is called the principle of microscopic reversibility, or more simply, *micro-reversibility.*

Examples of reversible processes

Elastic collision processes are clearly reversible: an atom or an electron undergoing a collision has an equal probability, statistically, of either gaining or losing energy.

On the contrary, inelastic collisions are not always easily reversible: to ensure their reversibility, the medium needs to be denser than for elastic collisions and, in the case where three particles are involved instead of only two, extremely dense. For example, consider the following:

- a *superelastic collision*, or *collision of the second kind*

$$\underline{\underline{\mathrm{e}}} + \mathrm{A}(0) \rightarrow \mathrm{A}(j) + \mathrm{e} \;\Leftrightarrow \mathrm{A}(j) + \mathrm{e} \rightarrow \mathrm{A}(0) + \underline{\underline{\mathrm{e}}}\,. \tag{1.4}$$

The symbol $\underline{\underline{\mathrm{e}}}$ denotes a high energy electron and e denotes a lower energy electron; A(0) designates the ground state of the atom and A(j) indicates an excited state of the same atom: the double arrow $\Leftrightarrow$ separates the two energy directions for the process considered. If the atom in the state j emits a photon before experiencing a collision, reversibility is not satisfied. A medium in which the number of collisions is sufficiently high, such that the mean time between collisions is smaller than the deexcitation time of the level considered, is therefore required.

- *collisional recombination*

$$\underline{\underline{\mathrm{e}}} + \mathrm{A}(0) \rightarrow \underline{\underline{\mathrm{e}}} + \mathrm{A}^+(j) + \mathrm{e} \Leftrightarrow \mathrm{A}^+(j) + \underline{\underline{\mathrm{e}}} + \mathrm{e} \rightarrow \mathrm{A}(0) + \underline{\underline{\mathrm{e}}}\,. \tag{1.5}$$

In this case, we see that reversibility requires a three-body interaction, which means that it is difficult to obtain complete thermodynamic equilibrium (TE) unless the medium is sufficiently dense to ensure that the three components interact simultaneously.

- *emission and absorption of photons*

$$\underset{\text{absorption}}{\mathrm{A}(i) + h\nu \rightarrow \mathrm{A}(j)} \;\Leftrightarrow\; \begin{cases} \mathrm{A}(j) \rightarrow \mathrm{A}(i) + h\nu & \text{spontaneous emission} \\ \mathrm{A}(j) + h\nu \rightarrow \mathrm{A}(i) + 2h\nu & \text{stimulated emission} \end{cases} \tag{1.6}$$

where h is the Planck constant, ν, the frequency of the emitted or absorbed photon; j and i denote the upper and lower energy levels of the transition respectively ($j > i$).

Consequences of complete TE

Complete thermodynamic equilibrium is obtained when the four equilibrium laws described below are simultaneously satisfied. To characterise the system, it is only necessary to know the temperature T and the density of nuclei N_n.

1. Maxwell-Boltzmann distribution of the microscopic velocities $\boldsymbol{w}$
For electrons, for the case of an isotropic distribution, we have (Appendix I):

$$f(w) = \left(\frac{m_e}{2\pi k_B T}\right)^{3/2} \exp\left(-\frac{m_e w^2}{2k_B T}\right) , \tag{1.7}$$

where k_B is the Boltzmann constant, m_e is the electron mass, and the temperature T is measured in kelvin. The most probable velocity of a particle in the Maxwell-Boltzmann distribution, v_{th}, is given by:

$$v_{th} = \left(\frac{2k_B T}{m_e}\right)^{1/2} , \tag{1.8}$$

which can be used to write (1.7) in a form which is simpler and easier to remember:

$$f(w) = \frac{\pi^{-3/2}}{v_{th}^3} \exp\left(-\frac{w^2}{v_{th}^2}\right) . \tag{1.9}$$

Remark: A sufficient condition for the velocity distribution of particles to be Maxwellian is that the plasma is in thermodynamic equilibrium.

2. Boltzmann's law relating the population of the excited states to the ground state
$$\frac{n_j}{n_0} = \left(\frac{g_j}{g_0}\right) \exp\left[-\frac{(\mathcal{E}_j - \mathcal{E}_0)}{k_B T}\right] , \tag{1.10}$$
where n_0 is the density of atoms in the ground state with energy $\mathcal{E}_0$ and n_j the density of atoms in the excited state of energy $\mathcal{E}_j$, with g_0 and g_j the corresponding statistical weights (or degeneracies)[16].
3. Planck's law, or *black body radiation*, relating the spectral intensity distribution of the EM radiation. This intensity, at frequency ν, is given by:
$$I_\nu = \frac{2h\nu^3}{c^2}\left[\exp\left(\frac{h\nu}{k_B T}\right) - 1\right]^{-1} , \tag{1.11}$$
where c is the speed of light in vacuum.
4. The *Saha's equation* describes the equilibrium between ionisation processes (creation of charged particles) and volume recombination (disappearance of charged particles by neutralisation of an ion by an electron (Sect. 1.8.1). This law allows us to calculate the density n_i of singly ionised (positive) ions, relative to the density n_0 of neutral atoms, from the plasma temperature. Assuming that the ions and neutral atoms are in their ground state[17], this equation takes the simple form:
$$\frac{n_e n_i}{n_0} = \frac{2g_i}{g_0}\frac{(2\pi m_e k_B T)^{3/2}}{h^3} \exp\left(-\frac{\mathcal{E}_i - \mathcal{E}_0}{k_B T}\right) , \tag{1.12}$$
where g_i and g_0 are the respective quantum degeneracies of the energy levels i and the ground state, n_e the electron density and $\mathcal{E}_i$ the energy level (at threshold) of the first ionisation.
To calculate the density relation between ions of charge Z (that is, those having lost Z electrons) and those of charge $(Z-1)$, we use the relation:
$$\frac{n_e n_i[Z]}{n_i[Z-1]} = \frac{2g_i[Z]}{g_i[Z-1]}\frac{(2\pi m_e k_B T)^{3/2}}{h^3} \exp\left(-\frac{\mathcal{E}_i[Z] - \mathcal{E}_i[Z-1]}{k_B T}\right) , \tag{1.13}$$
where this time, $\mathcal{E}_i$ is the energy of ionisation of the Zth electron with respect to the atom ionised $(Z-1)$ times; the symbol [ ] indicates the dependence on Z and $Z-1$ of n_i, g_i and $\mathcal{E}_i$; the values $n_i[Z]$ and $n_i[Z-1]$ are the densities of the ground states of the two types of ions.

[16] The degeneracy in energy of an atomic level is given by $2J+1$, where J is the quantum number of the total angular momentum of the level considered.

[17] To obtain, on the one hand, the total density of (singly and positively charged) ions, which includes not only those in the ground state but also those of all the excited states and, on the other hand, the total density of neutral atoms, including ground state and excited states, see Appendix II.

1.4.3 Different levels of departure from complete thermodynamic equilibrium

In most laboratory plasmas, the micro-reversibility of processes is not perfect, and the information then required to characterise the system increases, as the number of non-reversible processes increases[18]. We will examine this matter by describing situations in which micro-reversibility is less and less satisfied.

Local thermal equilibrium (LTE)

In an inhomogeneous plasma in which there is a density gradient of particles (inducing them to diffuse) or a temperature gradient (created, for example by a thermal flux to a wall), or in a homogeneous plasma from which photons are escaping (at least for certain lines or spectral regions), there is a net flux of energy across the system: the local decrease (or increase) of the energy of the system implies that micro-reversibility is not complete. However, if this local loss of energy is small with respect to the total energy at that point, or, equivalently, if the difference in energy between two neighbouring points in the system is small, then one can say it is in LTE.

The most common example of LTE is that of a plasma whose particle density is not large enough, or its volume is too small, to reabsorb the majority of the photons emitted: these photons, frequently limited to a given spectral region, thus escape from the system. This situation is often not deleterious to the equilibrium of the system, because additional processes occur to compensate for these reactions that, in complete TE, normally require the absorption of the photons.

Consider, for example, the reaction $\mathrm{A}(j) \rightarrow \mathrm{A}(0) + h\nu$, although not reversible in this case, is replaced by $\mathrm{A}(0) + \underline{\underline{\mathrm{e}}} \rightarrow \mathrm{A}(j) + \mathrm{e}$; this is called an *improper detailed balance*, in contrast to the proper balance under perfect micro-reversibility. The radiation in such a system does not obey Planck's law, but the flux which escapes is weak, and the three other equilibrium laws for TE apply locally: Maxwell-Boltzmann for the particle distribution function, Boltzmann for the densities of excited states of atoms (molecules) and Saha for ionisation-recombination: a single temperature $T(\boldsymbol{r})$, defined locally at $\boldsymbol{r}$, together with the densities $N_n(\boldsymbol{r})$ of the atomic (molecular) nuclei is sufficient to characterise the system.

In the case where a net particle flux traverses the system (diffusion, convection), the concept of LTE is applicable on condition that the time (referred to as the relaxation time) necessary for a constituent particle of a (thermodynamic) sub-system at temperature $T_1(\boldsymbol{r})$, at point $\boldsymbol{r}_1$, to reach equilibrium with the sub-system at temperature $T_2(\boldsymbol{r})$ at point $\boldsymbol{r}_2$, is very short. In this case, LTE is sustained locally.

---

[18] Recall that the TE system is simply and completely determined by its temperature T and density N_n of the atomic (molecular) nuclei.

Plasmas not in LTE: the particular case of a two-temperature plasma

When the plasma medium is less dense than was considered in the previous paragraph, the number of collisions between electrons and heavy particles is reduced. Since an electron transfers at most $4m_e/M$ of its energy to an ion or atom of mass M (demonstrated in Sect. 1.7.2), the collisional transfer of energy between electrons and heavy particles is insufficient for these particles to all have the same average energy. However, if the interactions between particles of the same type are sufficiently numerous, there is equipartition of energy within this population of particles, and these particles still have a Maxwell-Boltzmann distribution. In such a case, each species can be characterised by an appropriate temperature: electron temperature T_e, ion temperature T_i, and neutral particle temperature (or gas temperature) T_g.

A particular case of interest is that in which the electron temperature is much greater than that of the other particles in the plasma, in which case it is the electrons which introduce the energy into the system[19]. A frequently observed situation is that in which $T_e > T_i \approx T_g$ (called a *two-temperature plasma*). In such a two-temperature plasma, the population density of the different energy levels of neutral atoms (and ions) cannot be described by the Boltzmann equation (1.10). In fact, the time between two successive electron-neutral collisions for excitation or de-excitation of those levels close to the ground state is much longer than their radiative lifetime: these levels therefore populate and depopulate radiatively rather than by electron collisions, and their control by the electron kinetics is lost. On the other hand, the higher-lying levels, which are situated just below the ionisation energy (Fig. III.1 in Appendix III), are usually in collisional equilibrium, and the Boltzmann energy law gives their population density according to $T_{exc} \simeq T_e$. Such a system is said to be in *local partial equilibrium* (Appendix III) because the upper levels are in Boltzmann equilibrium with the electrons. To describe the system, it is therefore necessary to define many "temperatures" (the term "characteristic parameters" would be more correct) to distinguish it from LTE.

No thermodynamic equilibrium characteristics, but a stationary state

The energy distribution of particles are no longer Maxwellian: for example, inelastic collisions can strongly depopulate certain energy intervals which would be found in a Maxwell-Boltzmann distribution. In this case, we can no longer speak of a temperature, but only of an average energy, and it is

[19] When there is a preferred path for the introduction of energy, this raises the problem of the repartition of energy in the plasma. If there are insufficient interactions between the different types of particles, their average energy will not be the same.

again necessary to know the exact form of the distribution function in order to characterise the system.

In conclusion, the further from LTE, the greater the necessity to provide data to characterise the system.

1.5 Natural oscillation frequency of electrons in a plasma

1.5.1 Origin and description of the phenomenon

If a plasma, whose dimensions are much greater than the Debye length λ_D (the average distance below which there is no charge neutrality, Sect. 1.6), experiences a local perturbation from neutrality (resulting, for example, from the random movement of particles), this equilibrium will be re-established by a collective movement of charges (Sect. 1.1). If there are few or no collisions, this return to equilibrium will take the form of a pendular oscillation about the point where the initial disturbance occurred.

In order to understand this phenomenon, consider Fig. 1.7, which is an idealised representation of the distribution of ions and electrons in a plasma. Initially, the charges are distributed alternately and equidistant such that the electric field is zero at each point: the charged particles (supposedly with no thermal energy to set them in motion!) must remain immobile, in their state of equilibrium. Displacing a group of electrons by a distance x with respect to their initial equilibrium position will result in an electric field (this field is given by Poisson's equation (1.1) and is called the *space charge field*) which draws the electrons back to their original position, a motion which tends at the same time to reduce the electric field intensity. However, the accelerated electrons are unable to stop at their equilibrium position, but continue their motion from this point, thus creating a new departure from charge neutrality, with an electric field in the opposite direction to the initial field. The electrons thus continue their pendular motion about their equilibrium position, provided such a motion is not damped by collisions.

This collective motion of electrons produces a local oscillatory motion whose angular frequency (see following proof) is given by:

$$\omega_{pe} = \left(\frac{\bar{n}_e e^2}{m_e \epsilon_0} \right)^{1/2} , \tag{1.14}$$

where $\bar{n}_e$ is the unperturbed electron density, ϵ_0 is the permittivity of vacuum; $f_{pe} = \omega_{pe}/2\pi$ is the natural frequency of oscillation of electrons in a plasma (also referred to as Langmuir oscillation), or more commonly, the *electron plasma frequency*.

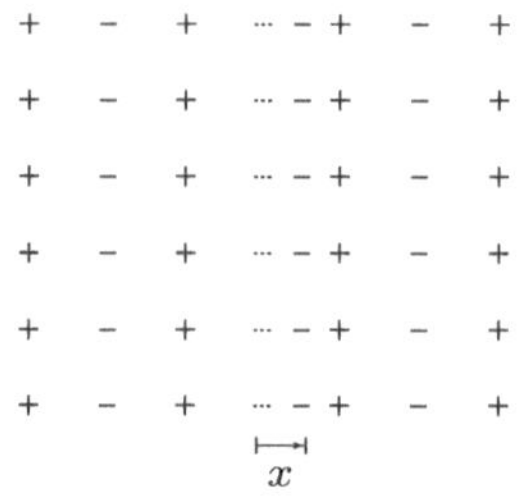

Fig. 1.7 (Highly) idealised representation of ions and electrons in a plasma, showing that a slight non-uniformity in the distribution, resulting from a displacement of a group of electrons by a distance x, creates an electric field in this region (referred to as the space charge field). The return of the displaced electrons to their initial positions, due to the electric field, leads to an oscillatory motion about their equilibrium position.

During these oscillations, the ions, which are much heavier than the electrons, remain practically immobile: they barely start to move in one direction, under the influence of the space charge field, when they are forced to move in the opposite direction.

1.5.2 Calculation of the electron plasma frequency

A simple hydrodynamic model, which describes the collective oscillatory motion of electrons as a fluid, allows us to obtain a value for the angular frequency ω_{pe}. The following approximations are made:

1. The ions are immobile, hence their density $\bar{n}_i$ remains unperturbed and uniform.
2. The thermal motion of the electrons is negligible: their velocity v_e due to the space charge field is such that $v_e \gg v_{th}$ (*cold plasma* approximation).
3. The electron-neutral collision frequency for momentum transfer ν is the most important collision frequency, but remains such that $\nu \ll \omega_{pe}$, in order to maintain the collective motion of the plasma.
4. The amplitude of the plasma oscillations is small.
5. There is no applied external magnetic field.

In terms of the hydrodynamic model (Sect. 3.5), we can describe the fluid electrons by the two following equations:

- Equation of conservation of particles:

$$\frac{\partial n_e}{\partial t} + \boldsymbol{\nabla} \cdot (n_e \boldsymbol{v}_e) = 0 \,. \tag{1.15}$$

- Equation of conservation of momentum transport[20]:

$$n_e m_e \left(\frac{\partial}{\partial t} + \boldsymbol{v}_e \cdot \boldsymbol{\nabla} \right) \boldsymbol{v}_e = -n_e e \boldsymbol{E} \ , \tag{1.16}$$

where $\boldsymbol{E}$ is the space charge field.

We can linearise the equations (approximation 4) by writing:

$$n_e(\boldsymbol{r}, t) = \bar{n}_e + \tilde{n}_e(\boldsymbol{r}, t) \ , \tag{1.17}$$

where $\tilde{n}_e(\boldsymbol{r}, t)$ is a perturbation on the density $\bar{n}_e$, uniform and constant in the absence of fluctuations ($\tilde{n}_e \ll \bar{n}_e$). We assume in addition, that the physical quantities vary with time, to first order, with a frequency $\omega/2\pi$ which we wish to calculate: thus we assume that $\boldsymbol{E} = \boldsymbol{E}_0 \exp(\mathrm{i}\omega t)$, $\boldsymbol{v}_e = \boldsymbol{v}_{e0} \exp(\mathrm{i}\omega t)$ and $\tilde{n}_e(\boldsymbol{r}, t) = \tilde{n}_{e0}(\boldsymbol{r}) \exp(\mathrm{i}\omega t)$. Equation (1.15) becomes (after cancelling the time-dependent exponential):

$$\mathrm{i}\omega \tilde{n}_{e0} + \bar{n}_e \boldsymbol{\nabla} \cdot \boldsymbol{v}_{e0} = 0 \ , \tag{1.18}$$

where we have ignored the term $\boldsymbol{\nabla} \cdot \tilde{n}_e \boldsymbol{v}_{e0}$, a second order term in a first order equation. From (1.16), we obtain:

$$\bar{n}_e m_e \mathrm{i}\omega \boldsymbol{v}_{e0} = -\bar{n}_e e \boldsymbol{E}_0 \ , \tag{1.19}$$

where $\boldsymbol{v}_{e0} \cdot \boldsymbol{\nabla} \boldsymbol{v}_{e0} \approx 0$, because it is also a second order term. To these two equations, we add Poisson's equation (1.1) which, for the present case, can be written:

$$\boldsymbol{\nabla} \cdot \boldsymbol{E} = \frac{\bar{n}_i e - n_e e}{\epsilon_0} \approx -\frac{\tilde{n}_e e}{\epsilon_0} \tag{1.20}$$

because charge neutrality requires $\bar{n}_e = \bar{n}_i$.

We can use Eqs. (1.18) to (1.20) to eliminate $\boldsymbol{v}_{e0}$ and n_{e0}. From (1.19):

$$\boldsymbol{v}_{e0} = -\left(\frac{e}{\mathrm{i}\omega m_e} \right) \boldsymbol{E}_0 \ , \tag{1.21}$$

and inserting (1.21) into (1.18), we obtain:

$$\tilde{n}_{e0} = -\frac{\bar{n}_e e}{\omega^2 m_e} \boldsymbol{\nabla} \cdot \boldsymbol{E}_0 \ . \tag{1.22}$$

Using the value of n_{e0} obtained simultaneously from (1.22) and (1.20), we find:

$$\tilde{n}_{e0} = -\frac{\bar{n}_e e}{\omega^2 m_e} \boldsymbol{\nabla} \cdot \boldsymbol{E}_0 = -\frac{\epsilon_0}{e} \boldsymbol{\nabla} \cdot \boldsymbol{E}_0 \ , \tag{1.23}$$

---

[20] We have neglected the collisional interaction terms (approximation 3) and used the cold plasma approximation (2); the field $\boldsymbol{E}$ is that of the space charge.

such that:

$$\nabla \cdot \boldsymbol{E}_0 \left(\frac{\bar{n}e^2}{m_e \epsilon_0 \omega^2} - 1 \right) = 0 \tag{1.24}$$

and, for $\nabla \cdot \boldsymbol{E}_0 \neq 0$, implying:

$$\omega = \omega_{pe} \,, \tag{1.25}$$

where $\omega_{pe} \equiv \left(\bar{n}_{e0} e^2 / \epsilon_0 m_e\right)^{\frac{1}{2}}$ as in (1.14).

Remarks:

1. In the cold plasma approximation ($T_e = 0$)[21], the collective oscillation of the plasma is restricted to the neighborhood of the perturbation which caused it; it does not propagate, nor is it a wave. In order for an electromagnetic wave to exist[22], we must be able to define a group velocity v_g[23], which is obtained from the dispersion equation. In the present case, from (1.25) where ω_{pe} is a constant, $v_g \equiv \partial\omega/\partial\beta = 0$.
   However, if we take into account the scalar pressure exerted by the thermal motion of the electrons under their own motion, (Sect. 3.5), whose average value is determined by their temperature, we obtain, for the same oscillatory motion (Quémada, Sect. 6.4.1):
   $$\omega^2 = \omega_{pe}^2 + \gamma\beta^2 \frac{k_B T_e}{m_e} \,, \tag{1.26}$$
   where $v_g = \gamma\beta^2 k_B T_e / m_e \omega$ is non zero if $T_e \neq 0$. In this equation, γ is the *ratio* c_p/c_v of the *specific heats* of the gas. For an adiabatic transformation, $\gamma = (2+\bar{\delta})/\bar{\delta}$, where $\bar{\delta}$ is the number of degrees of freedom for the gaseous species; in the case of a monoatomic gas, $\bar{\delta} = 1$ or $\gamma = 3$ (Sect. 3.6).
2. In a bounded plasma, i.e., where the boundaries must be taken into account, the oscillation frequency is:
   $$\omega = \omega_{pe}/\sqrt{2} \text{ for cylindrical geometry,} \tag{1.27}$$
   $$\omega = \omega_{pe}/\sqrt{3} \text{ for spherical geometry.} \tag{1.28}$$

---

[21] The cold plasma approximation neglects the thermal velocity compared to another characteristic velocity of the plasma, in this case v_e, by assuming $T_e = 0$.

[22] In order to propagate an electromagnetic wave, it is necessary to have a transport of energy from one point in space to another, i.e. the Poynting vector $\boldsymbol{S} = \boldsymbol{E} \wedge \boldsymbol{H}$ should be non-zero.

[23] In a propagating medium without attenuation, the modulus of the wave vector is expressed as $\beta \equiv \lambda/2\pi$ (also called the *wavenumber*), and the *group velocity* is given by $v_g \equiv \partial\omega/\partial\beta$.

3. A numerical approximation to the natural frequency for electrons is:

$$f_{pe}(\text{Hz}) \simeq 9000\sqrt{n_e(\text{cm}^{-3})}\ . \qquad (1.29)$$

4. One can calculate the natural frequency for ions in the plasma in the same way as for electrons, obtaining:

$$\omega_{pi} = \left(\frac{n_i e^2}{m_i \epsilon_0}\right)^{\frac{1}{2}}, \qquad (1.30)$$

where it should be noted that the ion plasma frequency, because it is a function of the inverse of the ion mass m_i, is much smaller than the electron plasma frequency.

1.6 Debye length: effect of screening in the plasma

1.6.1 Description of the phenomenon

If we introduce conducting electrodes into the plasma to create a source of potential, the electrons will be attracted to the positive terminal and the (positive) ions to the negative terminal. The excess of charges of a given sign thus created is concentrated, however, in a small region around the electrode, called the *sheath*, the rest of the plasma remaining macroscopically neutral. The sheath acts as a screen, spatially limiting the influence of the prevailing electric field on the plasma[24].

A similar screening mechanism is also found in the main body of the plasma itself, where the potential of any given particle is not felt beyond a distance λ_D, the Debye length. We will show that the electrostatic potential of a positive singly-charged ion in a plasma at a distance r from this ion is given by:

$$\phi(r) = \frac{e}{4\pi\epsilon_0 r}\exp\left(-\frac{r}{\lambda_D}\right), \qquad (1.31)$$

where the exponential term represents the *screening effect*, which strongly reduces the range of the potential of the ion compared to that in vacuum. In fact, for $r = \lambda_D$, the potential of the ion will have decreased by a factor 1/e compared to its value in vacuum (e is used here, exceptionally, to denote the base of the natural logarithm). The range of the screening depends on the energy of the thermal motion of the particles and their density, as we will see.

[24] In fact, if we introduce an object into a plasma which is made of dielectric or conducting material (which does not, however, act as an electrode) a sheath forms (Sect. 3.14) around this object, because its surface charges negatively; as we will see, this effect is due to the higher mobility of the electrons, relative to that of the ions.

1.6.2 Calculation of the potential exerted by an ion in a two-temperature plasma: definition of the Debye length

Consider the ion in question as a test-particle (positive): such a particle, hypothetically, acts on the other particles without being influenced by them. Placed in the plasma at the origin of a system of spherical coordinates, it will create a perturbation by its electrostatic field. We wish to know the potential $\phi(r)$ produced by the ion at a distance r, taking into account the cloud of electrons and ions which surround it. The electron and ion densities $n_e(r)$ and $n_i(r)$ at the origin of the system are different, but not at infinity, where they are equal, $n_{e\infty} = n_{i\infty}$ (the perturbation is not felt there).

We will suppose that, at a sufficient distance r, to be specified below, the electron and ion populations have a Maxwell-Boltzmann distribution, characterised, for greater generality, by different electron and ion temperatures, T_e and T_i respectively (two-temperature plasma, Sect. 1.4.3). The electron and ion densities, for such distributions, in the presence of a potential $\phi(r)$, satisfy the relation (I.15) (Appendix I), namely:

$$n_\alpha(r) = n_{\alpha\infty} \exp\left(-\frac{\Phi(r)}{k_B T_\alpha}\right) , \tag{1.32}$$

where the potential energy $\Phi(r) = +e\phi(r)$ for the case of a positive ion. For the two types of particles, we then have:

$$n_i(r) = n_{i\infty} \exp\left(-\frac{e\phi(r)}{k_B T_i}\right) , \tag{1.33}$$

$$n_e(r) = n_{e\infty} \exp\left(\frac{e\phi(r)}{k_B T_e}\right) . \tag{1.34}$$

Certainly, as other authors have noted [14], in taking account of the perturbation created by the test-particle, the hypothesis of a Maxwell-Boltzmann distribution is not valid in the immediate vicinity of this perturbation. We do not need to be concerned in the case of the present demonstration, because we suppose that such a distribution only occurs after a distance r sufficiently far from the test particle, such that its potential is strongly screened by the surrounding particles, or more exactly when $e\phi(r)/k_B T \ll 1$. These conditions allow us to expand Eqs. (1.33) and (1.34) to first order, to obtain:

$$n_i(r) = n\left(1 - \frac{e\phi(r)}{k_B T_i}\right) , \tag{1.35}$$

$$n_e(r) = n\left(1 + \frac{e\phi(r)}{k_B T_e}\right) , \tag{1.36}$$

since $n_{e\infty} = n_{i\infty} = n$.

Formulation of the differential equation defining $\phi(r)$

The local charge density at r is thus:

$$\rho(r) = en\left(1 - e\frac{\phi(r)}{k_B T_i}\right) - en\left(1 + e\frac{\phi(r)}{k_B T_e}\right),$$

i.e.:

$$\rho(r) = -\left(\frac{e^2\phi(r)}{k_B T_i} + \frac{e^2\phi(r)}{k_B T_e}\right) n\,. \tag{1.37}$$

Poisson's equation allows us to obtain a differential equation for $\phi(r)$ since:

$$\boldsymbol{\nabla} \cdot \boldsymbol{E} = \rho/\epsilon_0 \tag{1.38}$$

leads to:

$$\boldsymbol{\nabla} \cdot \boldsymbol{\nabla}\phi = -\rho/\epsilon_0\,. \tag{1.39}$$

Hence, using (1.37):

$$\nabla^2\phi = \phi(r)\left(\frac{ne^2}{\epsilon_0 k_B T_i} + \frac{ne^2}{\epsilon_0 k_B T_e}\right). \tag{1.40}$$

Denoting:

$$\lambda_{D\alpha}^2 \equiv \frac{\epsilon_0 k_B T_\alpha}{ne^2}, \tag{1.41}$$

the term in parenthesis in (1.40) can be written:

$$\frac{1}{\lambda_D^2} = \frac{1}{\lambda_{De}^2} + \frac{1}{\lambda_{Di}^2}, \tag{1.42}$$

and (1.40) then becomes:

$$\nabla^2\phi = \frac{\phi(r)}{\lambda_D^2}, \tag{1.43}$$

where λ_{De} and λ_{Di} are the Debye lengths for electrons and ions respectively, and λ_D is the *global* Debye length, or simply the Debye length.

Since (1.43) only depends on r, it is spherically symmetric and it can be written in spherical coordinates as:

$$\frac{1}{r^2}\frac{\mathrm{d}}{\mathrm{d}r}\left[r^2\frac{\mathrm{d}}{\mathrm{d}r}\phi(r)\right] = \frac{\phi(r)}{\lambda_D^2}\,. \tag{1.44}$$

Solution to the differential equation (1.44)

We will expand the potential $\phi(r)$ as a product of two contributions: $\phi_c(r)$ is applicable in the neighbourhood of the test particle, while $f(r)$ describes the behaviour at long distance.

- Solution for $r \approx 0$
  In this region, the test-ion potential is the most important, and it has spherical symmetry. After integrating Poisson's equation for this (+) ion

alone, we obtain:

$$\int_V \boldsymbol{\nabla} \cdot \boldsymbol{E} \, \mathrm{d}V = \int_V (\rho/\epsilon_0) \, \mathrm{d}V \equiv e/\epsilon_0 \; , \tag{1.45}$$

where the volume V is sufficiently small, such that it only contains the test-ion.
In addition, applying Gauss's law (an application of Green's theorem):

$$\int_V \boldsymbol{\nabla} \cdot \boldsymbol{E} \, \mathrm{d}V = \int_{S=\partial V} \boldsymbol{E} \cdot \mathrm{d}\boldsymbol{S} \; , \tag{1.46}$$

where S is the surface delimiting V. Spherical symmetry allows us to readily integrate across the surface:

$$\int_S \boldsymbol{E} \cdot \mathrm{d}\boldsymbol{S} = 4\pi r^2 E_r(r) \tag{1.47}$$

and from (1.45), (1.46) and (1.47) we obtain:

$$E(r) = \frac{e}{4\pi\epsilon_0 r^2} \tag{1.48}$$

and since:

$$E(r) = -\frac{\mathrm{d}\phi(r)}{\mathrm{d}r} \; , \tag{1.49}$$

we get the expected result for the potential $\phi(r)$ in the immediate vicinity of the test-ion, $\phi_c(r)$:

$$\phi_c(r) = \frac{e}{4\pi\epsilon_0 r} \; , \tag{1.50}$$

the potential of a positive ion in vacuum.

- Solution for large r
We can write $\phi(r)$ in (1.44) in the form

$$\phi(r) = \phi_c(r) f(r) \; , \tag{1.51}$$

where, a priori, we require $f(r) \to 1$ as $r \to 0$ and $f(r) \to 0$ as $r \to \infty$. In this case, substituting (1.51) in (1.44), we obtain the equation:

$$\frac{\mathrm{d}^2 f}{\mathrm{d}r^2} = \left(\frac{1}{\lambda_D}\right)^2 f(r) \; , \tag{1.52}$$

which has two possible solutions:

$$f_1(r) = \exp\left(-\frac{r}{\lambda_D}\right) \quad \text{and} \quad f_2(r) = \exp\left(+\frac{r}{\lambda_D}\right) \; , \tag{1.53}$$

where $f_2(r)$ is rejected, because we require $f(r) \to 0$ as $r \to \infty$. Inserting (1.50) and (1.53) in (1.51), we finally arrive at the expression for the potential of a test particle, at distance r, when it is immersed in a plasma:

$$\phi(r) = \frac{e}{4\pi\epsilon_0 r} \exp\left(-\frac{r}{\lambda_D}\right) . \tag{1.31}$$

Remarks:

1. The screening factor, expressed by the exponential factor in (1.31), is independent of the sign of the charge of the test particle.
2. The Debye length becomes shorter as the plasma density increases (1.41): i.e., the potential of the test particle is more rapidly screened as the density of charged particles surrounding it becomes more important.
3. In non-LTE plasmas, the temperature of the ions T_i being in general much smaller than that of the electrons T_e ($T_i \ll T_e$), the Debye length in the plasma can be approximated by the ion Debye length, which is much shorter than the electron Debye length, i.e. $\lambda_D \approx \lambda_{Di} \ll \lambda_{De}$: the screening effect, in this case, is governed principally by the ions.
4. For plasmas in thermodynamic equilibrium, ions and electrons have the same temperature ($T_i = T_e$), and the electron and ion Debye lengths are equal ($\lambda_{Di} = \lambda_{De}$). The Debye length is then given by $\lambda_D = \lambda_{De}/\sqrt{2}$.
5. Numerical expressions for $\lambda_{D\alpha}$:

$$\lambda_{D\alpha}(\text{cm}) = 6.9\,(T_\alpha/n)^{1/2} \quad \text{for } n \text{ in cm}^{-3}, \text{ and } T_\alpha \text{ in K}, \tag{1.54}$$

$$\lambda_{D\alpha}(\text{cm}) = 740\,(T_\alpha/n)^{1/2} \quad \text{for } n \text{ in cm}^{-3}, \text{ and } T_\alpha \text{ in eV}. \tag{1.55}$$

6. The Debye length can also be written in the form:

$$\lambda_{De} = \sqrt{\frac{k_B T_e \epsilon_0}{ne^2}} = \sqrt{\frac{1}{2}\frac{m_e \epsilon_0}{ne^2} v_{th}^2} = \frac{\sqrt{2}}{2}\frac{v_{th}}{\omega_{pe}} \approx \frac{v_{th}}{\omega_{pe}} , \tag{1.56}$$

which shows that an electron with the most probable thermal velocity, travels an electron Debye length λ_{De} in a time of the order of one period of the plasma electron oscillations. This relation summarises, to some extent, the way in which the collective motion of the electrons ensures macroscopic neutrality in the plasma.
7. The present derivation of the Debye length is idealised, because of the numerous hypotheses used, notably the test-particle concept, in which it is assumed that this particle is not influenced by other particles. It supposes, in addition, that electrons and ions which are situated at a sufficient distance form the test particle, have returned to a Maxwell-Boltzmann distribution.

8. In plasmas where the ions are considered solely as a continuous background to ensure charge neutrality (with the assumption, used in numerous calculations, that they do not deviate from a Maxwell-Boltzmann distribution), (1.35) reduces to $n_i(r) \approx n$, such that the screening of the electron (or ion) potential is due, in this case, uniquely to the electrons, i.e. $\lambda_D \approx \lambda_{De}$. This assumption is adopted in Appendix V, and also in problem 1.5, which proposes an alternative interpretation of the Debye length.
9. One condition for plasma neutrality to be restored following a perturbation (by a collision, for example), and for the different charged particles in the plasma to resume a Maxwell-Boltzmann distribution, is that the time between two collisions should be much longer than their natural oscillation period, i.e. $\nu \ll \omega_{p\alpha}$. This condition is easier to realise with electrons than with ions $\omega_{pe} \gg \omega_{pi}$), which justifies, in a number of cases, the assumption of a continuous background of ions to ensure plasma neutrality.
10. Conditions for the existence of a plasma
    - in order for macroscopic neutrality to be realised inside the plasma, it is necessary that L, the smallest dimension defining the volume occupied by the plasma, be much larger than the Debye length, i.e $L \gg \lambda_D$. Another condition, already mentioned in Sect. 1.5.2, is that $\nu \ll \omega_{pe}$.
    - The number of charges N_D in a Debye sphere should be much larger than 1, otherwise this is a "non-ideal" plasma, in which there is no screening effect: this condition can be written:

$$N_D \equiv n \left(\frac{4}{3} \pi \lambda_D^3 \right) \gg 1 \,. \tag{1.57}$$

1.7 Collision phenomena in plasmas

As we have observed in Sect 1.4.2, the repartition of energy between the different constitutive elements in a plasma is established by an ensemble of particle-particle and particle-photon interactions. We will use the term "collision" in future in a more general sense than simply an impact between two more or less rigid spheres, leading to an exchange of kinetic energy. In effect, the long range interaction (Coulomb force), as well as those interactions leading to the excitation of an atom by an electron collision (non-conservation of kinetic energy) lead us to consider, in the most general way, that there is a *collision* if the path or internal state of a particle has been modified by the presence of one or many other particles in its vicinity.

1.7.1 Types of collision

We can distinguish two main categories of collision, depending on whether the Coulomb force is directly involved, or not.

Collisions not involving the Coulomb force

These concern collisions between two neutral particles, and most collisions between a neutral particle and a charged particle. We can differentiate, in this case, elastic and inelastic collisions.

Elastic collisions

These can be represented by an impact between two hard spheres, with conservation of total kinetic energy. They are principally low energy electron-neutral collisions, for example, below the energy threshold of the first excited atomic (molecular) level.

Inelastic collisions

There is no conservation of total kinetic energy. For example, still considering electron neutral collisions, an inelastic collision can occur provided the electron energy is above the threshold for excitation or ionisation of atoms, dissociation of molecules, or even for chemical ion-molecule reactions[25]. The processes of charge capture are equally inelastic in nature, because the internal energy of the participating atoms (molecules) will be modified (see also Sect. 1.7.9).

Examples of inelastic collisions

1. Superelastic collisions (or collisions of the 2nd kind)
   An atom in an excited state can transfer its internal energy, either totally or in part, in the form of kinetic energy to an atom or an electron, by means of a collision. When the atom (molecule) A in an excited state j above the ground state collides with an electron, we have:

$$\mathrm{A}(j) + \mathrm{e} \rightarrow \mathrm{A}(0) + \underline{\underline{\mathrm{e}}} \,. \tag{1.58}$$

[25] Ion-molecule (and molecule-molecule) reactions produce the numerous chemical species present in certain reactive gas plasmas, for example in hydrocarbon plasmas.

This collision mechanism is especially favourable when the excited atom is in a state referred to as metastable, in which the lifetime[26] is much longer than for radiative states that undergo an electric-dipole transition. As an example, the impact between an electron and a mercury atom in a metastable state can result in an atom in the ground state and provide an energy of 4.7 or 5.6 eV (there are two possible metastable states) to the incident electron.

2. Transfer of charge (charge exchange)
   During a collision between a neutral atom B with an ion A^+, there is a strong probability that the neutral gives an electron to the ion, which is then neutralised:
   $$A^+ + B \rightarrow A + B^+ \,. \tag{1.59}$$
   Thus, an ion A^+ previously accelerated in a high electric field can be converted to a high energy neutral atom, unaffected by the presence of electric or magnetic fields.
3. Electron capture (attachment process)
   Negative ions are created by the capture of an electron by a neutral species. One very effective process is dissociative attachment:
   $$AB + e \rightarrow A + B^- \,, \tag{1.60}$$
   where the electron attaches to one of the fragments of the molecule dissociated during the collision.

Coulomb collisions

The interaction between charged particles is governed by the *Coulomb force* which, in the case of a "collision" between an ion (with Z positive charges) and an electron, may be expressed by:

$$F = \frac{Ze^2}{4\pi\epsilon_0 r^2} \,. \tag{1.61}$$

As in the case of non-Coulomb collisions, we can differentiate between elastic and inelastic collisions.

Elastic collisions

This is the case of electron-electron, electron-ion and ion-ion collisions when the electron energy is too low ($T_{eV} < 100\,\text{eV}$) for the emission or absorption of EM radiation. Elastic Coulomb collisions are discussed in detail in Appendix V.

[26] The lower the pressure of the carrier gas, the longer the life of the metastable states ($\approx$ µs to several hours). The lifetime of the electric-dipole radiative states is independent of pressure ($\approx 10^{-7}$–10^{-8} seconds).

Inelastic Coulomb collisions

Coulomb's collisions can also be inelastic, and lead to either recombination processes, or emission and absorption of EM radiation as mentioned earlier.

- Examples of recombination processes

  1. Electron-ion recombination
     An electron and a positive ion can neutralise each other. This is the case for *radiative recombination*:

     $$e + A^+ \rightarrow A + h\nu \tag{1.62}$$

     and for *dissociative recombination* of a molecular ion:

     $$e + AB^+ \rightarrow A + B\,. \tag{1.63}$$

     As in the case of charge exchange, the process of dissociative recombination is extremely effective.
  2. Mutual neutralisation
     In plasmas rich in negative ions, there is a very strong probability that a negative ion will give an electron to a positive ion. There is thus a *mutual neutralisation*:

     $$A^+ + B^- \rightarrow A + B\,. \tag{1.64}$$

     Positive or negative ions, previously accelerated in a high electric field, can thus be converted to high-energy neutrals, unaffected by electric or magnetic fields.
  3. Electron detachment
     Negative ions can also lose their electron during a collision with an electron, by *electron detachment*:

     $$e + A^- \rightarrow A + 2e\,. \tag{1.65}$$

     With mutual neutralisation, electron detachment is the most effective mechanism for the loss of negative ions.

- Examples of the emission or absorption of radiation

  1. Bremsstrahlung
     The emission or absorption of radiation can result from electron-electron, electron-ion and ion-ion collisions when the energies of the charged particles are sufficiently high ($T_{eV} > 100\,\text{eV}$). We encounter this type of interaction, for example, in high flux laser plasmas, in the case where the electron penetrates the electron shells, without displacing any of the electrons. We distinguish Bremsstrahlung (braking radiation) of the direct type (emission of energy in the form of photons)

from inverse Bremsstrahlung (absorption of photons). This radiation is distributed in the continuum across a spectral band, which is quite broad, generally in the X-ray region.

2. Inner-shell atomic line emission
Consider again the case of an electron-neutral interaction. This time, let the energy of the electron be such that it reaches the innermost shell (K shell) of the atom (heavy) and dislodges an electron there, which generates an X-ray on returning to its shell. The spectral emission is in the form of line radiation, which generally dominates the continuum spectral band of Bremsstrahlung radiation.

Remark:

The probability of these different collisions occurring can be characterised by a *reaction coefficient*. We will see (Sect. 1.7.9, remark 2) that this coefficient, written k_{ij}[27], equal to $\langle \hat{\sigma}_{ij}(w_{\alpha\beta}) w_{\alpha\beta} \rangle$ where $\hat{\sigma}_{ij}(w_{\alpha\beta})$ is the effective cross-section of the reaction considered, $w_{\alpha\beta}$ is the modulus of the relative velocity of the particles α and β in the interaction; the symbol $\langle\ \rangle$ represents an average taken over the velocity (or energy) distribution function of the particles. Henceforth, it is convenient to define the concept of effective collision frequency explicitly (Sect. 1.7.3 and after).

1.7.2 Momentum exchange and energy transfer during a collision between two particles

The considerations and results from this section will allow us to subsequently quantify the physical significance of the dependence of the effective cross-section for momentum transfer with respect to the angle of deflection of the particles after collision (Sect. 1.7.4). It also enables us to better understand the collisional term in the momentum transport equation.

We assume that all collisions are binary, including those between charged particles, on the understanding that in the last case, this is only a first approximation. Furthermore, as is commonly the case in the kinetic theory of gases, we consider that the trajectory of a particle can be separated into two parts: the part of the trajectory occurring between two collisions, during which each particle individually experiences the external forces, and that part of the trajectory which is principally affected by the (mutual) collisional interaction, during which the external forces are ignored.

---

[27] The subscripts i and j indicate that this involves either an interaction between particles of type i and those of type j, or that the particle (atom, molecule, ion) which experiences a collision moves from state i to state j.

Conservation equations and identification of independent variables (not determined by the conservation equations)

Consider two particles α and β, whose velocities $\boldsymbol{w}_\alpha$ and $\boldsymbol{w}_\beta$ are known, a priori, before the collision[28]. Following the assumption noted above that no external force is present for the duration of the collision, there is conservation of momentum and total energy[29]:

$$\boldsymbol{p} \equiv \boldsymbol{p}_\alpha + \boldsymbol{p}_\beta \equiv m_\alpha \boldsymbol{w}_\alpha + m_\beta \boldsymbol{w}_\beta = m_\alpha \boldsymbol{w}'_\alpha + m_\beta \boldsymbol{w}'_\beta \equiv \boldsymbol{p}' , \tag{1.66}$$

$$\frac{m_\alpha w_\alpha^2}{2} + \frac{m_\beta w_\beta^2}{2} = \frac{m_\alpha w_\alpha'^2}{2} + \frac{m_\beta w_\beta'^2}{2} + \Delta\mathcal{E} , \tag{1.67}$$

where the "prime" indicates the values after the collision. The energy term $\Delta\mathcal{E}$ allows us to include inelastic collisions; this quantity represents the difference in internal energy of the particles after the collision:

- $\Delta\mathcal{E} = 0$ for inelastic collisions
- $\Delta\mathcal{E} > 0$ for collisions of the 1st kind: excitation and ionisation
- $\Delta\mathcal{E} < 0$ for collisions of the 2nd kind: superelastic de-excitation

It should be noted that radiative phenomena (absorption and emission of photons) are not included in the present context.

For a given value of $\Delta\mathcal{E}$ (we use the published energy levels for excitation and ionisation), we have four equations: (1.66) is vectorial and (1.67) is scalar. Since we require six components to completely characterise the velocity vectors after the collision, $\boldsymbol{w}'_\alpha$ and $\boldsymbol{w}'_\beta$, we are left with two components which are not determined by the conservation equations (1.66) and (1.67): these two quantities are determined by the law of interaction governing the type of collision considered, taking into account the initial relative position of the particles.

We will now introduce a change of reference frame, in order to express the kinetic quantities in the centre of mass frame, in place of the laboratory frame. This leads us to expressions that better describe the physics of the collisional interactions.

Relative velocity of two particles and velocity of their centre of mass

By definition, the position $\boldsymbol{r}_0$ of the *centre of mass* (CM) (of two particles α and β, with positions $\boldsymbol{r}_\alpha$ and $\boldsymbol{r}_\beta$ in the laboratory frame, is given by:

---

[28] Thus making the supposition that the plasma particles are distinguishable and can therefore be described in a non quantum manner, which is generally correct.

[29] The contents of this section is a classical development of kinetic theory, which can be found, for example, in V.E. Golant et al.

$$\boldsymbol{r}_0 = \frac{m_\alpha \boldsymbol{r}_\alpha + m_\beta \boldsymbol{r}_\beta}{m_\alpha + m_\beta} \tag{1.68}$$

from which:

$$\boldsymbol{w}_0 = \frac{m_\alpha \boldsymbol{w}_\alpha + m_\beta \boldsymbol{w}_\beta}{m_\alpha + m_\beta}\,, \tag{1.69}$$

where $\boldsymbol{w}_0$ is the velocity of the CM frame. The CM is in *uniform motion* during the interaction, because total momentum is conserved (see (1.69)) throughout the collision, hence:

$$\boldsymbol{w}_0 = \boldsymbol{w}_0' \,. \tag{1.70}$$

The fact that the CM is in uniform motion allows us to use it to describe the motion of the particles during the interaction: their velocities in this frame, denoted by $\boldsymbol{w}_{\alpha 0}$ and $\boldsymbol{w}_{\beta 0}$ before the collision and $\boldsymbol{w}'_{\alpha 0}$ and $\boldsymbol{w}'_{\beta 0}$ after collision, are obtained by setting $\boldsymbol{w}_0 = 0$ in (1.69), where we have replaced $\boldsymbol{w}_\alpha$ and $\boldsymbol{w}_\beta$ by $\boldsymbol{w}_{\alpha 0}$ and $\boldsymbol{w}_{\beta 0}$. This leads us to the following simple relations:

$$\boldsymbol{w}_{\beta 0} = -\frac{m_\alpha}{m_\beta}\boldsymbol{w}_{\alpha 0}\,, \qquad \boldsymbol{w}'_{\beta 0} = -\frac{m_\alpha}{m_\beta}\boldsymbol{w}'_{\alpha 0}\,, \tag{1.71}$$

which shows that the velocities of the two particles, both before and after the collision, are anti-parallel in the CM frame (see also Appendix IV). This property suggests that we should introduce their relative velocity $\boldsymbol{w}_{\alpha\beta}$ in the calculations:

$$\boldsymbol{w}_{\alpha\beta} \equiv \boldsymbol{w}_\alpha - \boldsymbol{w}_\beta = \boldsymbol{w}_{\alpha 0} - \boldsymbol{w}_{\beta 0} \tag{1.72}$$

from which we can write the expression for the velocities of the particles in the CM frame:

$$\boldsymbol{w}_{\alpha 0} = \left(\frac{m_\beta}{m_\alpha + m_\beta}\right)\boldsymbol{w}_{\alpha\beta}\,, \qquad \boldsymbol{w}_{\beta 0} = -\left(\frac{m_\alpha}{m_\alpha + m_\beta}\right)\boldsymbol{w}_{\alpha\beta}\,. \tag{1.73}$$

These various transformations allow us to completely determine the motion of the particles in the laboratory frame as the superposition of the rectilinear motion of the CM and the relative motion of the particles in this frame. In effect:

$$\boldsymbol{w}_\alpha \equiv \boldsymbol{w}_0 + \boldsymbol{w}_{\alpha 0} = \boldsymbol{w}_0 + \left(\frac{m_\beta}{m_\alpha + m_\beta}\right)\boldsymbol{w}_{\alpha\beta}\,, \tag{1.74}$$

$$\boldsymbol{w}_\beta \equiv \boldsymbol{w}_0 + \boldsymbol{w}_{\beta 0} = \boldsymbol{w}_0 - \left(\frac{m_\alpha}{m_\alpha + m_\beta}\right)\boldsymbol{w}_{\alpha\beta}\,. \tag{1.75}$$

Remark: As we will see below, the CM is the frame in which we can describe binary collisions "naturally" (effective cross sections, collision frequencies, mean free paths are best introduced in this frame), the use of the relative velocity of particles experiencing a collision being an essential element of this description.

Expression for the conservation of total energy as a function of relative velocity only

Taking account of (1.74) and (1.75), we can write:

$$\frac{m_\alpha w_\alpha^2}{2} + \frac{m_\beta w_\beta^2}{2} \equiv \left(\frac{m_\alpha + m_\beta}{2}\right) w_0^2 + \frac{\mu_{\alpha\beta} w_{\alpha\beta}^2}{2} , \tag{1.76}$$

where $\mu_{\alpha\beta}$ is the *reduced mass*: $\mu_{\alpha\beta} \equiv \dfrac{m_\alpha m_\beta}{m_\alpha + m_\beta}$ (1.77)

and $\mu_{\alpha\beta} w_{\alpha\beta}^2/2$ is the kinetic energy related to the relative motion.

The conservation equation (1.67) can thus be written:

$$\left(\frac{m_\alpha + m_\beta}{2}\right) w_0^2 + \frac{\mu_{\alpha\beta} w_{\alpha\beta}^2}{2} = \left(\frac{m_\alpha + m_\beta}{2}\right) w_0'^2 + \frac{\mu_{\alpha\beta} {w'}_{\alpha\beta}^2}{2} + \Delta\mathcal{E} \tag{1.78}$$

and since $\boldsymbol{w}_0 = \boldsymbol{w}_0'$ (1.70), we find finally:

$$\frac{\mu_{\alpha\beta} w_{\alpha\beta}^2}{2} = \frac{\mu_{\alpha\beta} {w'}_{\alpha\beta}^2}{2} + \Delta\mathcal{E} . \tag{1.79}$$

Only the kinetic energy with respect to the relative motion can be transferred into internal energy (potential energy); the individual velocities do not enter into such a transfer.

Particular case of an electron-atom collision

The atom (particle β) is assumed to be at rest with respect to the electron (particle α): $\boldsymbol{w}_{\alpha\beta} = \boldsymbol{w}_\alpha - \boldsymbol{w}_\beta \approx \boldsymbol{w}_\alpha$. Taking account of the fact that $m_\beta \gg m_\alpha$, we have $\mu_{\alpha\beta} \approx m_\alpha$ and equation (1.79) then reduces to:

$$\frac{m_\alpha}{2}\left(w_\alpha^2 - w_\alpha'^2\right) \equiv \Delta\mathcal{E}_{c\alpha} = \Delta\mathcal{E} , \tag{1.80}$$

which signifies that the change of internal energy of an atom during an inelastic collision is equal to the change of kinetic energy of the electron, the kinetic energy of the atom remaining essentially unchanged.

Change of momentum of a particle following an elastic collision ($\Delta\mathcal{E} = 0$)

For the particle α, by definition:

$$\Delta\boldsymbol{p}_\alpha \equiv m_\alpha \boldsymbol{w}_\alpha' - m_\alpha \boldsymbol{w}_\alpha = m_\alpha \boldsymbol{w}_{\alpha 0}' - m_\alpha \boldsymbol{w}_{\alpha 0} , \tag{1.81}$$

and this can be expressed uniquely (following (1.73)) as a function of the difference in the relative velocities of particles α and β before and after the collision as[30]:

$$\Delta \boldsymbol{p}_\alpha = \frac{m_\alpha m_\beta}{m_\alpha + m_\beta}(\boldsymbol{w}'_{\alpha\beta} - \boldsymbol{w}_{\alpha\beta}) \tag{1.82}$$

and setting:

$$\boldsymbol{w}'_{\alpha\beta} - \boldsymbol{w}_{\alpha\beta} \equiv \Delta \boldsymbol{w}_{\alpha\beta} \,, \tag{1.83}$$

we obtain:

$$\Delta \boldsymbol{p}_\alpha = \mu_{\alpha\beta} \Delta \boldsymbol{w}_{\alpha\beta} \,, \tag{1.84}$$

a remarkable result that we will exploit next by developing the vector $\Delta \boldsymbol{w}_{\alpha\beta}$ in an appropriate frame of reference.

As we have seen at the beginning of the derivation of Eqs. (1.66) and (1.67), two components out of six of the velocity vectors after the collision depend on the laws of interaction. These two unknowns can be expressed, taking into account the relative direction of the velocities $\boldsymbol{w}_{\alpha\beta}$ and $\boldsymbol{w}'_{\alpha\beta}$, by means of the angles θ and ϕ of a spherical coordinate system, as shown in Fig. 1.8.

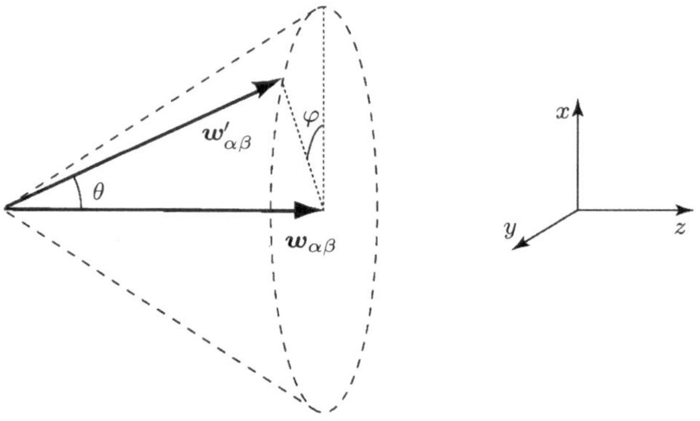

Fig. 1.8 Schematic of the relative velocities before and after collision, $\boldsymbol{w}_{\alpha\beta}$ and $\boldsymbol{w}'_{\alpha\beta}$, in a spherical coordinate system (θ, ϕ) tied to the centre of mass, with $\boldsymbol{w}_{\alpha\beta}$ directed along the z axis.

In this spherical coordinate system, ϕ is the angle made with the *interaction plane* (that is to say, the plane containing $\boldsymbol{w}_{\alpha\beta}$ and $\boldsymbol{w}'_{\alpha\beta}$), with a fixed plane (somewhere in space) including $\boldsymbol{w}_{\alpha\beta}$. The angle θ, between $\boldsymbol{w}_{\alpha\beta}$ and $\boldsymbol{w}'_{\alpha\beta}$, is the *scattering angle* of the particles in the CM frame. The angle θ depends only on the force law and the *impact parameter* (the distance of closest approach of the two particles if there is no interaction, Fig. IV.1 of Appendix IV). In the laboratory frame, the scattering angle $\theta_{\alpha L}$ is defined by the velocity of the "incident" particle before and after the collision, $\boldsymbol{w}_\alpha$ and $\boldsymbol{w}'_\alpha$ (see Fig. IV.2 of Appendix IV).

---

[30] Obviously $\Delta \boldsymbol{p}_\alpha = -\Delta \boldsymbol{p}_\beta$.

We will now project $\Delta \boldsymbol{w}_{\alpha\beta}$ on the three axes of the coordinate system in Fig. 1.8:

- Along $\boldsymbol{w}_{\alpha\beta}$ (z axis of the chosen frame):

$$(\Delta w_{\alpha\beta})_z = |\boldsymbol{w}'_{\alpha\beta}| \cos\theta - |\boldsymbol{w}_{\alpha\beta}| \,, \tag{1.85}$$

but $|\boldsymbol{w}_{\alpha\beta}| = |\boldsymbol{w}'_{\alpha\beta}|$ due to the conservation of kinetic energy ((1.79), with $\Delta\mathcal{E} = 0$), hence:

$$(\Delta w_{\alpha\beta})_z = |\boldsymbol{w}_{\alpha\beta}|(\cos\theta - 1) \,. \tag{1.86}$$

- Along the directions perpendicular to $\boldsymbol{w}_{\alpha\beta}$ (x and y axes):

$$(\Delta w_{\alpha\beta})_x = |\boldsymbol{w}_{\alpha\beta}| \sin\theta \cos\varphi \tag{1.87}$$

because the projection of $\Delta \boldsymbol{w}_{\alpha\beta}$ in the x direction, $\mathrm{Pr}_x(\Delta w_{\alpha\beta})$, is by definition equal to $\mathrm{Pr}_x(w'_{\alpha\beta}) - \mathrm{Pr}_x(w_{\alpha\beta})$ where here, $\mathrm{Pr}_x(w_{\alpha\beta}) = 0$ such that:

$$\mathrm{Pr}_x(\Delta \boldsymbol{w}_{\alpha\beta}) = |\boldsymbol{w}'_{\alpha\beta}| \sin\theta \cos\varphi = |\boldsymbol{w}_{\alpha\beta}| \sin\theta \cos\varphi \,. \tag{1.88}$$

For the same reason:

$$(\Delta w_{\alpha\beta})_y = |\boldsymbol{w}_{\alpha\beta}| \sin\theta \sin\varphi \,. \tag{1.89}$$

In the case of a central force, all values of ϕ are statistically equally probable; we therefore say that the particle scattering is isotropic (isotropic in ϕ). From this fact, if we have a sufficient number of particles, the average values of $\cos\phi$ (1.87) and $\sin\phi$ (1.89) are zero, and we can write:

$$\Delta \boldsymbol{w}_{\alpha\beta} = -(1 - \cos\theta)\boldsymbol{w}_{\alpha\beta} \,, \tag{1.90}$$

where, finally, the explicit form of equation (1.84) is:

$$\Delta \boldsymbol{p}_\alpha = -\frac{m_\alpha m_\beta}{m_\alpha + m_\beta}(1 - \cos\theta)(\boldsymbol{w}_\alpha - \boldsymbol{w}_\beta) \,. \tag{1.91}$$

This expression for the change in momentum of the particle α during an elastic collision with the particle β introduces a dependence on $(1 - \cos\theta)$ with scattering angle θ.

Particular case of an electron-atom collision

The atom (particle β) is assumed to be at rest relative to the electron (particle α) such that $w_\beta \ll w_\alpha$ in (1.91), from which:

$$\frac{\Delta \boldsymbol{p}_\alpha}{\boldsymbol{p}_\alpha} = -\frac{m_\beta}{m_\alpha + m_\beta}(1 - \cos\theta) \,. \tag{1.92}$$

Since the incident particle α is much lighter than the particle β, we have:

$$\frac{\Delta \boldsymbol{p}_\alpha}{\boldsymbol{p}_\alpha} = -(1-\cos\theta), \tag{1.93}$$

such that, for $\theta = \pi$, we obtain:

$$\frac{\Delta \boldsymbol{p}_\alpha}{\boldsymbol{p}_\alpha} = -2\,, \tag{1.94}$$

or alternatively:

$$\boldsymbol{p}'_\alpha = -\boldsymbol{p}_\alpha\,. \tag{1.95}$$

This result corresponds simply to a change in sign of the impulse of the incident particle, while the target particle remains immobile (assuming $m_\beta \approx \infty$); this is the largest possible value of $\Delta|\boldsymbol{p}_\alpha|$.

Change in the kinetic energy of a particle following an elastic collision ($\Delta\mathcal{E} = 0$)

In the case of the particle α, its change in kinetic energy is given by:

$$\begin{aligned}\Delta\mathcal{E}_{c\alpha} &= \frac{m_\alpha w_\alpha'^2}{2} - \frac{m_\alpha w_\alpha^2}{2} = \frac{m_\alpha}{2}\left[(\boldsymbol{w}_0' + \boldsymbol{w}_{\alpha 0}')^2 - (\boldsymbol{w}_0 + \boldsymbol{w}_{\alpha 0})^2\right] \\ &= \frac{m_\alpha}{2}\left[2\boldsymbol{w}_0\cdot(\boldsymbol{w}_{\alpha 0}' - \boldsymbol{w}_{\alpha 0}) + w_{\alpha 0}'^2 - w_{\alpha 0}^2\right]\end{aligned} \tag{1.96}$$

since $\boldsymbol{w}_0 = \boldsymbol{w}_0'$ (1.70).
Furthermore, $w_{\alpha 0}' = w_{\alpha 0}$. In effect, since $\boldsymbol{w}_{\alpha 0} = \boldsymbol{w}_{\alpha\beta} m_\beta/(m_\alpha + m_\beta)$ (see (1.73)), we can write:

$$w_{\alpha 0}^2 = \left(\frac{m_\beta}{m_\alpha + m_\beta}\right)^2 w_{\alpha\beta}^2 \text{ and } w_{\alpha 0}'^2 = \left(\frac{m_\beta}{m_\alpha + m_\beta}\right)^2 w'^2_{\alpha\beta} \tag{1.97}$$

and, since $w_{\alpha\beta}^2 = w'^2_{\alpha\beta}$ from (1.79) with $\Delta\mathcal{E} = 0$, we therefore have:

$$w_{\alpha 0}'^2 = w_{\alpha 0}^2\,. \tag{1.98}$$

The expression (1.96) then reduces to:

$$\Delta\mathcal{E}_{c\alpha} = m_\alpha \boldsymbol{w}_0\cdot(\boldsymbol{w}_{\alpha 0}' - \boldsymbol{w}_{\alpha 0}) = \boldsymbol{w}_0\cdot\Delta\boldsymbol{p}_\alpha \tag{1.99}$$

such that, from (1.69) and (1.91):

$$\Delta\mathcal{E}_{c\alpha} = -\left(\frac{m_\alpha \boldsymbol{w}_\alpha + m_\beta \boldsymbol{w}_\beta}{m_\alpha + m_\beta}\right)\frac{m_\alpha m_\beta}{m_\alpha + m_\beta}(1-\cos\theta)\cdot(\boldsymbol{w}_\alpha - \boldsymbol{w}_\beta) \tag{1.100}$$

or alternatively:

$$\Delta\mathcal{E}_{c\alpha} = -\frac{m_\alpha m_\beta}{(m_\alpha + m_\beta)^2}(1 - \cos\theta)\left[m_\alpha w_\alpha^2 - m_\beta w_\beta^2 + (m_\beta - m_\alpha)(\boldsymbol{w}_\beta \cdot \boldsymbol{w}_\alpha)\right] , \tag{1.101}$$

where the average of the scalar $\boldsymbol{w}_\beta \cdot \boldsymbol{w}_\alpha$ is zero, if all the initial relative orientations of the particles have the same probability density[31]. We can finally write (still in the CM frame):

$$\Delta\mathcal{E}_{c\alpha} = -2\frac{m_\alpha m_\beta}{(m_\alpha + m_\beta)^2}(1 - \cos\theta)\left[\frac{m_\alpha w_\alpha^2}{2} - \frac{m_\beta w_\beta^2}{2}\right] . \tag{1.102}$$

Remarks:

1. In (1.102), the term $$2\frac{m_\alpha m_\beta}{(m_\alpha + m_\beta)^2} \equiv \delta \tag{1.103}$$
   is called the *energy transfer coefficient*. This coefficient has a maximum value of $1/2$ for $m_\alpha = m_\beta$. Note that the cumulative value of the difference $(1 - \cos\theta)$ over the ensemble of values of the scattering angle θ (0 to π) is equal to unity.
   For an electron colliding with an atom, $\delta \approx 2m_e/M$, which is a very weak transfer of energy during the collision. In this case, the maximum transfer of energy from the electron to the atom, assuming that the atom (particle β) is at rest relative to the electron, occurs for a scattering angle $\theta = \pi$. The maximum fraction of energy transferred from the electron to the atom is then:
   $$\frac{\Delta\mathcal{E}_{ce}}{\mathcal{E}_{ce}} = -\frac{4m_e}{M} , \tag{1.104}$$
   the minus sign indicating an energy transfer from the electron to the atom.
2. The transfer of kinetic energy from one particle to another is, according to (1.102), proportional to the difference in kinetic energy between the two particles involved in the collision.
3. Expressed in the CM frame, the change of kinetic energy and momentum following a collision has the same dependence on the scattering angle, i.e. $(1 - \cos\theta)$: we will use this result in the definition of certain cross-sections (Sect. 1.7.4).
4. In the laboratory frame, the relations we have just developed are much more complicated. For instance, the fraction of energy lost by the incident particle following a collision is given by:

---

[31] Do not confuse this property (**before** collision) with that of isotropy in φ (angle **after** interaction) of interactions in the presence of central forces.

$$\frac{\Delta\mathcal{E}_{c\alpha}}{\mathcal{E}_{c\alpha}} = -2r_{\alpha/\beta}\frac{\left[1-\cos\theta_{\alpha L}(1-r_{\alpha/\beta}^2\sin\theta_{\alpha L})^{1/2}+r_{\alpha/\beta}\sin^2\theta_{\alpha L}\right]}{(1+r_{\alpha/\beta})^2} \tag{1.105}$$

where $r_{\alpha/\beta} \equiv m_\alpha/m_\beta$ (Heald and Wharton). We can verify that, for $m_\alpha \ll m_\beta$ (i.e. $r_{\alpha/\beta} \approx 0$), this expression becomes $-2r_{\alpha/\beta}(1-\cos\theta_{\alpha L})$, the laboratory frame thus coinciding with the CM (see also Appendix IV).

1.7.3 Microscopic differential cross-section

In practice, it is impossible to determine all the kinetic parameters of a collision occurring in a plasma: there are too many particles involved, and their motion, before collision, is random. To overcome this difficulty, we use a statistical description. One such description leads us to the concept of a cross-section.

Characterisation of the angular dispersion of a mono-energetic beam of particles by one scattering centre

Laboratory frame

Consider a mono-energetic beam of particles incident on a unique scattering centre at rest (Fig. 1.9)[32]. The flux of these particles, with velocity $\boldsymbol{w}$ is given by $\boldsymbol{\Gamma} = n\boldsymbol{w}$, where n is the number of particles per unit volume: this flux is a number of particles per unit surface area per second. As Fig. 1.9 suggests, the number of particles $\mathrm{d}N_d/\mathrm{d}t$ deviated by the scattering centre, per unit time and in the solid angle $\mathrm{d}\Omega(\theta_L, \varphi)$, is:

- proportional to the solid angle $\mathrm{d}\Omega$, where $\mathrm{d}\Omega = \sin\theta_L \mathrm{d}\theta_L \mathrm{d}\phi$ in spherical coordinates,
- proportional to the flux $\boldsymbol{\Gamma}$ of the incident particles,

such that we can set:

$$\frac{\mathrm{d}N_d}{\mathrm{d}t} = \hat{\sigma}(w, \theta_L)|\boldsymbol{\Gamma}|\,\mathrm{d}\Omega\,, \tag{1.106}$$

where the proportionality factor,$\hat{\sigma}$, referred to as the *microscopic scattering cross-section*, depends on θ_L and, in general[33], on the modulus of the particle velocity w.

Note that $\hat{\sigma}$ has the units of area (cm^2 are commonly used in plasma physics), as determined by comparing the dimensional analysis of the terms on the left and on the right side of (1.106):

[32] This is an idealised description, which isolates a particle as a unique target and, in addition, assumes that it is at rest.

[33] In the so called "billiard ball" model, where particles are assumed to be rigid spheres, the cross-sections do not depend on the velocity of the particles (see exercise 1.10).

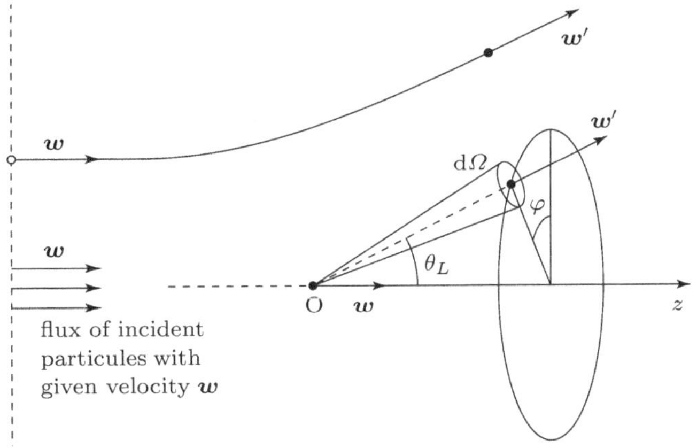

Fig. 1.9 Incident flux of particles with velocity $\boldsymbol{w}$, deviated by a scattering centre initially at rest at point O (laboratory frame).

$$\frac{\mathrm{d}N_d}{\mathrm{d}t} \equiv \text{number of particles (scattered)/s} ,$$

$$\hat{\sigma}|\boldsymbol{\Gamma}|\,\mathrm{d}\Omega \equiv \hat{\sigma}\left[\frac{\text{number of particles}}{\text{cm}^2\,\text{s}}\right] [\text{dimensionless}] .$$

From the physics standpoint, $\hat{\sigma}$ does not represent the real surface of the scattering centre, but rather that "seen" by the incident particles, depending on their velocity, for example, hence the term effective scattering surface or cross-section; the larger this value, the more probable the interaction.

Centre of mass frame

The situation described above corresponds perfectly to the case of a beam of electrons directed towards an atom (scattering centre) assumed at rest with respect to the electrons.

Due to the small mass of the electron, the description in the CM frame coincides with the laboratory frame (Appendix IV). In the general case, however, the study of binary collisions (Sect. 1.7.2) is more effectively treated in the CM frame. In effect, the description of the collisions is simplified (for example, a single angle θ is sufficient to characterise the scattering of particles, while in the laboratory frame it is necessary to include the angles $\theta_{\alpha L}$ and $\theta_{\beta L}$, as shown in Appendix IV) and more general (for example, the dependence on the individual velocities is replaced, in the centre of mass frame, by the relative velocities (modulus) of the particles before and after the collision).

Consider again the case of a mono-energetic beam of particles α with velocity $\boldsymbol{w}_\alpha$ and a scattering centre of initial velocity $\boldsymbol{w}_\beta = 0$ (laboratory frame). Since the velocity $\boldsymbol{w}_\alpha$ is equal to the relative velocity $\boldsymbol{w}_{\alpha\beta}$ in the laboratory frame, in the CM frame (Fig. 1.10), the particle flux can be expressed as

$\boldsymbol{\Gamma} = m\boldsymbol{w}_{\alpha\beta}$ [34]. The number of particles deviated by the scattering centre in the centre of mass frame can be expressed in an analogous fashion to (1.106):

$$\frac{\mathrm{d}N_d}{\mathrm{d}t} = \hat{\sigma}(w_{\alpha\beta}, \theta)|\boldsymbol{\Gamma}| \,\mathrm{d}\Omega \,, \tag{1.107}$$

where $\hat{\sigma}$ depends on $w_{\alpha\beta}$, the modulus of the relative velocity of the particles α and β. The relation thus obtained is of a completely general nature, precisely because the reference frame is the CM.

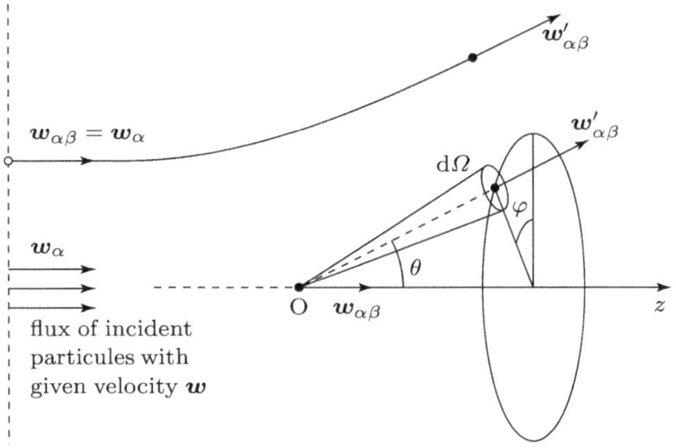

Fig. 1.10 Description in the centre of mass frame of an incident flux of particles α with velocity $\boldsymbol{w}_\alpha$, deviated by a scattering centre β initially at rest at point O ($\boldsymbol{w}_\beta = 0$) in the laboratory frame. The relative velocity $\boldsymbol{w}_{\alpha\beta}$ before the collision, identical in the two frames, leads, in the present case, to $\boldsymbol{w}_{\alpha\beta} = \boldsymbol{w}_\alpha$.

Remark: The differential cross-section can be expressed not only as a function of the relative velocity $w_{\alpha\beta}$, but also as a function of the kinetic energy $\mu_{\alpha\beta}w_{\alpha\beta}^2/2$ linked to the relative motion (1.79). In the case of electrons, we have $\mu_{\alpha\beta} \approx m_e$, such that the energy linked to the relative motion:

$$\frac{\mu_{\alpha\beta}w_{\alpha\beta}^2}{2} \simeq \frac{m_e w_e^2}{2} \tag{1.108}$$

is equal to the kinetic energy of the electrons.

Example of the measurement of a differential cross-section

Figure 1.11 shows the schematic view of an apparatus used to determine the angular dependence of the scattering of a beam of electrons by a gas.

[34] Note that in the centre of mass frame, the scattering centre is, in general, never at rest ($\boldsymbol{w}_{\beta 0} \neq 0$, see (1.73)).

Figure 1.12 shows the result of such a measurement for the case of elastic scattering by neon atoms, for different values of energy of the electron beam. The current obtained as a function of the scattering angle θ is proportional to the differential cross-section.

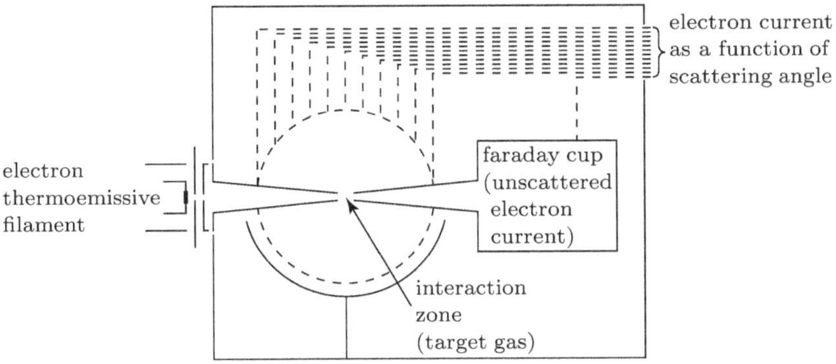

Fig. 1.11 Apparatus for measuring the differential cross-section of elastic collisions of electrons with a gas (from [31]).

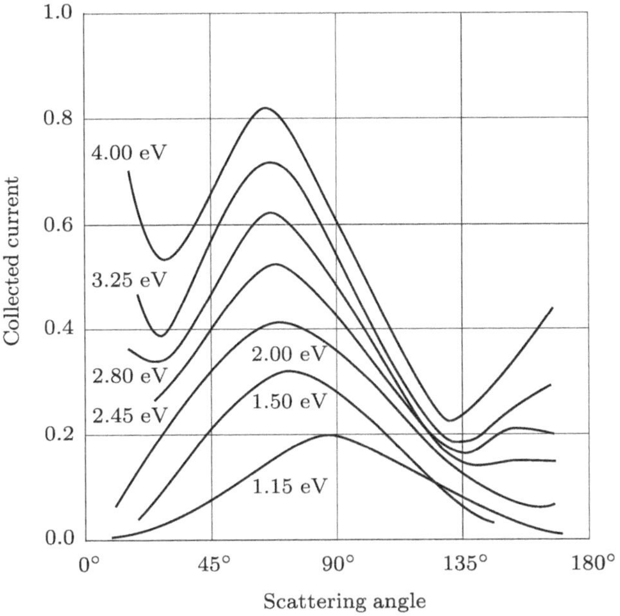

Fig. 1.12 Collected current (un-normalised differential cross section) in the case of elastic scattering of a beam of electrons of different energies by neon atoms (after [31]).

Remarks:

1. If the incident flux is sufficiently uniform and mono-energetic, and on condition that the scattering medium (interaction zone) is not favourable for multiple collisions of the same incident particle, the angular distribution of the scattered particles simply reflects the force law between the incident particles and the scattering particle at that energy.
   In the case of a Coulomb interaction, for example, we have (Rutherford scattering, see Appendix V):
$$\hat{\sigma}(w_{\alpha\beta}, \theta) = \frac{(e^2/8\pi\epsilon_0\mu_{\alpha\beta}w_{\alpha\beta}^2)^2}{\sin^4(\theta/2)} \,. \tag{1.109}$$
2. Recalling that $\boldsymbol{\Gamma}$ is a flux, relation (1.107) allows us to regard $\hat{\sigma}\mathrm{d}\Omega$ as the element of oriented surface (see Fig. 1.10) which, traversed by this flux, leads to $\mathrm{d}N_d/\mathrm{d}t$. This "effective" surface for capturing scattered particles varies with $w_{\alpha\beta}$ and θ, as shown in Fig. 1.12.
3. An incident beam of mono-energetic electrons can be represented in a quantum way by a planar monochromatic wave, partially dispersed by the target-particle during the interaction.
4. Some authors choose to define the differential cross-section as $\hat{\sigma}\mathrm{d}\Omega$ rather than $\hat{\sigma}$. On the other hand, we could equally regard our $\hat{\sigma}$ as being $\mathrm{d}\hat{\sigma}'/\mathrm{d}\Omega$, to emphasise the differential nature of this cross-section.

1.7.4 Total (integrated) microscopic cross-section

When integrating the *microscopic* (scattering by a single target) differential *cross-section* for collision $\hat{\sigma}$ over all the values of the scattering angle $\mathrm{d}\Omega$, we obtain the total microscopic cross-section; thus, assuming the scattering is isotropic in φ:
$$\hat{\sigma}_{tc}(w_{\alpha\beta}) = 2\pi \int_0^{\pi} \hat{\sigma}(w_{\alpha\beta}, \theta) \sin\theta \,\mathrm{d}\theta \,. \tag{1.110}$$

The total cross-section (1.110) is often divergent (this is the case for Rutherford scattering[35]). Note, in addition, that the value of $\hat{\sigma}_{tc}$ simply accounts for the number of particles scattered, while not taking into account the value of their momentum exchange; in fact, a collision in which $\theta = \pi$ will have zero contribution, while its contribution will be a maximum for $\theta = \pi/2$, for example, although in reality the change in momentum is a maximum for $\theta = \pi$. In

---

[35] In the case of Coulomb collisions, $\hat{\sigma}(\theta)$ is proportional to $\sin^{-4}(\theta/2)$ (see remark 1 in the previous section) and the integral (1.110) is therefore divergent for $\theta = 0$. This signifies that as $\theta \to 0$, there is a large probability of observing small deflections.

order to characterise the transfer of momentum, we use the following relation instead[36]:

$$\hat{\sigma}_{tm}(w_{\alpha\beta}) = 2\pi \int_0^\pi \hat{\sigma}(w_{\alpha\beta},\theta)(1-\cos\theta)\sin\theta \, d\theta \, . \qquad (1.111)$$

The weighting introduced by the factor $(1-\cos\theta)$ takes, in effect, account of the influence of the scattering angle in the exchange of momentum between particles (Sect. 1.7.2) during collisions; thus, the term ensures that the contribution to the integral (1.111) practically disappears for scattering in which $\theta \approx 0$.

The integral (1.111) converges for electron-neutral and ion-neutral collisions, but still diverges for Coulomb interactions (Appendix V): this divergence comes from the large contribution of long range collisions, which are very weak interactions in terms of transfer of energy ($\theta \approx 0$). These interactions, however, do not have any physical significance for distances longer than the Debye length (due to the effect of electrostatic screening), and it is therefore adequate to terminate the integral when the impact parameter becomes greater than the Debye length (Appendix V).

Remarks:

1. All total microscopic cross-sections will be designated by $\hat{\sigma}_{tx}$, where x can represent, inter alia, either c or m.
2. Experimental microscopic cross-sections are most often expressed in units of $\pi a_0^2 = 0.88 \times 10^{-16}\,\text{cm}^2$, where a_0 is the radius of the first orbit of the Bohr hydrogen atom.
3. Although in defining the concept of cross-section, we have considered an elastic collision, as we have remarked in (Sect. 1.7.1) above, we can also use the same concept to characterise all types of binary interaction: charge transfer, excitation, collisional de-excitation...

1.7.5 Total macroscopic cross-section

We have just defined the microscopic cross-section, where we have assumed the existence of a unique scattering centre, which is not truly realistic, but necessary to introduce the concept of a cross-section since it has only a physical meaning at the microscopic level. To measure a cross-section implies, in effect, that we consider an incident flux on a very large number of scattering

[36] Recall that, for two particles of given reduced mass $\mu_{\alpha\beta}$ and relative velocity $\boldsymbol{w}_{\alpha\beta}$, the factor $(1-\cos\theta)$ characterises the magnitude of the momentum exchange during a collision (1.91).

centres per unit volume. This leads us to define a cross section, referred to as macroscopic, and experimentally measurable, from which the corresponding microscopic cross-section can be deduced. In the following, we will establish the connection between the total microscopic and macroscopic cross-sections, starting with the formalism developed for the total microscopic cross-section (Sect. 1.7.4).

Consider a beam of particles with flux $\boldsymbol{\Gamma} = n\boldsymbol{w}$, incident on a semi-infinite medium (in y and z) this time containing not one but N scattering centres per unit volume, which are assumed to be at rest. We wish to calculate the remaining incident flux after it has traversed a distance x in the medium, under the assumption that the total microscopic cross-section is known. This may be schematically represented in the following manner (see Fig. 1.13):

- N: density of scattering centres
- A: surface of the slice considered, whose thickness is $\mathrm{d}x$, such that $NA\mathrm{d}x$ is the number of scattering centres in the slice and $(NA\mathrm{d}x)\hat{\sigma}_{tx}$ their total effective surface.

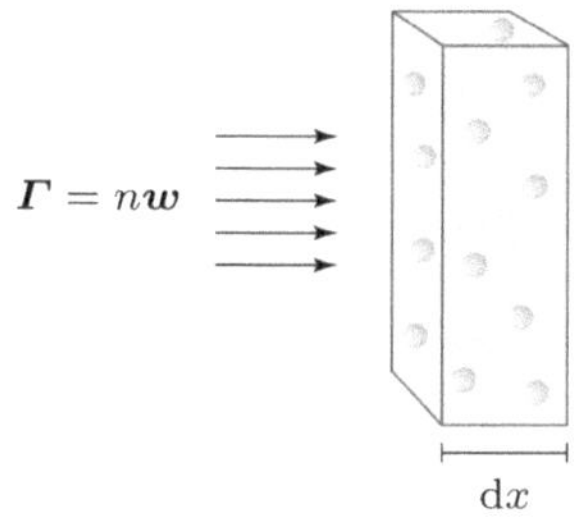

Fig. 1.13 Flux $\boldsymbol{\Gamma}$ of incident particles on a volume element of thickness $\mathrm{d}x$ and surface A, containing N target particles per unit volume.

The fraction $|\mathrm{d}\boldsymbol{\Gamma}|/|\boldsymbol{\Gamma}|$ of the flux intercepted by the scattering centres in the slice of thickness $\mathrm{d}x$ is equal to the ratio of the total effective surface over the surface A of the slice, i.e.:

$$\frac{|\mathrm{d}\boldsymbol{\Gamma}|}{|\boldsymbol{\Gamma}|} = -\frac{NA\hat{\sigma}_{tx}\mathrm{d}x}{A} \tag{1.112}$$

from which, after integration between $x = 0$ and x with $\boldsymbol{\Gamma}(x = 0) \equiv \boldsymbol{\Gamma}_0$:

$$\boldsymbol{\Gamma}(x) = \boldsymbol{\Gamma}_0 \exp\left(-N\hat{\sigma}_{tx}x\right) = \boldsymbol{\Gamma}_0 \exp(-P_x x) \ , \tag{1.113}$$

with:

$$N\hat{\sigma}_{tx} = P_x \ , \tag{1.114}$$

where P_x is the corresponding total *macroscopic cross-section.* Thus, if the subscript $x \equiv c$, then P_c represents the total cross-section for a simple collision:

$$P_c = N\hat{\sigma}_{tc} \tag{1.115}$$

while $x \equiv m$ refers to the total cross-section for momentum transfer:

$$P_m = N\hat{\sigma}_{tm} \,. \tag{1.116}$$

From (1.114), P_x is expressed[37] in cm^{-1}. Further, P_x represents the probability of collision in units of length. To demonstrate this, consider the expression (1.112):

$$\frac{(N\,\hat{\sigma}_{tx}A)\,\mathrm{d}x}{A} \equiv \frac{\begin{array}{c}\text{total cross-section of scattering centres}\\ \text{in a slice of thickness } \mathrm{d}x\end{array}}{\text{surface of the slice}}$$

$$= \begin{pmatrix}\text{probability of a collision}\\ \text{over a distance } \mathrm{d}x\end{pmatrix} \equiv P_x \mathrm{d}x \,. \tag{1.117}$$

Relation between P_x and its standard value

By convention, the published values of P_x are given for a reference pressure p_R, corresponding to 1 torr and 0°C, and denoted by:

$$P_{xo} = N_0\,\hat{\sigma}_{tx} \,, \tag{1.118}$$

where N_0 is the density of targets under these conditions (3.5377×10^{16} atoms or molecules per cm^{-3}). Knowing P_{x0}, we can calculate P_x at a particular temperature T_C (°C) and pressure p (torr). Consider N as the density of targets at T_C and p, by definition:

$$P_x \equiv N\,\hat{\sigma}_{tx} = \frac{N}{N_0}(N_0\,\hat{\sigma}_{tx}) = \frac{N}{N_0}P_{x0} \,. \tag{1.119}$$

Value of P_x at a particular pressure and temperature

It would be useful to replace the ratio N/N_0 in (1.119) by an expression that includes T_C and p directly, parameters which are easily measurable. From the perfect gas law, we have $N = p/(k_B T_K)$ and $N_0 = p_R/(k_B \times 273)$, from which:

$$\frac{N}{N_0} = \frac{p\times 273}{p_R T_K} = \frac{p \times 273}{p_R(273 + T_C)} \,, \tag{1.120}$$

where the subscripts K and C designate the temperatures expressed in kelvin and degrees Celsius respectively.

[37] According to the International System (SI), it would be more correct to express the density of particles in m^{-3}, but the cm^{-3} is well established as a practical unit, in particular in plasma physics. This leads to the microscopic cross-section being expressed in cm^{-2} and the macroscopic cross-section in cm^{-1}.

By convention, we write:

$$\frac{p}{p_R} \equiv \hat{p} \quad \text{and} \quad \frac{N}{N_0} \equiv p_0 \; , \tag{1.121}$$

where p_0 is the *reduced pressure* (note that this is a dimensionless quantity, as well as $\hat{p}$, and not really a pressure), and for the expression (1.120):

$$p_0 = \frac{\hat{p} \times 273}{273 + T_C} \; , \tag{1.122}$$

and we can verify that, for $\hat{p} = 1$ and $T_C = 0°\text{C}$, we have $p_0 = 1$ (without units) such that we can write (1.119) and (1.121) in a practical form:

$$P_x(T, p) = p_0 P_{x0} \; . \tag{1.123}$$

The flux of particles experiencing no collision over a distance x in the plasma (1.113) can now be expressed with the help of the macroscopic cross-section at standard conditions and the reduced pressure in the form:

$$\boldsymbol{\Gamma}(x) = \boldsymbol{\Gamma}_0 \exp(-p_0 P_{x0} x) \; . \tag{1.124}$$

Examples of cross-sections

Figure 1.14 compares the macroscopic electron-neutral collision cross section for momentum transfer P_m, and that for simple collisions P_c (i.e. only accounting for the number of collisions), for discharges in three rare gases. In general, we observe that $P_c > P_m$ (the behaviour of these cross-sections will be more fully discussed in Sect. 1.7.8). Note that, in these examples, the cross-sections are expressed as a function of the energy of the incident particles, rather than their velocity.

Remark:

The macroscopic cross-section for a simple collision P_c is obtained directly by measuring the attenuation of a beam of electrons (1.113): the value of P_c takes no account at all of the amount of momentum transfer, but of only the total number of collisions taking place in contrast to P_m. On the other hand, the macroscopic cross-section for P_m is determined by integrating the experimentally measured differential (macroscopic) cross-section over θ, following (1.111), shown, for example in Fig. 1.12. One can deduce the microscopic differential cross-section by a suitable normalisation of the macroscopic value.

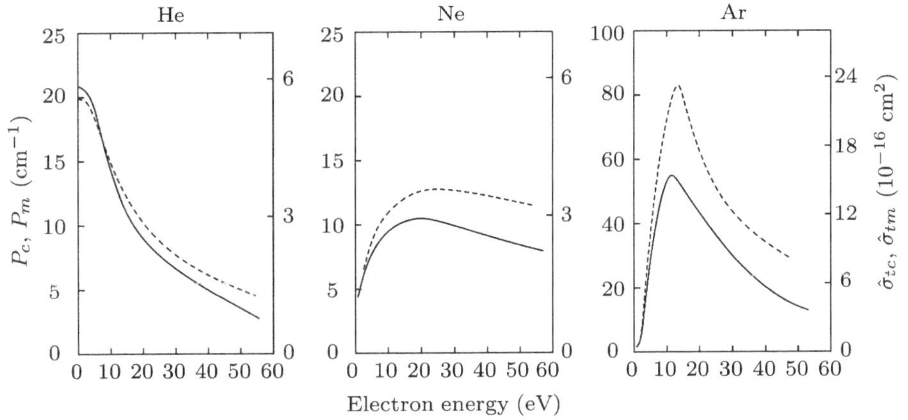

Fig. 1.14 Elastic electron-neutral cross-sections in three rare gases for standard conditions. The values of the macroscopic cross sections P_c (simple collision: dashes) and P_m (momentum transfer: full lines) are expressed in cm^{-1}, while the corresponding values for microscopic cross-sections are expressed in cm^2 (after Massey and Burhop, 1952).

1.7.6 Expression for the temperature of a plasma in electron-volt

The temperature T characterising the Maxwell-Boltzmann distribution for particles is normally expressed in kelvin (K). However, in plasma physics, this temperature is preferably expressed in electron-volt, T_{eV}. T can be converted to T_{eV} using the relation $k_BT/e = T_{eV}$, where $k_B = 1.38 \times 10^{-23}\,\mathrm{J\,K^{-1}}$ and $e = 1.6\times10^{-19}\,\mathrm{C}$. Since k_BT is an energy, $k_BT/e \equiv T_{eV}$ is an energy per unit charge (J/C) and should strictly be expressed in volt. Further, it should be noted that T_{eV} only represents 2/3 of the average energy of the particles in the plasma, $\frac{3}{2}k_BT$. Thus, for $T_{eV} = 1\,\mathrm{eV}$, the temperature in kelvin is about 11600 for an average energy which is really 1.5 eV.

This convention permits, inter alia, a quick estimation[38] of the values of the different cross sections participating in the various collision processes. We will show this by examining the case of electron collisions.

Velocity of an electron with energy of 1 eV

The velocity of an electron accelerated along the x direction, starting from a position of rest, by a potential difference U, obeys the expression:

[38] The cross-section evaluated at energy T_{eV} gives only an approximate value, for instance, of the average collision frequency (Sect. 1.7.8): the exact value is obtained by actually integrating this cross-section over the energy distribution function of the particles.

$$\frac{1}{2}m_e w_x^2 = eU \tag{1.125}$$

or:

$$w_x = \left(\frac{2eU}{m_e}\right)^{1/2} . \tag{1.126}$$

For $U = 1$ volt, the electron acquires an energy of 1 electron-volt (1.6×10^{-19} J) corresponding to a velocity:

$$w_x = 5.93 \times 10^5 \text{ m/s} . \tag{1.127}$$

The relation (1.126) remains valid for a beam of electrons provided that it is mono-energetic. It is also valid for a singly charged ion, replacing m_e by m_i.

Estimation of the value of a cross-section in a plasma

In a plasma, the motion of electrons is in three spatial directions and is not mono-energetic, which is different from the conditions of the measurement of a cross-section. If we wish nevertheless to estimate the value taken by a cross-section in a given plasma, one can imagine an "enlarged" beam of electrons with velocity v_{th}, the most probable velocity of the electrons in the plasma (Appendix I). The relation $\frac{1}{2}m_e v_{th}^2 = k_B T_e$, since it expresses the most probable kinetic energy of the electron, demonstrates the usefulness of the relation $T_{eV} = k_B T_e/e$ to obtain a good approximation of the value of the cross-section expressed as a function of energy, rather than velocity.

1.7.7 Collision frequency and mean free path between two collisions

Collision frequency

For a mono-energetic beam of particles with velocity w, the distance travelled by each particle in time t is $x = wt$, which allows us to write (1.124), with $\boldsymbol{\Gamma} = n\boldsymbol{w}$, in the form:

$$n(t) = n(0) \exp\left[-(p_0 P_{x0} w)t\right] . \tag{1.128}$$

$n(t)$ being the density of particles in the beam at time t and $n(0)$ its value at $t = 0$.

The characteristic decay-time of the exponential (1.128):

$$\tau = (p_0 P_{x0} w)^{-1} \tag{1.129}$$

represents the time taken by the beam to decay, as a result of collisions experienced, to 1/e (e being in the present case the base of natural logarithms) of its initial value.

To define the collision frequency, note from (1.128) that:

$$\frac{\mathrm{d}n}{\mathrm{d}t} = -\frac{n}{\tau} \tag{1.130}$$

and introducing $\tau = \nu_x^{-1}$:

$$\frac{\mathrm{d}n}{\mathrm{d}t} = -\nu_x n \,, \tag{1.131}$$

where $\mathrm{d}n/\mathrm{d}t$ represents the number of incident particles which exit the beam (as a result of collisions with the target particles) per unit volume per second: $\mathrm{d}n/\mathrm{d}t$ is thus, in its most general form, the total number of collisions per unit volume per second experienced by the incident particles. Dividing $\mathrm{d}n/\mathrm{d}t$ by n in (1.131) shows that ν_x is the number of collisions per second per incident particle (subscripts $x = c$: simple collision; $x = m$: collision for momentum transfer), in other words the *collision frequency* of a particle. From (1.129), (1.130) and (1.131), for a beam of particles with velocity w incident on targets at rest, we can identify:

$$\nu_x = p_0 P_{x0} w = N \hat{\sigma}_{tx} w \,. \tag{1.132}$$

For incident electrons whose energy is expressed in eV, $eU \equiv U_{eV}$, from (1.126) and (1.132) we have the numerical expression:

$$\nu_x(\mathrm{s}^{-1}) = 5.93 \times 10^7 \sqrt{U_{eV}}\, p_0 P_{x0}\, (\mathrm{cm}^{-1}) \,. \tag{1.133}$$

Remark: Note that ν_x refers to the collision frequency for a given velocity of incident particles: only the average frequency (defined in the following section) correctly represents the number of collisions per second in a medium in which there is a velocity distribution.

Free path between two collisions

Taking account of the meaning of ν_x, τ appears as the time between two collisions, so that the probable free path between two collisions ℓ_x is clearly:

$$\ell_x = w\tau = \frac{w}{\nu_x} \,. \tag{1.134}$$

From (1.132) and (1.134) and, taking account of (1.118) and (1.119), we find:

$$\ell_x = \frac{1}{p_0 P_{x0}} = \frac{1}{N \hat{\sigma}_{tx}} \,. \tag{1.135}$$

Knowing that τ is also the time necessary for the number of particles in the beam to decrease to 1/e of their initial value, ℓ_x is thus the distance that must be traversed by a beam for its flux to be attenuated to 1/e of its initial value as a result of collisions.

1.7.8 Average collision frequency and mean free path

Average collision frequency

In general, cross-sections vary, often greatly, as a function of the energy of the incident particles, depending on the type of collisions which they represent.

As an example, consider the electron-neutral collision cross-section in a rare gas with high mass such as argon (Fig. 1.14). For this gas (as for krypton and xenon), the value of the energy has a great influence on the collision probability; thus, we observe a pronounced minimum in P_x for an energy a little less than 1 eV: this is the so called Ramsauer's *minimum*, a purely quantum effect resulting from the diffraction of the wave function of the incident electron on the outer electrons of the target atom, with destructive (minimum) or constructive (maximum) interference after interaction. For an incident electron having an energy U_{eV} of 1 eV, the de Broglie wavelength is close to the diameter of the atom, hence the diffraction. On the other hand, for $U_{eV} \to \infty$, $\hat{\sigma}_{tx} \to 0$, because the fraction of time the particles are close together approaches zero (to give a picture of this small probability, consider the case of a car crossing a red light at extremely high speed: the probability of an accident, proportional to the crossing time, is extremely small).

Particles in a plasma are not mono-energetic, their velocity $\boldsymbol{w}$ forming a distribution $f(\boldsymbol{w})$. Knowing that the cross-section $\hat{\sigma}_{tx}$ varies with w, how do we define a collision frequency representative of what is occurring in the plasma? In this case, it is necessary to consider the *average collision frequency* $\langle \nu_x \rangle$, defined by:

$$\langle \nu_x \rangle = \frac{N \displaystyle\int_{\boldsymbol{w}} \hat{\sigma}_{tx}(w)\, n\, w f(\boldsymbol{w})\, \mathrm{d}\boldsymbol{w}}{\displaystyle\int_{\boldsymbol{w}} n\, f(\boldsymbol{w})\, \mathrm{d}\boldsymbol{w}} . \tag{1.136}$$

In effect, while for an incident particle with velocity $\boldsymbol{w}$ we have $\nu_x = N\hat{\sigma}_{tx}w$ (1.132), for an ensemble of incident particles for which the velocity is in the interval $\langle \boldsymbol{w}, \boldsymbol{w} + \mathrm{d}\boldsymbol{w} \rangle$, the number of collisions per second per unit volume, assuming the normalisation condition (I.4), is $\nu_x f(\boldsymbol{w})\, \mathrm{d}\boldsymbol{w}$. The denominator in (1.136) corresponds to the density n of incident particles, including all velocities (again normalisation condition (I.4)). The ratio of the total number of collisions per unit volume (nominator) to the total number of particles per unit volume (denominator) gives an average number of collisions, per unit volume, per electron. The expression (1.136) can equally be written in a condensed form:

$$\langle \nu_x \rangle = N \langle \hat{\sigma}_{tx}(w) w \rangle \,, \tag{1.137}$$

where the brackets symbolically represent an integration over the distribution function (in velocity or energy) of the particles.

Example:

In an argon plasma with an electron temperature $T_{eV} = 4\,\text{eV}$ (assuming a Maxwell-Boltzmann distribution) $v_{th} = 1.19 \times 10^6\,\text{m}\,\text{s}^{-1}$ ($\langle w \rangle = 1.34 \times 10^6\,\text{m}\,\text{s}^{-1}$), the integration of the distribution over the electron-neutral momentum transfer cross-section (Fig. 1.14) gives $\langle \nu_m \rangle = 4 \times 10^9\,\text{s}^{-1}$ at 1 torr, 0°C. At 0.1 torr and 0°C, we obviously have $\langle \nu_m \rangle = 4 \times 10^8\,\text{s}^{-1}$. In the following, to simplify the notation, ν will denote the average value of the electron momentum transfer frequency and ν_c that for simple collisions (related to $\hat{\sigma}_{tc}$).

We have defined the average collision frequency of particles in a plasma, satisfying a distribution $f(\boldsymbol{w})$, with the other particles assumed to be initially at rest (1.137). In the more general case of collisions between particles α and particles β (both moving)[39] and correlated among themselves (Sect. 3.2), the expression for the average collision frequency $\langle \nu_{\alpha\beta} \rangle$ of particles α on particles β can be written, in the centre of mass frame:

$$\langle \nu_{\alpha\beta} \rangle = \frac{\displaystyle\int_{\boldsymbol{w}_\alpha}\int_{\boldsymbol{w}_\beta} \hat{\sigma}_{\alpha\beta}(|\boldsymbol{w}_\alpha - \boldsymbol{w}_\beta|)|\boldsymbol{w}_\alpha - \boldsymbol{w}_\beta|\, n_\alpha n_\beta\, f_{\alpha\beta}(\boldsymbol{w}_\alpha, \boldsymbol{w}_\beta)\, \mathrm{d}\boldsymbol{w}_\alpha \mathrm{d}\boldsymbol{w}_\beta}{\displaystyle\int_{\boldsymbol{w}_\alpha} n_\alpha f_\alpha(\boldsymbol{w}_\alpha)\, \mathrm{d}\boldsymbol{w}_\alpha} \,, \tag{1.138}$$

where the function $f_{\alpha\beta}(\boldsymbol{w}_\alpha, \boldsymbol{w}_\beta)$ is a pair (two-point) correlation function in the velocities $\boldsymbol{w}_\alpha$ and $\boldsymbol{w}_\beta$, expressing a correlated binary interaction (Sect. 3.2). Note the explicit presence of the modulus of the relative velocity of the two particles, a characteristic of the description in the centre of mass frame. The denominator in (1.138) corresponds to the density of incident particles α, including all velocities. The expression (1.138) is a generalisation of the relation (1.136).

If there is no correlation between the particles (Sect. 3.2), we can replace the pair correlation function by two functions depending separately on the velocity of each of the two kinds of species, by setting $f_{\alpha\beta}(\boldsymbol{w}_\alpha, \boldsymbol{w}_\beta) = f_\alpha(\boldsymbol{w}_\alpha) f_\beta(\boldsymbol{w}_\beta)$, such that:

[39] When the target particles are considered at rest, their density is denoted by N but n_β when both particles α and β are moving.

$$\langle \nu_{\alpha\beta} \rangle = \frac{n_\beta \int\limits_{\boldsymbol{w}_\alpha} \int\limits_{\boldsymbol{w}_\beta} \hat{\sigma}_{\alpha\beta}(w_{\alpha\beta}) w_{\alpha\beta} f_\alpha(\boldsymbol{w}_\alpha) f_\beta(\boldsymbol{w}_\beta) \, \mathrm{d}\boldsymbol{w}_\alpha \mathrm{d}\boldsymbol{w}_\beta}{\int\limits_{\boldsymbol{w}_\alpha} f_\alpha(\boldsymbol{w}_\alpha) \, \mathrm{d}\boldsymbol{w}_\alpha} \,. \tag{1.139}$$

The average value $\langle \nu_{\alpha\beta} \rangle$ can be written:

$$\langle \nu_{\alpha\beta} \rangle = n_\beta \langle \hat{\sigma}_{\alpha\beta}(w_{\alpha\beta}) w_{\alpha\beta} \rangle \,, \tag{1.140}$$

where the brackets symbolically represent an integration over the distribution function (in velocity or energy) of the particles α and β. Note that the frequency $\nu_{\alpha\beta}$ is, as a rule, different from $\nu_{\beta\alpha}$ (see, for instance, (3.126)). One can see again that (1.140) is a generalisation of the frequency $\langle \nu_x \rangle$ given by (1.137).

Mean free path

Following an analogous method to that which allowed us to define $\langle \ell_{\alpha\beta} \rangle$, the *mean free path of particles* α for collisions with particles β can be expressed, in the more general case, as the average value of the quotient of the relative velocity $w_{\alpha\beta}$ on the collision frequency $\nu_{\alpha\beta}$, i.e.:

$$\langle \ell_{\alpha\beta} \rangle = \left\langle \frac{w_{\alpha\beta}}{\nu_{\alpha\beta}} \right\rangle = \frac{1}{n_\beta} \left\langle \frac{1}{\hat{\sigma}_{\alpha\beta}(w_{\alpha\beta})} \right\rangle \,. \tag{1.141}$$

Returning to the preceding numerical example ($T_{eV} = 4\,\mathrm{eV}$ at 0.1 torr, 0°C in argon), we obtain an electron-neutral mean free path for electrons of $\langle \ell_m \rangle \approx 3\,\mathrm{mm}$. Henceforth, the mean free path for momentum transfer will be simply denoted by ℓ.

1.7.9 Examples of collision cross-sections

In the preceding section, to establish our ideas, we started by considering elastic electron-neutral collisions. As we have already remarked, the concept of cross-sections is more general. In the following sections, we will consider the cases of inelastic electron-neutral collisions (ionisation, excitation, dissociation) and those of elastic and inelastic ion-neutral collisions (charge exchange).

Electron-neutral collisions leading to the ionisation of an atom (molecule)

In laboratory plasmas, ionisation of a gas usually occurs as a result of electron impact. The probability of impact ionisation by atom-atom collisions is, in fact, small in plasmas with less than a few atmospheres pressure. If we take an argon atom, its step-wise ionisation through the metastable states as a relay, which is the lowest energy pathway to achieve ionisation, requires a minimum energy of 11.5 eV (Appendix VI). Thus, assuming a Maxwell-Boltzmann particle energy distribution function, this would require that the most energetic atoms in the plasma reach a temperature of more than 100,000 K, which is not realistic in laboratory plasmas.

The ionisation cross-section for electron-neutral collisions generally exhibits the following characteristics (Fig. 1.15):

- a very precise energy threshold $\mathcal{E}_i$, below which the cross-section is zero. For atoms, the *ionisation threshold* corresponds to the *ionisation potential*. For molecules, a number of ionisation thresholds co-exist (dissociative and non dissociative ionisation).
- immediately above the energy threshold $\mathcal{E}_i$, an (almost) linear growth of the cross-section as a function of the energy U_{eV} of the electron[40],
- then the cross-section passes through a maximum for an energy $\mathcal{E}_m$, followed by a slow diminution.

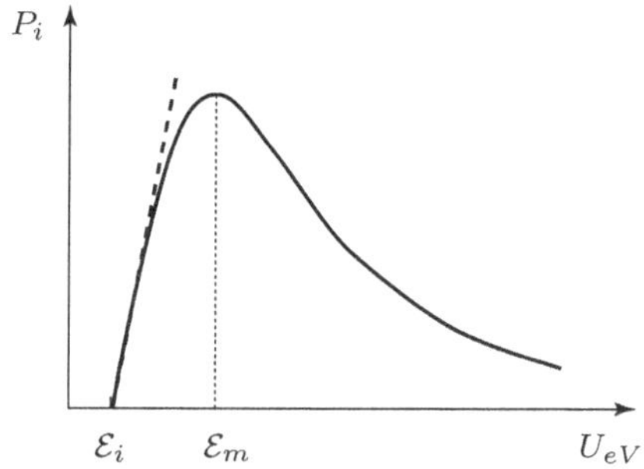

Fig. 1.15 General form of the ionisation cross-section of an atom by electron collisions.

The dashed line in Fig. 1.15, with slope a_i, drawn from the energy threshold $\mathcal{E}_i$ of the cross-section, represented by the expression:

$$P_i = a_i(U_{eV} - \mathcal{E}_i) \text{ for } U_{eV} \geq \mathcal{E}_i \,, \tag{1.142}$$

is a good approximation for the initial portion ($\mathcal{E}_i < U_{eV} < \mathcal{E}_m$) of the ionisation cross-section.

[40] Since the target-particles are not compelled to be at rest, it is more correct to speak of their relative energy (velocity) at the moment of collision. In the case of electron-neutral interactions, this distinction is generally negligible above a fraction of an electron volt in the case where the gas is not very warm (see also remark 2 at the end of this section).

Electron-neutral collisions leading to the excitation of an atom (molecule)

For ionisation processes, there is an energy threshold $\mathcal{E}_j$ below which the cross-section is zero. Its growth after $\mathcal{E}_j$ is almost linear with U_{eV} before passing through a maximum, as illustrated in Fig. 1.16.

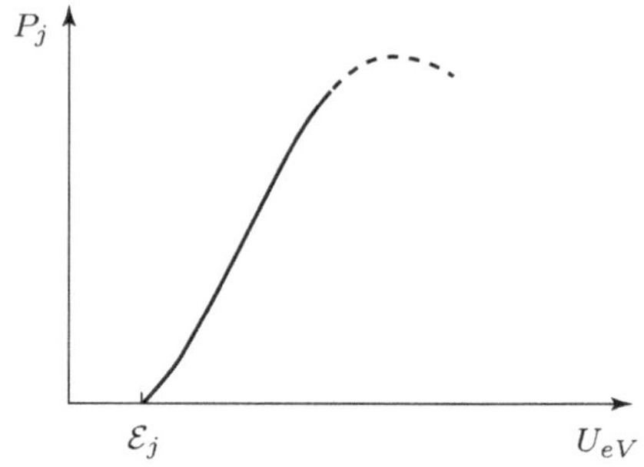

Fig. 1.16 General form of a cross-section for excitation of level j of an atom by electron collisions.

Electron-neutral collisions leading to the dissociation of a molecule

Here also, there is an energy threshold $\mathcal{E}_d$ below which the cross-section is zero. For a complex molecule, many dissociation thresholds may coexist, depending on the nature of the fragments resulting from the collision.

Elastic ion-atom collisions

Cross-sections for the scattering of ions by their own atoms (molecules) all have exactly the same form: interaction is more probable at low velocity. Figure 1.17 shows, as an example, the case for different hydrogen ions incident on H_2 molecules, after having been accelerated by a potential difference U (1.126).

Ion-atom collisions leading to a transfer of charge

Consider an atomic or molecular ion A^+ and an atom or a molecule B. During their interaction, there can be an exchange of an electron, which gives rise to the reaction[41]:

$$A^+ + B \rightarrow A + B^+ \,. \tag{1.143}$$

[41] This represents an inelastic collision: the internal energy of the atom (molecule) is modified when losing or recovering an electron.

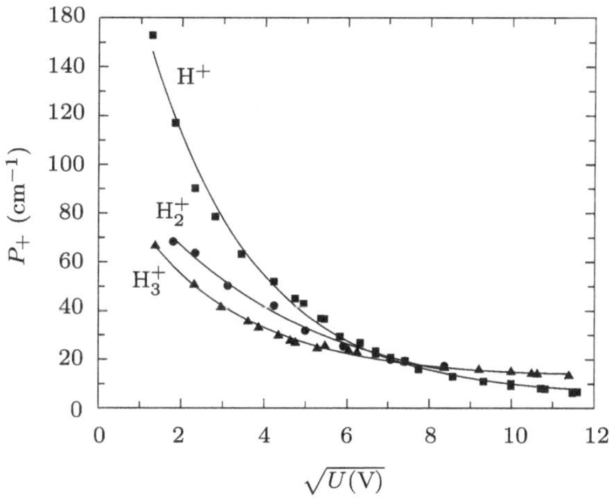

Fig. 1.17 Cross-sections for elastic collisions of different hydrogen ions of energy eU with H_2 (after [37] and [38]).

One interesting case is that of *resonant transfer* (A = B) where an atomic ion has a collision with an atom of the same species. Fig. 1.18 shows that this cross-section (denoted by t) reaches its greatest value for very low energies (general rule).

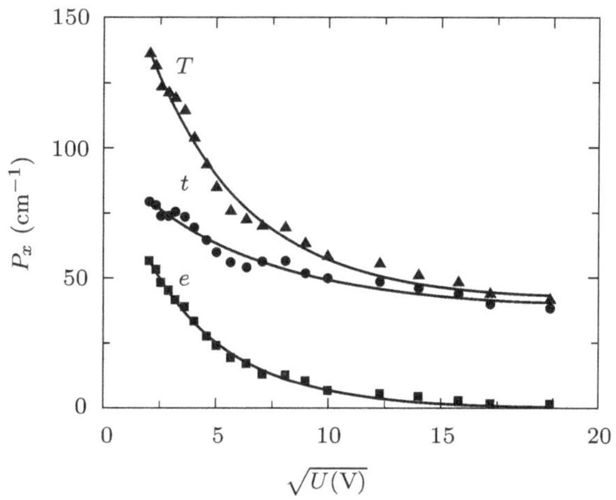

Fig. 1.18 Cross-section for charge exchange (index $x = t$) of a helium atom and its ion with energy eU, and the resulting total cross-section ($x = T$) (after [9], with permission of the American Institute of Physics).

To complete the comparison, the elastic interaction between the ion A^+ and the atom A is also shown. The value of the scattering cross section (denoted by e) always has a lower value than for resonant transfer: this difference is large in helium and neon, for example, but small for argon.

Atom-atom collisions

Although difficult to measure experimentally, because they involve the interaction between neutral particles, these cross-sections can be calculated from the analytical expressions for the interaction potential between two atoms. They enable the thermal conductivity of the gas to be calculated. As an example, this type of interaction can be modelled by the Lennard-Jones potential, illustrated in Fig. 1.19:

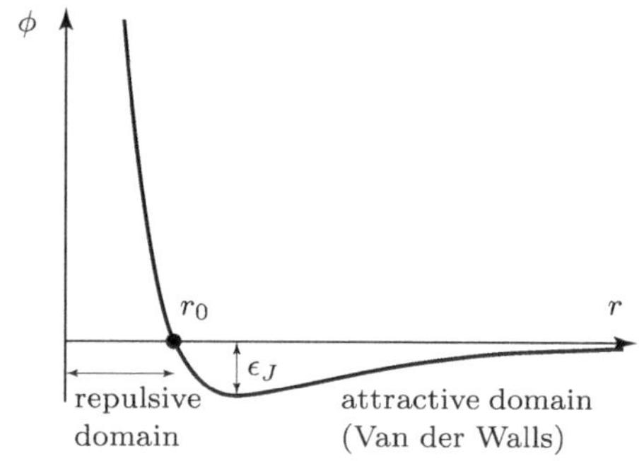

Fig. 1.19 Lennard-Jones potential to model atom-atom interaction.

$$\phi(r) = 4\epsilon_J \left[(r_0/r)^{12} - (r_0/r)^6\right] , \tag{1.144}$$

where the first term represents the repulsive potential when the two particles are sufficiently in contact and the second term, the long distance attractive potential, r_0 being the value of r (internuclear distance) for which $\phi(r) = 0$ and ϵ_J is the depth of the potential well.

Remarks:

1. In the case of excitation and ionisation by electron collisions, the amount of energy lost by the electron and transferred to the internal energy of the atom is quantified, whatever the energy of the electron at the moment of interaction: the energy of the electron after the collision is its initial energy minus the threshold energy of the excited level (Franck and Hertz experiment, 1914).
2. Reaction coefficient: definition and general expression
   Consider a particular reaction, symbolically represented by:

   $$\mathrm{A} + \mathrm{B} \xrightarrow[k_{\mathrm{AB}}]{} \mathrm{C} + \mathrm{D} , \tag{1.145}$$

   where A and B are the particles before collision and C and D are the particles after collision. The probability of this occurring is characterised by a *reaction coefficient* (expressed at 0°C, 1 torr) given by the expression:

$$k_{\mathrm{AB}} = \langle \hat{\sigma}_{\mathrm{AB}}(w_{\mathrm{AB}})\, w_{\mathrm{AB}} \rangle \qquad \text{units: cm}^3\,\text{s}^{-1}\,, \tag{1.146}$$

where $w_{\mathrm{AB}} = |\boldsymbol{w}_{\mathrm{A}} - \boldsymbol{w}_{\mathrm{B}}|$ is the modulus of the difference in velocity before collision (see problem 1.9 and footnote 40). The use of the reaction coefficient, instead of a "reaction frequency" ν_{AB}, is more general because it does not involve the density of targets N_{B}, which varies according to the operating conditions. We know, in fact, that $\nu_{\mathrm{AB}} = N_{\mathrm{B}} k_{\mathrm{AB}}$ and the published values are those of k_{AB} (in place of a more detailed cross-section).

3. Hydrodynamic momentum equation and nature of the collision term
   Consider a weakly ionised plasma (dominated by electron-neutral collisions) submitted to an oscillating electric field $E_0 \exp(\mathrm{i}\omega t)$, whose frequency is sufficiently high for the motion of ions to be neglected. Suppose also that the velocity of thermal motion of the electrons is negligible compared to its speed $\mathrm{d}x/\mathrm{d}t$ in the HF field (cold plasma approximation)).[23] These considerations lead us to the momentum equation for an electron in an HF field (in one dimension)[42]:

$$m_e \frac{\mathrm{d}^2 x}{\mathrm{d}t^2} = -eE_0 \exp(\mathrm{i}\omega t) - \nu m_e \frac{\mathrm{d}x}{\mathrm{d}t}\,. \tag{1.147}$$

   Since the term $\nu m_e \mathrm{d}x/\mathrm{d}t$ has the dimensions of a force, $\nu m_e v \mathrm{d}t$ represents the change in momentum during $\mathrm{d}t$ and $\nu m_e v$ its variation per second; furthermore, a dimensional analysis shows that ν is in numbers per second. In this context, the term $-\nu m_e \mathrm{d}x/\mathrm{d}t \equiv -\nu p_e$ represents the number of times per second that an electron loses its momentum. The frequency ν thus appears as the (average) collision frequency for the transfer of momentum from the electron to the atom (ion). The solution of the differential equation (1.147) gives the (complex) amplitude of the electron in the HF field as:

$$x_0 = \frac{eE_0}{m_e \omega(\omega - \mathrm{i}\nu)}\,, \tag{1.148}$$

   where we note that ν has the same units as the angular frequency ω of the HF field.

In (1.147), νm_e appears as the classic hydrodynamic coefficient of viscosity, generally assumed independent of the velocity of the particles. The influence of ions and atoms on the motion of electrons under the influence of an HF field can be regarded as a viscosity term impeding the electron motion.

---

[42] We will obtain the relation (1.143) in the framework of the Lorentz hydrodynamic model for plasma electrons (Sect. 3.7).

1.8 Mechanisms for creation and loss of charged particles in a plasma and their conservation equation

1.8.1 Loss mechanisms

We can distinguish two principal loss mechanisms of charged particles through neutralisation (also called recombination).

Diffusion to the walls, where the charge particles are neutralised

Diffusion is a collisional process that occurs if the medium is inhomogeneous in density or in (charged or neutral) particle energy. The role of collisions in this process is to randomise the directions of the particles after equi-probable collisions. However, in a medium of inhomogeneous density, there are more particles in the regions of high density than in the regions of low density so this will create a net flux towards the low-density region.

Diffusion of charged particles from the centre of the plasma to the walls can constitute an extremely efficient method for removing these particles, by their recombination at the walls into neutral particles. Such a "sink" at the walls, as it provides a much lower density than on the axis, gives rise to a flux of charged particles $\boldsymbol{\Gamma} = n\boldsymbol{v}$ where $\boldsymbol{v}$, the diffusion velocity, is directed from the centre towards the walls. The value of this flux is given by the relation[43]:

$$\boldsymbol{\Gamma} = -D\boldsymbol{\nabla}n \,, \tag{1.149}$$

which indicates that the flux of particles to the wall increases with the density gradient of the particles $\boldsymbol{\nabla}n$; the factor D is called the *diffusion coefficient*, free or ambipolar, depending on whether the plasma density is low or high, as we shall see in Sect. 3.8.

Volume recombination

Charged particles can also be neutralised in the plasma volume itself, by collision, and not at the walls.

The recombination of an **atomic ion**, assuming that its radiative recombination (1.62) is negligible, requires the presence of a third particle (*three-body recombination*), in this case an electron, to ensure the conservation of energy and momentum for the interaction. This reaction can be written:

$$\mathrm{A}^{+} + \mathrm{e} + \mathrm{e} \rightarrow \mathrm{A} + \mathrm{e} \,. \tag{1.150}$$

---

[43] Generally, the diffusion coefficient D depends on position and then one should write $\boldsymbol{\Gamma} = -\boldsymbol{\nabla}(Dn)$: see Sect. 3.8, Remark 4.

Note that this is an inverse reaction, in the sense of micro-reversibility (Sect. 1.4.2), for the ionisation process of atom A. This is the process of charge recombination that occurs in the ionisation-recombination equilibrium governed by the Saha equation. The number of atomic ions which recombine per unit volume per second in the plasma, is thus proportional to $n_{i1}n_e^2$, where n_{i1} is the density of atomic ions and n_e is that of the electrons. In the case where there is only one type of atomic ion, $n_{i1} = n_e$, and the number of ions that have recombined per unit volume per second ($\mathrm{cm}^{-3}\,\mathrm{s}^{-1}$) can be written $\alpha_{ar}n_e^3$, where α_{ar}, a reaction coefficient, is the three body *recombination coefficient* (units: $\mathrm{cm}^6\,\mathrm{s}^{-1}$).
By formally writing:

$$\alpha_{ar}n_e^3 \equiv \nu_{ar}n_e \ , \tag{1.151}$$

we can define ν_{ar}, the corresponding atomic recombination frequency.

The recombination of a **molecular ion (two-body recombination)** follows as:

$$\mathrm{A}_2^+ + \mathrm{e} \rightarrow \mathrm{A} + \mathrm{A} \ . \tag{1.152}$$

This recombination is referred to as *dissociative recombination*. If the energy liberated during this neutralisation is greater than that of the first excited state of the atom A, one of these atoms will be in an excited state, unless the excess energy is transformed into kinetic energy of the atoms[44]. The number of molecular ions recombining per unit volume per second is therefore proportional to $n_{i2}n_e$, where n_{i2} is the density of molecular ions. If there is only one kind of molecular ion ($n_{i2} = n_e$), in an analogous fashion to (1.151):

$$\alpha_{mr}n_e^2 \equiv \nu_{mr}n_e \ , \tag{1.153}$$

where α_{mr} is the molecular dissociative recombination coefficient and ν_{mr} the corresponding frequency.

In the case of plasmas rich in **negative ions**, the recombination of positive ions with negative ions follows the reaction:

$$\mathrm{A}^+ + \mathrm{B}^- \rightarrow \mathrm{A} + \mathrm{B} \tag{1.154}$$

referred to as *mutual recombination*. This type of interaction is independent of whether the ions are atomic or molecular. The number of positive (and negative) ions recombining per unit volume per second is proportional to n_in_{i-}, where n_i is the density of positive ions (atomic or molecular) and n_{i-} is that of the negative ions. In an analogous fashion to (1.151) and (1.153), we can write, for example:

$$\alpha_{r\pm}n_in_{i-} \equiv \nu_{r\pm}n_{i-} \ , \tag{1.155}$$

[44] In the case of rare gas atoms, one of the two atoms is necessarily excited into a metastable state.

where $\alpha_{r\pm}$ is the mutual recombination coefficient for a negative ion and a positive ion, and $\nu_{r\pm}$ is the corresponding frequency.

The order of magnitude of $\alpha_{r\pm}$ is $10^{-7}\,\mathrm{cm}^3\,\mathrm{s}^{-1}$, a value which only weakly varies with the nature of the ions and their relative energy. It should also be noted that the value of $\alpha_{r\pm}$ is lower by a factor of 2 or 3 than the coefficient α_{mr} for dissociative recombination of molecular ions.

Predominance of the removal mechanism for charged particles according to their pressure region (plasmas without negative ions)

As a general rule, in the pressure interval between 10 mtorr ($\sim$ 1 Pa) and around 10 torr ($\approx 10^3$ Pa) and for a discharge in an atomic gas, diffusion dominates, provided the smallest dimension (e.g. the radius for a long plasma column) is sufficiently small for the volume losses to be negligible: in the same pressure interval, but for a molecular gas, volume recombination predominates. At higher pressures, volume recombination becomes the dominant mechanism for the loss of charges, even for atomic gases.

1.8.2 Creation mechanisms

When ionisation results from a single electron collision with an atom in its ground state (*direct ionisation*), the corresponding ionisation frequency (number of ions created per second per electron) ν_{id}, can be written (Sect. 1.7.8):

$$\nu_{id} = \langle P_i(w_e) w_e \rangle \equiv \mathcal{N}_0 \langle \hat{\sigma}_{ti}(w_e) w_e \rangle \;, \tag{1.156}$$

where P_i is the macroscopic cross-section and $\hat{\sigma}_{ti}$, the total microscopic cross-section, both corresponding to direct ionisation, and $\mathcal{N}_{\prime}$ the density of the ground state of the atom.

However, when the electron density is sufficiently high, direct ionisation is no longer the only path possible. *Ionisation* can then take place in *steps*, using the excited states of the atom as relay states. This ionisation process is advantageous, because it can take place with electrons of lower energy than for direct ionisation. A frequent case of collisional ionisation in two steps is that where successively:

$$\mathrm{e} + \mathrm{A}(0) \rightarrow \mathrm{A}(j) + \mathrm{e} \;, \tag{1.157}$$

$$\mathrm{e} + \mathrm{A}(j) \rightarrow \mathrm{A}^+ + \mathrm{e} + \mathrm{e} \;, \tag{1.158}$$

the excited state $\mathrm{A}(j)$ being metastable (a state weakly susceptible to radiative de-excitation between two collisions: see footnote 26). In this case, the number of atoms ionised per unit volume per electron, taking into account de-excitation of the intermediate levels by electron collisions[45] (losses proportional to n_e) and collision-free losses (by diffusion, therefore independent of n_e) can be written in the form (Appendix VI):

$$\nu_{ie} = \frac{\rho_{ie} n_e}{1 + \eta n_e} , \tag{1.159}$$

where the coefficients ρ_{ie} and η characterise ionisation in two steps and the effect of the intermediate steps respectively. When n_e is very large (saturation), (1.159) reduces to:

$$\nu_{ie} \simeq \rho_{ie}/\eta \, . \tag{1.160}$$

1.8.3 Conservation equation for charged particles

If we take into account the two loss mechanisms we have just discussed, together with ionisation, the conservation equation for particles can be most generally written as (Sect. 3.5):

$$\frac{\partial n}{\partial t} + \boldsymbol{\nabla} \cdot n\boldsymbol{v} = (\nu_i - \nu_r) n \, , \tag{1.161}$$

where $\nu_i = \nu_{ie} + \nu_{id}$ and $\nu_r = \nu_{ar} + \nu_{im}$. The term on the right-hand side of the equation includes the number of particles created by ionisation and the number lost by volume recombination, per unit volume per second respectively, the diffusion losses being taken into account in the term on the left-hand side $\boldsymbol{\nabla} \cdot n\boldsymbol{v}$.

In the stationary state, and in the absence of volume recombination[46], (1.161), including (1.149), reduces to:

$$\boldsymbol{\nabla} \cdot (-D \boldsymbol{\nabla} n) = \nu_i n \tag{1.162}$$

and, provided D does not depend on position, we have:

$$\nabla^2 n = (-\nu_i / D) \, n \, . \tag{1.163}$$

In the case where ν_i and D are independent of n, this is an eigenvalue problem[47] (Sect. 3.8), fixed by the boundary conditions (imposed by the geometry of the discharge chamber): for a cylindrical tube, by symmetry

[45] The de-excitation (quenching) of metastable atoms by collision with heavy particles (ground state and metastable state atoms) becomes significant in discharges with a low degree of ionisation (Appendix VI).

[46] In the stationary state and in the absence of diffusion losses, (1.161) obviously reduces to $\nu_i = \nu_r$.

[47] This is not so, for example, for a two-step ionisation where the problem is non-linear (1.159).

$(\mathrm{d}n/\mathrm{d}r)_{r=0} = 0$, and we often take $n(r = R) = 0$, where R is the internal radius of the tube. In the case of a very long cylindrical tube ($L \gg R$) and for $n(r = R) = 0$, we have $n(r) = n(0)J_0(2.405\, r/R)$, where J_0 is the Bessel function of the first kind and of order zero (Fig. 3.4).

Problems

1.1.

a) Find the vertical distribution, in equilibrium, of the density and velocity of neutral particles of mass M, in the terrestrial atmosphere, under the influence of gravity. Take the origin of the vertical coordinate z to be the surface of the Earth, where the density of particles is $\hat{n}_0$. Assume that the particles have a Maxwell-Boltzmann velocity distribution and, to a first approximation, that the temperature T of the particles and the gravitational constant g do not vary vertically.

b) Calculate the average velocities $\langle w \rangle$, $\langle \boldsymbol{w} \rangle$ and the average kinetic energy as functions of the height z.

It will be useful here to refer to Appendix I and the table of formulae and integrals (see Appendix XX).

Answer

a) The system is subjected to a conservative force ($\boldsymbol{F} = -\boldsymbol{\nabla}\Phi$) since the gravitational force $F = -Mg$ is the derivative of a potential (more correctly a potential energy), in this case $\Phi(z) = Mgz$. Assuming, to a first approximation, that the particles have a Maxwell-Boltzmann velocity distribution, then for the present system (1.17):

$$f(z, w) = \hat{n}_0 \exp\left(-\frac{Mgz}{k_B T}\right) f(w) \,, \tag{1.164}$$

where the scalar function:

$$f(w) = \left(\frac{M}{2\pi k_B T}\right)^{\frac{3}{2}} \exp\left(-\frac{Mw^2}{2k_B T}\right) \tag{1.165}$$

represents an isotropic distribution in velocity (remark 2 below). Clearly:

$$f(z, w) = n(z) f(w) \,, \tag{1.166}$$

and the vertical distribution of the particle density is then given by:

$$n(z) = \hat{n}_0 \exp\left(-\frac{Mgz}{k_B T}\right) . \qquad (1.167)$$

Remarks:

1. Since we can write $f(z, w) = n(z)f(w)$, the function $f(z, w)$ is said to be separable (Sect. 3.3).
2. The fact that we have assumed $\partial T/\partial z \simeq 0$ allows us, in effect, to assume that there is no significant loss of particles or energy in any given direction, such that the velocity distribution function is isotropic

b) Generally speaking (see Sect. 3.3 for further details), the average value of a variable $A(\boldsymbol{r}, \boldsymbol{w})$ taken over the distribution function $f(\boldsymbol{r}, \boldsymbol{w})$ can be written:

$$\langle A(\boldsymbol{r}, \boldsymbol{w})\rangle = \frac{\int\limits_{-\infty}^{\infty} A(\boldsymbol{r}, \boldsymbol{w}) f(\boldsymbol{r}, \boldsymbol{w})\, \mathrm{d}^3 w}{\int\limits_{-\infty}^{\infty} f(\boldsymbol{r}, \boldsymbol{w})\, \mathrm{d}^3 w} . \qquad (1.168)$$

Because the function $f(\boldsymbol{r}, \boldsymbol{w})$ is separable and isotropic (1.166), expression (1.168) as a function of z becomes:

$$\langle A(z, \boldsymbol{w})\rangle = n(z) \frac{\int\limits_{-\infty}^{\infty} A(z, \boldsymbol{w}) f(w)\, \mathrm{d}^3 w}{n(z) \int\limits_{-\infty}^{\infty} f(w)\, \mathrm{d}^3 w} = \int\limits_{-\infty}^{\infty} A(z, \boldsymbol{w}) f(w)\, \mathrm{d}^3 w , \qquad (1.169)$$

where the denominator, as we can verify, is equal to unity since the function $f(w)$ (1.165) is normalized:

$$\int\limits_{-\infty}^{\infty} f(w)\, \mathrm{d}^3 w = \int\limits_{0}^{\infty} \left(\frac{M}{2\pi k_B T}\right)^{\frac{3}{2}} \exp\left(-\frac{M w^2}{2 k_B T}\right) 4\pi w^2\, \mathrm{d}w = 1 . \qquad (1.170)$$

Finally, from (1.169), $\langle w\rangle$, the average scalar velocity has the value:

$$\langle w\rangle \equiv \int\limits_{0}^{\infty} w f(w)\, 4\pi w^2\, \mathrm{d}w = \sqrt{\frac{8 k_B T}{\pi M}} , \qquad (1.171)$$

and thus is independent of z.
To obtain the average vector velocity $\langle \boldsymbol{w}\rangle$, we can calculate its components, for example in the x direction:

$$\langle w_x \rangle = \left(\frac{M}{2\pi k_B T}\right)^{\frac{3}{2}} \underbrace{\int_{-\infty}^{\infty} w_x \exp\left(-\frac{M w_x^2}{2k_B T}\right) \mathrm{d}w_x}_{\text{odd integrand}} \int_{-\infty}^{\infty} \exp\left(-\frac{M w_y^2}{2k_B T}\right) \mathrm{d}w_y \cdots \tag{1.172}$$

Since the integrand over w_x is odd, its integral from $-\infty$ to $+\infty$ is zero, and $\langle w_x \rangle = 0$.
This is the same for the velocity components in y and z, such that $\langle \boldsymbol{w} \rangle = 0$.
Finally, the average value of $\langle w^2 \rangle$ can be expressed as:

$$\langle w^2 \rangle \equiv \int_{-\infty}^{\infty} w^2 f(w)\, \mathrm{d}^3 w = \int_{0}^{\infty} 4\pi w^4 f(w)\, \mathrm{d}w = \frac{3k_B T}{M}\,, \tag{1.173}$$

and the average kinetic energy is given by:

$$\langle \mathcal{E}_c \rangle = \frac{3}{2} k_B T\,. \tag{1.174}$$

Remark:

Since there are neither sources or sinks of particles, it is not possible to have a net movement of particles in a given direction, from which, inter alia, $\langle w_z \rangle = 0$. The density gradient in the z direction was established as the gravitational force "was applied", before the stationary state was reached!

1.2.

a) The *random particle flux* is defined as the average value of the flux crossing a surface in one single direction (in the positive z direction, for example) thus:

$$\Gamma_z = \langle n w_z \rangle\,. \tag{1.175}$$

Calculate this flux of particles of mass m and density n, assuming thermodynamic equilibrium at a temperature T. Perform the calculation in Cartesian coordinates, then in spherical coordinates ("useful" integrals can be found in Appendix XX).

b) Calculate the corresponding *random energy flux*. Perform the calculation in Cartesian coordinates, then in spherical coordinates.

Answer

a) In the case in which the velocity distribution function is Maxwellian and isotropic (necessary conditions for thermodynamic equilibrium), the random flux of particles is written, in Cartesian coordinates (remember, in this context, that n is independent of position):

$$\Gamma_z = n\left(\frac{m}{2\pi k_B T}\right)^{\frac{3}{2}} \int_0^{\infty} w_z \exp\left(-\frac{m w_z^2}{2k_B T}\right) \mathrm{d}w_z \int_{-\infty}^{\infty} \exp\left(-\frac{m w_x^2}{2k_B T}\right) \mathrm{d}w_x$$

$$\times \int_{-\infty}^{\infty} \exp\left(-\frac{m w_y^2}{2k_B T}\right) \mathrm{d}w_y \,. \tag{1.176}$$

Therefore:

$$\Gamma_z = n\left(\frac{m}{2\pi k_B T}\right)^{\frac{3}{2}} \times \frac{1}{2}\left(\frac{2k_B T}{m}\right) \times \sqrt{\pi}\left(\frac{2k_B T}{m}\right)^{\frac{1}{2}} \times \sqrt{\pi}\left(\frac{2k_B T}{m}\right)^{\frac{1}{2}} ,$$

$$\Gamma_z = \frac{n}{4}\sqrt{\frac{8k_B T}{\pi m}} = \frac{n\langle w\rangle}{4} \,. \tag{1.177}$$

Expressing the velocities in spherical coordinates, where $w_z = w\cos\theta$ and $\mathrm{d}^3 w = w^2 \mathrm{d}w \, \sin\theta \, \mathrm{d}\theta \, \mathrm{d}\varphi$:

$$\Gamma_z = n\left(\frac{m}{2\pi k_B T}\right)^{\frac{3}{2}} \int_0^{2\pi} \mathrm{d}\varphi \int_0^{\frac{\pi}{2}} \sin\theta\cos\theta \, \mathrm{d}\theta \int_0^{\infty} w^3 \exp\left(-\frac{m w^2}{2k_B T}\right) \mathrm{d}w \,, \tag{1.178}$$

thus:

$$\Gamma_z = n\left(\frac{m}{2\pi k_B T}\right)^{\frac{3}{2}} \times 2\pi \times \frac{1}{2} \times \frac{1}{2}\left(\frac{2k_B T}{m}\right)^2 = \frac{n\langle w\rangle}{4} \,. \tag{1.179}$$

Comparing (1.177) and (1.179), the value of Γ_z is clearly independent of the coordinate system in which the velocities are expressed.

Remark:

If n is the particle density, only half of this density will contribute to the random flux in a given direction! We can therefore write the random flux in the form:

$$\Gamma_z = \frac{n}{2}\langle w_z(w_z > 0)\rangle \,, \tag{1.180}$$

from which we obtain, based on (1.177) or (1.179):

$$\langle w_z(w_z > 0)\rangle = \frac{\langle w\rangle}{2} \,. \tag{1.181}$$

This example illustrates the difficulty of separating the different contributions to an average quantity when the molecular properties themselves are the product (here: n times w_z) or quotient of different quantities.

b) The random kinetic energy flux along z, in Cartesian coordinates, can be written:

$$P_z = \left\langle n w_z \left(\frac{m w^2}{2} \right) \right\rangle , \tag{1.182}$$

and:

$$P_z = n \left(\frac{m}{2\pi k_B T} \right)^{\frac{3}{2}} \int_{-\infty}^{\infty} \int_{-\infty}^{\infty} \int_{0}^{\infty} \frac{m w_z w^2}{2} \exp\left(-\frac{m w^2}{2 k_B T} \right) \mathrm{d}w_x \mathrm{d}w_y \mathrm{d}w_z , \tag{1.183}$$

which, after decomposing the terms, becomes:

$$\begin{aligned} P_z = {} & \frac{nm}{2} \left(\frac{m}{2\pi k_B T} \right)^{\frac{3}{2}} \times \\ & \left[\int_{-\infty}^{\infty} w_x^2 \exp\left(-\frac{m w_x^2}{2 k_B T} \right) \mathrm{d}w_x \int_{-\infty}^{\infty} \exp\left(-\frac{m w_y^2}{2 k_B T} \right) \mathrm{d}w_y \int_0^{\infty} w_z \exp\left(-\frac{m w_z^2}{2 k_B T} \right) \mathrm{d}w_z \right. \\ & + \int_{-\infty}^{\infty} \exp\left(-\frac{m w_x^2}{2 k_B T} \right) \mathrm{d}w_x \int_{-\infty}^{\infty} w_y^2 \exp\left(-\frac{m w_y^2}{2 k_B T} \right) \mathrm{d}w_y \int_0^{\infty} w_z \exp\left(-\frac{m w_z^2}{2 k_B T} \right) \mathrm{d}w_z \\ & \left. + \int_{-\infty}^{\infty} \exp\left(-\frac{m w_x^2}{2 k_B T} \right) \mathrm{d}w_x \int_{-\infty}^{\infty} \exp\left(-\frac{m w_y^2}{2 k_B T} \right) \mathrm{d}w_y \int_0^{\infty} w_z^3 \exp\left(-\frac{m w_z^2}{2 k_B T} \right) \mathrm{d}w_z \right] , \end{aligned} \tag{1.184}$$

from which we obtain:

$$P_z = 2 k_B T \frac{n \langle w \rangle}{4} = 2 k_B T \, \Gamma_z . \tag{1.185}$$

In spherical coordinates, the random flux is written:

$$P_z = \frac{nm}{2} \left(\frac{m}{2\pi k_B T} \right)^{\frac{3}{2}} \int_0^{2\pi} \mathrm{d}\varphi \int_0^{\frac{\pi}{2}} \sin\theta \cos\theta \, \mathrm{d}\theta \int_0^{\infty} w^5 \exp\left(-\frac{m w^2}{2 k_B T} \right) \mathrm{d}w , \tag{1.186}$$

thus:

$$P_z = \frac{nm}{2} \left(\frac{m}{2\pi k_B T} \right)^{\frac{3}{2}} \times 2\pi \times \frac{1}{2} \times \left(\frac{2 k_B T}{m} \right)^3 , \tag{1.187}$$

$$P_z = 2k_B T \, \Gamma_z \, . \tag{1.188}$$

Once again, we have verified that the value of P_z is independent of the coordinate system in which the velocities are expressed.

Remarks:

1. P_z has the dimensions of an energy flux or, equally, of a power density per unit surface ($\mathrm{W\,m^{-2}}$).
2. We know that the average kinetic energy of these particles is given by $\frac{3}{2}k_B T$. However, the random energy flux introduces the factor $2k_B T$ and not $\frac{3}{2}k_B T$. This difference is due to the much greater weight of the velocities appearing in the calculation of the random energy flux ($w^3\,\mathrm{d}^3w$ than in the calculation of average kinetic energy ($w^2\,\mathrm{d}^3w$).

1.3. Consider a helium plasma, in which the density of nuclei is $10^{20}\,\mathrm{m^{-3}}$. Calculate the density of neutral atoms, of electrons and of both singly and doubly ionised atoms, when the plasma temperature is:

- $T_{eV} = 1\,\mathrm{eV}$
- $T_{eV} = 10\,\mathrm{eV}$

assuming thermodynamic equilibrium. The energy threshold for ionisation of helium, with respect to the ground state of the neutral atom, is $\mathcal{E}_{i1} = 24.59\,\mathrm{eV}$ for the first ionisation and $\mathcal{E}_{i2} = 54.4\,\mathrm{eV}$ for the second. Perform the calculation analytically and not using the computer.

Answer

The equations for conservation of nuclei[48] and of charges are respectively:

$$N_n = n_0 + n_{i1} + n_{i2} \, , \tag{1.189}$$

$$n_e = n_{i1} + 2n_{i2} \, , \tag{1.190}$$

where N_n, n_0, n_{i1}, n_{i2} and n_e denote the densities of nuclei, neutral atoms, singly ($Z = 1$) and doubly charged ($Z = 2$) ions and electrons respectively. The assumption of thermodynamic equilibrium allows us to apply the Saha equation for the two ionisation levels (see Sect. 1.4.2):

[48] The conservation of nuclei expresses the way helium atoms at 0 K become distributed into neutral atoms and ions when subjected to a temperature $T \neq 0$.

$$\frac{n_e n_{i1}}{n_0} = \frac{2g_{i1}}{g_0}\frac{(2\pi m_e k_B T)^{\frac{3}{2}}}{h^3}\exp\left(-\frac{\mathcal{E}_{i1}}{k_B T}\right) , \tag{1.191}$$

$$\frac{n_e n_{i2}}{n_{i1}} = \frac{2g_{i2}}{g_{i1}}\frac{(2\pi m_e k_B T)^{\frac{3}{2}}}{h^3}\exp\left(-\frac{(\mathcal{E}_{i2}-\mathcal{E}_{i1})}{k_B T}\right) . \tag{1.192}$$

We thus have four equations to resolve a problem with four unknowns. The quantum degeneracy $g[Z]$ (statistical weights) of the three electronic states of helium, given by $2J+1$, are listed in the table below ($\boldsymbol{L}$ is the total orbital angular momentum and $\boldsymbol{S}$ the total spin under ground state conditions; $\boldsymbol{J}$ the total angular momentum, is the modulus of the vectorial sum (in a quantum sense) of $\boldsymbol{L}$ and $\boldsymbol{S}$).

	L	S	J	$g[Z]=2J+1$
He	0	0	0	1
He^+	0	$\frac{1}{2}$	$\frac{1}{2}$	2
He^{++}	0	0	0	1

a) $T_{eV} = 1\,\mathrm{eV}$

Although an analytical calculation is possible (see b), we will proceed by successive approximation (iterative method) to solve the first part of the problem. In the first iteration, we will neglect the number of doubly charged ions with respect to those that are singly charged. This can be justified by the large gap between the thermal energy of the particles (1 eV) and the energy of the second ionisation of helium (54.4 eV). In consequence, the conservation equations for charges (1.190) and nuclei (1.189) reduce to:

$$n_e \simeq n_{i1} , \tag{1.193}$$

$$N_n \simeq n_0 + n_{i1} . \tag{1.194}$$

Let us write the relevant Saha equation for the equilibrium of the singly charged ions with respect to the neutral state, and the equilibrium of the doubly charged ions with respect to singly charged ones with $T_{eV} = 1\,\mathrm{eV}$:

$$\frac{n_e n_{i1}}{n_0} = \frac{2\times 2}{1}\times 3.02\times 10^{27}\times \mathrm{e}^{-24.5}$$

$$= 2.76\times 10^{17}\,\mathrm{m}^{-3} = A_1 , \tag{1.195}$$

$$\frac{n_e n_{i2}}{n_{i1}} = \frac{2\times 1}{2}\times 3.02\times 10^{27}\times \mathrm{e}^{-(54.4-24.5)}$$

$$= 3.12\times 10^{14}\,\mathrm{m}^{-3} = A_2 . \tag{1.196}$$

Combining equations (1.193), (1.194) and (1.195), we obtain a second order equation in n_{i1}:

$$n_{i1}^2 + A_1 n_{i1} - N_n A_1 = 0 \tag{1.197}$$

from which:

$$n_{i1} = -\frac{A_1}{2} \pm \sqrt{\left(\frac{A_1}{2}\right)^2 + N_n A_1} \ . \tag{1.198}$$

Only the case $n_{i1} > 0$ is physically realistic, so we deduce $n_{i1} = 5.12 \times 10^{18}\,\mathrm{m}^{-3}$.
Knowing n_{i1}, we can then deduce n_e and n_0 with the help of Eqs. (1.193) and (1.194): $n_e = 5.12 \times 10^{18}\,\mathrm{m}^{-3}$ and $n_0 = 9.49 \times 10^{19}\,\mathrm{m}^{-3}$.
A second iteration, using equation (1.196) with the n_e and n_{i1} values just determined, gives $n_{i2} = 3.12 \times 10^{14}\,\mathrm{m}^{-3}$

b) $T_{eV} = 10\,\mathrm{eV}$
This time we will calculate the exact solution, by solving a 3rd order equation in n_e. Writing the relevant Saha equation for the equilibrium of the singly charged ions with respect to the neutral state, and the equilibrium of the doubly charged ions with respect to singly charged helium ions with $T_{eV} = 10\,\mathrm{eV}$:

$$\begin{aligned} \frac{n_e n_{i1}}{n_0} &= \frac{2 \times 2}{1} \times 9.55 \times 10^{28} \times \mathrm{e}^{-2.45} \\ &= 3.29 \times 10^{28}\,\mathrm{m}^{-3} = B_1 \ , \end{aligned} \tag{1.199}$$

$$\begin{aligned} \frac{n_e n_{i2}}{n_{i1}} &= \frac{2 \times 1}{2} \times 9.55 \times 10^{28} \times \mathrm{e}^{-(5.44-2.45)} \\ &= 4.79 \times 10^{27}\,\mathrm{m}^{-3} = B_2 \ . \end{aligned} \tag{1.200}$$

Equations (1.190) and (1.189) can be written in the form:

$$n_{i1} = 2(N_n - n_0) - n_e \ , \tag{1.201}$$

$$n_{i2} = n_e - (N_n - n_0) \ . \tag{1.202}$$

Introducing expressions for n_{i1} and n_{i2} in (1.199) and (1.200), we find:

$$B_1 = \frac{n_e(2(N_n - n_0) - n_e)}{n_0} \tag{1.203}$$

from which:

$$n_0 = \frac{n_e(2N_n - n_e)}{B_1 + 2n_e} \ , \tag{1.204}$$

and:

$$B_2 = \frac{n_e(n_e - (N_n - n_0))}{2(N_n - n_0) - n_e} \ , \tag{1.205}$$

hence:

$$n_0 = \frac{n_e(N_n - n_e) + B_2(2N_n - n_e)}{2B_2 + n_e} \ . \tag{1.206}$$

We can eliminate n_0, using relations (1.204) and (1.206), which results in a third degree polynomial in n_e:

$$n_e^3 + (B_1)n_e^2 + [B_1(B_2 - N_n)]n_e - 2B_1B_2N_n = 0\,. \tag{1.207}$$

Retaining the positive real root of this equation, we obtain $n_e \simeq 2.00 \times 10^{20}\,\mathrm{m}^{-3}$. The density of the neutrals can be obtained from either equation (1.204) or equation (1.206): $n_0 \simeq 2.53 \times 10^4\,\mathrm{m}^{-3}$. Finally, using equations (1.201) and (1.202), we have $n_{i1} = 4.17 \times 10^{12}\,\mathrm{m}^{-3}$ and $n_{i2} = 1.00 \times 10^{20}\,\mathrm{m}^{-3}$, and we observe that, this time, $n_{i1} \ll n_{i2}$.

1.4. Calculate the electron density in an oven containing sodium vapour at 2000 K. The density of sodium nuclei is $10^{28}\,\mathrm{m}^{-3}$.

Energy threshold of the first ionisation of sodium: $\mathcal{E}_{i1} = 5.14\,\mathrm{eV}$
Energy threshold of the second ionisation of sodium (with respect to neutral ground state): $\mathcal{E}_{i2} = 47.29\,\mathrm{eV}$
Statistical weights: $g_0 = 2$, $g_{i1} = 1$, $g_{i2} = 4$.

Answer

Assuming the system is in thermodynamic equilibrium, we can use the Saha equation (see Sect. 1.4.2). In this case, for the singly charged sodium ion with respect to neutral sodium:

$$\frac{n_{i1}n_e}{n_0} = \frac{(2\pi m_e k_B T)^{\frac{3}{2}}}{h^3}\, 2\, \frac{B'(T)}{B(T)} \exp\left(-\frac{\mathcal{E}_{i1}}{k_B T}\right)\,, \tag{1.208}$$

where n_0 is the density of neutral atoms (in the ground state), n_{i1} is the density of singly charged ions (in the ground state) and n_e, the electron density; $B(T)$ is the partition function (Appendix II), defined by:

$$B(T) = \sum_k g_{0k} \exp\left(-\frac{\mathcal{E}_{0k}}{k_B T}\right)\,, \tag{1.209}$$

where the summation is over the different excited states of the neutral atom ($k = 0$ designates the ground state) and:

$$B'(T) = \sum_j g_{i1j} \exp\left(-\frac{\mathcal{E}_{i1j}}{k_B T}\right)\,, \tag{1.210}$$

where the summation is over the different excited states of the singly charged ion ($j = 0$ designates its ground state); g_{0k}, g_{i1j} represent the level degeneracies; $\mathcal{E}_{0k}$, $\mathcal{E}_{i1j}$ are the threshold energies of the excited levels with respect to the ground state of the atom and the ground state of the singly charged ion, respectively.

It is easy to see that when $k_B T$ is small compared to $\mathcal{E}_{0k}$ and $\mathcal{E}_{i1j}$, the partition functions reduce to $B(T) \simeq g_0$ and $B'(T) \simeq g_{i10}$ respectively (Appendix II).
If the atom loses another electron, the density of the doubly charged ions will be linked to that of the singly charged ions by the Saha equation:

$$\frac{n_{i2} n_e}{n_i} = \frac{(2\pi m_e k_B T)^{\frac{3}{2}}}{h^3} 2 \frac{g_{i2}}{g_{i1}} \exp\left[-\frac{\mathcal{E}_{i2} - \mathcal{E}_{i1}}{k_B T}\right] , \tag{1.211}$$

where n_{i2} is the density of the doubly charged ions.
It is obvious, in the present case, that since the particle temperature is 2000 K (i.e. in eV 2000/11600 $\simeq$ 0.17 eV) and the energy threshold of the first ionisation is 5.14 eV, the gas will be very weakly ionised and we can set $n_{2i} = 0$ (in effect, the value of $\exp(-5.14/0.17)$ in (1.208) relative to that of $\exp[-(47.29 - 5.14)/0.17]$ in (1.211) results in a comparison between 1.13×10^{-13} and a number 10^{-100}!) Finally, because $n_e \simeq n_i$, the Saha equation (1.208) reduces to:

$$\frac{n_e^2}{n_0} = \frac{(2\pi m_e k_B T)^{\frac{3}{2}}}{h^3} 2 \frac{1}{2} \exp -\frac{5.14}{0.17} . \tag{1.212}$$

The additional equation for conservation of nuclei, in the case where $n_{i2} \simeq 0$, can be written:

$$n_0 + n_{i1} \equiv n_0 + n_e = N_n , \tag{1.213}$$

where N_n is the density of the sodium nuclei. Thus from (1.212) and (1.213), expressing the energies in Joule:

$$n_e^2 = (N_n - n_e) \left[\frac{(2\pi \times 9.11 \times 10^{-31} \times 1.38 \times 10^{-23} \times 2000)^{\frac{3}{2}}}{(6.62 \times 10^{-34})^3}\right] \times \exp\left[-\frac{5.14 \times 1.6 \times 10^{-19}}{1.38 \times 10^{-23} \times 2000}\right] , \tag{1.214}$$

or:

$$n_e^2 = (N_n - n_e) A , \tag{1.215}$$

where we have set:

$$A = \frac{6.28 \times 10^{-74}}{290 \times 10^{-102}} (1.13 \times 10^{-13}) = 2.4 \times 10^{13} \text{ (m}^{-3}\text{)} . \tag{1.216}$$

We now need to resolve a quadratic equation of the form:

$$n_e^2 + A n_e - N_n A = 0 , \tag{1.217}$$

hence:

$$n_e = \frac{-A \pm \sqrt{A^2 + 4 N_n A}}{2} = 4.9 \times 10^{15} \text{ m}^{-3} . \tag{1.218}$$

The degree of ionisation $n_{i1}/(n_{i1} + n_0)$ is extremely small: 0.5%.

1.5.

a) Consider a plasma at a given instant (one dimensional configuration, see figure) with an electron density n_e that is 1% higher than the ion density n_i, in a slice of the plasma from $x = -\ell/2$ to $x = +\ell/2$. Derive the expressions for $E(x)$, the electric field and for $V(x)$, the potential in the region of non-neutrality.
Assume:

- that the field E is zero in the plasma surrounding the slice (because it is uniform and macroscopically neutral);
- that the potential at $x = -\ell/2$ and $x = +\ell/2$ (the plasma potential) is taken as the potential reference ($V(-\ell/2) = V(+\ell/2) = 0$).

Determine the direction of the space charge field.
Evaluate the electric field intensity at the borders of the region of non-neutrality at $x = -\ell/2$ and $x = +\ell/2$ for $n_i = 10^{16}\,\mathrm{m}^{-3}$ and for a slice width of $\ell = 2\,\mathrm{cm}$. Analyse this relation dimensionally, then calculate the potential at $x = 0$.

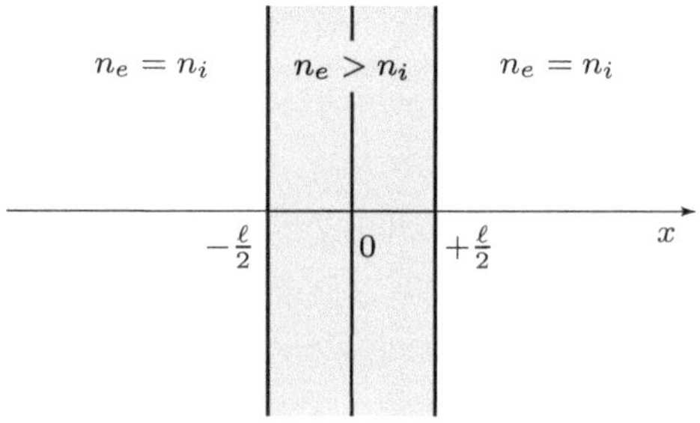

Fig. 1.20 One dimensional representation of a slice of non-neutrality of width ℓ in a plasma.

b) What is the energy (in eV) necessary for an electron, incident at $x = -\ell/2$ (from $x < -\ell/2$) to overcome the potential barrier (imposed by the non-neutral field) and cross the non-neutral zone to arrive at $x = +\ell/2$?

c) Using the preceding relations, derive an expression for the maximum distance that an electron with average thermal energy ($k_BT/2$ in one dimension) can travel away from its position of neutrality in the macroscopically neutral region, due to its thermal motion. We assume that the ions are immobile. Evaluate this distance for $k_BT/e = 1\,\mathrm{eV}$ and $n_e = 10^{16}\,\mathrm{m}^{-3}$.

d) Knowing that the electrostatic potential energy of an ensemble of charges in a volume V is:

$$\mathcal{W}_E = \frac{1}{2}\epsilon_0 \int_V E^2 \,\mathrm{d}V\ , \tag{1.219}$$

derive an expression for the electrostatic energy (or potential energy) associated with the presence of charges in the zone extending from $x = -\ell/2$ to $x = +\ell/2$. Evaluate this potential for the conditions given in a).

Answer

a) This type of problem can be treated with the Poisson equation $\nabla \cdot \boldsymbol{E} = \rho/\epsilon_0$ (1.1), a self-consistent relation since the density of charges ρ is the source of the field $\boldsymbol{E}$, ϵ_0 being the permittivity of vacuum.
In the case in which the non-neutrality of charges is to be resolved in a single dimension x, as illustrated in the figure above, we have, from Poisson's equation:

$$\frac{\mathrm{d}E}{\mathrm{d}x} = \frac{\rho}{\epsilon_0} , \tag{1.220}$$

$$\int \mathrm{d}E = \int \frac{\rho}{\epsilon_0} \, \mathrm{d}x , \tag{1.221}$$

$$E(x) = \frac{\rho}{\epsilon_0} x + C_1 . \tag{1.222}$$

Further, the potential ϕ, defined by $\boldsymbol{E} = -\nabla\phi$, can be written in one dimension:

$$\frac{\mathrm{d}\phi}{\mathrm{d}x} = -E(x) \tag{1.223}$$

and:

$$\int \mathrm{d}\phi = \int -E \, \mathrm{d}x , \tag{1.224}$$

which, by substituting (1.222), becomes:

$$\int \mathrm{d}\phi = \int -\left(\frac{\rho}{\epsilon_0} x + C_1\right) \mathrm{d}x , \tag{1.225}$$

$$\phi(x) = -\frac{\rho}{\epsilon_0}\frac{x^2}{2} - C_1 x + C_2 . \tag{1.226}$$

Because of symmetry, $\phi(x) = \phi(-x)$, so $C_1 = 0$, and:

$$E(x) = \frac{\rho}{\epsilon_0} x \tag{1.227}$$

in the interval $-\ell/2 \le x \le \ell/2$. Further, since $\phi(-\ell/2) = \phi(\ell/2) = 0$, the expression for C_2 is:

$$C_2 = \frac{\rho}{2\epsilon_0}\frac{\ell^2}{4} , \tag{1.228}$$

hence:

$$\phi(x) = +\frac{\rho}{2\epsilon_0}\left(\frac{\ell^2}{4} - x^2\right) . \tag{1.229}$$

The variations of the electric field and the potential $\phi(x)$ are drawn below, as a function of x, for $\rho < 0$. There is a discontinuity in the electric field $E(x)$ and the derivative $\mathrm{d}\phi(x)/\mathrm{d}x$ of the potential at $x = \pm\ell/2$ due to the discontinuity of the space charge at each side of the boundary $x = \pm\ell/2$. For $\rho < 0$ ($n_e > n_i$), according to (1.227), the space charge electric field is positive (directed to the right, Sect. 2.2.1) for negative values of x such that $-\ell/2 \leq x \leq 0$ and it is negative (directed to the left) for positive values of x such that $0 \leq x \leq \ell/2$. The direction of the electric field is such that the field tends to attract the ions into the space charge region and repel the electrons. Consequently, the non-neutrality cannot remain for very long, at most a time of order ω_{pe}^{-1}.

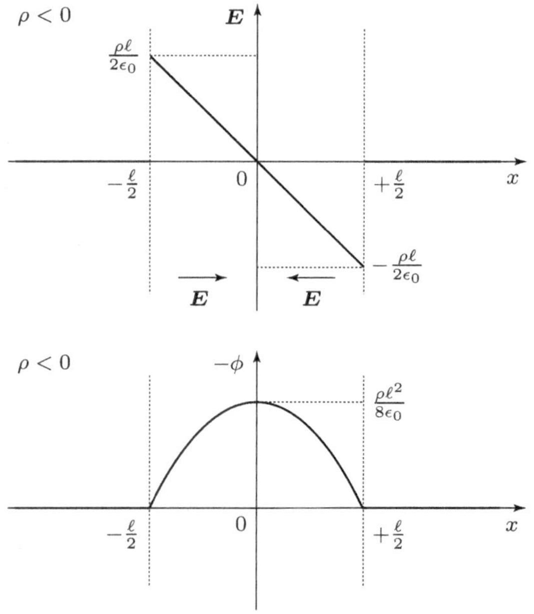

In the present case, $\rho \equiv (n_i - n_e)e = (\Delta n)e$. Therefore, from (1.227) the local field intensity at x is:

$$E = \frac{(\Delta n)e}{\epsilon_0}x , \tag{1.230}$$

where $|\Delta n|$ is, from the initial assumptions, $0.01n_i$.

We can make the dimensional analysis at the same time as we continue the numerical calculation, $x = \ell/2 = 10^{-2}$ m, $\Delta n = 10^{14}$ m^{-3}, $\epsilon_0 = 8.85 \times$

$10^{-12}\,\mathrm{F\,m^{-1}}$, $e = 1.6 \times 10^{-19}\,\mathrm{C}$, where the coulomb C is linked, as a unit, to the farad F by the relation C = F V.
From equation (1.230), at $x = -\ell/2$:

$$E = \frac{10^{14}\ 1.6 \times 10^{-19}\ 10^{-2}}{8.85 \times 10^{-12}} = \frac{\mathrm{C m}}{\mathrm{m^3(F\,m^{-1})}} = \frac{\mathrm{FV}}{\mathrm{mF}} = \frac{\mathrm{V}}{\mathrm{m}} = 18\,\mathrm{kV\,m^{-1}}\,, \tag{1.231}$$

the units $\mathrm{V\,m^{-1}}$ being those of the electric field.
In the same manner, the potential at $x = 0$ becomes, from (1.229):

$$\phi(0) = -\frac{e\Delta n \ell^2}{8\epsilon_0}\,, \tag{1.232}$$

giving $\phi(0) = -90\,\mathrm{V}$.

b) In order for an electron coming from $x < -\ell/2$ to cross the space charge region, it is necessary for the initial kinetic energy U in the direction x to be greater than the energy necessary to break through the potential due to the space charge (caused by electrons!), which is:

$$U = e\phi(0) = -90\,\mathrm{eV}\,. \tag{1.233}$$

c) We will show that the maximum distance x over which an electron with average thermal energy can travel from its neutral position in a macroscopically neutral region, due to this thermal energy, is the Debye length (Sect. 1.6). This maximum distance is fixed by the equality between the thermal energy of this "average" electron, and the potential energy linked to the space charge electric field produced by the deviation of the electron with respect to its neutral position, namely:

$$\frac{1}{2}k_B T = |U|\,. \tag{1.234}$$

However, the work exerted by the electric field $\boldsymbol{E}$ on an electron is given by the expression (Sect. 2.1):

$$\mathrm{d}U = F\,\mathrm{d}x = -|\boldsymbol{E}|e\,\mathrm{d}x\,, \tag{1.235}$$

Hence from (1.227) and since the electron is moving from $x = 0$ to x:

$$|U| = \int_0^x |\boldsymbol{E}|e\,\mathrm{d}x = \frac{e\rho x^2}{2\epsilon_0}\,. \tag{1.236}$$

Assuming that the ions are at rest[49], only the electrons move into the space charge field, and we obtain, following (1.236):

[49] Delcroix and Bers (Sect. 1.4.1), in their case, assumed a complete absence of ions.

$$|U| \equiv \frac{n_e e^2 x^2}{2\epsilon_0} , \tag{1.237}$$

then, making use of (1.234):

$$\frac{n_e e^2 x^2}{2\epsilon_0} = \frac{1}{2} k_B T , \tag{1.238}$$

and finally:

$$x = \sqrt{\frac{\epsilon_0 k_B T}{e^2 n_e}} , \tag{1.239}$$

which is, in effect, the expression for λ_{De} (1.41).

Application: for $T_{eV} = 1\,\text{eV}$ and $n_e = 10^{16}\,\text{m}^{-3}$, we find $\lambda_{De} = 74\,\mu\text{m}$.

d) If we express an element of volume as $\mathrm{d}V = S\,\mathrm{d}x$ where S is a surface, the general expression for the electrostatic potential energy being:

$$\mathcal{W}_E = \frac{1}{2}\epsilon_0 \int_V E^2\,\mathrm{d}V , \tag{1.219}$$

we can write:

$$\mathcal{W}_E = \frac{1}{2}\epsilon_0 \int_{-\ell/2}^{\ell/2} E^2 S\,\mathrm{d}x . \tag{1.240}$$

We can then deduce the electrostatic energy density per unit surface:

$$\frac{\mathcal{W}_E}{S} = \frac{1}{2}\epsilon_0 \int_{-\ell/2}^{\ell/2} E^2\,\mathrm{d}x = \frac{(e\Delta n_e)^2}{24\epsilon_0}\, l^3 . \tag{1.241}$$

For the conditions indicated in a), we find numerically:

$$\frac{\mathcal{W}_E}{S} = 9.6 \times 10^{-6}\,\text{J}\,\text{m}^{-2} . \tag{1.242}$$

1.6. Consider two parallel conducting plates extending to infinity in y and z, and placed at $x = \pm d$. Their potential ϕ is assumed to be zero. The space between the plates is occupied by a gas of charged particles of one single type, of uniform density n and charge q.

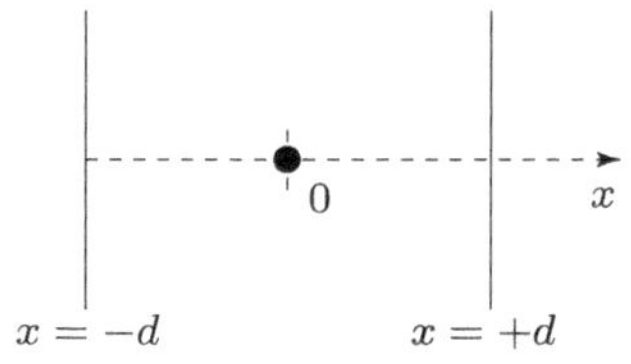

a) Show that the potential distribution between the plates is given by:

$$\phi(x) = \frac{nq}{2\epsilon_0}(d^2 - x^2) \,. \tag{1.243}$$

b) Assuming that the distance d is greater than the Debye length λ_D, what is the probability of finding a particle with a kinetic energy sufficiently high to overcome the potential energy $\mathcal{E}$ and travel from one plate to the mid-distance between the plates? Consider that the particles have a Maxwell-Boltzmann velocity distribution and discuss to what extent the hypothesis that a particle has such an energy is generally acceptable?

c) Calculate the Debye length in the case of an electric discharge in which the plasma density is $10^{16}\,\mathrm{m}^{-3}$ and the temperature is $2\,\mathrm{eV}$.

Answer

a) Poisson's equation $\boldsymbol{\nabla}\cdot\boldsymbol{E} = \rho/\epsilon_0$ and the relation $\boldsymbol{E} = -\boldsymbol{\nabla}\phi$ allows us to write the equation for the potential ϕ in the form:

$$\nabla^2\phi = -\frac{\rho}{\epsilon_0} \tag{1.244}$$

and, in the present case, since there is only one type of charged particle, $\rho = nq$, we can thus write:

$$\nabla^2\phi = -\frac{nq}{\epsilon_0} \,. \tag{1.245}$$

In the present configuration (one single dimension), a first integration gives:

$$\boldsymbol{\nabla}\phi = -\frac{nqx}{\epsilon_0} + C_1 \,. \tag{1.246}$$

The position $x = 0$ represents the axis of symmetry of the system, such that:

$$\boldsymbol{\nabla}\phi(x=0) = 0 \,, \tag{1.247}$$

therefore $C_1 = 0$.
A second integration leads to:

$$\phi(x) = -\frac{nqx^2}{2\epsilon_0} + C_2 \,, \tag{1.248}$$

but $\phi(x = \pm d) = 0$, from which $C_2 = nqd^2/2\epsilon_0$ and, finally:

$$\phi(x) = \frac{nq}{2\epsilon_0}(d^2 - x^2) \,, \tag{1.249}$$

recalling that we have assumed that n is independent of position.

b) The potential difference between the axis and one of the plates is, following (1.249):

$$\phi(0) - \phi(\pm d) = \frac{nqd^2}{2\epsilon_0} , \tag{1.250}$$

and the energy given to the particle in moving from $x = d$ to $x = 0$ is obtained by multiplying the potential difference by q:

$$\mathcal{E} = \frac{nq^2d^2}{2\epsilon_0} . \tag{1.251}$$

Assuming that $d > \lambda_D$, we then have:

$$\mathcal{E} = \frac{nq^2d^2}{2\epsilon_0} > \frac{nq^2}{2\epsilon_0}\lambda_D^2 \equiv \frac{nq^2}{2\epsilon_0}\frac{k_BT\epsilon_0}{nq^2} = \frac{k_BT}{2} , \tag{1.252}$$

i.e. the electrostatic energy of the particles would be higher than their average kinetic energy (Maxwell-Boltzmann velocity distribution), $m\langle w_x^2\rangle/2 = k_BT/2$. In such a situation, is it possible to find a particle with a kinetic energy higher than the potential energy $\mathcal{E}$? To answer this question, we first consider the case where d is equal to or slightly larger than λ_D ($d \geq \lambda_D$), and then the case where d is much larger than λ_D ($d \gg \lambda_D$).
Since the velocity distribution of the particles is Maxwellian, and as long as d remains of the order of λ_D, there are particles with velocities such that their kinetic energy is much higher than the average kinetic energy (see Fig. I.2 in Appendix I). However, in the case where $d \gg \lambda_D$, considering, on the one hand, that the electrostatic energy $\mathcal{E}$ increases with the square of d, and, on the other hand, that the number of high energy particles decreases exponentially with velocity in a Maxwell-Boltzmann distribution, the proposed hypothesis of finding a particle with enough energy to travel from one plate to the mid-distance between the plates becomes unrealistic.

c) We have the numerical relation (Sect. 1.6):

$$\lambda_D = 740\left(\frac{T_{eV}}{n}\right)^{\frac{1}{2}} , \tag{1.55}$$

where λ_D is in cm and n in cm^{-3}. In the present case, n is 10^{10} cm^{-3} and $T_{eV} = 2$ eV, from which:

$$\lambda_D = 740\left(\frac{2}{10^{10}}\right)^{\frac{1}{2}} = 740\sqrt{2}10^{-5}\,\text{cm} \simeq 1.05\cdot10^{-2}\,\text{cm} = 105\,\mu\text{m} . \tag{1.253}$$

1.7.

a) In a plasma, charged particles move randomly due to their thermal energy. One of the conditions for the existence of a plasma is that the energy of the thermal motion of the charged particles (characterised here by a temperature $T_e = T_i = T$) should be much greater than the average energy of Coulomb attraction exerted between ions ($Z = 1$) and electrons.

Show that this assumption leads to the relation:

$$n_e \lambda_D^3 \gg 1 , \tag{1.254}$$

an inequality expressing the neutrality condition in a plasma with electron density n_e and Debye's length λ_D.

N.B. The average distance d between an electron and an ion can be deduced from the trivial relation $\left(\frac{4}{3}\pi d^3\right) n_e = 1$, where d is the radius of the sphere.

b) Transform the equilibrium equation for ionisation (Saha's equation) to the form:

$$\frac{n_e}{n_0} = \frac{2^{5/2} g_i}{g_0} \frac{\left(n_e \lambda_D^3\right)}{\sqrt{n_e a_0^3}} \exp\left(-\frac{\mathcal{E}_i}{k_B T}\right) , \tag{1.255}$$

where:

$$a_0 = \frac{4\pi\epsilon_0 \hbar^2}{m_e e^2} \tag{1.256}$$

is the radius of the first orbit of the Bohr hydrogen atom, n_0 the density of neutrals. g_i, g_0 the statistical weights of the singly-charged ion (ground state) and neutral atom (ground state), respectively, and $\mathcal{E}_i$, the threshold energy for the first ionisation of helium.

c) Show that (1.254) is neither a sufficient, nor a necessary condition for n_e/n_0 to be large (high degree of ionisation).

Constant: $a_0 = 0.05\,\text{nm}$. Application: $\mathcal{E}_i = 24.59\,\text{eV}$ (energy threshold of the first ionisation of the helium atom).

Answer

a) The energy of interaction between an electron and an ion can be written:

$$\phi(d) = \frac{e^2}{4\pi\epsilon_0 d} , \tag{1.257}$$

where d is the average distance between the two types of charged particles. The assumption stated above implies that:

$$\frac{e^2}{4\pi\epsilon_0 d} \ll k_B T_e . \tag{1.258}$$

To estimate the average distance d, we turn to the obvious relation:

$$\left(\frac{4}{3}\pi d^3\right) n_e = 1 , \tag{1.259}$$

which states that the spherical volume around an ion, of radius d, only contains one particle (on average). From (1.259) and (1.257), we obtain:

$$\left(\frac{4}{3}\pi n_e\right)^{\frac{1}{3}} \frac{e^2}{4\pi\epsilon_0} \frac{1}{k_B T_e} \ll 1 \,. \tag{1.260}$$

From the definition (1.41):

$$\lambda_{De} = \left(\frac{\epsilon_0 k_B T_e}{ne^2}\right)^{\frac{1}{2}} \,, \tag{1.261}$$

we obtain the inequality (1.260) raised to the cubic power:

$$\frac{4\pi n_e}{3}\left(\frac{n_e e^2}{4\pi\epsilon_0 k_B T_e}\right)^3 \frac{1}{n_e^3} \ll 1 \,, \tag{1.262}$$

from which:

$$\frac{4\pi}{3(4\pi)^3}\frac{1}{n_e^2}\frac{1}{\lambda_D^6} \ll 1 \,. \tag{1.263}$$

The factor $1/3(4\pi)^2$ simply reinforces the fact that:

$$n_e \lambda_D^3 \gg 1 \,. \tag{1.264}$$

b) The Saha relation, for singly ionised atoms, from Sect. 1.4.2, is:

$$\frac{n_e n_i}{n_0} = \frac{2g_i}{g_0}\frac{(2\pi m_e k_B T)^{\frac{3}{2}}}{h^3} \exp\left(-\frac{\mathcal{E}_i}{k_B T}\right) \,. \tag{1.12}$$

Noting $n_e = n_i$ (the helium ion is assumed to be singly ionised) and introducing λ_D in (1.12) via a ratio equal to unity:

$$\frac{n_e^2}{n_0} = \frac{2g_i}{g_0}\frac{(2\pi m_e k_B T)^{\frac{3}{2}}}{h^3}\left(\frac{\lambda_D^2 n_e e^2}{\epsilon_0 k_B T}\right)^{\frac{3}{2}} \exp\left(-\frac{\mathcal{E}_i}{k_B T}\right) \,, \tag{1.265}$$

such that:

$$\frac{n_e^2}{n_0} = \frac{2g_i}{g_0}\frac{n_e^2}{n_e^{\frac{1}{2}}}\lambda_D^3\left(\frac{2\pi m_e e^2}{h^2\epsilon_0}\right)^{\frac{3}{2}} \exp\left(-\frac{\mathcal{E}_i}{k_B T}\right) \,, \tag{1.266}$$

$$\frac{n_e}{n_0} = \frac{2g_i}{g_0}\frac{n_e\lambda_D^3}{\sqrt{n_e}}\left(\frac{2m_e e^2}{4\pi\hbar^2\epsilon_0}\right)^{\frac{3}{2}} \exp\left(-\frac{\mathcal{E}_i}{k_B T}\right) \,, \tag{1.267}$$

and finally:

$$\frac{n_e}{n_0} = \frac{2^{\frac{5}{2}}g_i}{g_0}\frac{n_e\lambda_D^3}{\sqrt{n_e a_0^3}} \exp\left(-\frac{\mathcal{E}_i}{k_B T}\right) \,. \tag{1.268}$$

c) In the case where T is low, for n_e/n_0 to be large, the relation (1.268) requires that n_0 be small, that is to say the gas pressure should be low. For this, $n_e\lambda_D^3 \gg 1$ is, a priori, insufficient.
In the case where T is high, n_e/n_0 is a priori large: the exponential decay is weak and $n_e a_0^3 \ll 1$ (which is the case when $n_e \approx 10^{15}\,\text{cm}^{-3}$, $a_0 = 5\times 10^{-9}$ cm, from which $n_e a_0^3$: $10^{15}(5)^3 10^{-27}$, clearly an expression always $\ll 1$). It is therefore not necessary for $n_e\lambda_D^3 \gg 1$ to achieve a high degree of ionisation.

1.8.

a) Consider a lithium ion with kinetic energy $\mathcal{E}_c = 60\,\text{eV}$, colliding with a helium atom at rest. Will the helium be ionised? If this is the case, indicate if it will produce a single or double ionisation.
b) What would happen if the incident lithium ions were replaced by an electron with the same energy.

Mass of lithium: $\simeq 6.9\,m_p$
Mass of helium: $\simeq 4.0\,m_p$
Mass of proton: $m_p = 1.67\times 10^{-27}$ kg
Energy threshold for single ionisation of helium: 24.59 eV
Energy threshold for second ionisation of helium: 54.4 eV

Answer

This concerns a collision between an ion and an atom. The laws of conservation of energy and momentum indicate (Sect. 1.7.2) that only the kinetic energy linked to the relative motion can be transferred into internal energy. The expression for conservation of energy can be effectively written (1.79) as:

$$\frac{\mu_{hl} w_{hl}^2}{2} = \frac{\mu_{hl} {w'}_{hl}^2}{2} + \Delta\mathcal{E} \tag{1.269}$$

with $\boldsymbol{w}$ and $\boldsymbol{w}'$ the relative helium-lithium velocities before and after collision, and μ_{hl} the reduced mass:

$$\mu_{hl} = \frac{m_l m_h}{m_l + m_h}\,, \tag{1.270}$$

where m_l and m_h are the masses of the lithium and helium atom respectively. For a given initial relative velocity $\boldsymbol{w}$, the maximum energy that can be transferred into internal energy, from (1.269), is obtained for $\boldsymbol{w}' = 0$, i.e.:

$$\Delta\mathcal{E}_{\max} = \frac{\mu_{hl} w_{hl}^2}{2}\,. \tag{1.271}$$

Given that the helium atom is at rest, the initial relative velocity $\boldsymbol{w}_{hl}$ is equal to $\boldsymbol{w}_l$, the initial velocity of the lithium ion, whose kinetic energy is:

$$\mathcal{E}_c = \frac{m_l w_l^2}{2} \ . \tag{1.272}$$

Explicitly, (1.306) can be written:

$$\Delta\mathcal{E}_{\max} = \mathcal{E}_c \left(\frac{m_h}{m_h + m_l} \right) \ . \tag{1.273}$$

Remarks:

1. If the kinetic energy linked to the relative motion is completely transferred to internal energy (in other words, there is complete equality between the energy threshold of the inelastic reaction and the transferable kinetic energy), then $\boldsymbol{w}'_{hl} = 0$ ($\boldsymbol{w}'_h = \boldsymbol{w}'_l$). Following (1.74) and (1.75) and including (1.70):
$$\boldsymbol{w}'_h = \boldsymbol{w}'_l = \boldsymbol{w}'_0 = \boldsymbol{w}_0 \ . \tag{1.274}$$
Since $\boldsymbol{w}_h = 0$, we find from (1.75):
$$\boldsymbol{w}_0 = \boldsymbol{w}_l \, \frac{m_l}{m_l + m_h} \ . \tag{1.275}$$
Equation (1.274) shows that, after the collision, the helium and lithium nuclei both follow a uniform, straight path, with a velocity equal to that of the centre of mass $\boldsymbol{w}_0$ and given by (1.275).
If the energy threshold of the inelastic reaction is below the maximum transferable energy, the remaining energy will be repartitioned as elastic energy between the helium and the lithium. If the relative velocity after collision is $\boldsymbol{w}'_{hl}$, the velocities $\boldsymbol{w}'_h$ and $\boldsymbol{w}'_l$ are given by (1.74) and (1.75).
2. Assuming that the kinetic energy linked to the relative motion is integrally transferred into internal energy, we have $\boldsymbol{w}'_{hl} = 0$. In this case, from (1.274) and (1.275), the velocity $\boldsymbol{w}_l$ of the lithium particle before collision and the velocities $\boldsymbol{w}'_l$ and $\boldsymbol{w}'_h$ of the lithium and helium particles after collision are all co-linear; the collision is thus necessarily frontal.

Numerical application:

a) The mass of the lithium ion being $\simeq 6.9 m_p$, and that of the helium atom $\simeq 4.0 m_p$, the numerical value of (1.273) is then:

$$\Delta\mathcal{E}_{\max} = \mathcal{E}_c \, \frac{4.0 m_p}{6.9 m_p + 4.0 m_p} = 60 \, \frac{4.0}{6.9 + 4.0} = 22.02 \, \mathrm{eV} \ .$$

This energy is not sufficient to induce even a single ionisation of the helium atom.

b) If the lithium ion is replaced by an electron as incident particle, we obtain from (1.273):

$$\Delta\mathcal{E}_{\max} = 60\frac{4.0m_p}{m_e + 4.0m_p} = 59.99\,\mathrm{eV}\ .$$

In this case, we can achieve a double ionisation of helium.

Another solution

We could also derive the solution directly from the conservation equations:

1. Conservation of kinetic energy (1.67):

$$\frac{m_l w_l^2}{2} + \frac{m_h w_h^2}{2} = \frac{m_l w_l'^2}{2} + \frac{m_h w_h'^2}{2} + \Delta\mathcal{E}\ , \tag{1.276}$$

   where m_l is the mass of the lithium ion, $\boldsymbol{w}_l$ its velocity before collision, $\boldsymbol{w}_l'$ its velocity after collision, m_h is the mass of the helium atom, $\boldsymbol{w}_h$ its velocity before collision, $\boldsymbol{w}_h'$ its velocity after collision, and$\Delta\mathcal{E}$ is the internal energy of the helium atom after collision.
2. Conservation of momentum (1.66):

$$m_l\boldsymbol{w}_l + m_h\boldsymbol{w}_h = m_l\boldsymbol{w}_l' + m_h\boldsymbol{w}_h'\ . \tag{1.277}$$

   In the present case, the helium ion is at rest before the collision: $\boldsymbol{w}_h = 0$. Using the law of conservation of momentum (1.277), the velocity of the helium atom after collision reduces to:

$$\boldsymbol{w}_h' = \frac{m_l(\boldsymbol{w}_l - \boldsymbol{w}_l')}{m_h}\ , \tag{1.278}$$

   and the law of conservation of total energy can be written in the following form:

$$\frac{m_l w_l^2}{2} = \frac{m_l w_l'^2}{2} + \frac{m_l^2}{2m_h}(\boldsymbol{w}_l - \boldsymbol{w}_l')^2 + \Delta\mathcal{E}\ . \tag{1.279}$$

   Under these conditions, the initial velocity of the lithium is known (the kinetic energy $\mathcal{E}_c$ being given), the change of internal energy $\mathcal{E}_c$ is simply a function of $\boldsymbol{w}_l'$, the velocity of the lithium ion after collision.
3. Calculation of the maximum possible $\Delta\mathcal{E}$ (to be compared with the ionisation energy thresholds).
   The energy available for transfer to internal energy will be a maximum when:

$$\frac{\mathrm{d}(\Delta\mathcal{E})}{\mathrm{d}\boldsymbol{w}_l'} = 0\ , \tag{1.280}$$

   which allows us to determine the maximum value of $\Delta\mathcal{E}$.

From (1.279):

$$\Delta\mathcal{E} = \frac{m_l w_l^2}{2} - \frac{m_l w_l'^2}{2} - \frac{m_l^2}{2m_h}(\boldsymbol{w}_l - \boldsymbol{w}_l')^2 \,, \tag{1.281}$$

$$\Delta\mathcal{E} = \frac{m_l w_l^2}{2} - \frac{m_l w_l'^2}{2} - \frac{m_l^2}{2m_h} w_l^2 + \frac{m_l^2}{m_h}\boldsymbol{w}_l \cdot \boldsymbol{w}_l' - \frac{m_l^2}{2m_h} w_l'^2 \,. \tag{1.282}$$

This becomes:

$$\frac{\mathrm{d}(\Delta\mathcal{E})}{\mathrm{d}\boldsymbol{w}_l'} = -m_l \boldsymbol{w}_l' + \frac{m_l^2}{m_h}\boldsymbol{w}_l - \frac{m_l^2}{m_h}\boldsymbol{w}_l' = 0 \,, \tag{1.283}$$

$$\boldsymbol{w}_l' \left(m_l + \frac{m_l^2}{m_h} \right) = \boldsymbol{w}_l \frac{m_l^2}{m_h} \,, \tag{1.284}$$

and finally:

$$\boldsymbol{w}_l' = \boldsymbol{w}_l \frac{m_l}{m_l + m_h} \,, \tag{1.285}$$

which shows that the maximum occurs for a frontal collision (use (1.91) with $\theta = \pi$ and $w_h = 0$).
In this case, the value of $\Delta\mathcal{E}_{\max}$, starting from (1.279) and using (1.285), is:

$$\Delta\mathcal{E}_{\max} = \frac{m_l w_l^2}{2} - \frac{m_l^3 w_l^2}{2(m_l + m_h)^2} - \frac{m_l^2}{2m_h} w_l^2 \left(1 - \frac{m_l}{m_l + m_h} \right)^2 \,, \tag{1.286}$$

$$\Delta\mathcal{E}_{\max} = \frac{m_l w_l^2}{2} - \frac{m_l w_l^2}{2} \frac{m_l^2}{(m_l + m_h)^2} - \frac{m_l w_l^2}{2} \frac{m_l}{m_h} \left(\frac{m_h}{m_l + m_h} \right)^2 \,, \tag{1.287}$$

and:

$$\Delta\mathcal{E}_{\max} = \mathcal{E}_c \left(1 - \frac{m_l^2}{(m_l + m_h)^2} - \frac{m_l}{m_h} \frac{m_h^2}{(m_l + m_h)^2} \right) \,, \tag{1.288}$$

which finally leads to:

$$\Delta\mathcal{E}_{\max} = \mathcal{E}_c \left(\frac{m_h}{m_l + m_h} \right) \,, \tag{1.289}$$

and we recover the numerical implementation already treated above.

A further solution

Instead of looking for the maximum possible value of $\Delta\mathcal{E}$, we can calculate the velocities after the collision as a function of $\Delta\mathcal{E}$. If the value of $\Delta\mathcal{E}$ does not satisfy the conservation laws, then the values of the velocities after collision so obtained will be either negative or imaginary.

In (1.282) of variant 1 of the solution, $\Delta\mathcal{E}$ is a maximum when the scalar product $\boldsymbol{w}_l \cdot \boldsymbol{w}'_l$ is a maximum, that is when $\boldsymbol{w}_l$ and $\boldsymbol{w}'_l$ are colinear (frontal collision) and in the same direction (see (1.285)). Thus we have:

$$\Delta\mathcal{E} = \mathcal{E}_c - \frac{m_l w_l'^2}{2} - \mathcal{E}_c \frac{m_l}{m_h} + \frac{m_l^2}{m_h}\sqrt{\frac{2\mathcal{E}_c}{m_l}}\, w'_l - \frac{m_l w_l'^2}{2}\frac{m_l}{m_h} , \tag{1.290}$$

$$\Delta\mathcal{E} = \mathcal{E}_c \left(1 - \frac{m_l}{m_h}\right) - \frac{m_l w_l'^2}{2}\left(1 + \frac{m_l}{m_h}\right) + \frac{m_l^2}{m_h}\sqrt{\frac{2\mathcal{E}_c}{m_l}}\, w'_l , \tag{1.291}$$

$$\frac{m_l w_l'^2}{2}\left(1 + \frac{m_l}{m_h}\right) - \frac{m_l^2}{m_h}\sqrt{\frac{2\mathcal{E}_c}{m_l}}\, w'_l + \left[\Delta\mathcal{E} - \mathcal{E}_c\left(1 - \frac{m_l}{m_h}\right)\right] = 0 , \tag{1.292}$$

and finally:

$$w'_l = \frac{-\mathcal{B} \pm \sqrt{\mathcal{B}^2 - 4\mathcal{A}\mathcal{C}}}{2\mathcal{A}} , \tag{1.293}$$

where ($\mathcal{E}_c$ and $\Delta\mathcal{E}$ are now in eV and $\mathcal{B}^2$ in kg J):

$$\mathcal{A} = \frac{m_l}{2}\left(1 + \frac{m_l}{m_h}\right) \ (\text{kg}) , \tag{1.294}$$

$$\mathcal{B} = -\frac{m_l^2}{m_h}\sqrt{\frac{2e\mathcal{E}_c}{m_l}} , \tag{1.295}$$

$$\mathcal{C} = \left[e\Delta\mathcal{E} - e\mathcal{E}_c\left(1 - \frac{m_l}{m_h}\right)\right] \ (\text{J}) . \tag{1.296}$$

The velocity of the ion after collision w'_l (1.293), must be a real quantity. In addition, the law of conservation of momentum should give a single value for the velocity of the lithium ion after the collision. Under these conditions, the value of $\Delta\mathcal{E}$ that leads to a single root of equation (1.293) will be the portion of energy transferred by the lithium ion to the helium atom during the collision.

To allow a single root, the discriminant of the quadratic equation is necessarily zero:

$$\mathcal{D} = \mathcal{B}^2 - 4\mathcal{A}\mathcal{C} = 0 . \tag{1.297}$$

From (1.297) we obtain:

$$\frac{m_l^4}{m_h^2}\frac{2e\mathcal{E}_c}{m_l} = 4\frac{m_l}{2}\left(1 + \frac{m_l}{m_h}\right)\left[e\Delta\mathcal{E} - e\mathcal{E}_c\left(1 - \frac{m_l}{m_h}\right)\right] , \tag{1.298}$$

$$\frac{m_l^2}{m_h}\frac{\mathcal{E}_c}{(m_h + m_l)} + \mathcal{E}_c\frac{(m_h - m_l)}{m_h} = \Delta\mathcal{E} . \tag{1.299}$$

Thus:

$$\Delta\mathcal{E} = \mathcal{E}_c \left(\frac{m_l^2}{m_h(m_h + m_l)} + \frac{(m_h - m_l)(m_h + m_l)}{m_h(m_h + m_l)} \right) , \tag{1.300}$$

$$\Delta\mathcal{E} = \mathcal{E}_c \left(\frac{m_l^2 + m_h^2 - m_l^2}{m_h(m_h + m_l)} \right) , \tag{1.301}$$

and finally:

$$\Delta\mathcal{E} = \mathcal{E}_c \left(\frac{m_h}{m_h + m_l} \right) , \tag{1.302}$$

the numerical application for which has already been treated above.
If we choose to give $\Delta\mathcal{E}$ in (1.276) a value equal to the first ionisation of helium ($\Delta\mathcal{E} = 24.59\,\mathrm{eV}$), we obtain from (1.297):

$$\mathcal{A} = \frac{6.9m_p}{2}\left(1 + \frac{6.9m_p}{4.0m_p}\right) = 1.57 \times 10^{-26}\ (\mathrm{kg}) ,$$

$$\mathcal{B} = -\frac{6.9^2 m_p}{4}\sqrt{\frac{2}{6.9}\frac{1.6 \times 10^{-19}\,60}{1.67 \times 10^{-27}}} = -8.11 \times 10^{-22}\ (\mathcal{B}^2 \text{ in kg J}) ,$$

$$\mathcal{C} = \left[1.6 \times 10^{-19}\left(24.59 - 60\left(1 - \frac{6.9m_p}{4.0m_p}\right)\right)\right] = 1.089 \times 10^{-17}\ (\mathrm{J}) ,$$

$$\mathcal{D} = \mathcal{B}^2 - 4\mathcal{A}\mathcal{C} = 6.58 \times 10^{-43} - 6.84 \times 10^{-43} = -2.6 \times 10^{-44} < 0 ,$$

which is an unacceptable solution, because it would require an imaginary value for the velocity of the lithium ion after the collision, w_l' (1.293).
For $\Delta\mathcal{E} = 22.02\,\mathrm{eV}$,

$$\mathcal{A} = 1.57 \times 10^{-26}\ (\mathrm{kg}) ,$$

$$\mathcal{B} = -8.11 \times 10^{-22}\ (\mathcal{B}^2 \text{ en kg J}) ,$$

$$\mathcal{C} = \left[1.6 \times 10^{-19}\left(22.02 - 60\left(1 - \frac{6.9m_p}{4.0m_p}\right)\right)\right] = 1.048 \times 10^{-17}\ (\mathrm{J}) ,$$

$$\mathcal{D} = \mathcal{B}^2 - 4\mathcal{A}\mathcal{C} = 6.58 \times 10^{-43} - 6.58 \times 10^{-43} = 0 ,$$

and:

$$w_l' = \frac{1}{2}\frac{1.81 \times 10^{-22}}{1.57 \times 10^{-26}} = 2.58 \times 10^4\ \mathrm{m\,s^{-1}} . \tag{1.303}$$

The solution is valid (satisfying the conservation laws), but the internal energy is less than the energy threshold for the first ionisation of helium, as was shown in the original solution.

1.9. Consider two populations of particles, α and β, with masses m_α and m_β, whose velocity distribution functions $f_\alpha(\boldsymbol{w}_\alpha)$ and $f_\beta(\boldsymbol{w}_\beta)$ are isotropic, with temperatures T_α and T_β, respectively.

a) Calculate the average value $\langle|\boldsymbol{w}_\alpha - \boldsymbol{w}_\beta|\rangle$ of the modulus of the relative velocities between particles α and β, using the following steps:

1. Write the relation expressing the average of the modulus of the relative velocities by integration over the distribution functions f_α and f_β.
2. Introduce the following change of variables:

$$\boldsymbol{w}_{\alpha\beta} = \boldsymbol{w}_\alpha - \boldsymbol{w}_\beta \qquad \text{and} \qquad \boldsymbol{w}_0 = \frac{a\boldsymbol{w}_\alpha + b\boldsymbol{w}_\beta}{a+b} \tag{1.304}$$

   and calculate the coefficients a and b such that the product of the distribution functions $f_\alpha f_\beta$ can be written in the form:

$$f_\alpha(\boldsymbol{w}_\alpha)f_\beta(\boldsymbol{w}_\beta) = f_{\alpha\beta}(\boldsymbol{w}_{\alpha\beta})f_0(\boldsymbol{w}_0)\,. \tag{1.305}$$

   For this, express $\boldsymbol{w}_\alpha$ and $\boldsymbol{w}_\beta$ as functions of $\boldsymbol{w}_0$ and $\boldsymbol{w}_{\alpha\beta}$, and proceed by identification.
3. Verify that the Jacobian of the transformation of coordinates is equal to unity and thus $\mathrm{d}^3w_\alpha\,\mathrm{d}^3w_\beta = \mathrm{d}^3w_{\alpha\beta}\,\mathrm{d}^3w_0$. Then integrate over $\boldsymbol{w}_0$ and $\boldsymbol{w}_{\alpha\beta}$ to obtain the average value of $\langle|\boldsymbol{w}_\alpha - \boldsymbol{w}_\beta|\rangle$.

b) Calculate $v_{\alpha\beta}$, the most probable relative velocity.

Answer

a)

1. To obtain the average value of $|\boldsymbol{w}_\alpha - \boldsymbol{w}_\beta|$, it is necessary to integrate this term over the velocities of the population ensemble of the particles α and β, i.e.:

$$\langle|\boldsymbol{w}_\alpha - \boldsymbol{w}_\beta|\rangle = \int\limits_{\boldsymbol{w}_\alpha}\int\limits_{\boldsymbol{w}_\beta} |\boldsymbol{w}_\alpha - \boldsymbol{w}_\beta|\, f_\alpha f_\beta\, \mathrm{d}^3w_\alpha \mathrm{d}^3w_\beta \tag{1.306}$$

   with:

$$f_\alpha f_\beta = \left(\frac{m_\alpha}{2\pi k_B T_\alpha}\right)^{\frac{3}{2}}\left(\frac{m_\beta}{2\pi k_B T_\beta}\right)^{\frac{3}{2}} \exp\left[-\left(\frac{m_\alpha w_\alpha^2}{2k_B T_\alpha} + \frac{m_\beta w_\beta^2}{2k_B T_\beta}\right)\right]\,. \tag{1.307}$$

   Equation (1.306) is justified by considering the general definition of a hydrodynamic quantity (3.39) where the function $f_{\alpha\beta}(\boldsymbol{w}_\alpha, \boldsymbol{w}_\beta)$ is an uncorrelated pair-correlation distribution function (Sect. 3.2) such that $f_{\alpha\beta}(\boldsymbol{w}_\alpha, \boldsymbol{w}_\beta) = f_\alpha(\boldsymbol{w}_\alpha)f_\beta(\boldsymbol{w}_\beta)$.

2. The calculation of $\boldsymbol{w}_\alpha$ and $\boldsymbol{w}_\beta$ as a function of the new coordinates $\boldsymbol{w}_{\alpha\beta}$ and $\boldsymbol{w}_0$ given by (1.74) and (1.75) is:

$$\boldsymbol{w}_\alpha = \boldsymbol{w}_0 + \frac{b\boldsymbol{w}_{\alpha\beta}}{a+b} , \tag{1.308}$$

$$\boldsymbol{w}_\beta = \boldsymbol{w}_0 - \frac{a\boldsymbol{w}_{\alpha\beta}}{a+b} . \tag{1.309}$$

Substituting $\boldsymbol{w}_\alpha$ and $\boldsymbol{w}_\beta$ by $\boldsymbol{w}_{\alpha\beta}$ and $\boldsymbol{w}_0$ in the exponential term of (1.307), we obtain:

$$\begin{aligned}\frac{m_\alpha w_\alpha^2}{2k_BT_\alpha}+\frac{m_\beta w_\beta^2}{2k_BT_\beta} &= \frac{m_\alpha}{2k_BT_\alpha}\left[w_0^2+\frac{b^2}{(a+b)^2}w_{\alpha\beta}^2+\frac{2b}{a+b}\boldsymbol{w}_0\cdot\boldsymbol{w}_{\alpha\beta}\right]\\ &+\frac{m_\beta}{2k_BT_\beta}\left[w_0^2+\frac{a^2}{(a+b)^2}w_{\alpha\beta}^2-\frac{2a}{a+b}\boldsymbol{w}_0\cdot\boldsymbol{w}_{\alpha\beta}\right] .\end{aligned} \tag{1.310}$$

To eliminate the cross terms in (1.310) by addition, a and b must be chosen such that:

$$\frac{a}{b} = \frac{m_\alpha T_\beta}{m_\beta T_\alpha} , \tag{1.311}$$

and we will take $a = m_\alpha T_\beta$ and $b = m_\beta T_\alpha$.
The argument of the exponential given by equation (1.310) can thus be written:

$$\begin{aligned}\frac{m_\alpha w_\alpha^2}{2k_BT_\alpha}+\frac{m_\beta w_\beta^2}{2k_BT_\beta} &= w_0^2\frac{(m_\alpha+m_\beta)(m_\alpha T_\beta+m_\beta T_\alpha)}{2k_BT_\alpha T_\beta(m_\alpha+m_\beta)}\\ &+w_{\alpha\beta}^2\frac{m_\alpha m_\beta}{m_\alpha+m_\beta}\,\frac{m_\alpha+m_\beta}{2k_B(m_\alpha T_\beta+m_\beta T_\alpha)}\\ &=\frac{m_0w_0^2}{2k_BT_0}+\frac{\mu_{\alpha\beta}w_{\alpha\beta}^2}{2k_BT_{\alpha\beta}}\end{aligned} \tag{1.312}$$

after setting:

$$m_0 = m_\alpha + m_\beta , \qquad T_0 = \frac{T_\alpha T_\beta}{T_{\alpha\beta}} ,$$

$$\mu_{\alpha\beta} = \frac{m_\alpha m_\beta}{m_\alpha+m_\beta} , \qquad T_{\alpha\beta} = \frac{m_\alpha T_\beta + m_\beta T_\alpha}{m_\alpha+m_\beta} .$$

Further, noting that the product $m_\alpha m_\beta/T_\alpha T_\beta$ in the pre-exponential terms of equation (1.307) can be written in the form:

$$\frac{m_\alpha m_\beta}{T_\alpha T_\beta} = \frac{m_\alpha m_\beta}{m_\alpha+m_\beta}\,\frac{(m_\alpha+m_\beta)T_{\alpha\beta}}{T_{\alpha\beta}T_\alpha T_\beta} = \frac{\mu_{\alpha\beta}m_0}{T_{\alpha\beta}T_0} . \tag{1.313}$$

We can thus verify that:

$$f_\alpha(\boldsymbol{w}_\alpha)f_\beta(\boldsymbol{w}_\beta) = f_{\alpha\beta}(\boldsymbol{w}_{\alpha\beta})f_0(\boldsymbol{w}_0) \,, \tag{1.314}$$

where $f_{\alpha\beta}$, characterising the relative velocity distribution, and f_0 the velocity distribution for $\boldsymbol{w}_0$ are defined by:

$$f_{\alpha\beta} = \left(\frac{\mu_{\alpha\beta}}{2\pi k_B T_{\alpha\beta}}\right)^{\frac{3}{2}} \exp\left(-\frac{\mu_{\alpha\beta} w_{\alpha\beta}^2}{2k_B T_{\alpha\beta}}\right) \,, \tag{1.315}$$

$$f_0 = \left(\frac{m_0}{2\pi k_B T_0}\right)^{\frac{3}{2}} \exp\left(-\frac{m_0 w_0^2}{2k_B T_0}\right) \,. \tag{1.316}$$

3. Consider the change of frame for the velocities defined by (1.308) and (1.309). The Jacobian $\mathcal{J}$ for this transformation has the value:

$$\mathcal{J} \equiv \begin{vmatrix} 1 & -\dfrac{a}{a+b} \\ 1 & \dfrac{b}{a+b} \end{vmatrix} = 1 \,. \tag{1.317}$$

Since $\mathrm{d}\boldsymbol{w}_\alpha \mathrm{d}\boldsymbol{w}_\beta = \mathcal{J} \mathrm{d}\boldsymbol{w}_{\alpha\beta}\, \mathrm{d}\boldsymbol{w}_0$, we finally have:

$$\mathrm{d}\boldsymbol{w}_\alpha \mathrm{d}\boldsymbol{w}_\beta = \mathrm{d}\boldsymbol{w}_{\alpha\beta}\, \mathrm{d}\boldsymbol{w}_0 \,. \tag{1.318}$$

The average value of the modulus of the relative velocities can thus be written:

$$\langle|\boldsymbol{w}_\alpha - \boldsymbol{w}_\beta|\rangle = \langle|\boldsymbol{w}_{\alpha\beta}|\rangle = \int\limits_{w_{\alpha\beta}} \int\limits_{w_0} |\boldsymbol{w}_{\alpha\beta}| f_{\alpha\beta} f_0 \, \mathrm{d}\boldsymbol{w}_{\alpha\beta} \mathrm{d}\boldsymbol{w}_0 \,, \tag{1.319}$$

hence:

$$\int\limits_{w_{\alpha\beta}} |\boldsymbol{w}_{\alpha\beta}| f_{\alpha\beta} \, \mathrm{d}\boldsymbol{w}_{\alpha\beta} \underbrace{\int\limits_{w_0} f_0 \, \mathrm{d}\boldsymbol{w}_0}_{1} = \int\limits_{w_{\alpha\beta}} |\boldsymbol{w}_{\alpha\beta}| f_{\alpha\beta} \, \mathrm{d}\boldsymbol{w}_{\alpha\beta} \,. \tag{1.320}$$

The function $f_{\alpha\beta}$ being isotropic, using (1.315) we can develop (1.320) in the following way:

$$\langle|\boldsymbol{w}_\alpha - \boldsymbol{w}_\beta|\rangle = 4\pi \int\limits_0^\infty w_{\alpha\beta}^3 \left(\frac{\mu_{\alpha\beta}}{2\pi k_B T_{\alpha\beta}}\right)^{\frac{3}{2}} \exp\left(-\frac{\mu_{\alpha\beta} w_{\alpha\beta}^2}{2k_B T_{\alpha\beta}}\right) \mathrm{d}w_{\alpha\beta} \,, \tag{1.321}$$

$$\langle|\boldsymbol{w}_\alpha - \boldsymbol{w}_\beta|\rangle = 4\pi\left(\frac{\mu_{\alpha\beta}}{2\pi k_B T_{\alpha\beta}}\right)^{\frac{3}{2}} \times \frac{1}{2}\left(\frac{2k_B T_{\alpha\beta}}{\mu_{\alpha\beta}}\right)^2 , \tag{1.322}$$

$$= \frac{2}{\sqrt{\pi}}\left(\frac{\mu_{\alpha\beta}}{2k_B T_{\alpha\beta}}\right)^{\frac{3}{2}}\left(\frac{2k_B T_{\alpha\beta}}{\mu_{\alpha\beta}}\right)^2 , \tag{1.323}$$

$$\langle|\boldsymbol{w}_\alpha - \boldsymbol{w}_\beta|\rangle = \frac{2}{\sqrt{\pi}}\left(\frac{2k_B T_{\alpha\beta}}{\mu_{\alpha\beta}}\right)^{\frac{1}{2}} = \sqrt{\frac{8k_B T_{\alpha\beta}}{\pi\mu_{\alpha\beta}}} , \tag{1.324}$$

$$= \sqrt{\frac{8k_B(m_\alpha T_\beta + m_\beta T_\alpha)(m_\alpha + m_\beta)}{\pi(m_\alpha + m_\beta)m_\alpha m_\beta}} , \tag{1.325}$$

$$\langle|\boldsymbol{w}_\alpha - \boldsymbol{w}_\beta|\rangle = \sqrt{\frac{8k_B}{\pi}\left(\frac{T_\beta}{m_\beta} + \frac{T_\alpha}{m_\alpha}\right)} . \tag{1.326}$$

Remarks:

a. If one of the temperatures is zero ($T_\beta = 0$), we recover the value of the average particle velocity for temperature T_α ($T_\alpha \neq 0$) because $\langle|\boldsymbol{w}_\alpha - \boldsymbol{0}|\rangle = \langle|\boldsymbol{w}_\alpha|\rangle = (8k_B T_\alpha/\pi m_\alpha)^{\frac{1}{2}}$.
b. If $\alpha = \beta$, $\langle|\boldsymbol{w}_\alpha - \boldsymbol{w}'_\alpha|\rangle = \sqrt{2}\langle w_\alpha\rangle$. This second case corresponds to a collision between two particles from the same population.
c. If the cross-section $\hat{\sigma}_{\alpha\beta}(|\boldsymbol{w}_\alpha - \boldsymbol{w}_\beta|)$ is a simple analytic function of $|\boldsymbol{w}_\alpha - \boldsymbol{w}_\beta|$, it is then possible to calculate the reaction coefficient $k_{\alpha\beta} = \langle\hat{\sigma}_{\alpha\beta}(|\boldsymbol{w}_\alpha - \boldsymbol{w}_\beta|)|\boldsymbol{w}_\alpha - \boldsymbol{w}_\beta|\rangle$ (1.146) by using the same method.

b) Assuming that the distribution functions are isotropic, the distribution function $f_{\alpha\beta}(\boldsymbol{w}_{\alpha\beta})$ is described in spherical coordinates by:

$$g(w_{\alpha\beta}) \equiv 4\pi w_{\alpha\beta}^2 f_{\alpha\beta}(w_{\alpha\beta}) ,$$

hence:

$$g(w_{\alpha\beta}) \equiv 4\pi w_{\alpha\beta}^2\left(\frac{\mu_{\alpha\beta}}{2\pi k_B T_{\alpha\beta}}\right)^{\frac{3}{2}} \exp\left(-\frac{\mu_{\alpha\beta} w_{\alpha\beta}^2}{2k_B T_{\alpha\beta}}\right) . \tag{1.327}$$

The most probable velocity $v_{\alpha\beta}$ is obtained when:

$$\frac{\partial g}{\partial w_{\alpha\beta}} = 0 . \tag{1.328}$$

The derivation of (1.327) shows that relation (1.328) is satisfied when:

$$w_{\alpha\beta}^2 = \frac{2k_B T_{\alpha\beta}}{\mu_{\alpha\beta}} . \tag{1.329}$$

The expression for the most probable velocity is then:

$$v_{\alpha\beta} = \sqrt{\frac{2k_B T_{\alpha\beta}}{\mu_{\alpha\beta}}} \, ,$$

which, when we expand $T_{\alpha\beta}$ and $\mu_{\alpha\beta}$ as before, is:

$$v_{\alpha\beta} = \sqrt{\frac{2k_B T_\alpha}{m_\alpha} + \frac{2k_B T_\beta}{m_\beta}} \, ,$$

which is the square root of the sum of the most probable velocity of each type of particle.

1.10. Consider "billiard ball" binary elastic collisions, where the species α and β are rigid spheres of radius r_α and r_β.

a) Draw the geometric representation of a collision in the centre of mass frame at the moment of impact: indicate the particle velocities before and after the collision, the impact parameter s and the scattering angle θ.
b) Calculate the microscopic differential scattering cross-section $\hat{\sigma}(w_{\alpha\beta}, \theta)$.
c) Deduce the total microscopic scattering cross-sections $\hat{\sigma}_{tc}(w_{\alpha\beta})$ and $\hat{\sigma}_{tm}(w_{\alpha\beta})$.
d) Consider ion-neutral and electron-neutral cross-sections in a "billiard ball" model, assuming that the electron radius is zero ($r_\alpha = 0$) and that the radius of the ions is equal to the radius r_β of the neutrals. Deduce the relationship of the cross-sections. Compare this relation with that obtained for collisions of electrons and helium ions on helium atoms:

$$\langle \hat{\sigma}_{en}(w) \rangle (T_{eV} = 2\,\text{eV}, T_n = 300\,\text{K}) = 5 \times 10^{-16}\,\text{cm}^2 \, ,$$
$$\langle \hat{\sigma}_{in}(w) \rangle (T_i = T_n = 300\,\text{K}) = 3.5 \times 10^{-15}\,\text{cm}^2 \, .$$

Answer

a) The geometry of an elastic "billiard ball" collision, whose interaction is by nature repulsive, is represented in the figure below. The velocities $\boldsymbol{w}_{\alpha 0}$ and $\boldsymbol{w}_{\beta 0}$, and $\boldsymbol{w}'_{\alpha 0}$ and $\boldsymbol{w}'_{\beta 0}$, are the velocities of the particles α and β before and after collision respectively. The distance s between the two pairs of asymptotes is the impact parameter (the distance of closest approach in the absence of an interaction). The scattering angle θ is related to the angle $\chi_{\max}$, the maximum angle between $\boldsymbol{r}$ (the relative position of the centre of the particles) and the relative velocity $\boldsymbol{w}_{\alpha\beta} = \boldsymbol{w}_{\alpha 0} - \boldsymbol{w}_{\beta 0}$ before collision by:

$$2\chi_{\max} + \theta = \pi \, . \qquad (1.330)$$

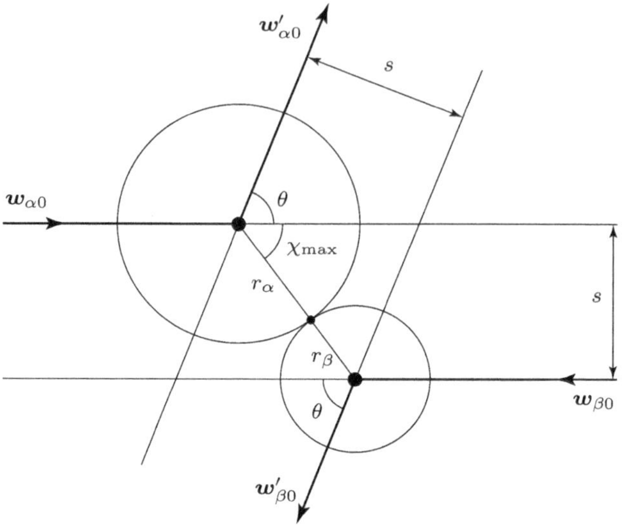

Fig. 1.21 Geometric representation of a "billiard ball" interaction in the centre of mass frame.

b) The differential relation between the total microscopic cross-section $\hat{\sigma}_{tc}$ and the microscopic differential scattering cross section $\hat{\sigma}(w_{\alpha\beta}, \theta)$ is deduced from (1.110):

$$\mathrm{d}\hat{\sigma}_{tc} = 2\pi\hat{\sigma}(w_{\alpha\beta}, \theta) \sin\theta \, \mathrm{d}\theta \, . \tag{1.331}$$

The microscopic cross-section element $\mathrm{d}\hat{\sigma}_{tc}$ can be expressed in a simple way as a function of the impact parameter s (V.34), as:

$$\mathrm{d}\hat{\sigma}_{tc} = 2\pi s \, \mathrm{d}s \, . \tag{1.332}$$

It is therefore sufficient to determine the relation between s and θ, then to deduce the corresponding differential cross-section and finally to identify (1.332) with (1.331).
From the figure, we have:

$$s = (r_\alpha + r_\beta) \sin\chi_{\mathrm{max}} \, , \tag{1.333}$$

then, taking account of (1.330):

$$s = (r_\alpha + r_\beta) \cos\frac{\theta}{2} \, , \tag{1.334}$$

and thus:

$$\mathrm{d}s = -\frac{r_\alpha + r_\beta}{2} \sin\frac{\theta}{2} \, \mathrm{d}\theta \, . \tag{1.335}$$

The $(-)$ sign signifies that an increase in the impact parameter results in a reduction in the scattering angle θ. Including (1.334) and (1.335), (1.332) can be written:

$$2\pi s \, \mathrm{d}s = -\frac{(r_\alpha + r_\beta)^2}{4} \sin\theta \, \mathrm{d}\theta \tag{1.336}$$

and, by identification with (1.331), we obtain:

$$\hat{\sigma}(w_{\alpha\beta}, \theta) = \left| -\frac{(r_\alpha + r_\beta)^2}{4} \right| . \tag{1.337}$$

This verifies that, for the hard sphere "billiard ball" model, the microscopic differential cross-section depends neither on the relative velocities, nor on the scattering angle.

Remark: Relation (1.333) can also be obtained by integration of (V.18) from $r = \infty$ to $r_{\text{min}} = r_\alpha + r_\beta$, by considering the interaction potential $\phi(r)$ to be zero between $r = \infty$ and r_{min} and becoming infinite for r less than r_{min}. The integral (V.18) can then be written:

$$\chi_{\text{max}} = \int_\infty^{r_{\text{min}}} \frac{s \, \mathrm{d}r}{r^2 \sqrt{1 + \left(\frac{s}{r}\right)^2}} , \tag{1.338}$$

and by change of variable:

$$u = \frac{s}{r} , \tag{1.339}$$

we obtain:

$$\chi_{\text{max}} = \int_0^{s/(r_\alpha + r_\beta)} -\frac{\mathrm{d}u}{\sqrt{1 - u^2}} , \tag{1.340}$$

hence:

$$\chi_{\text{max}} = \arcsin \frac{s}{r_\alpha + r_\beta} \tag{1.341}$$

and finally:

$$s = (r_\alpha + r_\beta) \sin \chi_{\text{max}} . \tag{1.333}$$

c) The total microscopic collision cross-section (1.110) can be written:

$$\hat{\sigma}_{tc} = 2\pi \int_0^\pi \frac{(r_\alpha + r_\beta)^2}{4} \sin\theta \, \mathrm{d}\theta \tag{1.342}$$

and, after integration:

$$\hat{\sigma}_{tc} = \pi (r_\alpha + r_\beta)^2 , \tag{1.343}$$

which is a value corresponding to a tangential collision ($s = r_\alpha + r_\beta$) and which can be obtained directly by multiplying (1.337) by 4π steradian, the solid angle over a sphere.
The total microscopic momentum transfer cross-section (1.111) can be written:

$$\hat{\sigma}_{tm} = 2\pi \int_0^\pi \frac{(r_\alpha + r_\beta)^2}{4} \sin\theta(1 - \cos\theta)\,\mathrm{d}\theta \,, \tag{1.344}$$

$$\hat{\sigma}_{tm} = \frac{\pi}{2}(r_\alpha + r_\beta)^2 \int_0^\pi (1 - \cos\theta)\,\mathrm{d}(-\cos\theta) \,, \tag{1.345}$$

then, after integration:

$$\hat{\sigma}_{tm} = \pi(r_\alpha + r_\beta)^2 \,, \tag{1.346}$$

which is identical to the value obtained for the total microscopic cross section for simple collisions (1.343).

d) In the case of ion-neutral collisions $r_\alpha = r_\beta$, such that:

$$\hat{\sigma}_{in} = 4\pi r_\beta^2 \,, \tag{1.347}$$

while for electron-neutral collisions ($r_\alpha = 0$):

$$\hat{\sigma}_{en} = \pi r_\beta^2 \,. \tag{1.348}$$

Therefore, within the billiard-ball model, the cross-section for electron-neutral collisions is a factor 4 times smaller than the cross section for ion-neutral collisions. For helium, from experiments this time, there is a factor of 7 between these two cross-sections but they were not determined for the same relative velocities or relative energies.

Chapter 2
Individual Motion of a Charged Particle in Electric and Magnetic Fields

There are three distinct levels of modelling of the action of $\boldsymbol{E}$ and $\boldsymbol{B}$ fields on the charged particles in a plasma. Starting with the simplest and moving to the most complicated, we have:

The single trajectory model

In this description, the fields $\boldsymbol{E}$ and $\boldsymbol{B}$ are given, imposed from the exterior: no account is taken of the fields created by the motion of the particles. Further, collisions are completely neglected, including Coulomb interactions: this model only describes the motion of an isolated particle.

The hydrodynamic model

In this case, the plasma consists either of two fluids (that of the electrons and that of the ions), or of a single fluid (for instance, that of the electrons, the ions remaining at rest and forming a continuous background, providing an effective viscosity to the electron motion). The motion of each fluid is characterised locally by an average velocity $\boldsymbol{v}$ whose value results from an integration of the velocity distribution of the particles contained in the volume element considered (Sect. 3.3). The motion of the charged particles creates the fields $\boldsymbol{E}$ and $\boldsymbol{B}$ (for which the average local value is retained (macroscopic fields)) which are included in a self-consistent manner in the equations of motion[50]. In addition, the model includes collisions, which modify the pre-determined motion defined by the superposition of the external and induced fields.

In order to establish self-consistency between the charged particle motion and the fields they produce, we need to consider first the velocity of the

[50] The coupling of the $\boldsymbol{E}$ and $\boldsymbol{B}$ fields with the charged particles is said to be self-consistent because the motion of the particles creating the fields $\boldsymbol{E}$ and $\boldsymbol{B}$ is itself influenced by the fields that it produces.

M. Moisan, J. Pelletier, *Physics of Collisional Plasmas*,
DOI 10.1007/978-94-007-4558-2_2,

fluid elements. This is obtained from the equation of motion, in which the Lorentz' force (Sect. 2.1) is included, assuming values for the $\boldsymbol{E}$ and $\boldsymbol{B}$ fields for the first iteration. Once $\boldsymbol{v}$ has been determined, we can calculate the total current density $\boldsymbol{J}$ from the component fluids involved ($\boldsymbol{J} = \sum_\alpha n_\alpha q_\alpha \boldsymbol{v}_\alpha$). We can then complete the loop in two ways to obtain iterated values for $\boldsymbol{E}$ and $\boldsymbol{B}$:

- from $\boldsymbol{J}$, recover $\boldsymbol{E}$ from the electromagnetic relation:

$$\boldsymbol{J} = \sigma \boldsymbol{E} \ , \tag{2.1}$$

  where σ is the electrical conductivity from the fluids involved, and from the known value of $\boldsymbol{E}$, calculate $\boldsymbol{B}$ by one or other of Maxwell's curl equations:

$$\boldsymbol{\nabla} \wedge \boldsymbol{E} = -\frac{\partial \boldsymbol{B}}{\partial t} \ , \tag{2.2}$$

$$\boldsymbol{\nabla} \wedge \boldsymbol{B} = \mu_0 \epsilon_0 \frac{\partial \boldsymbol{E}}{\partial t} + \mu_0 \boldsymbol{J} \ , \tag{2.3}$$

- from the density $\boldsymbol{J}$, calculate the charge density ρ from the continuity equation (for example $\partial\rho/\partial t + \nabla \cdot \boldsymbol{J} = 0$) and obtain $\boldsymbol{E}$ from Poisson's equation:

$$\boldsymbol{\nabla} \cdot \boldsymbol{E} = \rho/\epsilon_0 \ , \tag{1.1}$$

  then, determine $\boldsymbol{B}$ through (2.2) or (2.3).

Remark: Note that the conductivity σ, which relates $\boldsymbol{J}$ and $\boldsymbol{E}$, plays a key role in obtaining field-particle self-consistence: we shall calculate the expression for σ in the framework of various models.

The kinetic or microscopic model

This is the description with the highest resolution. It uses the individual velocity distributions of the particles: this allows us to include certain phenomena that escape the hydrodynamic model, such as, for example, the Landau damping (resonance effect between a wave propagating in the plasma and particles with velocities within a certain interval). This model includes the fields and collisions self consistently, this time on the microscopic scale (individual particles), a more refined approach than that provided by the macroscopic values (average values over the velocity distribution of the particles).

The present chapter is devoted to the study of the individual motion of a charged particle in given $\boldsymbol{E}$ and $\boldsymbol{B}$ fields. This model gives a first glimpse of the complex phenomena taking place at the heart of a plasma, with the assumption that there are no collisions in the body of the plasma or at the walls. In the first place, we will examine the solution of the equation of motion through a series of particular cases, to finally determine the general solution[51].

2.1 The general equation of motion of a charged particle in E and B fields and properties of that equation

Suppose q_a is the charge of a particle of mass m_a, moving with a velocity $\boldsymbol{w} = \mathrm{d}\boldsymbol{r}/\mathrm{d}t$ and suppose $\boldsymbol{E}(\boldsymbol{r},t)$ and $\boldsymbol{B}(\boldsymbol{r},t)$ are the external fields: the particle is subject to the action of the Lorentz' *force* that, in the non-relativistic case, takes the form[52]:

$$\boldsymbol{F} \equiv q_a \left[\boldsymbol{E}(\boldsymbol{r},t) + \boldsymbol{w} \wedge \boldsymbol{B}(\boldsymbol{r},t)\right] \,. \tag{2.4}$$

This equation is the result of observation. It is valid if the particle is sufficiently small to be taken as a point (this therefore avoids the need to consider the problem of repartition of charges in the particle volume).

2.1.1 The equation of motion

From (2.4), we can write:

$$m_\alpha \frac{\mathrm{d}^2\boldsymbol{r}}{\mathrm{d}t^2} = q_\alpha \left[\boldsymbol{E}(\boldsymbol{r},t) + \frac{\mathrm{d}\boldsymbol{r}}{\mathrm{d}t} \wedge \boldsymbol{B}(\boldsymbol{r},t)\right] \,. \tag{2.5}$$

This equation leads to a second order differential equation for each axial component of the coordinate system. For example, in Cartesian coordinates:

$$m_\alpha \frac{\mathrm{d}^2 x}{\mathrm{d}t^2} = q_\alpha \left[E_x + \left(B_z \frac{\mathrm{d}y}{\mathrm{d}t} - B_y \frac{\mathrm{d}z}{\mathrm{d}t}\right)\right] \,, \tag{2.6}$$

$$m_\alpha \frac{\mathrm{d}^2 y}{\mathrm{d}t^2} = q_\alpha \left[E_y + \left(B_x \frac{\mathrm{d}z}{\mathrm{d}t} - B_z \frac{\mathrm{d}x}{\mathrm{d}t}\right)\right] \,, \tag{2.7}$$

$$m_\alpha \frac{\mathrm{d}^2 z}{\mathrm{d}t^2} = q_\alpha \left[E_z + \left(B_y \frac{\mathrm{d}x}{\mathrm{d}t} - B_x \frac{\mathrm{d}y}{\mathrm{d}t}\right)\right] \,. \tag{2.8}$$

---

[51] The principal reference for this section is Electrodynamics of Plasmas by Jancel and Kahan, Chap. 4. See also Delcroix, Physique des plasmas, Vol. I, Sect. 12.3, Delcroix and Bers, Physique des plasmas, Vol. I, Sect. 2.3, and Allis, Motions of Ions and Electrons [2].

[52] The relativistic equation is:

$$m_a \frac{\mathrm{d}\boldsymbol{w}}{\mathrm{d}t} = q_a \left(1 - \frac{w^2}{c^2}\right)^{\frac{1}{2}} \left[\boldsymbol{E}(\boldsymbol{r},t) + \boldsymbol{w} \wedge \boldsymbol{B}(\boldsymbol{r},t) - \frac{w^2}{c^2}(\boldsymbol{w} \cdot \boldsymbol{E})\right] \,,$$

where c is the speed of light in vacuum.

2.1.2 The kinetic energy equation

Taking the scalar product of (2.5) with $\boldsymbol{w} = \mathrm{d}\boldsymbol{r}/\mathrm{d}t$, we obtain the *kinetic energy equation*:

$$\frac{m_\alpha}{2}\frac{\mathrm{d}}{\mathrm{d}t}\left|\frac{\mathrm{d}\boldsymbol{r}}{\mathrm{d}t}\right|^2 = q_\alpha \boldsymbol{E}(\boldsymbol{r},t)\cdot\frac{\mathrm{d}\boldsymbol{r}}{\mathrm{d}t} + q_\alpha\left(\frac{\mathrm{d}\boldsymbol{r}}{\mathrm{d}t}\wedge\boldsymbol{B}(\boldsymbol{r},t)\right)\cdot\frac{\mathrm{d}\boldsymbol{r}}{\mathrm{d}t}\,, \tag{2.9}$$

where the second term on the RHS vanishes, since $(\boldsymbol{A}\wedge\boldsymbol{B})\cdot\boldsymbol{A} = \boldsymbol{0}$: the resulting equation is in scalar form and constitutes an invariant in any frame of reference. After integration of the equation over time t from t_0 to t (in position, from $\boldsymbol{r}_0$ to $\boldsymbol{r}$), we have:

$$\frac{m_\alpha}{2}\left[\left|\frac{\mathrm{d}\boldsymbol{r}}{\mathrm{d}t}\right|_{\boldsymbol{r}}^2 - \left|\frac{\mathrm{d}\boldsymbol{r}}{\mathrm{d}t}\right|_{\boldsymbol{r}_0}^2\right] = q_\alpha\int_{t_0}^{t}\boldsymbol{E}\cdot\mathrm{d}\boldsymbol{r}\,, \tag{2.10}$$

where the RHS of the equation represents the *work* done on the particle *by the electric field.* From this, we can draw the following important conclusions:

1. *The magnetic field* does "no work" because the force it exerts on the particle is perpendicular to its velocity[53]. It follows that the magnitude of the velocity of a charged particle is not affected by the presence of a magnetic field. However, the magnitudes of the velocity components perpendicular to $\boldsymbol{B}$ can vary, as we will show for the cyclotron motion (Sect. 2.2.2). Supposing that $\boldsymbol{B}$ is directed along Ox, this implies that:

$$w_\perp^2 = w_{y0}^2 + w_{z0}^2 = w_y^2(t) + w_z^2(t)\,, \tag{2.11}$$

   where the subscript 0 denotes the velocity at $t = 0$: in other words, a magnetic field can only change the direction of the velocity, not its magnitude. However, the application of a magnetic field to a plasma makes it possible, among other things, to conserve the energy of the system by reducing the diffusion losses of the charged particles to the walls, as we shall see (Sect. 3.8).
2. Only the electric field can "heat" the charged particles, i.e., give them energy.

2.2 Analysis of particular cases of E and B

We will successively treat the following cases: only an electric field acting on the particle (Sect. 2.2.1); the particle is subjected to a constant, uniform

[53] Heating by magnetic pumping, where $\boldsymbol{B}$ varies periodically, can be considered as resulting from the action of the $\boldsymbol{E}$ field through the Maxwell equation $\nabla\wedge\boldsymbol{E} = -\partial\boldsymbol{B}/\partial t$.

magnetic field, with or without an electric field $\boldsymbol{E}$ (Sect. 2.2.2); and finally, the most complex situation, the particle moves in a magnetic field that is (slightly) non uniform or (slowly) varying in time (Sect. 2.2.3). We will see that the different solutions obtained for the particular cases can be included in a general equation describing the particle motion in such $\boldsymbol{E}$ and $\boldsymbol{B}$ fields.

2.2.1 Electric field only ($B = 0$)

From (2.6), (2.7) and (2.8), we obtain:

$$\frac{\mathrm{d}^2x}{\mathrm{d}t^2} = \frac{q_a}{m_a}E_x(\boldsymbol{r},t)\,, \quad \frac{\mathrm{d}^2y}{\mathrm{d}t^2} = \frac{q_a}{m_a}E_y(\boldsymbol{r},t)\,, \quad \frac{\mathrm{d}^2z}{\mathrm{d}t^2} = \frac{q_a}{m_a}E_z(\boldsymbol{r},t)\,. \tag{2.12}$$

We can now treat the following cases.

Constant and uniform electric field $\boldsymbol{E}$

By direct integration of (2.12) in vectorial form, we deduce:

$$\boldsymbol{w} = \frac{q_\alpha}{m_\alpha}\boldsymbol{E}t + \boldsymbol{w}_0\,, \tag{2.13}$$

$$\boldsymbol{r} = \frac{q_\alpha}{m_\alpha}\boldsymbol{E}\frac{t^2}{2} + \boldsymbol{w}_0 t + \boldsymbol{r}_0\,, \tag{2.14}$$

which describe uniformly accelerated motion.

Remarks:

1. From (2.13), one can see that the component of motion along a direction perpendicular to $\boldsymbol{E}$ is not affected by the presence of this field; this can be shown by decomposing $\boldsymbol{w}$ in directions parallel and perpendicular to $\boldsymbol{E}$. The situation is completely different with $\boldsymbol{B}$, because the corresponding force acts perpendicularly to $\boldsymbol{B}$ (and to $\boldsymbol{w}$) (2.4).
2. Since the field $\boldsymbol{E}$ selectively accelerates the component of velocity parallel to it, we could say that it tends, if not to confine, at least to orient the particle in this direction.
3. From (2.13) and (2.14), we can conclude that the velocity, as well as the distance travelled by an ion of mass m_i under the effect of a field $\boldsymbol{E}$ during a given time is m_e/m_i times smaller than that of an electron of mass m_e in the same field, which justifies the commonly used assumption that the ion is at rest with respect to the electron.

Conservative field $\boldsymbol{E}(\boldsymbol{r}, t)$

Since the electric field is conservative, we can write:

$$\boldsymbol{E} = -\boldsymbol{\nabla}\phi(\boldsymbol{r}, t) \ , \tag{2.15}$$

where ϕ is the potential acting on the particle. The vectorial equation of motion:

$$m_\alpha \frac{\mathrm{d}^2\boldsymbol{r}}{\mathrm{d}t^2} = -q_\alpha \boldsymbol{\nabla}\phi \ , \tag{2.16}$$

scalar multiplied by $\mathrm{d}\boldsymbol{r}/dt$ shows, after integration over time t, that the variation of kinetic energy is equal to the (negative) variation of the potential energy, such that the total energy is, of course, conserved:

$$\frac{m_\alpha}{2}\left[\left|\frac{\mathrm{d}\boldsymbol{r}}{\mathrm{d}t}\right|^2_{\boldsymbol{r}} - \left|\frac{\mathrm{d}\boldsymbol{r}}{\mathrm{d}t}\right|^2_{\boldsymbol{r}_0}\right] = -q_\alpha[\phi(\boldsymbol{r}, t) - \phi(\boldsymbol{r}_0, t_0)] \ . \tag{2.17}$$

Equation (2.17) is a variant of (2.10).

Application to the case where ϕ is time independent

The motion of an electron in an electrostatic potential is similar to the propagation of a luminous wave in a medium of refractive index n_r, as shown below.

Consider the case of two media where ϕ, moreover, does not depend on $\boldsymbol{r}$, thus $\boldsymbol{E}$ is zero (2.15). The crossing of a discontinuity in potential ($\phi_1 \neq \phi_2$, Fig. 2.1) determines the existence of a field $\boldsymbol{E}$ (at the interface only) and, as a result, the particle experiences an instantaneous acceleration (or deceleration), the velocity thus changing from $\boldsymbol{w}_1$ to $\boldsymbol{w}_2$.

However, the components of the velocities parallel to the interface between the two media remain the same from one side to the other, because the electric field $\boldsymbol{E}$ is perpendicular to this interface (Remark 1 above) from which, noting $\boldsymbol{p} = m_e\boldsymbol{w}$:

$$|\boldsymbol{p}_1| \sin\theta_1 = |\boldsymbol{p}_2| \sin\theta_2 \ , \tag{2.18}$$

which, when written in the form:

$$\frac{|\boldsymbol{p}_1|}{|\boldsymbol{p}_2|} = \frac{\sin\theta_2}{\sin\theta_1} \ , \tag{2.19}$$

appears as the well known geometrical optics law of Descartes, if θ_1 and θ_2 are considered as the angle of incidence and refraction respectively, and where the momentum p_i of the particle is proportional to the index of the medium[54].

---

[54] Doing this, one finds that $n_r = A\sqrt{\mathcal{E} - q_\alpha\phi}$, where A is a constant and $\mathcal{E}$ the total energy of the particle.

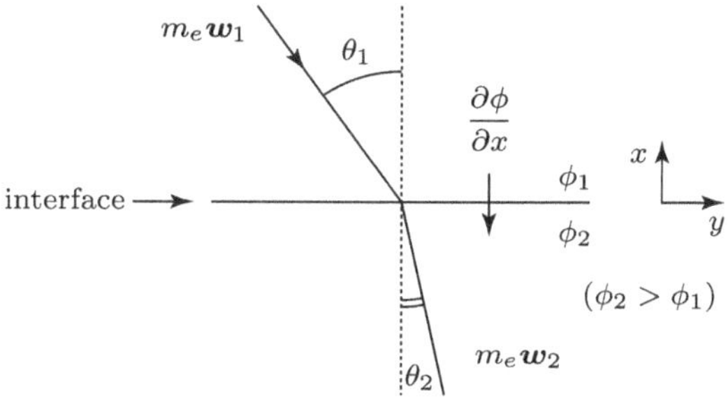

Fig. 2.1 Description of the refraction path in optical electronics.

The field E is uniform, but oscillates periodically as a function of time

This case corresponds to that in which the charged particles are present either in a plasma created by a high frequency field (HF), or in a plasma produced by other means (for example, a continuous current discharge) onto which a significant HF field has been superimposed.

The equation of motion is, in this case:

$$\frac{\mathrm{d}^2 \boldsymbol{r}}{\mathrm{d}t^2} = \frac{q_\alpha}{m_\alpha} \boldsymbol{E}_0 \mathrm{e}^{\mathrm{i}\omega t} \tag{2.20}$$

and, after successive integrations from $t = 0$ to t, and supposing that the initial velocity of the particle is $\boldsymbol{w}_0$ (taking $w_0 \neq 0$, to remain completely general), we obtain:

$$\boldsymbol{w} = \frac{\mathrm{d}\boldsymbol{r}}{\mathrm{d}t} = \frac{1}{\mathrm{i}\omega}\left[\frac{q_\alpha \boldsymbol{E}_0}{m_\alpha}\mathrm{e}^{\mathrm{i}\omega t} - \frac{q_\alpha \boldsymbol{E}_0}{m_\alpha}\right] + \boldsymbol{w}_0 \, , \tag{2.21}$$

or:

$$\boldsymbol{w} = \frac{q_\alpha \boldsymbol{E}_0}{\mathrm{i} m_\alpha \omega}\mathrm{e}^{\mathrm{i}\omega t} + \left(\boldsymbol{w}_0 - \frac{q_\alpha \boldsymbol{E}_0}{\mathrm{i}\omega m_\alpha}\right) , \tag{2.22}$$

and:

$$\boldsymbol{r} = -\frac{q_\alpha \boldsymbol{E}_0}{m_\alpha \omega^2}\mathrm{e}^{\mathrm{i}\omega t} + \left(\boldsymbol{w}_0 - \frac{q_\alpha \boldsymbol{E}_0}{\mathrm{i}\omega m_\alpha}\right) t + \boldsymbol{r}_c \, , \tag{2.23}$$

where $\boldsymbol{r}_c$ is a constant of integration, the initial position of the particle being

$$\boldsymbol{r}_0 = -\frac{q_\alpha \boldsymbol{E}_0}{m_\alpha \omega^2} + \boldsymbol{r}_c \, . \tag{2.24}$$

Examination of the relative phases of E, w and r

We consider a charged particle (taken to be a positive ion), with zero initial velocity, in an electric field $E_0 \cos \omega t$ of period $\mathcal{T}$, and examine the detailed

behaviour of its velocity and trajectory[55] as a function of time, with the aid
of Fig. 2.2. To simplify this presentation, we ignore the non-periodic term in
velocity in (2.22).

convention :
field direction $\boldsymbol{E}$, $+ \to -$
ion motion $\oplus \to$

$$\Re(E) = E$$
$$E = E_0 \cos \omega t$$

$$\Re(w) = w_E$$
$$w_E = \frac{q_\alpha E_0}{m_\alpha \omega} \sin \omega t$$

$$\Re(r) = x_E$$
$$x_E = -\frac{q_\alpha E_0}{m_\alpha \omega^2} \cos \omega t$$

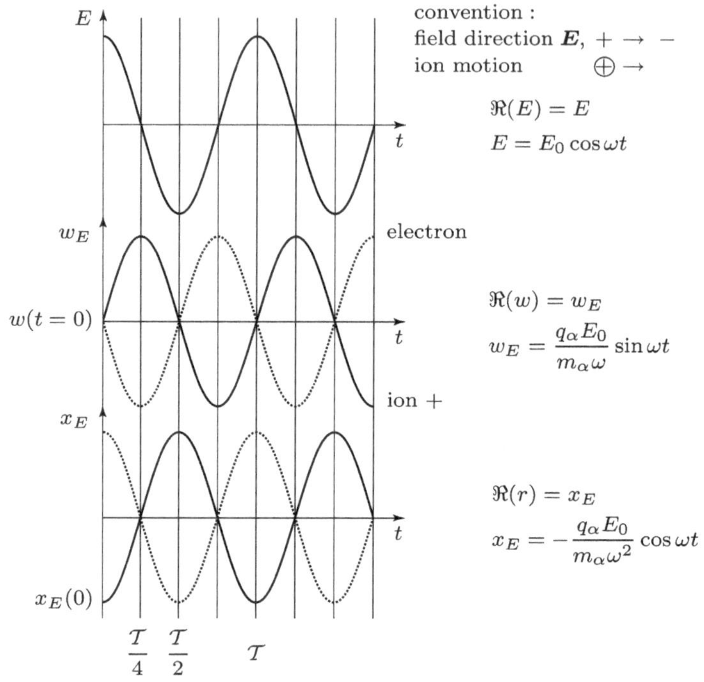

Fig. 2.2 Velocity and trajectory of a positive ion (full curve) and of an electron (dotted)
in an alternating electric field of period $\mathcal{T}$.

1. Velocity: the velocity of the charged particle is in phase with the field $\boldsymbol{E}$.
   From $t = 0$ to $t = \mathcal{T}/4$, the positive ion is accelerated in the positive
   direction of the field: its velocity increases during the entire quarter period
   and reaches its maximum value at $t = \mathcal{T}/4$, when the electric field passes
   through zero.
   Between $t = \mathcal{T}/4$ and $\mathcal{T}/2$, the field $\boldsymbol{E}$ is in the opposite direction to the
   velocity of the positive ion, so it can only be retarded. The velocity passes
   through zero at the same time as the electric field reaches its maximum,
   the situation being symmetric to $t = 0$: in order to return to zero velocity,
   a field of the same amplitude but in the opposite direction is required.
   Between $t = \mathcal{T}/2$ and $3\mathcal{T}/4$, by symmetry, the velocity of the particle
   reaches its maximum opposite to its initial direction at the same time as

---

[55] By convention, the electric field existing between a positive charge and a negative charge
is directed towards the negative charge. As a result, a positive ion is accelerated in the
direction of the field (see Fig. 2.2).

the electric field passes through zero, just before changing sign, and so on. The velocity of the ion is then $\pi/2$ behind the phase of the field $\boldsymbol{E}$. This de-phasing with respect to the field $\boldsymbol{E}$, as we shall see, is such that the transfer of energy from the field to the charged particle is zero over a complete period.

2. Trajectory: in the case of a positive ion, the phase of the trajectory lags by π behind that of the electric field (in opposite phase), while an electron is in phase with the field. The amplitude of motion *of a charged particle* in a HF field $\boldsymbol{E}$ is referred to as the *extension of the periodic motion* and denoted by x_E.
   For a positive ion (initial position $x_E(0)$ in Fig. 2.2), since the initial velocity is assumed to be zero, the direction of motion, according to our convention, is in the direction of the field and only changes direction when the velocity $\boldsymbol{w}_E$ passes through zero (at $t = \mathcal{T}/2$): at this time, the field has its maximum in the opposite direction: there is clearly a lag in phase of π in the motion of the ion in the field.
   In contrast, the spatial oscillation of the electron motion is in phase with the HF field (following the convention of the direction of the field that we have adopted).

Transfer of energy from an oscillating electric field $\boldsymbol{E}$ to a charged particle

The kinetic energy resulting from the work done by an electric field $\boldsymbol{E}$ on the charge, in the time interval t_0 to t can be written (see (2.10)):

$$W \equiv \frac{m_\alpha}{2} w^2 \bigg|_{r_0}^{r} = q_\alpha \int_{r_0}^{r} \boldsymbol{E} \cdot \mathrm{d}\boldsymbol{r} = q_\alpha \int_{t_0}^{t} \boldsymbol{E} \cdot \boldsymbol{w} \mathrm{d}t \,, \tag{2.25}$$

and, following (2.22):

$$\begin{aligned} W &= \Re\left[q_\alpha \int_{t_0}^{t} \boldsymbol{E}_0 \mathrm{e}^{\mathrm{i}\omega t} \cdot \left(\frac{q_\alpha \boldsymbol{E}_0}{m_\alpha \mathrm{i}\omega} \mathrm{e}^{\mathrm{i}\omega t} + \boldsymbol{w}_0 - \frac{q_\alpha \boldsymbol{E}_0^2}{\mathrm{i}\omega m_\alpha} \right) \mathrm{d}t \right] , \\ &= \Re\left[\frac{q_\alpha^2 E_0^2}{i m_\alpha \omega} \int_{t_0}^{t} \mathrm{e}^{\mathrm{i}2\omega t} \mathrm{d}t + \left(q_\alpha \boldsymbol{E}_0 \cdot \boldsymbol{w}_0 - \frac{q_\alpha^2 E_0^2}{i m_\alpha \omega} \right) \int_{t_0}^{t} \mathrm{e}^{\mathrm{i}\omega t} \mathrm{d}t \right] , \\ &= \Re\left[- \frac{q_\alpha^2 E_0^2}{2 m_\alpha \omega^2} \mathrm{e}^{\mathrm{i}2\omega t} \bigg|_{t_0}^{t} + \left(\frac{q_\alpha \boldsymbol{E}_0 \cdot \boldsymbol{w}_0}{\mathrm{i}\omega} + \frac{q_\alpha^2 E_0^2}{m_\alpha \omega^2} \right) \mathrm{e}^{\mathrm{i}\omega t} \bigg|_{t_0}^{t} \right] , \\ &= - \frac{q_\alpha^2 E_0^2}{2 m_\alpha \omega^2} \cos 2\omega t \bigg|_{t_0}^{t} + \frac{q_\alpha^2 E_0^2}{m_\alpha \omega^2} \cos \omega t \bigg|_{t_0}^{t} + \frac{q_\alpha \boldsymbol{E}_0 \cdot \boldsymbol{w}_0}{\omega} \sin \omega t \bigg|_{t_0}^{t} , \end{aligned} \tag{2.26}$$

where $\Re(A)$ denotes the real part of a complex quantity A. In the scalar product under the integral, $\boldsymbol{w}$ reduces to $\boldsymbol{w}_E$ (2.22), the component of the velocity parallel to $\boldsymbol{E}$ (there is no work done in the direction perpendicular to $\boldsymbol{E}$).

The value of the integral (2.26) over a period $\mathcal{T} = 2\pi/\omega$, i.e. between the times t_0 and $t_0 + 2\pi/\omega$, is zero. The total kinetic energy acquired during a period is actually zero, because during the first half-period the work is done in one direction and in the opposite direction during the second half-period.

However, if the oscillatory motion of the particle is interrupted by a collision before the repetition of a complete period starting from t_0, when the field has been applied, the integral (2.26) is non-zero and the corresponding energy taken from the field will be acquired by the particle[56]. In order to demonstrate this, we must leave the very simplified model of individual trajectories (collisionless plasma model) for a moment and consider the hydrodynamic model including collisions.

Transfer of energy from an oscillating field E to electrons via collisions: power absorbed by the electrons and plasma permittivity (a digression from individual trajectories)

Consider an electron fluid, coupled to ions and neutrals via collisions. Assuming that the thermal motion of electrons is negligible compared to their motion resulting from the field $\boldsymbol{E}$ ($v_{th} \ll v_E$, *cold plasma approximation*), the corresponding hydrodynamic equation for momentum transport (Sect. 3.7) can then be written:

$$m_e \frac{\mathrm{d}\boldsymbol{v}}{\mathrm{d}t} = -e\boldsymbol{E}_0 \mathrm{e}^{\mathrm{i}\omega t} - m_e \nu \boldsymbol{v} , \tag{2.27}$$

where $\boldsymbol{v}$ is the (macroscopic) velocity of electrons and ν the average electron-neutral momentum transfer collision frequency. The physical meaning of this equation has already been discussed (1.147).

In fact, we are not very far from the context of individual trajectories in the sense that we can consider that (2.27) describes the motion of a single particle in a medium where it is subject to a friction force.

In the cold plasma approximation, the electron velocity is purely periodic, such that:

$$\boldsymbol{v} = \boldsymbol{v}_0 \mathrm{e}^{\mathrm{i}\omega t} , \tag{2.28}$$

and, substituting $\boldsymbol{v}$ in (2.27), we obtain:

---

[56] The particle "acquires" this energy at the moment of collision, this energy being totally or partially shared with the particle with which it interacts. Recall that in the case of an electron-neutral collision, the electron only partially transfers its energy; more exactly, a fraction of the order of m_e/M of that energy (Sect. 1.7.2).

$$\boldsymbol{v} = -\frac{e\boldsymbol{E}(t)}{m_e(\nu + \mathrm{i}\omega)} , \tag{2.29}$$

which determines $\boldsymbol{v}_0$.
Since $\mathrm{d}\boldsymbol{r}/\mathrm{d}t \equiv \boldsymbol{v}$, again neglecting thermal motion, we have:

$$\boldsymbol{r} = \frac{\boldsymbol{v}}{\mathrm{i}\omega} , \tag{2.30}$$

that is:

$$\boldsymbol{r} = \frac{e\boldsymbol{E}(t)}{m_e\omega(\omega - \mathrm{i}\nu)} . \tag{2.31}$$

- Average HF power absorbed per electron
  The work per unit time and per electron in the field $\boldsymbol{E}$ can be written:

$$-\, e\boldsymbol{E}\cdot\boldsymbol{v} , \tag{2.32}$$

  which thus represents the instantaneous power taken from the field. The average value of the product of two complex variables A and B over a period, each varying sinusoidally with the same frequency, is $\Re(AB^*)/2$ (B^* is the complex conjugate of B). The power taken from the field over a period, or the average power, per electron, is then:

$$\theta_a \equiv \Re\left(\frac{-e\boldsymbol{E}\cdot\boldsymbol{v}^*}{2}\right) = \Re\left[\frac{e^2E_0^2}{2m_e}\frac{1}{(\nu - \mathrm{i}\omega)}\right] = \frac{e^2}{m_e}\frac{\nu}{\nu^2+\omega^2}\overline{E^2} , \tag{2.33}$$

  where $\sqrt{\overline{E^2}} = E_0/\sqrt{2}$ is the mean squared value of the electric field.
  If $\nu/\omega \ll 1$ (HF discharge approximation), we have (2.33):

$$\theta_a \approx \frac{e^2}{m_e}\frac{\nu}{\omega^2}\overline{E^2} , \tag{2.34}$$

  and we can verify that, for $\nu = 0$, the transfer of energy from the field $\boldsymbol{E}$ is zero, $\theta_a = 0$, conforming to the result we have already obtained above in the case of individual trajectories.
  In the opposite case of $\nu/\omega \gg 1$ (low-frequency discharge approximation), we obtain:

$$\theta_a \approx \frac{e^2}{m_e}\frac{\overline{E^2}}{\nu} . \tag{2.35}$$

  Expressions (2.34) and (2.35) are essential to the understanding of HF plasmas (Sect. 4.2).
- Electrical conductivity and permittivity in the presence of collisions
  The motion of charged particles in the field $\boldsymbol{E}$ creates a *current*, called the *conduction current*. For an electron density n_e, the current density can be written:

$$\boldsymbol{J} = -n_e e\boldsymbol{v} \tag{2.36}$$

and in complex notation, following (2.29):

$$\boldsymbol{J} = \frac{n_e e^2}{m_e(\nu + \mathrm{i}\omega)} \boldsymbol{E}(t) \; . \tag{2.37}$$

Since from electromagnetism:

$$\boldsymbol{J} = \sigma \boldsymbol{E} \; , \tag{2.38}$$

where σ is the (scalar) electrical conductivity of electrons, we find by identification from (2.37) and (2.38):

$$\sigma = \frac{n_e e^2}{m_e(\nu + \mathrm{i}\omega)} \; . \tag{2.39}$$

Note that in the case where there are no collisions ($\nu = 0$), σ is purely imaginary and the plasma then behaves as a **perfect dielectric**.
The permittivity ϵ_p of the plasma relative to vacuum in a field $E_0 \mathrm{e}^{\mathrm{i}\omega t}$ is related to the conductivity σ (demonstrated in Remark 2 below):

$$\epsilon_p = 1 + \frac{\sigma}{\mathrm{i}\omega\epsilon_0} \; , \tag{2.40}$$

where ϵ_0 is the permittivity of vacuum. Substituting σ from (2.39), we find:

$$\epsilon_p = 1 - \frac{\omega_{pe}^2}{\omega(\omega - \mathrm{i}\nu)} \; , \tag{2.41}$$

which, in the absence of collisions, reduces to:

$$\epsilon_p = 1 - \frac{\omega_{pe}^2}{\omega^2} \; , \tag{2.42}$$

an expression which shows that the exact case where $\omega = \omega_{pe}$ represents a singular value for the propagation of a wave, since $\epsilon_p = 0$.

Remarks:

1. Note that the value of θ_a (2.33) is inversely proportional to the mass of the particles, which means that we can usually neglect the power transferred to the ions in assessing the HF-particle power balance. We can also verify that for constant ω, θ_a passes through a maximum when $\nu = \omega$[57]; this is the case in which the transfer of energy is the most efficient.
2. The use of the conductivity σ in the preceding pages corresponds to the representation of *charges in vacuum*, as distinct from the *dielectric description* expressed by ϵ_p where, from the beginning, we prefer to consider the

[57] Recall that the collision frequency ν depends on the average energy of the electrons (and on the energy distribution function) and gas pressure (Sect. 1.7.8).

displacement current rather than the conduction current to describe the motion of charged particles in a HF field.

In effect, in the case of a purely dielectric description of the plasma, (2.3) can be expressed in the form:

$$\nabla \wedge \boldsymbol{B} = \mu_0 \frac{\partial \boldsymbol{D}}{\partial t} \equiv \mu_0 \epsilon_0 \epsilon_p \frac{\partial \boldsymbol{E}}{\partial t} \ . \tag{2.43}$$

Assuming a periodic variation $\mathrm{e}^{\mathrm{i}\omega t}$ in the electromagnetic field with angular frequency ω, we obtain the terms on the RHS of (2.3) and (2.43) respectively:

$$\mu_0 \epsilon_0 \frac{\partial \boldsymbol{E}}{\partial t} + \mu_0 \boldsymbol{J} = \mu_0 \epsilon_0 \mathrm{i}\omega \boldsymbol{E}_0 \mathrm{e}^{\mathrm{i}\omega t} + \mu_0 \sigma \boldsymbol{E}_0 \mathrm{e}^{\mathrm{i}\omega t} \ , \tag{2.44}$$

$$\mu_0 \epsilon_0 \epsilon_p \frac{\partial \boldsymbol{E}}{\partial t} = \mu_0 \epsilon_0 \epsilon_p \mathrm{i}\omega \boldsymbol{E}_0 \mathrm{e}^{\mathrm{i}\omega t} \ , \tag{2.45}$$

which, by identification, leads to:

$$\mathrm{i}\omega \epsilon_0 \epsilon_p = \mathrm{i}\omega \epsilon_0 + \sigma \ , \tag{2.46}$$

from which we obtain the complex relative permittivity of the plasma given by (2.40).

2.2.2 Uniform static magnetic field

MAGNETIC FIELD ONLY ($\boldsymbol{E} = 0$)

The study of this simple case will allow us to introduce the concepts of cyclotron gyration and helical motion. Cyclotron motion of particles produces a magnetic field $\boldsymbol{B}'$, in the opposite direction to the externally applied field $\boldsymbol{B}$, giving the plasma a diamagnetic character.

We will use Cartesian coordinates, such that Ox is oriented in the direction of $\boldsymbol{B}$. From the general equations of motion (2.6) and (2.8), setting $\boldsymbol{E} = (0, 0, 0)$ and $\boldsymbol{B} = (B, 0, 0)$, we obtain:

$$\frac{\mathrm{d}^2 x}{\mathrm{d}t^2} = 0 \ , \tag{2.47}$$

$$\frac{\mathrm{d}^2 y}{\mathrm{d}t^2} = \frac{q_\alpha B}{m_\alpha} \frac{\mathrm{d}z}{\mathrm{d}t} \ , \tag{2.48}$$

$$\frac{\mathrm{d}^2 z}{\mathrm{d}t^2} = -\frac{q_\alpha B}{m_\alpha} \frac{\mathrm{d}y}{\mathrm{d}t} \ . \tag{2.49}$$

These equations can be rewritten by introducing the *cyclotron* (angular) *frequency*:

$$\omega_{c\alpha} = -\frac{q_\alpha B}{m_\alpha} , \tag{2.50}$$

the sign convention being such that $\omega_{c\alpha}$ is positive for electrons[58].
Ignoring the subscript α for simplicity, (2.47) to (2.49) take the form:

$$\ddot{x} = 0 , \tag{2.51}$$

$$\ddot{y} = -\omega_c \dot{z} , \tag{2.52}$$

$$\ddot{z} = \omega_c \dot{y} . \tag{2.53}$$

We will solve these equations, using the initial conditions ($t = 0$): $x = y = z = 0$ (the particle is initially at the origin of the coordinate system), $\dot{x} = w_{x0} = w_{\parallel 0}$, $\dot{y} = w_{y0}$ and $\dot{z} = w_{z0}$: for complete generality, the components of the initial velocity parallel and perpendicular to $\boldsymbol{B}$ are non zero.
Integrating (2.53), we obtain:

$$\dot{z} = \omega_c y + C_1 = \omega_c y + w_{z0} , \tag{2.54}$$

where the constant of integration C_1, in view of our initial conditions, is equal to w_{z0}. Introducing this value of $\dot{z}$ into (2.52) for $\ddot{y}$;

$$\ddot{y} = -\omega_c^2 y - \omega_c w_{z0} . \tag{2.55}$$

This equation can be rearranged such that the LHS is homogeneous in y:

$$\ddot{y} + \omega_c^2 y = -\omega_c w_{z0} , \tag{2.56}$$

which has the form of a "forced" harmonic oscillator. The solution to this equation is given by the sum of the general solution without the RHS, and a particular solution of the differential equation including the RHS, thus:

$$y = A_1 \cos \omega_c t + A_2 \sin \omega_c t - \frac{w_{z0}}{\omega_c} . \tag{2.57}$$

We will now determine the constants A_1 and A_2 in (2.57):

$$y(t=0) \equiv A_1 - \frac{w_{z0}}{\omega_c} = 0 \text{ from which } A_1 = \frac{w_{z0}}{\omega_c} , \tag{2.58}$$

$$\dot{y}(t=0) \equiv w_{y0} = A_2 \omega_c \text{ from which } A_2 = \frac{w_{y0}}{\omega_c} . \tag{2.59}$$

We now need to calculate $z(t)$: from (2.54) with (2.57)–(2.59),

---

58 Some authors prefer to write $\omega_{c\alpha} = |q_\alpha| B/m_\alpha$, but it is still necessary to define the direction in which the respective positively and negatively charged particles rotate around a line of force of the field $\boldsymbol{B}$.

$$\dot{z} = \omega_c \left[\frac{w_{z0}}{\omega_c} \cos \omega_c t + \frac{w_{y0}}{\omega_c} \sin \omega_c t - \frac{w_{z0}}{\omega_c} \right] + w_{z0} \, , \tag{2.60}$$

and, after integrating over t:

$$z = \frac{w_{z0}}{\omega_c} \sin \omega_c t - \frac{w_{y0}}{\omega_c} \cos \omega_c t + C_2 \, , \tag{2.61}$$

and since $z(t=0) = 0$, we find $C_2 = w_{y0}/\omega_c$.
The three equations describing the orbit of a charged particle can finally be written:

$$x = w_{x0} t = w_{\|0} t \, , \tag{2.62}$$

$$y = \frac{w_{z0}}{\omega_c} \cos \omega_c t + \frac{w_{y0}}{\omega_c} \sin \omega_c t - \frac{w_{z0}}{\omega_c} \, , \tag{2.63}$$

$$z = \frac{w_{z0}}{\omega_c} \sin \omega_c t - \frac{w_{y0}}{\omega_c} \cos \omega_c t + \frac{w_{y0}}{\omega_c} \, . \tag{2.64}$$

In the yOz plane, the particle motion describes a circle[59], for which the centre is fixed by the constants of integration, in this case $Y, Z = -w_{z0}/\omega_c, -w_{y0}/\omega_c$. To demonstrate this, we will write the equation of the corresponding circular trajectory:

$$\begin{aligned}
(y-Y)^2 + (z-Z)^2 &\equiv \left(y + \frac{w_{z0}}{\omega_c} \right)^2 + \left(z - \frac{w_{y0}}{\omega_c} \right)^2 \\
&= \frac{w_{z0}^2}{\omega_c^2} \cos^2 \omega_c t + \frac{w_{y0}^2}{\omega_c^2} \sin^2 \omega_c t + \frac{2 w_{z0} w_{y0}}{\omega_c^2} \cos \omega_c t \sin \omega_c t \\
&+ \frac{w_{z0}^2}{\omega_c^2} \sin^2 \omega_c t + \frac{w_{y0}^2}{\omega_c^2} \cos^2 \omega_c t - \frac{2 w_{z0} w_{y0}}{\omega_c^2} \cos \omega_c t \sin \omega_c t \\
&= \frac{w_{z0}^2 + w_{y0}^2}{\omega_c^2} \equiv \frac{w_{\perp 0}^2}{\omega_c^2} = r_B^2 \, ,
\end{aligned} \tag{2.65}$$

from which we can define a radius whose value is:

$$r_B = \frac{w_{\perp 0}}{\omega_c} = \frac{m_e}{eB} w_{\perp 0} \, . \tag{2.66}$$

In summary, in the plane perpendicular to $\boldsymbol{B}$, we observe a periodic circular motion with an angular frequency ω_c, the cyclotron frequency[60], whose

[59] The relations (2.63) and (2.64) which describe a periodic motion have the same amplitude and the same frequency, with a difference of phase $\pi/2$. In the framework of Lissajous curves, this gives rise to a circle. Note that, in English, the distinction between frequency and angular frequency is often ignored.

[60] Equivalently, the gyro-frequency of particles α ($\alpha = e, i$).

radius r_B is called the *Larmor radius*[61], $w_{\perp 0}$ being the initial speed of the particle in the yOz plane. To determine the direction of rotation of particles of mass m_α and of charge q_α, we ignore the constant, initial velocity of the particle in the yOz plane. For the electron, since by convention $\omega_{ce} > 0$, we see from (2.63) and (2.64) that for $\omega_c t = 0$, $y = w_{z0}/\omega_c$ and $z = -w_{y0}/\omega_c$, while for $\omega_c t = \pi/2$ ($t = \mathcal{T}_c/4$, where $\mathcal{T}_c$ is the cyclotron period), $y = w_{y0}/\omega_c$ and $z = w_{z0}/\omega_c$. It follows that, for a field $\boldsymbol{B}$ away from the reader, the gyration of the electron is in the clockwise direction (towards the right), as is shown in Fig. 2.3a, while the positive ion rotates in the anti-clockwise direction (towards the left). In the direction parallel to $\boldsymbol{B}$, the velocity is constant, equal to $w_{\parallel 0}$, and the motion is uniform, since this velocity is not modified by $\boldsymbol{B}$. The combination of the cyclotron motion and uniform motion gives rise to a trajectory in the form of a helix (Fig. 2.3b), which rotates around the magnetic field line (referred to as the *guiding centre*).

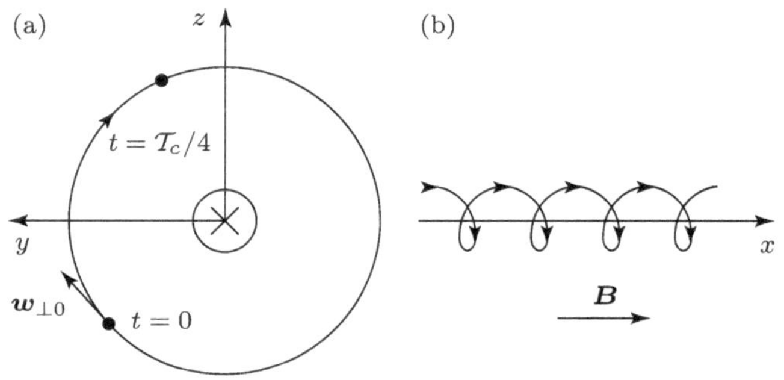

Fig. 2.3 **a** Cyclotron motion of an electron in the plane perpendicular to $\boldsymbol{B}$, the field directed along the Ox axis, away from the reader. The points on the circle show the position of the electron at $t = 0$ and $t = \mathcal{T}_c/4$. **b** Helical motion of the electron along the B field.

Interesting particular cases:

- If $w_{\parallel 0} = 0$, the helical trajectory degenerates into a circular orbit. The radius of the orbit is then dependent on the total velocity w_0 of the particle, and $r_B = w_0 m_e/eB$.
- If $w_{\perp 0} = 0$, the trajectory is rectilinear and parallel to $\boldsymbol{B}$.

Remarks:

1. The decrease in the diameter of the helix with increasing $\boldsymbol{B}$ results in a confinement of charged particles in the direction perpendicular to $\boldsymbol{B}$. In fact, as $\boldsymbol{B}$ tends to infinity, $r_B \to 0$, such that transverse motion is not

[61] Equivalently, the cyclotron radius or radius of gyration.

possible: we will see in Sect. 3.8 that this effect reduces the particle diffusion perpendicular to $\boldsymbol{B}$, towards the walls.

2. A uniform field $\boldsymbol{B}$ cannot affect $w_{\parallel}$ so $w_{\parallel}(t) = w_{\parallel 0}$ where the subscript zero corresponds to the time $t = 0$: this is a property of the Lorentz force in the case $\boldsymbol{E} = 0$. If $\boldsymbol{E} = 0$, from conservation of kinetic energy: $w_{\perp}^2(t) + w_{\parallel}^2(t) \equiv w^2(t) = w_0^2$. Since we have just seen that $w_{\parallel} = w_{\parallel 0}$, then $w_{\perp}^2 = w_{\perp 0}^2$ and, thus $w_{\perp}^2(t) \equiv w_y^2(t) + w_z^2(t) = w_{\perp 0}^2$. Thus, in a field $\boldsymbol{B}$, the components w_y and w_z can vary, as was mentioned in Sect. 2.1 (Remark 1).
3. The *pitch of the helix* is obtained by calculating the axial distance travelled during one revolution. If this pitch is p_h, and $\mathcal{T}_c$ is the cyclotron period, then $p_h = w_{\parallel 0}\mathcal{T}_c = w_{\parallel 0}/f_c = 2\pi w_{\parallel 0}/\omega_c$, and we obtain:

$$p_h = 2\pi \left(\frac{w_{\parallel 0}}{w_{\perp 0}} \right) r_B \ . \tag{2.67}$$

4. A useful way to represent the helical motion is:

$$\boldsymbol{w} = \boldsymbol{w}_{\parallel 0} + \boldsymbol{\omega}_c \wedge \boldsymbol{r}_B \ , \tag{2.68}$$

where $\boldsymbol{w}_{\parallel 0}$ describes the motion of the guiding centre and the second term, the circular cyclotron motion of the particle; the vector $\boldsymbol{\omega}_c$ is directed along $\boldsymbol{B}$ and defines the axis of rotation and its direction; the vector $\boldsymbol{r}_B$, the orbit radius, has its origin at the guiding centre.
5. Since the Larmor radius is proportional to the mass of the particles, (see (2.66)), it follows that for ions of mass m_i, $r_{Bi} = r_{Be}m_i/m_e$, i.e. $r_{Bi} \gg r_{Be}$.
6. The cyclotron frequency (2.50) or gyration frequency does not depend on the velocity of the particles, but only on their mass and charge. This property allows energy to be given uniquely to particles of a given mass and charge by means of an electric field oscillating at $\omega = \omega_{c\alpha}$, independently of their velocity distribution: we can therefore obtain a form of selective heating by means of cyclotron resonance, which will be treated in detail later (2.146).
7. A usefull numerical relation to calculate the cyclotron frequency for electrons is:

$$f_{ce}(\text{Hz}) = 2.799 \times 10^{10} B \ (\text{tesla}) \ . \tag{2.69}$$

Thus for $B = 0.1\,\text{T}$ (10^3 gauss), $f_{ce} = 2.8\,\text{GHz}$. The corresponding frequency for ions of mass m_i is m_i/m_e times smaller.
8. The *diamagnetic field* created by the circulating cyclotron current is given by the Biot-Savart Law (Lorrain et al):

$$\boldsymbol{B}' = \frac{\mu_0}{4\pi} \int\limits_V \frac{\boldsymbol{J} \wedge \boldsymbol{r}}{r^3} \, \mathrm{d}V \ . \tag{2.70}$$

In this expression, $\boldsymbol{r}$ points from the source (charge) towards the guiding centre axis (Fig. 2.4). Note that $\boldsymbol{B}$ and $\boldsymbol{B}'$ are calculated at the same $\boldsymbol{r}$

position for comparison purposes. The diamagnetic field $\boldsymbol{B}'$ points in the same direction for electrons and ions: particles of opposite charge revolve in opposite directions around $\boldsymbol{B}$, such that their respective currents rotate in the same direction, as is shown in Fig. 2.4. The vectorial product $\boldsymbol{J} \wedge \boldsymbol{r}$ from (2.70) indicates that $\boldsymbol{B}'$ is in the opposite direction to the field $\boldsymbol{B}$ responsible for the cyclotron motion (this cannot be otherwise!). The magnetic field in the plasma is given by the vectorial sum of $\boldsymbol{B}$ and $\boldsymbol{B}'$ (see exercise 2.2).

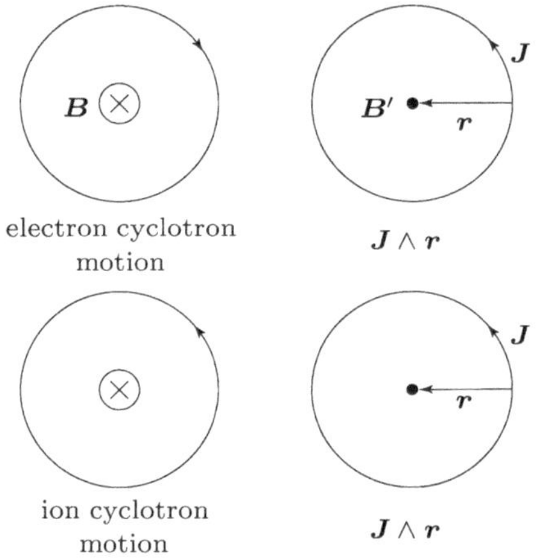

Fig. 2.4 Determining the orientation of the diamagnetic field $\boldsymbol{B}'$ created by the cyclotron motion in a field $\boldsymbol{B}$ imposed into the page: $\boldsymbol{B}'$ comes out of the page towards the reader.

STATIC UNIFORM ELECTRIC AND MAGNETIC FIELDS

In this section, we will show that the effect of uniform, constant fields $\boldsymbol{E}$ and $\boldsymbol{B}$ leads to a motion, called the electric field drift (also known as the $\boldsymbol{E} \wedge \boldsymbol{B}$ drift), of ions and electrons in the plasma, perpendicular to both $\boldsymbol{E}$ and $\boldsymbol{B}$. After this, we will derive an equation that incorporates all the fundamental motions studied to date. As a further application, we will calculate the electric conductivity, for the same $\boldsymbol{E}$ and $\boldsymbol{B}$ fields, and show that it is a tensor.

In the first case, the superposition of electric and magnetic fields modifies the magnitude of the velocity, such that the part of the Lorentz force tied to the magnetic field, $q_\alpha \boldsymbol{w} \wedge \boldsymbol{B}$, is continuously varying. It is noteworthy that, since the two fields are uniform and constant in time, the orbits can be calculated analytically and are easily represented graphically.

The case where $\boldsymbol{E}$ and $\boldsymbol{B}$ are arbitrarily oriented (with $w_0 = 0$)

The Cartesian frame is once again constructed such that $\boldsymbol{B}$ is directed along the Ox axis. Since the orientation of $\boldsymbol{E}$ in this frame is independent of $\boldsymbol{B}$, the $\boldsymbol{E}$ field has a component along each of the axes. We suppose that the charged particle, at $t = 0$, is situated at the origin of the frame $x = y = z = 0$, and, in contrast to the previous case ($\boldsymbol{B}$ only), at rest $\dot{x} = \dot{y} = \dot{z} = 0$. This last condition implies that $w_{\perp 0} = 0$, removing the contribution of the cyclotron motion to the particle trajectory entirely, allowing us to examine the effect of the electric field drift alone (the case $w_{\perp 0} \neq 0$ is treated further in the text, for $\boldsymbol{E}$ perpendicular and parallel to $\boldsymbol{B}$.)

1. The equations of motion
   From (2.6)–(2.8), we obtain:

$$\ddot{x} = \frac{q_\alpha}{m_\alpha} E_x \, , \tag{2.71}$$

$$\ddot{y} = \frac{q_\alpha}{m_\alpha} E_y - \omega_c \dot{z} \, , \tag{2.72}$$

$$\ddot{z} = \frac{q_\alpha}{m_\alpha} E_z + \omega_c \dot{y} \, . \tag{2.73}$$

2. Calculation of the trajectories
   The equations of motion are integrated analogously to the previous case.
   Calculation of y: Integration of (2.73) gives:

$$\dot{z} = \frac{q_\alpha}{m_\alpha} E_z t + \omega_c y \, . \tag{2.74}$$

   Substituting $\dot{z}$ in (2.72):

$$\ddot{y} = \frac{q_\alpha}{m_\alpha} E_y - \omega_c \left[\frac{q_\alpha}{m_\alpha} E_z t + \omega_c y \right] \, . \tag{2.75}$$

   This equation can be rearranged such that the LHS is homogeneous:

$$\ddot{y} + \omega_c^2 y = -\frac{\omega_c q_\alpha}{m_\alpha} E_z t + \frac{q_\alpha}{m_\alpha} E_y \, , \tag{2.76}$$

   for which the solution is:

$$y = A_1 \cos \omega_c t + A_2 \sin \omega_c t - \frac{q_\alpha}{m_\alpha \omega_c} E_z t + \frac{q_\alpha}{m_\alpha \omega_c^2} E_y \, . \tag{2.77}$$

   The constants A_1 and A_2 are fixed by the initial conditions.
   Since $y(t = 0) = 0$, (2.77) yields:

$$A_1 + \frac{q_\alpha}{m_\alpha \omega_c^2} E_y = 0 \, , \tag{2.78}$$

from which:

$$A_1 = -\frac{q_\alpha}{m_\alpha \omega_c^2} E_y \tag{2.79}$$

and since $\dot{y}(t=0) = 0$, $A_2\omega_c - (q_\alpha/m_\alpha\omega_c)E_z = 0$, such that:

$$A_2 = \frac{q_\alpha}{m_\alpha \omega_c^2} E_z \ . \tag{2.80}$$

Calculation of z. Substituting the value of y obtained from (2.77), together with (2.79) and (2.80), in (2.74):

$$\dot{z} = \frac{q_\alpha E_z t}{m_\alpha} + \omega_c \left[-\frac{q_\alpha E_y}{m_\alpha \omega_c^2} \cos\omega_c t + \frac{q_\alpha E_z}{m_\alpha \omega_c^2} \sin\omega_c t - \frac{q_\alpha E_z t}{m_\alpha \omega_c} + \frac{q_\alpha E_y}{m_\alpha \omega_c^2} \right] , \tag{2.81}$$

and, after integrating:

$$z = -\frac{q_\alpha E_y}{\omega_c^2 m_\alpha} \sin\omega_c t - \frac{q_\alpha E_z}{\omega_c^2 m_\alpha} \cos\omega_c t + \frac{q_\alpha E_y t}{\omega_c m_\alpha} + C_3 \ . \tag{2.82}$$

Since $z(t=0) = 0 = -(q_\alpha/\omega_c^2 m_\alpha)E_z + C_3$:

$$C_3 = \frac{q_\alpha E_z}{m_\alpha \omega_c^2} \ . \tag{2.83}$$

Calculation of x. Two successive integrations of (2.71) lead to:

$$x = \frac{q_\alpha}{m_\alpha} E_x \frac{t^2}{2} \ . \tag{2.84}$$

Finally, the equations for the trajectory as a function of time (for $\boldsymbol{B} \parallel \hat{\mathbf{e}}_x$) can be written:

$$x = \frac{q_\alpha}{m_\alpha} E_x \frac{t^2}{2} \ , \tag{2.85}$$

$$y = -\frac{q_\alpha}{\omega_c^2 m_\alpha} E_y \cos\omega_c t + \frac{q_\alpha}{\omega_c^2 m_\alpha} E_z \sin\omega_c t - \frac{q_\alpha}{\omega_c m_\alpha} E_z t + \frac{q_\alpha}{\omega_c^2 m_\alpha} E_y \ , \tag{2.86}$$

$$z = -\frac{q_\alpha}{\omega_c^2 m_\alpha} E_y \sin\omega_c t - \frac{q_\alpha}{\omega_c^2 m_\alpha} E_z \cos\omega_c t + \frac{q_\alpha}{\omega_c m_\alpha} E_y t + \frac{q_\alpha}{\omega_c^2 m_\alpha} E_z \ . \tag{2.87}$$

3. Study of the motion described by (2.85) to (2.87)
   The presence of the uniform and constant fields $\boldsymbol{E}$ and $\boldsymbol{B}$ results in a *drift motion* (called the *electric field drift*) of the charged particle perpendicular to $\boldsymbol{B}$ and $\boldsymbol{E}_\perp$, the component of $\boldsymbol{E}$ perpendicular to $\boldsymbol{B}$. In fact, if $w_0 = 0$, as is the case here, the non-periodic part of the motion in the plane yOz is as follows: the particle initially moves in the direction of $\boldsymbol{E}_\perp$ (for a positive

ion, Fig. 2.5) or in the opposite direction (electron). Due to the velocity $\boldsymbol{w}_\perp$ thus acquired, the magnetic part of the Lorentz' force $\boldsymbol{F}_{Lm}$ produces a motion perpendicular to $\boldsymbol{E}_\perp$ and $\boldsymbol{B}$, precisely in the direction of the drift motion, since $\boldsymbol{F}_{Lm} = q_\alpha \boldsymbol{w}_\perp \wedge \boldsymbol{B}$.

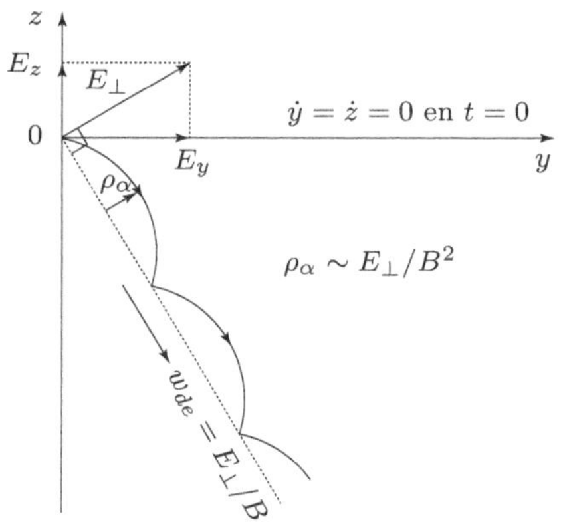

Fig. 2.5 Cycloidal motion of the drift for a positive ion (the field $\boldsymbol{B}$ is out of the page). The ion is initially ($t = 0$) at the origin of the frame and at rest, then it moves, on average, along the drift axis represented by the dotted line.

The projection of the motion in the yOz plane (the plane perpendicular to $\boldsymbol{B}$) is thus a cycloidal trajectory, as is shown in Fig. 2.5: the non-periodic terms $(q_\alpha/m_\alpha\omega_c)E_it$ $[i = y, z]$ push the particle in a direction perpendicular to $\boldsymbol{E}_\perp$ and $\boldsymbol{B}$ along a virtual straight line, whose parametric equation is given by:

$$y_d = -\frac{q_\alpha}{m_\alpha\omega_c}E_zt\,, \tag{2.88}$$

and:

$$z_d = \frac{q_\alpha}{m_\alpha\omega_c}E_yt\,. \tag{2.89}$$

These relations can be combined to give:

$$z_d = -\frac{E_y}{E_z}y_d\,. \tag{2.90}$$

The average velocity of this shifting motion, called the *electric field drift velocity*, taken from (2.88) and (2.89), is:

$$w_{de} = \sqrt{\left(\frac{q_\alpha E_z}{m_\alpha\omega_c}\right)^2 + \left(\frac{q_\alpha E_y}{m_\alpha\omega_c}\right)^2} = \frac{E_\perp}{B}\,. \tag{2.91}$$

This velocity is independent of the mass of the particle, and of its charge. Further, because the motion is directed perpendicular[62] to $\boldsymbol{E}$ (to both $\boldsymbol{E}_\perp$ and $\boldsymbol{E}_\parallel$ components, see Fig. 2.5), the particle in its drift motion does no work in the field $\boldsymbol{E}$: the drift velocity thus remains constant.
A uniformly accelerated motion in the direction perpendicular to the yOz, plane, following the E_x component of the electric field, must be added to the motion in the yOz plane.

4. Comparative study of the cycloidal motion of electrons and ions.
We will ignore the motion due to $\boldsymbol{E}_\parallel$. Recall the convention: the motion of positive ions is in the direction of the electric field. At $t = 0$, the electron and the ion are at the origin of the frame, with zero velocity. Immediately afterwards, the ion starts to move in the direction of $\boldsymbol{E}_\perp$ but its trajectory is instantly curved, by the magnetic component of the Lorentz force, following $\boldsymbol{w}_{de}$ (Fig. 2.5). The electron is initially accelerated in the opposite direction, but the Lorentz force leads it to follow the same drift direction as the ion because of the opposite sign of its charge ($\boldsymbol{F}_{Lm} = -e\boldsymbol{w}_e \wedge \boldsymbol{B}$): the two trajectories (if we ignore the influence of $\boldsymbol{E}_\parallel$) are confined in the plane ($\boldsymbol{w}_{de}, \boldsymbol{E}_\perp$), as is shown in Fig. 2.6.
In (2.86) where $y = -(q_\alpha E_y/\omega_{c\alpha}^2 m_\alpha)\cos\omega_{c\alpha}t + \cdots$, the amplitude of the periodic motion of the particle is proportional to m_α ($\omega_{c\alpha}^2 m_\alpha \propto m_\alpha^{-1}$): the electrons describe much smaller arcs than those of the ions but their number per second is much larger (Fig. 2.6) since the ratio of the masses $m_i/m_e \gg 1$ leads to $\omega_{ce}/\omega_{ci} \gg 1$.

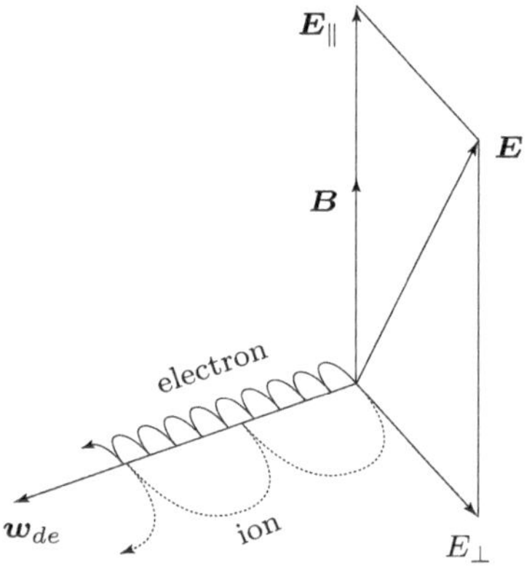

Fig. 2.6 Schematic representation of the motion of electrons and ions in the electric field drift, showing that the arcs described by the electrons have much smaller amplitudes but are more numerous.

[62] To see that $\boldsymbol{w}_{de}$ is perpendicular to $\boldsymbol{E}$, note that the slope of the trajectory describing the particle motion $z = f(y)$ is given by $\Delta x/\Delta y = -E_y/E_z$ (2.90) while the orientation of $\boldsymbol{E}_\perp$ in the same frame (y, z) is expressed by E_z/E_y: these slopes are therefore orthogonal. To distinguish it from the present drift velocity, the drift in a field $\boldsymbol{E}$ including collisions (Sect. 3.8.2) will be called the *collisional drift velocity*.

Remarks:

1. $E_\perp/B$ has the units of velocity (the proof is left to the reader)
2. The maximum amplitude ρ_α of the cycloid of a particle of type α with respect to the drift axis is proportional to $E_\perp/B^2$ (Fig. 2.5). The calculation of this expression is also left to the reader.

The preceding discussion can be treated in a more complete manner by considering more generally that $\boldsymbol{w}_0 \neq 0$: then the influence of the cyclotron gyration is superimposed on the drift velocity in the total motion of the particle. Nonetheless to simplify the calculation, we will assume $\boldsymbol{E} \perp \boldsymbol{B}$.

Perpendicular $\boldsymbol{E}$ and $\boldsymbol{B}$ fields with $\boldsymbol{w}_0 \neq 0$: combined drift and cyclotron motion

The $\boldsymbol{B}$ field is still along Ox but this time $\boldsymbol{E}$ is entirely along Oz. This leads to the following equations for the trajectory of the charged particle:

$$x = w_{\parallel 0} t \ , \tag{2.92}$$

$$y = \frac{w_{z0}}{\omega_c}\cos\omega_c t + \left(\frac{w_{y0}}{\omega_c} + \frac{q_\alpha E}{m_\alpha \omega_c^2}\right)\sin\omega_c t - \frac{q_\alpha E}{m_\alpha \omega_c} t - \frac{w_{z0}}{\omega_c} \ , \tag{2.93}$$

$$z = \frac{w_{z0}}{\omega_c}\sin\omega_c t - \left(\frac{w_{y0}}{\omega_c} + \frac{q_\alpha E}{m_\alpha \omega_c^2}\right)\cos\omega_c t + \left(\frac{w_{y0}}{\omega_c} + \frac{q_\alpha E}{m_\alpha \omega_c^2}\right) . \tag{2.94}$$

To illustrate the various forms of the trajectories, one needs to consider the ratio w_{y0}/w_{de}, where $w_{de} = E_\perp/B$ (we will assume $w_{y0} = w_{z0}$) and distinguish three particular cases.
To do this, consider the term:

$$\frac{w_{y0}}{\omega_c} + \frac{q_\alpha E}{m_\alpha \omega_c^2}$$

appearing in the expressions (2.93) and (2.94) for y and z. Taking into account the convention on the sign of $\omega_{c\alpha}$ (2.50), this term can be transformed in terms of the ratio w_{y0}/w_{de} such that:

$$\frac{1}{\omega_c}\left[w_{y0} - \frac{q_\alpha E}{m_\alpha}\frac{m_\alpha}{q_\alpha B}\right] = \frac{1}{\omega_c}\left[w_{y0} - \frac{E}{B}\right] = \frac{1}{\omega_c}\left[w_{y0} - w_{de}\right] . \tag{2.95}$$

If $w_{de} \gg w_{y0}$, then $w_{y0} \simeq 0$ and $w_{z0} \simeq 0$ (no cyclotron motion because $w_{\perp 0} \simeq 0$) and Eq. (2.93) for y reduces to:

$$y = -\frac{q_\alpha E}{m_\alpha \omega_c^2}\left(\omega_c t - \sin\omega_c t\right) , \tag{2.96}$$

which obviously leads to (2.86) in the case where $E_y = 0$.
For the same approximation ($w_{y0} \simeq 0$ and $w_{z0} \simeq 0$), Eq. (2.94) for z becomes:

$$z = \frac{q_\alpha E}{m_\alpha \omega_c^2}(1 - \cos\omega_c t) , \tag{2.97}$$

the expression obtained when $E_y = 0$ in (2.87).
We will now consider the three following typical cases:

- $w_{y0}/w_{de} = 100$ (Fig. 2.7a)
- $w_{y0}/w_{de} = 2$ (Fig. 2.7b)
- $w_{y0}/w_{de} \leq 1$ (Fig. 2.7c)

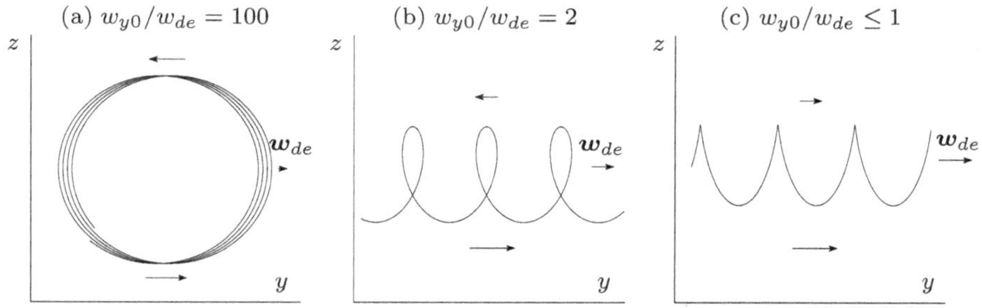

Fig. 2.7 Trajectory of a positive ion in uniform static $\boldsymbol{E}$ and $\boldsymbol{B}$ fields, with the respective components E_z and B_x, for different values of the ratio w_{y0}/w_{de} where $w_{z0} = w_{y0}$ (the $\boldsymbol{B}$ field is directed towards the reader).

Figure 2.7a shows that the cyclotron motion is hardly affected by a weak $\boldsymbol{E}$ field, the guiding centre being slightly displaced in the direction of the electric field drift. Figure 2.7b describes what happens to the cyclotron motion when it is strongly modified by the drag along y due to the electric field drift. Finally, Fig. 2.7c shows that all traces of cyclotron motion disappear when $w_{de} \geq w_{y0}$.

To obtain a simple analytic form for the resulting trajectories, suppose $w_{z0} = 0$ (in Fig. 2.7, note that $w_{z0} = w_{y0} \neq 0$). The resultant trajectory for $w_{y0}/w_{de} = 2$ is that of a quasi trochoid[63], for which the mathematical expression is:

$$y = a\tau - b\sin\tau \qquad z = b(\cos\tau - 1) \tag{2.98}$$

with, following (2.93) and (2.94) and assuming $w_{z0} = 0$:

$$a = \frac{E}{B\omega_c} , \qquad b = -\frac{1}{\omega_c}\left[w_{y0} - \frac{E}{B}\right] \qquad \text{and} \qquad \tau = \omega_c t .$$

[63] A true trochoid requires $y = a\tau - b\sin\tau$ and $z = a - b\cos\tau$.

In the case $w_{y0}/w_{de} < 1$ ($a \simeq b$), the trajectory is that of a cycloid (with a sign inversion):

$$y = a(\tau - \sin\tau) \quad z = a(\cos\tau - 1) \quad \text{with } a = \frac{E}{B\omega_c} \,. \tag{2.99}$$

Note that setting $w_{z0} = 0$ while $w_{y0}/w_{de} = 1$ (Eqs. (2.93) and (2.94)) suppresses all periodic motion in the y and z ($b = 0$) directions: all that remains is a rectilinear trajectory along y due to the electric field drift.

Remark: In the case $w_{de} \ll w_{\perp 0} = \omega_c r_B$ (weak $E_\perp$ field), as shown in Fig. 2.7a, the trajectories are quasi cyclotronic, with a weak drift velocity of their guiding centres in the direction perpendicular to $\boldsymbol{B}$ and $\boldsymbol{E}_\perp$. The guiding centre of the cyclotron trajectory of a positive ion moves slowly in the direction of the drift, because the cyclotron curvature is smaller when the ion moves in the direction of $\boldsymbol{E}_\perp$ ($w_\perp$ increases, as does r_B) than when it moves in the opposite direction to $\boldsymbol{E}_\perp$. This deformation of the cyclotron motion leads to a shift of the guiding centre and, accordingly, to the particle drift.

Parallel $\boldsymbol{E}$ and $\boldsymbol{B}$ fields: no drift motion

Assume the Ox axis is in the direction of the fields: It is then useful to distinguish two cases:

- The initial velocity is zero.
  From (2.85) to (2.87), we find:

$$x(t) = \frac{q_\alpha}{m_\alpha}\frac{E_x t^2}{2} \,, \tag{2.100}$$

$$y(t) = 0 \,, \tag{2.101}$$

$$z(t) = 0 \,. \tag{2.102}$$

  The motion is only along Ox and uniformly accelerated: since the $\boldsymbol{B}$ field is in the direction of motion, it plays no role on the trajectory of the particle ($\boldsymbol{F}_{Lm} \equiv q_\alpha \boldsymbol{w} \wedge \boldsymbol{B} = 0$ since $\boldsymbol{w} \parallel \boldsymbol{B}$).
- The initial velocity normal to $\boldsymbol{B}$ is non zero ($w_{y0} \neq 0$, $w_{z0} \neq 0$).
  Under these conditions, we can resume the development from (2.71)–(2.73). We then obtain a helical trajectory, as in the previous case of a magnetic field only, but the pitch of the helix increases (or decreases) because the E_x field gives rise to a velocity component w_x:

$$p_h = w_\parallel \mathcal{T}_c = \frac{2\pi}{|\omega_c|} w_\parallel = \frac{2\pi}{q_\alpha B} m_\alpha w_\parallel = \frac{2\pi m_\alpha}{q_\alpha B}\left(\frac{q_\alpha}{m_\alpha} E_x t\right) = \frac{2\pi}{B} E_x t \,. \tag{2.103}$$

The general solution

By combining the results of the preceding cases, it is possible to obtain the general characteristics of the motion of a charged particle in uniform, static fields, $\boldsymbol{E}$ and $\boldsymbol{B}$. The charged particle describes a trajectory which, in the most general form, consists of:

1. A cyclotron gyration in the plane perpendicular to $\boldsymbol{B}$, provided that $\boldsymbol{w}_{\perp 0} \neq \boldsymbol{0}$. If in addition $\boldsymbol{w}_{\parallel 0} \neq 0$, the particle motion develops in three dimensions, leading to a helical motion, with constant pitch if $\boldsymbol{E} = \boldsymbol{0}$ or increasing (decreasing) pitch if the $\boldsymbol{E}$ field has a component parallel to the $\boldsymbol{B}$ field.
2. A net motion perpendicular to both $\boldsymbol{E}$ and $\boldsymbol{B}$, referred to as the electric field drift trajectory, which is independent of both m_α and q_α, and has a constant velocity $w_{de} = E_\perp / B$.

Examination of the general equation of motion (2.5) will enable us to recover these results. For that purpose, we regroup the terms homogenous in $\boldsymbol{w}$ on the LHS:

$$\dot{\boldsymbol{w}} - \frac{q_\alpha}{m_\alpha} \boldsymbol{w} \wedge \boldsymbol{B} = \frac{q_\alpha}{m_\alpha} \boldsymbol{E} \,, \tag{2.104}$$

The solution of this differential equation consists of the general solution $\boldsymbol{w}_1$ of the homogeneous equation without the RHS (helical motion with constant pitch) to which is added a particular solution $\boldsymbol{w}_2$ that includes the RHS. We want to determine $\boldsymbol{w}$ such that:

$$\boldsymbol{w} = \boldsymbol{w}_1 + \boldsymbol{w}_2 \,. \tag{2.105}$$

- General solution without the RHS ($\boldsymbol{E} = 0$)
  The value of $\boldsymbol{w}_1$ has already been obtained (2.68) in the form:

$$\boldsymbol{w}_1 = \boldsymbol{w}_{\parallel 0} + \boldsymbol{\omega}_c \wedge \boldsymbol{r}_B \,, \tag{2.106}$$

  describing a helical motion, where $\boldsymbol{w}_{\parallel 0}$ is the initial velocity parallel to $\boldsymbol{B}$. Therefore, we only need to calculate $\boldsymbol{w}_2$.
- Particular solution including the RHS: the expression for $\boldsymbol{w}_2$
  We can construct this solution in a completely arbitrary way, provided that the result obtained is a true solution. To guide us in this process, we know that this particular solution must reproduce the drift motion. Because of this, we express $\boldsymbol{w}_2$ in a trihedral coordinate system, whose Cartesian axes are defined (Fig. 2.8) such that:

$$\hat{\mathrm{e}}_z \parallel \boldsymbol{B} \,, \quad \hat{\mathrm{e}}_y \parallel \boldsymbol{E}_\perp \,, \quad \hat{\mathrm{e}}_x \parallel (\boldsymbol{E}_\perp \wedge \boldsymbol{B}) \,.$$

  This method was proposed by J.L. Delcroix.

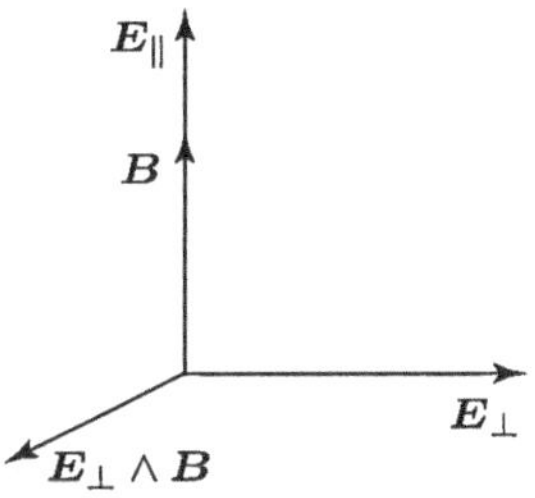

Fig. 2.8 Trihedral coordinate system used to calculate the particular solution (after J.L. Delcroix).

We are thus looking for a solution of the form:

$$\boldsymbol{w}_2 = a\boldsymbol{E}_\parallel + b\boldsymbol{E}_\perp + c(\boldsymbol{E}_\perp \wedge \boldsymbol{B}) \, , \tag{2.107}$$

$$\dot{\boldsymbol{w}}_2 = \dot{a}\boldsymbol{E}_\parallel + \dot{b}\boldsymbol{E}_\perp + \dot{c}(\boldsymbol{E}_\perp \wedge \boldsymbol{B}) \, , \tag{2.108}$$

which we can substitute in the equation of motion (2.5) including the RHS:

$$\begin{aligned} \dot{a}\boldsymbol{E}_\parallel + \dot{b}\boldsymbol{E}_\perp + \dot{c}(\boldsymbol{E}_\perp \wedge \boldsymbol{B}) &= \frac{q_\alpha}{m_\alpha}\left[a\boldsymbol{E}_\parallel + b\boldsymbol{E}_\perp + c(\boldsymbol{E}_\perp \wedge \boldsymbol{B})\right] \wedge \boldsymbol{B} \\ &= \frac{q_\alpha}{m_\alpha}(\boldsymbol{E}_\parallel + \boldsymbol{E}_\perp) \, . \end{aligned} \tag{2.109}$$

Noting that[64] $(\boldsymbol{E}_\perp \wedge \boldsymbol{B}) \wedge \boldsymbol{B} = -\boldsymbol{E}_\perp B^2$ and regrouping the terms along the different axes:

$$\left(\dot{a} - \frac{q_\alpha}{m_\alpha}\right)\boldsymbol{E}_\parallel + \left(\dot{b} + \frac{q_\alpha c B^2}{m_\alpha} - \frac{q_\alpha}{m_\alpha}\right)\boldsymbol{E}_\perp + \left(\dot{c} - \frac{b q_\alpha}{m_\alpha}\right)\boldsymbol{E}_\perp \wedge \boldsymbol{B} = 0 \, , \tag{2.110}$$

we obtain:

$$\dot{a} = \frac{q_\alpha}{m_\alpha} \, , \quad \dot{b} = \frac{q_\alpha}{m_\alpha} - \frac{q_\alpha c B^2}{m_\alpha} \, , \quad \dot{c} = \frac{q_\alpha}{m_\alpha} b \, , \tag{2.111}$$

for which a particular solution is obviously $\dot{a} = q_\alpha/m_\alpha$ and $\dot{b} = \dot{c} = 0$ such that:

$$a = \frac{q_\alpha t}{m_\alpha} \, , \quad b = 0 \, , \quad c = \frac{1}{B^2} \, . \tag{2.112}$$

This shows that we have actually chosen as particular solution that for which the initial velocity of the particle in the plane ($\boldsymbol{B}$, $\boldsymbol{E}_\perp$) is zero. We then have:

$$\boldsymbol{w}_2 = \frac{q_\alpha t}{m_\alpha}\boldsymbol{E}_\parallel + \frac{\boldsymbol{E}_\perp \wedge \boldsymbol{B}}{B^2} \, , \tag{2.113}$$

where the first term on the RHS is a uniformly accelerated motion along $\boldsymbol{B}$, the second term represents the electric drift in the direction perpendicular to both $\boldsymbol{E}_\perp$ and $\boldsymbol{B}$, for which the modulus of the velocity is $E_\perp/B$.

[64] Double vectorial product rule: $\boldsymbol{A} \wedge (\boldsymbol{B} \wedge \boldsymbol{C}) = \boldsymbol{B}(\boldsymbol{C} \cdot \boldsymbol{A}) - \boldsymbol{C}(\boldsymbol{A} \cdot \boldsymbol{B})$.

- Solution of the general equation of motion
  By adding $\boldsymbol{w}_1$ (2.106) (noting that $\boldsymbol{\omega}_c \wedge \boldsymbol{r}_B = -(q_\alpha/m_\alpha)\boldsymbol{B} \wedge \boldsymbol{r}_B$ and $\boldsymbol{w}_2$ (2.113), we obtain the full general solution:

$$\boldsymbol{w} = \underbrace{\boldsymbol{w}_{\parallel 0} + \frac{q_\alpha}{m_\alpha}\boldsymbol{r}_B \wedge \boldsymbol{B}}_{\text{Helical motion}} + \underbrace{\frac{q_\alpha t}{m_\alpha}\boldsymbol{E}_\parallel}_{\substack{\uparrow \\ \text{Uniformly} \\ \text{accelerated motion} \\ \text{along } \boldsymbol{E}_\parallel}} + \underbrace{\frac{\boldsymbol{E}_\perp \wedge \boldsymbol{B}}{B^2}}_{\text{Electric drift}} . \tag{2.114}$$

Electrical conductivity in the presence of a magnetic field: the need for a tensor representation (a digression from individual trajectories)

In Sect. 2.2.1, we calculated the electrical conductivity of charged particles in a periodic electric field ($\boldsymbol{B} = 0$). We now want to obtain an expression for the conductivity when the particles are subjected to uniform, static magnetic and electric fields.

In order to calculate the current created by the charged particles in the $\boldsymbol{E}$ an $\boldsymbol{B}$ fields, we will move from the trajectory of one particle to an ensemble of individual particle trajectories per unit volume. For this ensemble of particles, we will again make the assumption that their initial velocities are isotropic, such that on average, at $t = 0$, there is no directed motion: $\langle \boldsymbol{w}_{\perp 0} \rangle = 0$, $\langle \boldsymbol{w}_{\parallel 0} \rangle = 0$. In (2.114), it follows that $\boldsymbol{w}_{\parallel 0} = 0$ and $\boldsymbol{r}_B \wedge \boldsymbol{B} = 0$, $(\boldsymbol{r}_B \propto \boldsymbol{w}_{\perp 0})$[65]. The current density $\boldsymbol{J}_\alpha$ of charged particles of type α then reduces to:

$$\boldsymbol{J}_\alpha \equiv n_\alpha q_\alpha \boldsymbol{w}_\alpha = \frac{n_\alpha q_\alpha^2 t}{m_\alpha}\boldsymbol{E}_\parallel + \frac{n_\alpha q_\alpha}{B^2}(\boldsymbol{E}_\perp \wedge \boldsymbol{B}) . \tag{2.115}$$

In the following discussion, until equation (2.121), we shall omit the index α in $\boldsymbol{J}$ and σ.

Conductivity is now a tensor quantity: we will show that, if it is considered a priori as a scalar, it cannot satisfy (2.115). In fact, in the case where $\boldsymbol{J} = \sigma \boldsymbol{E}$, we would have the following components:

$$\boldsymbol{J} = \sigma E_x \hat{\mathbf{e}}_x + \sigma E_y \hat{\mathbf{e}}_y + \sigma E_z \hat{\mathbf{e}}_z , \tag{2.116}$$

but in developing (2.115), and since $\boldsymbol{E}_\perp = E_x \hat{\mathbf{e}}_x + E_y \hat{\mathbf{e}}_y$ ($\boldsymbol{B}$ is taken to be along z)[66], we obtain:

[65] The value of r_B, initially fixed by $\boldsymbol{w}_{\perp 0}$ in the case of the solution to (2.104) without the RHS ($\boldsymbol{E} = 0$), is not affected by the inclusion of the particular solution ($\boldsymbol{E} \neq 0$) because $\boldsymbol{w}_{2\perp} = 0$ ($b = 0$ in (2.112)).

[66] We have not decomposed equation (2.115) following the trihedral coordinate system of Fig. 2.8 because this, being vectorial, can be developed in any chosen coordinate system.

$$\boldsymbol{J} = \frac{n_\alpha q_\alpha}{B^2}(B)\, E_y \hat{\mathbf{e}}_x - \frac{n_\alpha q_\alpha}{B^2}(B)\, E_x \hat{\mathbf{e}}_y + \frac{n_\alpha q_\alpha^2}{m_\alpha} t\, E_z \hat{\mathbf{e}}_z \ , \tag{2.117}$$

because:

$$\boldsymbol{E}_\perp \wedge \boldsymbol{B} = \begin{vmatrix} \hat{\mathbf{e}}_x & \hat{\mathbf{e}}_y & \hat{\mathbf{e}}_z \\ E_x & E_y & 0 \\ 0 & 0 & B \end{vmatrix} . \tag{2.118}$$

Note that in (2.117) there is no E_x component along $\hat{\mathbf{e}}_x$ and no E_y component along $\hat{\mathbf{e}}_y$, as is required by (2.116). In fact, in (2.117), for example J_x has the form:

$$J_x = \left(\frac{n_\alpha q_\alpha}{B}\right) E_y \ , \tag{2.119}$$

from which we can conclude that σ cannot be a scalar in the presence of $\boldsymbol{B}$.

We will now seek to write the components of a tensor $\underline{\boldsymbol{\sigma}}$ explicitly, supposing it to be of order 2 (see Appendix VII for a brief introduction to tensors and Appendix VIII for tensor operations), defined by the relation:

$$\boldsymbol{J} = \underline{\boldsymbol{\sigma}} \cdot \boldsymbol{E} \ , \tag{2.120}$$

which can be written explicitly as:

$$J^i = \sigma^{ij} E_j \ , \tag{2.121}$$

where σ^{ij} is a tensor element with two (order 2) superscript (contravariant) indices. Note that the vector $\boldsymbol{J}$ is also contravariant but that $\boldsymbol{E}$ is (by nature) covariant: by convention, there is a summation over the same index when it appears in both the covariant and contravariant positions, and this index is said to be mute. In the following, however, we will not distinguish between the variance of the quantities. Expanding (2.121), we find:

$$\begin{aligned} \boldsymbol{J} = {} & (\sigma_{xx} E_x + \sigma_{xy} E_y + \sigma_{xz} E_z)\hat{\mathbf{e}}_x + (\sigma_{yx} E_x + \sigma_{yy} E_y + \sigma_{yz} E_z)\hat{\mathbf{e}}_y \\ & + (\sigma_{zx} E_x + \sigma_{zy} E_y + \sigma_{zz} E_z)\hat{\mathbf{e}}_z \ . \end{aligned} \tag{2.122}$$

By identification of (2.122) with (2.117),

$$\sigma_{xy} = \frac{n_\alpha q_\alpha}{B} \ , \qquad \sigma_{yx} = -\frac{n_\alpha q_\alpha}{B} \ , \qquad \sigma_{zz} = \frac{n_\alpha q_\alpha^2 t}{m_\alpha} \ , \tag{2.123}$$

such that the tensor can be represented by the matrix:

$$\underline{\boldsymbol{\sigma}} = n_\alpha q_\alpha \begin{pmatrix} 0 & 1/B & 0 \\ -1/B & 0 & 0 \\ 0 & 0 & q_\alpha t/m_\alpha \end{pmatrix} . \tag{2.124}$$

In the present case, and assuming a macroscopically neutral plasma ($n_e = n_i$), the total electric current due to the positive ions and the electrons (subscripts i and e respectively) is such that only its component along the direction of the $\boldsymbol{B}$ field is non zero, because along x and y, $\sigma^i_{xy} + \sigma^e_{xy} = (en_i/B) - (en_e/B) = 0$, etc. In fact, the electric field drift motion cannot give rise to a net current because the drift of the ions and electrons takes place in the same direction, so that the net transport of charge is zero[67].

Remarks:

1. In (2.121), the element σ_{ij} of the tensor $\underline{\boldsymbol{\sigma}}$ expresses the fact that the component E_j of the electric field (a force) in a given direction induces a current J^i (an action) in another direction.
2. The reader can calculate the corresponding relative permittivity tensor corresponding to $\underline{\boldsymbol{\sigma}}$ and introduce therein the electron plasma frequency, by generalising (2.40):
$$\underline{\epsilon}_p = \underline{\boldsymbol{I}} + \frac{\underline{\boldsymbol{\sigma}}}{\mathrm{i}\omega\epsilon_0} , \tag{2.125}$$
where $\underline{\boldsymbol{I}}$ is the unit tensor (represented by the unit matrix).

UNIFORM STATIC MAGNETIC FIELD AND UNIFORM PERIODIC ELECTRIC FIELD

The problem to be resolved is not very different from that of Eq. (2.104), which led to the general solution of the preceding case ($\boldsymbol{E}$ constant) because now:
$$\dot{\boldsymbol{w}} - \frac{q_\alpha}{m_\alpha}(\boldsymbol{w} \wedge \boldsymbol{B}) = \frac{q_\alpha}{m_\alpha}\boldsymbol{E}_0 \mathrm{e}^{\mathrm{i}\omega t} . \tag{2.126}$$

We are left to find a particular solution including the RHS[68], still with the trihedral coordinate system of Fig. 2.8, but this time setting:
$$\boldsymbol{w}_2 = a\boldsymbol{E}_{0\|}\mathrm{e}^{\mathrm{i}\omega t} + b\boldsymbol{E}_{0\perp}\mathrm{e}^{\mathrm{i}\omega t} + c(\boldsymbol{E}_{0\perp} \wedge \boldsymbol{B})\mathrm{e}^{\mathrm{i}\omega t} . \tag{2.127}$$

Substituting this expression into (2.126), we obtain:
$$\begin{aligned} &\left[\dot{a}\boldsymbol{E}_{0\|} + \dot{b}\boldsymbol{E}_{0\perp} + \dot{c}(\boldsymbol{E}_{0\perp} \wedge \boldsymbol{B})\right] \mathrm{e}^{\mathrm{i}\omega t} \\ &+\mathrm{i}\omega \left[a\boldsymbol{E}_{0\|} + b\boldsymbol{E}_{0\perp} + c(\boldsymbol{E}_{0\perp} \wedge \boldsymbol{B})\right] \mathrm{e}^{\mathrm{i}\omega t} \\ &-\frac{q_\alpha}{m_\alpha}\left[\left(a\boldsymbol{E}_{0\|} + b\boldsymbol{E}_{0\perp} + c(\boldsymbol{E}_{0\perp} \wedge \boldsymbol{B})\right) \wedge \boldsymbol{B}\right] \mathrm{e}^{\mathrm{i}\omega t} = \frac{q_\alpha}{m_\alpha}\left[\boldsymbol{E}_{0\|} + \boldsymbol{E}_{0\perp}\right] \mathrm{e}^{\mathrm{i}\omega t} . \end{aligned} \tag{2.128}$$

---

[67] It constitutes a neutral beam of charged particles!

[68] Remember that this solution $\boldsymbol{w}_2$ is related to the drift motion in $E_\perp$ and $\boldsymbol{B}$.

Noting that $\boldsymbol{E}_{0\parallel} \wedge \boldsymbol{B} = 0$, we obtain, along the different base vectors of the trihedral coordinate system, by identification:

$$\boldsymbol{E}_{0\parallel}\left(\dot{a} + \mathrm{i}\omega a - \frac{q_\alpha}{m_\alpha}\right) = 0 \rightarrow \dot{a} + \mathrm{i}\omega a = \frac{q_\alpha}{m_\alpha}\,, \tag{2.129}$$

$$\boldsymbol{E}_{0\perp}\left(\dot{b} + \mathrm{i}\omega b + \frac{q_\alpha c B^2}{m_\alpha} - \frac{q_\alpha}{m_\alpha}\right) = 0 \rightarrow \dot{b} + \mathrm{i}\omega b = -\frac{q_\alpha B^2}{m_\alpha}c + \frac{q_\alpha}{m_\alpha}\,, \tag{2.130}$$

$$\boldsymbol{E}_{0\perp} \wedge \boldsymbol{B}\left(\dot{c} + \mathrm{i}\omega c - \frac{q_\alpha b}{m_\alpha}\right) = 0 \rightarrow \dot{c} + \mathrm{i}\omega c = \frac{q_\alpha b}{m_\alpha}\,. \tag{2.131}$$

To find the solution, we must distinguish two situations:

1. Off-resonance case ($\omega \neq \omega_c$)

   - Solution of (2.129)–(2.131)
     A simple particular solution is then $\dot{a} = \dot{b} = \dot{c} = 0$; the value of the coefficients in this case are:

$$a = \frac{q_\alpha}{\mathrm{i}\omega m_\alpha}\,,\; b = \frac{q_\alpha}{\mathrm{i}\omega m_\alpha}(1 - B^2 c) \text{ and } c = \frac{q_\alpha b}{\mathrm{i}\omega m_\alpha}\,, \tag{2.132}$$

such that:

$$b = \frac{q_\alpha}{\mathrm{i}\omega m_\alpha}\left(1 - \frac{B^2 q_\alpha b}{\mathrm{i}\omega m_\alpha}\right)\,, \quad \text{i.e.} \quad b\left(1 - \frac{q_\alpha^2 B^2}{m_\alpha^2 \omega^2}\right) = \frac{q_\alpha}{\mathrm{i}\omega m_\alpha}\,, \tag{2.133}$$

where again:

$$b = -\frac{i q_\alpha}{\omega m_\alpha}\frac{1}{\left(1 - \dfrac{\omega_c^2}{\omega^2}\right)}\,. \tag{2.134}$$

Note that the coefficient b is finite on condition that $\omega \neq \omega_c$. Finally:

$$a = -\frac{i q_\alpha}{\omega m_\alpha}\,,\; b = \frac{i q_\alpha}{m_\alpha}\frac{\omega}{(\omega_c^2 - \omega^2)} \text{ and } c = \frac{q_\alpha^2}{m_\alpha^2}\frac{1}{(\omega_c^2 - \omega^2)}\,, \tag{2.135}$$

such that the general motion, off cyclotron resonance, can be written:

$$\begin{aligned} \boldsymbol{w} = \underset{\substack{\uparrow \\ \text{Helical motion} \\ +\text{ all initial} \\ \text{conditions}}}{\boldsymbol{w}_1} + \Bigg(& -\frac{i q_\alpha}{\omega m_\alpha}\underset{\substack{\uparrow \\ (-i)}}{\boldsymbol{E}_{0\parallel}} + \frac{\mathrm{i}\omega q_\alpha}{m_\alpha(\omega_c^2 - \omega^2)}\underset{\substack{\uparrow \\ (+i)}}{\boldsymbol{E}_{0\perp}} \\ & + \frac{q_\alpha^2}{m_\alpha^2(\omega_c^2 - \omega^2)}\overbrace{(\boldsymbol{E}_{0\perp} \wedge \boldsymbol{B})}^{\substack{(+1) \\ \downarrow}}\Bigg)\mathrm{e}^{\mathrm{i}\omega t}\,. \end{aligned} \tag{2.136}$$

Because this describes a periodic motion with the same frequency along the 3 axes and because of the particular phase relations between the three components of $\boldsymbol{w}_2$, namely (for a positive ion) $-\pi/2$ for $\boldsymbol{E}_{0\parallel}$ and $\pi/2$ for $\boldsymbol{E}_{0\perp}$ with respect to the axis $(\boldsymbol{E}_{0\perp} \wedge \boldsymbol{B})$ in the case where $\omega_c > \omega$, the trajectory obtained from (2.136) is closed on itself, corresponding to a helical motion, depending on the initial conditions superimposed on a three dimensional elliptical motion (which is difficult to represent graphically!).
In the particular case where $\omega = 0$ (constant field $\boldsymbol{E}$), we have seen that the velocity $\boldsymbol{w}_2$ describes the motion (axial and lateral) of the guiding centre[69]. In the presence of a harmonically varying $\boldsymbol{E}$ field, the drift motion does not occur: the term containing $\boldsymbol{E}_{0\perp} \wedge \boldsymbol{B}$ in (2.136) is not constant and when integrated, cannot yield a linear dependence on t, as is the case in (2.86) and (2.87) where $\boldsymbol{E}$ is constant. This drift is in fact "annihilated", because the $\boldsymbol{E}_{0\perp}$ component and, as a result the drift velocity, oscillate periodically. On the other hand, if ω tends to zero, the term $\boldsymbol{E}_{0\perp}$ in (2.136) disappears and the term in $\boldsymbol{E}_{0\parallel}$ reduces to $(q_\alpha/m_\alpha)E_{0\parallel}t$ because $\sin\omega t \to \omega t$, in complete agreement with the expression (2.113) for $\boldsymbol{w}_2$ obtained for constant $\boldsymbol{E}$.

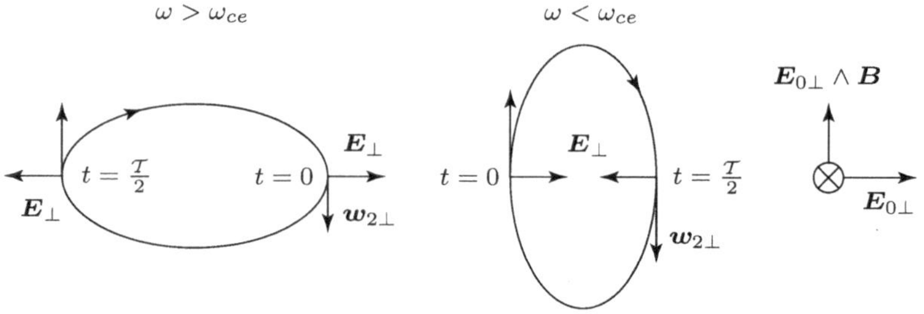

Fig. 2.9 Orientation of $\boldsymbol{w}_{2\perp}$ with respect to the reference frame $(\boldsymbol{E}_{0\perp} \wedge \boldsymbol{B}, \boldsymbol{E}_{0\perp}, \boldsymbol{B})$ for the case of a non-resonant electron cyclotron frequency. See Appendix IX for details.

- Representation of the $\boldsymbol{w}_{2\perp}$ component of the particular solution of (2.136)
Returning to the coordinate frame in Fig. 2.8, we find, in the plane perpendicular to $\boldsymbol{B}$, an ellipse whose major axis varies according to $\boldsymbol{E}_{0\perp}$ or $\boldsymbol{E}_{0\perp} \wedge \boldsymbol{B}$, depending on whether $\omega > \omega_c$ or $\omega < \omega_c$ (Fig. 2.9). To show this, we rewrite the two corresponding components of $\boldsymbol{w}_2$ in (2.136) in the form:

[69] In fact, for constant $\boldsymbol{E}$, $\boldsymbol{w}_2$ (2.114) includes the drift motion (perpendicular to $E_\perp$ and $\boldsymbol{B}$) and the uniformly accelerated motion along $\boldsymbol{B}$, which together describe the cyclotron motion around the guiding centre.

$$\frac{q_\alpha}{m_\alpha(\omega_c^2-\omega^2)}\left\{\mathrm{i}\omega \boldsymbol{E}_{0\perp}-\omega_c\frac{(\boldsymbol{E}_{0\perp}\wedge \boldsymbol{B})}{B}\right\}\mathrm{e}^{\mathrm{i}\omega t}\,, \tag{2.137}$$

noting that the term $\boldsymbol{E}_{0\perp}\wedge \boldsymbol{B}/B$ has the same modulus as $\boldsymbol{E}_{0\perp}$. We can then conclude that for $\omega > \omega_c$, the velocity $\boldsymbol{w}_{2\perp}$ is mainly[70] in phase quadrature (in advance for electrons because $q_\alpha = -e$) with the field $\boldsymbol{E}_\perp$ while for $\omega < \omega_c$, $\boldsymbol{w}_{2\perp}$ is principally in phase: this leads to the representation in Fig. 2.9.

2. Resonant case ($\omega = \omega_c$)
The particular solution can no longer have $\dot{b} = \dot{c} = 0$ because, following (2.135), the coefficients b and c would then tend to infinity. We can, however, retain the solution that corresponds to $\dot{a} = 0$, from (2.132):

$$a = \frac{q_\alpha}{\mathrm{i}\omega m_\alpha}\,. \tag{2.138}$$

To find the value of the coefficient c, we substitute the value of b given by (2.130) in (2.131) and obtain:

$$\dot{c}+\mathrm{i}\omega c = \frac{q_\alpha}{m_\alpha}\left[-\frac{q_\alpha B^2 c}{m_\alpha}+\frac{q_\alpha}{m_\alpha}-\dot{b}\right]\frac{1}{\mathrm{i}\omega} \tag{2.139}$$

and, to eliminate $\dot{b}$, we differentiate (2.131), and rearrange the result to write $\dot{b}$ in the form:

$$\dot{b} = (\ddot{c}+\mathrm{i}\omega\dot{c})\frac{m_\alpha}{q_\alpha}\,, \tag{2.140}$$

which, substituted into (2.139), gives:

$$\mathrm{i}\omega(\dot{c}+\mathrm{i}\omega c) = \frac{q_\alpha}{m_\alpha}\left[-\frac{q_\alpha B^2 c}{m_\alpha}+\frac{q_\alpha}{m_\alpha}-(\ddot{c}+\mathrm{i}\omega\dot{c})\frac{m_\alpha}{q_\alpha}\right]. \tag{2.141}$$

By regrouping the terms in (2.141), we obtain:

$$\ddot{c}+2\mathrm{i}\omega\dot{c} = \frac{q_\alpha^2}{m_\alpha^2}-\omega_c^2 c+\omega^2 c\,, \tag{2.142}$$

such that for resonance ($\omega = \omega_c$):

$$\ddot{c}+2\mathrm{i}\omega\dot{c} = \frac{q_\alpha^2}{m_\alpha^2}\,. \tag{2.143}$$

A valid particular solution for (2.143) is $\ddot{c} = 0$, which leads to $\dot{c} = q_\alpha^2/2\mathrm{i}\omega m_\alpha^2$, from which finally:

---

[70] The adverb mainly is used to emphasise that the weakest amplitude in (2.137) is not completely negligible, depending on the ratio ω/ω_c.

$$c = \frac{q_\alpha^2 t}{2\mathrm{i}\omega m_\alpha^2} \; . \tag{2.144}$$

The expression (2.144) for c substituted into (2.131) gives for b:

$$b = \frac{m_\alpha}{q_\alpha}\left[\frac{q_\alpha^2}{2\mathrm{i}\omega m_\alpha^2} + \frac{q_\alpha^2 t}{2m_\alpha^2}\right] = \frac{q_\alpha}{2m_\alpha\omega}[\omega t - \mathrm{i}] \; . \tag{2.145}$$

Ultimately, the particular solution can be written:

$$\boldsymbol{w}_2 = \left[-\frac{\mathrm{i}q_\alpha}{m_\alpha\omega}\boldsymbol{E}_{0\parallel} + \frac{q_\alpha}{2m_\alpha\omega}(\omega t - \mathrm{i})\boldsymbol{E}_{0\perp} - \frac{\mathrm{i}q_\alpha^2 t}{2\omega m_\alpha^2}(\boldsymbol{E}_{0\perp}\wedge\boldsymbol{B})\right]\mathrm{e}^{\mathrm{i}\omega t} \; . \tag{2.146}$$

Discussion of the solution

- the motion parallel to $\boldsymbol{B}$ is the same as that for non-resonance (and it is obviously independent of $\boldsymbol{B}$).
- the motion in the plane perpendicular to $\boldsymbol{B}$ is completely different. The terms involving $\boldsymbol{E}_{0\perp}$ and $(\boldsymbol{E}_{0\perp}\wedge\boldsymbol{B})$ increase indefinitely with time, and this motion tends towards an infinite amplitude: this is the phenomenon of gyro-magnetic resonance or *cyclotron resonance*.

The motion in the plane perpendicular to $\boldsymbol{B}$ can, in fact, be decomposed into 2 parts:

- a motion along $\boldsymbol{E}_{0\perp}$, purely oscillatory, with limited amplitude;
- a motion along $\boldsymbol{E}_{0\perp}$ and a motion along $\boldsymbol{E}_{0\perp}\wedge\boldsymbol{B}$, $\pi/2$ out of phase with respect to each other and with increasing amplitude: the result is a spiral of increasing radius r_B, as can readily be verified, but with constant rotation frequency (because $\omega_{c\alpha} = -q_\alpha B/m_\alpha$ is independent of the particle velocities).

Remarks:

1. If the $\boldsymbol{E}_\perp$ component of the electric field rotates in the opposite direction to the particle cyclotron motion, and at the same frequency, i.e. $\omega = -\omega_c$, there can be no resonance (see exercise 2.7).
2. It is obvious that the amplitude of the cyclotron motion cannot increase indefinitely because:

   - collisions can interrupt the electron (ion) motion, limiting the gain in energy,
   - in any case, the increase of the electron (ion) gyro-radius is limited by the dimensions of the vessel.

2.2.3 Magnetic field either (slightly) non uniform or (slightly) varying in time

The treatment of the equations of motion until now has been purely analytical, with no approximation. To deal with cases where particles are subjected to magnetic fields which are no longer uniform or no longer static, we must limit ourselves to $\boldsymbol{B}$ fields which are only slightly spatially non uniform, or slowly varying in time. This restriction allows us to consider a helical trajectory about an initial line of force, which imperceptibly modifies the orbit during a cyclotron rotation: in other words, a number of complete gyrations are required before the axial velocity of the guiding centre or its initial position in the direction perpendicular to $\boldsymbol{B}$ is significantly modified[71]. This slow variation of the guiding centre motion allows us to introduce the *guiding centre approximation*, also called the *adiabatic approximation* (in the sense that the particle energy varies very slowly), this concept being developed using a perturbation method.

Characteristics of the guiding centre approximation

- To zeroth order in this approximation, the trajectory in the plane perpendicular to $\boldsymbol{B}$ is circular. At a given point on the line of the field $\boldsymbol{B}$ defining the guiding centre axis, the field $\boldsymbol{B}$ is assumed to be uniform both in the plane containing the cyclotron trajectory and axially: this is the *local uniformity approximation*. At another point on this field line, the field $\boldsymbol{B}$ can be different, but it is once again assumed to be uniform transversely and axially. In the absence of an applied electric field, the motion in the direction of $\boldsymbol{B}$ is uniform. The complete trajectory is helical.
- To first order, the "inhomogeneties" (spatial or temporal) introduce variations in the guiding centre motion in both the direction of $\boldsymbol{B}$ (we are looking in particular for the axial velocity) and that perpendicular to $\boldsymbol{B}$ (of particular interest is the lateral position). These inhomogeneities occur locally, transversally as well as axially, as perturbations in the $\boldsymbol{B}$ field, assumed to be uniform to zeroth order.

The orbital magnetic moment associated with the cyclotron motion as a constant of motion defining the guiding centre approximation

The local uniformity approximation method that we have just introduced can be justified physically, and developed using a simple mathematical method,

---

[71] Recall that the *guiding centre axis* is defined instantaneously by the line of force of the field $\boldsymbol{B}$ around which the cyclotron motion occurs.

making use of the *orbital magnetic moment*, an invariant associated with the cyclotron component of the helical motion of the charged particles.

Definition: The magnetic moment μ of a current loop of intensity I bounding a surface S is equal to SI. In the context of our approximation, to order zero, we have $S = \pi r_B^2$ and $I = q_\alpha N_{\mathcal{T}_c}$, where $N_{\mathcal{T}_c}$ is the number of turns per second which are effected by the charged particle on its cyclotron orbit. Since $N_{\mathcal{T}_c} \equiv f_c = \omega_c/2\pi$, the modulus of μ is given by:

$$|\mu| = \pi r_B^2 \frac{q_\alpha|\omega_c|}{2\pi} \tag{2.147}$$

and:

$$|\mu| = \pi \left(\frac{w_\perp^2}{\omega_c^2}\right) \frac{q_\alpha|\omega_c|}{2\pi} = \frac{w_\perp^2 q_\alpha}{2|\omega_c|} = \frac{1}{2}\frac{m_\alpha w_\perp^2}{B} = \frac{\mathcal{E}_{\text{kin}\perp}}{B} , \tag{2.148}$$

where $\mathcal{E}_{\text{kin}\perp}$ is the kinetic energy of the particle in the plane perpendicular to $\boldsymbol{B}$. Since the magnetic field created by the cyclotron motion of the particle tends to oppose the applied field $\boldsymbol{B}$ (see p. 117, and the remark on diamagnetism), $\boldsymbol{\mu}$ is a vector anti-parallel to $\boldsymbol{B}$.

The magnetic moment is a constant of motion (to order zero)

Consider the case where the variation in $\boldsymbol{B}$ is simply a function of time[72]. From Maxwell's equations, this leads to the appearance of an electric field:

$$\boldsymbol{\nabla} \wedge \boldsymbol{E} = -\frac{\partial \boldsymbol{B}}{\partial t} , \tag{2.149}$$

which can accelerate (decelerate) the particles (without modifying the total kinetic energy). Thus, in the direction perpendicular to $\boldsymbol{B}$, we can write (2.10) such that:

$$\frac{\mathrm{d}}{\mathrm{d}t}\left(\frac{1}{2} m_\alpha w_\perp^2\right) \equiv q_\alpha \boldsymbol{E} \cdot \boldsymbol{w}_\perp , \tag{2.150}$$

where $\boldsymbol{E}$ is the field induced by the variation of $\boldsymbol{B}$ with time ($\partial \boldsymbol{B}/\partial t$). In this case, the variation in kinetic energy over a period $2\pi/\omega_c$ is given by:

$$\delta\left(\frac{1}{2} m_\alpha w_\perp^2\right) = \int_0^{2\pi/\omega_c} q_\alpha \boldsymbol{E} \cdot \frac{\mathrm{d}\boldsymbol{\ell}}{\mathrm{d}t}\, \mathrm{d}t , \tag{2.151}$$

[72] We could equally define the adiabaticity of $\boldsymbol{\mu}$ considering a spatial inhomogeneity: this is a question of reference frame. If $\boldsymbol{B}$ is inhomogeneous in the laboratory frame, in the frame of the particle, $\boldsymbol{B}$ varies with time.

where $\mathrm{d}\boldsymbol{\ell}/\mathrm{d}t$ is the instantaneous curvilinear velocity vector, tangent to the trajectory at each point. If we now suppose that the velocity parallel to $\boldsymbol{B}$ is not very large and that the guiding centre is only slightly displaced perpendicular to $\boldsymbol{B}$, notably because the field $\boldsymbol{B}$ does not greatly vary (the basic assumption for this calculation method), we can replace the integral over the helical trajectory by a line integral along the circular orbit (not perturbed by the inhomogeneity). Then, calling on Stokes theorem, which states that "the line integral of a vector along a closed contour is equal to the rotational flux of this vector traversing any surface bounded by this contour", we obtain:

$$\delta\left(\frac{1}{2}m_\alpha w_\perp^2\right) = \oint q_\alpha \boldsymbol{E}\cdot\mathrm{d}\boldsymbol{\ell} = q_\alpha \iint_S (\boldsymbol{\nabla}\wedge\boldsymbol{E})\cdot\mathrm{d}\boldsymbol{S} \tag{2.152}$$

and:

$$\delta\left(\frac{1}{2}m_\alpha w_\perp^2\right) = -q_\alpha \iint_S \frac{\partial \boldsymbol{B}}{\partial t}\cdot\mathrm{d}\boldsymbol{S} = \pm q_\alpha \frac{\partial B}{\partial t}\pi r_B^2\,, \tag{2.153}$$

since $\partial\boldsymbol{B}/\partial t$ is a flux perpendicular to the plane of the cyclotron motion (adiabatic approximation) and therefore to the surface element $\mathrm{d}\boldsymbol{S}$. The sign of the cosine of the angle between the direction of the normal to the elementary surface and the vector $\partial\boldsymbol{B}/\partial t$ determines the sign of the integrand.

The variation of the kinetic energy **per unit time** then takes the form ($\mathcal{T}_c$ being the period of gyration):

$$\frac{\mathrm{d}}{\mathrm{d}t}\left(\frac{1}{2}m_\alpha w_\perp^2\right) = \pm q_\alpha \frac{\partial B}{\partial t}\frac{\pi r_B^2}{\mathcal{T}_c} \equiv \frac{\partial B}{\partial t}\frac{\pi r_B^2 q_\alpha |\omega_c|}{2\pi} \tag{2.154}$$

and from (2.49), by definition, we find simply that:

$$\frac{\mathrm{d}}{\mathrm{d}t}\left(\frac{1}{2}m_\alpha w_\perp^2\right) = \mu\frac{\partial B}{\partial t}\,. \tag{2.155}$$

Also, following (2.148), it is equally possible to write:

$$\frac{\mathrm{d}}{\mathrm{d}t}\left(\frac{1}{2}m_\alpha w_\perp^2\right) = \frac{\mathrm{d}}{\mathrm{d}t}(\mu B) \equiv \frac{\partial \mu}{\partial t}B + \mu\frac{\partial B}{\partial t}\,, \tag{2.156}$$

such that, by comparing (2.155) and (2.156), it is obvious that $\partial\mu/\partial t = 0$, which shows that the moment μ is a constant in time.

This constant of motion is called the *first adiabatic invariant*. Remember that the magnetic moment is strictly constant only if $\boldsymbol{B}$ is completely uniform and $\boldsymbol{w}_{0\parallel} = 0$; it is constant, to a first approximation, if the change in $\boldsymbol{B}$ is slow, that is to say adiabatic.

Remark: In so far as one can consider the moment μ to be constant, the corresponding ratio $\mathcal{E}_{\text{kin}\perp}/B$ also remains constant and therefore whenever $\boldsymbol{B}$ varies, $\mathcal{E}_{\text{kin}\perp}$ should also vary in the same way and proportionally. Since the total kinetic energy is conserved (in the absence of an applied field $\boldsymbol{E}$), the values of $w_{\parallel}$ and $w_{\perp}$ will be modified in such a way that $w_{\perp}$ decreases and $w_{\parallel}$ increases and vice versa.

Static magnetic field, but non uniform in the direction parallel to $\boldsymbol{B}$ ($\boldsymbol{E} = \mathbf{0}$)

We will continue to suppose that there is no applied field $\boldsymbol{E}$ [73]. A priori, we are led to represent the magnetic field as being purely axial:

$$\boldsymbol{B} = B(z)\hat{\mathbf{e}}_z \ , \tag{2.157}$$

which will be proved to be incorrect: the gradient in $\boldsymbol{B}$ along z necessarily requires the existence of a component B_r. To see this, we assume a field $\boldsymbol{B}$ which is axially symmetric, as is shown, as an example, in Fig. 2.10.

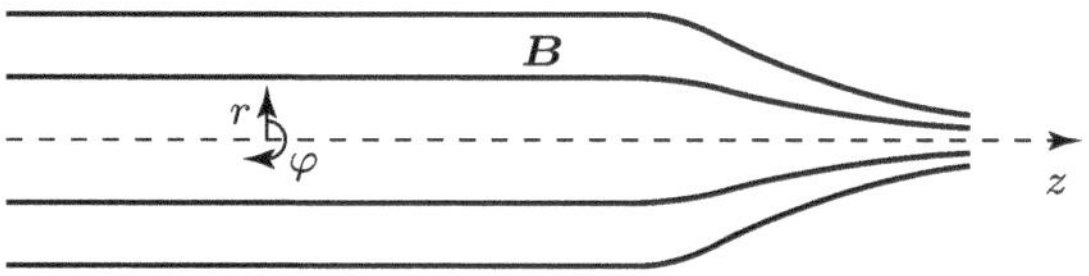

Fig. 2.10 Approximate representation of the lines of force in the case where the field $\boldsymbol{B}$ is axially symmetric and axially non uniform. The contraction of the lines of force indicates an increase in the intensity of B.

We need simply to consider the Maxwell equation:

$$\boldsymbol{\nabla} \cdot \boldsymbol{B} = 0 \tag{2.158}$$

(which signifies that the magnetic field lines should close) and to expand it in cylindrical coordinates as suggested by the symmetry of the problem. The units of local length are $e_1 = 1$, $e_2 = 1$ et $e_3 = r$, for the coordinates z, r, φ respectively[74]. We then obtain:

---

[73] Since $\boldsymbol{B}$ is constant in the laboratory frame, $\boldsymbol{\nabla} \wedge \boldsymbol{E} = -\partial \boldsymbol{B}/\partial t$ is zero and there is no electric field, which is not the case in the frame of the particle!

[74] Quite generally, the divergence of a vector can be expressed as (see Appendix XX):

$$\boldsymbol{\nabla} \cdot \boldsymbol{B} = \frac{1}{e_1 e_2 e_3} \left[\partial_1(e_2 e_3 B_1) + \partial_2(e_1 e_3 B_2) + \partial_3(e_1 e_2 B_3)\right] \ ,$$

where $\boldsymbol{\nabla} \cdot \boldsymbol{B}$ is in fact a pseudo-scalar (see Appendix VII).

$$\boldsymbol{\nabla}\cdot\boldsymbol{B}=\frac{\partial}{\partial z}B_z+\frac{1}{r}\frac{\partial}{\partial r}(rB_r)+\frac{1}{r}\frac{\partial}{\partial\varphi}B_\varphi=0. \tag{2.159}$$

By construction, Fig. 2.10 shows an axial symmetry of the $\boldsymbol{B}$ field, that is to say $\partial B_\varphi/\partial\varphi=0$, such that‘:

$$\frac{1}{r}\frac{\partial}{\partial r}(rB_r)=-\frac{\partial}{\partial z}B_z\ , \tag{2.160}$$

which implies that the inhomogeneity of the field $\boldsymbol{B}$ in its own direction cannot exist without the presence of a transverse component, which is B_r in the present case.

1. The expression for $\boldsymbol{B}$ in the neighbourhood of its axis of symmetry, for a weakly non-uniform field
   Assume that we know a priori the expression for $B_z(z)$ and its gradient $(\partial B_z/\partial z)_{r=0}$ at $r=0$. In addition, we can use Fig. 2.10 to see that B_z passes radially through a maximum on the axis of symmetry and that at $r=0$, $\partial B_z/\partial r=0$. Based on this, we assume that in the region close to the axis, $(\partial B/\partial z)_{r\simeq 0}\simeq$ constant, such that the B_z component is independent of r to second order. Under these conditions, by integration of (2.160) over r in the neighbourhood of the axis:

$$rB_r\approx-\int_0^r r'\left(\frac{\partial B_z}{\partial z}\right)_{r'=0}\mathrm{d}r'=-\frac{1}{2}r^2\left(\frac{\partial B_z}{\partial z}\right)_{r=0}. \tag{2.161}$$

   The complete and correct expression for the field $\boldsymbol{B}$ when it is non uniform in its own direction, and with the assumption of axial symmetry, is not (2.157), but rather:

$$\boldsymbol{B}=\hat{\mathbf{e}}_zB_z(z)-\hat{\mathbf{e}}_r\frac{r}{2}\left(\frac{\partial B_z}{\partial z}\right)_{r=0}. \tag{2.162}$$

   Note that the correction introduced by the B_r component becomes more important when the axial gradient is large, and as we move away from the axis. Under the basic assumptions of our calculation, this correction is of first order, and is in fact linear in r in the vicinity of the axis.
   Because the B_φ component is zero, and therefore $\boldsymbol{B}=\hat{\mathbf{e}}_rB_r+\hat{\mathbf{e}}_zB_z$, we can express $\boldsymbol{B}$ in Cartesian coordinates in the following way:

$$\boldsymbol{B}=-\frac{1}{2}x\left(\frac{\partial B_z}{\partial z}\right)_{x=y=0}\hat{\mathbf{e}}_x-\frac{1}{2}y\left(\frac{\partial B_z}{\partial z}\right)_{0,0}\hat{\mathbf{e}}_y+B_z\hat{\mathbf{e}}_z\ . \tag{2.163}$$

2. The trajectory of a charged particle in the calculated field $\boldsymbol{B}$
   We must solve:

$$m_\alpha\dot{\boldsymbol{w}}=q_\alpha(\boldsymbol{w}\wedge\boldsymbol{B})\ . \tag{2.164}$$

From our assumptions, the component of velocity perpendicular to $\boldsymbol{B}$ can be obtained, to first approximation, by supposing that the cyclotron motion takes place in a locally uniform field. All that remains is to calculate $\boldsymbol{w}_{\parallel}$.

3. The equation of motion in the direction of B_z
Since the field $\boldsymbol{B}$ is not completely uniform along z, the velocity of the guiding centre in the same direction does not remain constant.
To calculate this, set $\boldsymbol{w} = w_x\hat{\mathbf{e}}_x + w_y\hat{\mathbf{e}}_y + w_z\hat{\mathbf{e}}_z$, and consider (2.164):

$$m_\alpha \dot{\boldsymbol{w}}_{\parallel} = \hat{\mathbf{e}}_z q_\alpha [B_y w_x - B_x w_y] \,. \tag{2.165}$$

The variation of the guiding centre axial velocity described by (2.165) stems from the first order of our calculation method. It is therefore correct to use the zero order velocities in the plane perpendicular to the z axis to develop (2.165):

$$m_\alpha \dot{\boldsymbol{w}}_{\parallel} \approx \hat{\mathbf{e}}_z q_\alpha \left[-\frac{1}{2} y \left(\frac{\partial B_z}{\partial z}\right)_{0,0} w_x + \frac{1}{2} x \left(\frac{\partial B_z}{\partial z}\right)_{0,0} w_y \right] , \tag{2.166}$$

where the term $(\partial B_z/\partial z)_{0,0}$ is, by assumption, of first order while x, y, w_x and w_y are of order zero; the term on the RHS of (2.166) is thus of first order.

4. Solution of the equation of motion
The expressions for the position and velocity in the plane perpendicular to $\boldsymbol{B}$ are, from the assumptions of the approximation method, those already obtained in a uniform field $\boldsymbol{B}$ (Sect. 2.2.2, $\boldsymbol{E} = 0$). They can be written more succinctly:

$$w_x = A \sin(\omega_c t - \varphi) \,, \qquad x = -\frac{A}{\omega_c} \cos(\omega_c t - \varphi) \,, \tag{2.167}$$

$$w_y = A \cos(\omega_c t - \varphi) \,, \qquad y = \frac{A}{\omega_c} \sin(\omega_c t - \varphi) \,. \tag{2.168}$$

Setting $w_x(0) = 0$ and $w_y(0) = w_{y0}$, which leads to $\varphi = 0$ and $A = w_{y0}$, respectively, we obtain:

$$w_x = w_{y0} \sin \omega_c t \,, \qquad x = -\frac{w_{y0}}{\omega_c} \cos \omega_c t \,, \tag{2.169}$$

$$w_y = w_{y0} \cos \omega_c t \,, \qquad y = \frac{w_{y0}}{\omega_c} \sin \omega_c t \,. \tag{2.170}$$

This solution is such that, with $\omega_c > 0$ and $\boldsymbol{B}$ entering the page, the electrons are seen to rotate in the anti-clockwise direction; to check it, consider the values of x and y at $t = 0$ and $t = \pi/2\omega_c$. There is thus a change in convention, and to re-establish the motion in the true direction, we need to set $\omega_{ce} = -eB/m$ instead of $\omega_{ce} = eB/m$.

In order to come back to our initial conventions (Sect. 2.2.2, $\boldsymbol{E} = 0$), we must take $w_x = A\cos(\omega_c t - \varphi)$ and $w_y = A\sin(\omega_c t - \varphi)$ with $w_y(0) = 0$ and $w_x(0) = w_{x0}$ at $t = 0$. This yields:

$$w_x = w_{x0}\cos\omega_c t\ , \qquad x = \frac{w_{x0}}{\omega_c}\sin\omega_c t\ , \tag{2.171}$$

$$w_y = w_{x0}\sin\omega_c t\ , \qquad y = -\frac{w_{x0}}{\omega_c}\cos\omega_c t\ . \tag{2.172}$$

We can easily verify that (2.169) and (2.170) lead to $x^2+y^2 = (w_{y0}/\omega_c)^2 = r_B^2$. Thus, by substituting (2.169) and (2.170) into (2.166):

$$m_\alpha \dot{\boldsymbol{w}}_\parallel = \hat{\mathbf{e}}_z \frac{q_\alpha}{2}\left(\frac{\partial B_z}{\partial z}\right)_{0,0}\left[-\frac{w_{y0}^2}{\omega_c}\sin^2\omega_c t - \frac{w_{y0}^2}{\omega_c}\cos^2\omega_c t\right]\ , \tag{2.173}$$

$$\begin{aligned} m_\alpha \dot{w}_\parallel &= -\frac{q_\alpha}{2}\left(\frac{\partial B_z}{\partial z}\right)_{0,0}\left(\frac{w_{y0}^2}{\omega_c}\right) \\ &= -\frac{q_\alpha}{2}\left(\frac{\partial B_z}{\partial z}\right)_{0,0}\left(\frac{r_B^2\omega_c^2 m_\alpha}{q_\alpha B_\parallel}\right)\ , \end{aligned} \tag{2.174}$$

where, to allow for the sign of ω_c, we have chosen, exceptionally, $\omega_c = (q_\alpha/m_\alpha)B_\parallel$[75]. Simplifying:

$$\dot{w}_\parallel = -\frac{1}{2}\frac{r_B^2\omega_c^2}{B_\parallel}\left(\frac{\partial B_z}{\partial z}\right)_{0,0} \tag{2.175}$$

from which, finally, after integration:

$$\boldsymbol{w}_\parallel(t) = \boldsymbol{w}_\parallel(0) - \frac{\hat{\mathbf{e}}_z}{2} r_B^2\omega_c^2\frac{1}{B_\parallel}\left(\frac{\partial B_z}{\partial z}\right)_{0,0} t\ . \tag{2.176}$$

This is the velocity, entirely parallel to $\boldsymbol{B}$, of the guiding centre in the case where the gradient in $\boldsymbol{B}$ is principally in the direction of the field.
From (2.174), we can also derive an expression that will be useful later:

$$F_z = m_\alpha \dot{w}_\parallel = -\frac{1}{2} m_\alpha w_{\perp 0}^2 \frac{1}{B_\parallel}\left(\frac{\partial B_z}{\partial z}\right)_{0,0} \equiv -\mu\left(\frac{\partial B_z}{\partial z}\right)_{0,0}\ . \tag{2.177}$$

Appendix X suggests another demonstration of expression (2.177). In addition, Appendix XI uses (2.177) to show, with a different method than that developed from (2.149) to (2.156), that μ is a constant of motion in the guiding centre approximation.

---

[75] $B_\parallel$ represents the value of $B_z(z)$ along $z = 0$ (region of uniform $\boldsymbol{B}$).

5. Analysis of the motion $\boldsymbol{w}_{\|}$: retardation or acceleration of charged particles along an axial gradient in $\boldsymbol{B}$
   Following (2.176), the gradient $\partial B_z/\partial z$ subjects the charged particles to:

   - either a retardation if $\partial B_z/\partial z > 0$ because in this case $w_{\|}(t)$ slows down as a function of time, and finally changes the sign of the RHS of (2.176) with respect to the LHS. If B_0 is the value in the uniform $\boldsymbol{B}$ region and B_{max} the maximum value of B (Fig. 2.11), the region $B_0 < B < B_{\text{max}}$ where the particles are subject to reflection is called a *magnetic mirror*.
   - or an acceleration if $\partial B_z/\partial z < 0$, as is the case after reflection by a mirror, for example.

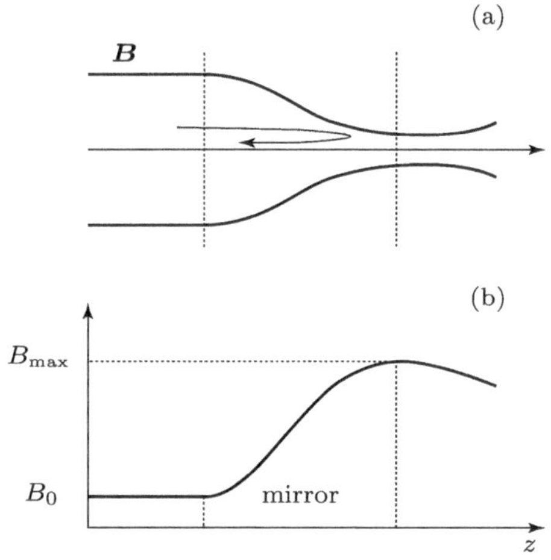

Fig. 2.11 a Magnetic field for the confinement of charged particles showing the mirror zone where they are reflected. **b** The value of B increases as the lines of force (figure a) get closer together.

The type of action exercised by $\partial B_z/\partial z$ on the velocity depends neither on the charge of the particle or its mass, because from (2.176):

$$\boldsymbol{w}_{\|} = \boldsymbol{w}_{\|0} - \frac{\hat{\mathbf{e}}_z}{2} \frac{w_{\perp}^2}{B_{\|}} \left(\frac{\partial B_z}{\partial z} \right)_{0,0} t \,, \tag{2.178}$$

and there is thus the possibility of confining all the charged particles. The efficiency of the confinement depends, finally, on the ratio $w_{\|}(0)/w_{\perp}(0)$: if it is too large, the mirror cannot play its role, as we will show below.

Remark: The role of the magnetic mirror (Fig. 2.11) can also be understood from the fact that, in the absence of an applied $\boldsymbol{E}$ field and within our guiding centre approximation, the total kinetic energy of the particle is conserved:

$$W_{\perp} + W_{\|} = \text{constant} \tag{2.179}$$

and only the ratio $W_\parallel / W_\perp$ can vary, thus:

$$\mathrm{d}W_\parallel = -\mathrm{d}W_\perp \,. \tag{2.180}$$

In addition, from (2.177), we can write the infinitesimal element of work effected by the particle on the field $\boldsymbol{B}$ in terms of the kinetic energy parallel to $\boldsymbol{B}$[76]:

$$F\mathrm{d}z \equiv \mathrm{d}W_\parallel = -\mu \mathrm{d}B_\parallel \,. \tag{2.181}$$

Inserting (2.180) in (2.181) and because $\mu = W_\perp / B$ (2.148), we have:

$$\mathrm{d}W_\perp = \mu \mathrm{d}B_\parallel = \frac{W_\perp}{B_\parallel} \mathrm{d}B_\parallel \tag{2.182}$$

or also:

$$\frac{\mathrm{d}W_\perp}{\mathrm{d}B_\parallel} = \frac{W_\perp}{B_\parallel} \equiv \mu \,. \tag{2.183}$$

This result signifies that if $B_\parallel$ increases, $W_\perp$ must increase, such that the ratio $W_\perp / B_\parallel$ remains constant. When the particle enters into the mirror zone, its energy $W_\parallel$ will decrease, if need be to zero, after which it will increase again after being "reflected". Since $W_\perp$ increases in the mirror neck (Fig. 2.11a), and because $r_B = W_\perp / B$, the question is whether the value of r_B could become so large that the particle reaches the wall. In fact, the value of r_B in the mirror zone decreases because the value of $\boldsymbol{B}$ increases more rapidly[77] than $W_\perp$.

6. The loss cone in the magnetic mirror of a linear machine
Consider the typical configuration of a linear magnetic confinement machine, with a mirror at each extremity such as that shown in Fig. 2.12. We are looking for the conditions such that the incident particles "cross the mirror", i.e. are lost.
Consider a particle traversing the uniform zone with a velocity $\boldsymbol{w}_0$ (making an angle α_0 with $\boldsymbol{B}$), as is shown in Fig. 2.13a. Let us now separate the velocity of this particle into parallel and perpendicular components with respect to the field $\boldsymbol{B}$. Thus in the region of uniform field (Fig. 2.13b), $\boldsymbol{w}_0 = \boldsymbol{w}_{0\parallel} + \boldsymbol{w}_{0\perp}$ (the subscript 0 indicates that the particle is in the homogeneous field region of the machine) where:

---

[76] In (2.177), we found $F = -\mu \partial B_z / \partial z$, from which $F\mathrm{d}z \simeq -\mu \mathrm{d}B_\parallel$.

[77] To verify this assertion, it is sufficient to differentiate $r_B^2 = w_\perp^2 / \omega_c^2$. Taking (2.182) into account, we find

$$\mathrm{d}r_B = -\frac{m_\alpha W_\perp}{r_B q_\alpha^2 B_\parallel^2} \left(\frac{\mathrm{d}B_\parallel}{B_\parallel} \right) .$$

In consequence, if the gradient of $B_\parallel$ is positive (mirror zone), the Larmor radius effectively decreases when $B_\parallel$ increases ($\mathrm{d}r_B$ is negative).

$$w_{0\parallel} = w_0 \cos\alpha_0 \; , \tag{2.184}$$

$$w_{0\perp} = w_0 \sin\alpha_0 \; , \tag{2.185}$$

with $w_0 = \sqrt{w_{0\parallel}^2 + w_{0\perp}^2}$.

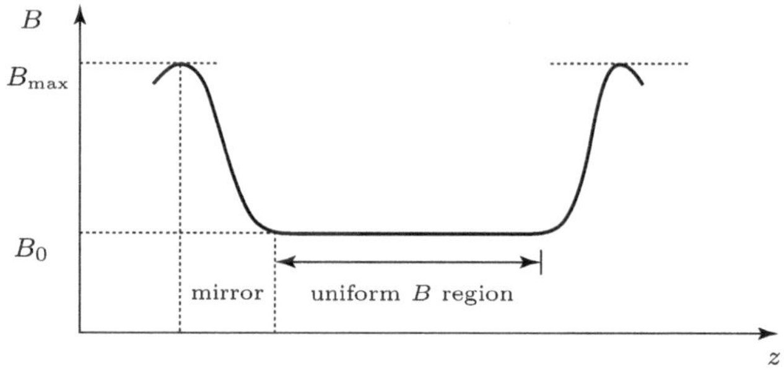

Fig. 2.12 Typical configuration of the confining magnetic field of a linear discharge in which each extremity is closed by a magnetic mirror (a configuration referred to as "minimum B").

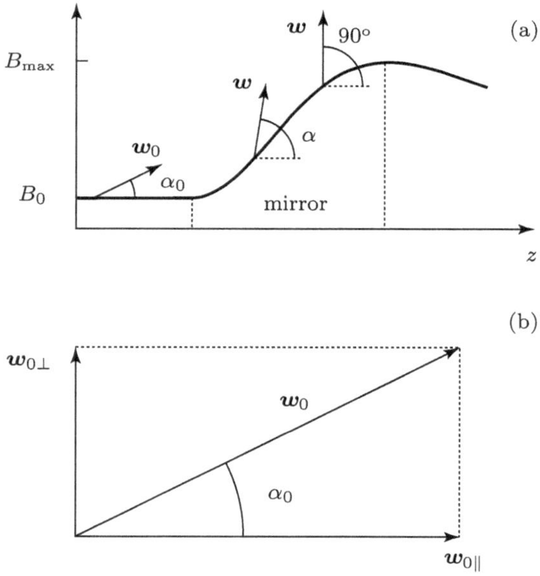

Fig. 2.13 **a** Orientation of the velocity vector with respect to the z axis in the zone of uniform B (α_0) and the mirror zone (α). **b** Decomposition of the velocity $\boldsymbol{w}_0$ along the z axis ($\boldsymbol{w}_{0\parallel}$) and perpendicular to it ($\boldsymbol{w}_{0\perp}$).

In the absence of an applied $\boldsymbol{E}$ field and with the assumption that $\boldsymbol{B}$ varies slowly along z, we know that $m_\alpha w_0^2/2$ = constant (where only the ratio $w_\perp/w_\parallel$ can vary) and that the magnetic moment μ is constant to first order.

We can thus establish a relation between the velocity in the uniform field region and that in the mirror, noting that from (2.168):

$$\mu = \frac{\frac{1}{2} m_\alpha w_0^2 \sin^2 \alpha_0}{B_0} = \frac{\frac{1}{2} m_\alpha w_0^2 \sin^2 \alpha}{B} , \tag{2.186}$$

where $w_\perp = w_0 \sin \alpha$ in the mirror region, such that:

$$\sin \alpha = \sin \alpha_0 \sqrt{\frac{B}{B_0}} . \tag{2.187}$$

There is a reflection of the particle in the case when $\alpha > \pi/2$. Equation (2.187) shows that if α_0 is sufficiently small (corresponding to a large enough "parallel" component of velocity of the particle in the homogeneous field region), the value of B/B_0 cannot be large enough to reach at least $\alpha = \pi/2$ ($\sin \alpha = 1$); it is certainly true for $\alpha_0 = 0$! When this is the case, the particle will cross the mirror and be neutralised on the end walls, and it will be "lost" for the plasma. We will denote α_{0m} as the minimum angle of α_0 for which there is still a reflection of particles at the maximum of the field $B_{\max}$. If we define the *mirror ratio* by:

$$\mathcal{R} \equiv B_{\max}/B_0 , \tag{2.188}$$

the value α_{0m} is obtained for $\sin \alpha = 1$ in (2.187):

$$1 = \sin \alpha_{0m} \sqrt{\frac{B_{\max}}{B_0}} \tag{2.189}$$

and finally:

$$\sin \alpha_{0m} = \frac{1}{\sqrt{\mathcal{R}}} . \tag{2.190}$$

The angle α_{0m} defines a cone, in the interior of which the particles leave the plasma at the end of the machine. Note that the efficiency of a magnetic mirror to reflect charged particles is independent of the modulus of the velocity of the particles (w_0) as well as their charge and mass.

7. The percentage of incident particles reflected by a magnetic mirror
We will consider the preceding magnetic field configuration (Fig. 2.13a) and suppose that the angular distribution of the particle velocities is isotropic in the uniform region: in other words, the density $n(\alpha_0)$ of particles with an angle α_0 is the same for each value of α_0. We wish to calculate $C_r = \Gamma_r/\Gamma_{\text{inc}}$, the fraction of incident flux Γ_{inc} reflected by the mirror, knowing that there is a reflection if $\alpha_0 > \alpha_{0m}$.
To do this, we must calculate the number of particles per second that are directed towards the mirror, Γ_{inc}, and then subtract the number of them for which $\alpha_0 < \alpha_{0m}$ (and which are not reflected), which will lead us to Γ_r. It is sufficient to establish such a balance for each value of α_0 on an elementary solid angle $d\Omega$, independently of the value of the azimuthal

angle φ owing to axial symmetry. We therefore consider the solid angle $\mathrm{d}\Omega(\alpha_0, \varphi)$ in which the particles enter (Fig. 2.14). To this angle $\mathrm{d}\Omega(\alpha_0, \varphi)$, there corresponds an elementary surface $\mathrm{d}\sigma(\alpha_0)$ directed along α_0, whose projection perpendicular to the mirror axis, $\mathrm{d}\sigma(\alpha_0)\cos\alpha_0$[78], constitutes the effective surface traversed by the incident flux in the direction of the mirror.

Fig. 2.14 Elementary surface $\mathrm{d}\sigma(\alpha_0)$ collecting particles of velocity $\boldsymbol{w}_0$ directed along α_0 and entering the solid angle $\mathrm{d}\Omega$.

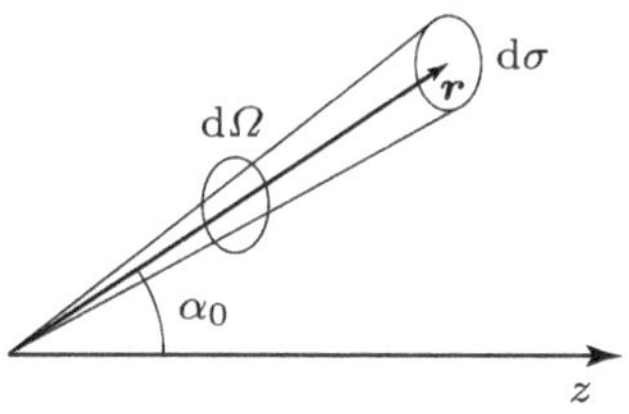

We then have:

$$\Gamma_{\mathrm{inc}}(\alpha_0) = n w_0 \, \mathrm{d}\sigma(\alpha_0) \cos\alpha_0 \ , \tag{2.191}$$

where, as we have seen, n does not depend on α_0.
By definition $\mathrm{d}\sigma = r^2 \mathrm{d}\Omega$ where $\mathrm{d}\Omega$ can be expressed in spherical coordinates $(r, c\alpha_0, \varphi)$ as:

$$\mathrm{d}\Omega = \sin\alpha_0 \, \mathrm{d}\alpha_0 \, \mathrm{d}\varphi \ . \tag{2.192}$$

The axial symmetry implies that the integration over φ yields 2π. We can then write:

$$C_r \equiv \frac{\Gamma_r}{\Gamma_{\mathrm{inc}}} = \frac{n w_0 r^2 (2\pi) \int_{\alpha_{0m}}^{\pi/2} \cos\alpha_0 \sin\alpha_0 \mathrm{d}\alpha_0}{n w_0 r^2 (2\pi) \int_0^{\pi/2} \cos\alpha_0 \sin\alpha_0 \mathrm{d}\alpha_0} \ . \tag{2.193}$$

The result is independent of the magnitude of the velocity, thus it is valid for all particle energy distributions.
Simplifying, and after a trigonometric transformation:

$$C_r \equiv \frac{\int_{\alpha_{0m}}^{\pi/2} \sin 2\alpha_0 \mathrm{d}\alpha_0}{\int_0^{\pi/2} \sin 2\alpha_0 \mathrm{d}\alpha_0} = \frac{-\cos 2\alpha_0 \Big|_{\alpha_{0m}}^{\pi/2}}{-\cos 2\alpha_0 \Big|_0^{\pi/2}} \ , \tag{2.194}$$

which gives:

$$\begin{aligned} C_r &= \frac{1 + \cos 2\alpha_{0m}}{2} = \frac{\left[1 + (1 - 2\sin^2\alpha_{0m})\right]}{2} = 1 - \sin^2\alpha_{0m} \\ &= 1 - \frac{B_0}{B_{\mathrm{max}}} \ , \end{aligned} \tag{2.195}$$

[78] Recall that a flux is by definition always evaluated normal to the surface that it traverses.

from which, finally:

$$C_r = 1 - \frac{1}{\mathcal{R}} . \tag{2.196}$$

Remarks:

1. The fraction of reflected particles becomes larger as $\mathcal{R}$ increases, that is to say as $B_{\max}$ becomes more important relative to B_0.
2. Satellite measurements have provided evidence for the existence of belts (layers) of high energy charged particles surrounding the earth. These particles, essentially electrons and protons from the solar wind, are trapped in the earth magnetic field and reflected at the poles: the lines of force of the $\boldsymbol{B}$ field become tighter at the poles, forming a mirror.
3. The particles confined in a system with a mirror at each extremity will oscillate between the two mirrors (see exercises 2.15 and 2.16).

Constant magnetic field, but non uniform in the direction perpendicular to *B*

The following section is divided into two parts: 1) the field lines are assumed rectilinear; 2) the curvature of the field lines is taken into account.

1. Field lines assumed rectilinear
We consider $\boldsymbol{B}$ entirely directed along the z axis and uniform along this axis. The gradient which affects it is, by hypothesis, perpendicular to it and uniquely directed along the y axis: $\boldsymbol{\nabla} B = (\partial B/\partial y)\hat{\mathbf{e}}_y$ and thus $\partial B/\partial x = 0$. In this case, we will assume that B increases slowly with y such that $\boldsymbol{B}$ can be expressed by:

$$\boldsymbol{B}(y) = \hat{\mathbf{e}}_z B_0(1 + \beta y) , \quad 0 < \beta \ll 1 . \tag{2.197}$$

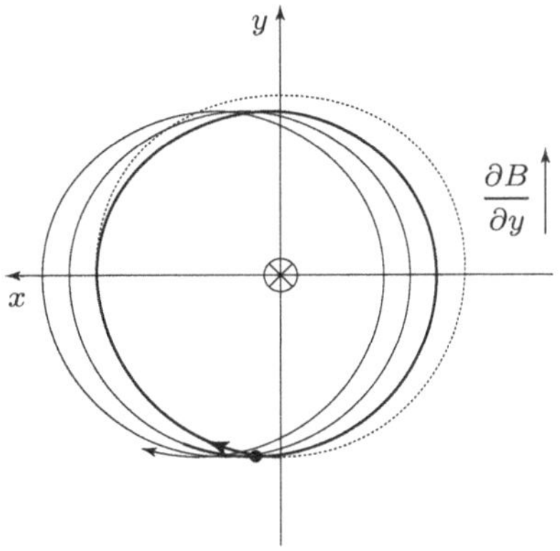

Fig. 2.15 (Trochoidal) trajectory of an electron in the plane perpendicular to the field $B\hat{\mathbf{e}}_z$, which is non uniform in the direction Oy (2.197). There is a magnetic field drift along x.

If the field were uniform ($\beta = 0$), we would have a cyclotron gyration of constant radius in the plane xOy (the dotted trajectory in Fig. 2.15). Due to the inhomogemeity of the field in this plane ($\beta \neq 0$), the trajectory is no longer an exact circle, and it does not close on itself, as is shown in Fig. 2.15[79]: this is due to the fact that the Larmor radius decreases, and with it, the radius of curvature of the trajectory, whenever the particle is moving towards increasing values of y (in the example considered), with the result that the guiding centre shifts. The guiding centre drifts, on average, along increasing x if the particle rotates in the clockwise direction as shown in Fig. 2.15; this average motion (over many periods) is called the *magnetic field drift.* It occurs in the direction perpendicular to $\boldsymbol{B}$ and to $\boldsymbol{\nabla}|\boldsymbol{B}|$, hence its alternative designation as the $\boldsymbol{\nabla}|\boldsymbol{B}|$ *drift.* We will now calculate the velocity $\boldsymbol{w}_{dm}$ of this magnetic field drift.

- The instantaneous velocity of the guiding centre
  To find $\mathrm{d}\boldsymbol{R}_g/\mathrm{d}t$, where $\boldsymbol{R}_g$ is the instantaneous position of the guiding centre (Fig. 2.16), we will call on our adiabatic approximation: the motion of the particle is determined to zeroth order by the cyclotron gyration in the field $\boldsymbol{B}$, when the effects of its non-uniformity are ignored: this motion is perturbed, to first order, by the magnetic field drift.

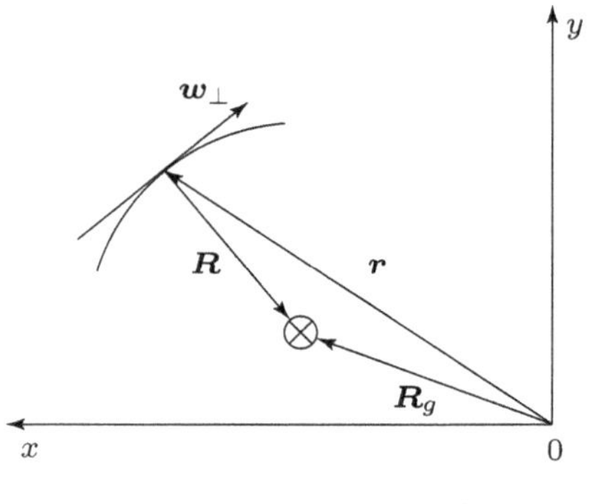

Fig. 2.16 The vector $\boldsymbol{R}$ describes the position of the guiding centre in the frame of the particle (in this case an electron), which itself is at position $\boldsymbol{r}$ in the laboratory frame. Note that $\boldsymbol{R}$ is perpendicular to the cyclotron trajectory at the point considered and that $\boldsymbol{R}_g = \boldsymbol{r} + \boldsymbol{R}$.

Zeroth order motion: calculation of $\boldsymbol{R}$

The radius of the gyration vector $\boldsymbol{R}$ gives the position of the guiding centre with respect to the particle, as illustrated in Fig. 2.16, and we will show that:

$$\boldsymbol{R} = \frac{m_\alpha}{q_\alpha B^2}(\boldsymbol{w} \wedge \boldsymbol{B}) \, . \tag{2.198}$$

To demonstrate this expression, we need only recall that, in general, for a particle situated at $\boldsymbol{r}'$ with respect to the axis about which it is

[79] According to our adiabatic approximation, many complete gyrations are required for this phenomenon to manifest itself.

rotating with a frequency ω, the tangential velocity obeys $\boldsymbol{w} = \boldsymbol{\omega} \wedge \boldsymbol{r}'$. In the present case, this translates into:

$$\boldsymbol{w} = +\frac{q_\alpha \boldsymbol{B}}{m_\alpha} \wedge \boldsymbol{R} . \tag{2.199}$$

Multiplying this expression vectorially on the right by $\boldsymbol{B}$, yields:

$$\boldsymbol{w} \wedge \boldsymbol{B} = \frac{q_\alpha}{m_\alpha}(\boldsymbol{B} \wedge \boldsymbol{R}) \wedge \boldsymbol{B} . \tag{2.200}$$

Recalling that the double vectorial product obeys the following rule:

$$\boldsymbol{P} \wedge (\boldsymbol{Q} \wedge \boldsymbol{T}) = \boldsymbol{Q}(\boldsymbol{T} \cdot \boldsymbol{P}) - \boldsymbol{T}(\boldsymbol{P} \cdot \boldsymbol{Q}) , \tag{2.201}$$

hence:

$$(\boldsymbol{Q} \wedge \boldsymbol{T}) \wedge \boldsymbol{P} = \boldsymbol{T}(\boldsymbol{P} \cdot \boldsymbol{Q}) - \boldsymbol{Q}(\boldsymbol{T} \cdot \boldsymbol{P}) , \tag{2.202}$$

we find that:

$$\boldsymbol{w} \wedge \boldsymbol{B} = \frac{q_\alpha}{m_\alpha} \left[\boldsymbol{R}(\boldsymbol{B} \cdot \boldsymbol{B}) - \boldsymbol{B}(\boldsymbol{R} \cdot \boldsymbol{B})\right] , \tag{2.203}$$

where the term $\boldsymbol{R} \cdot \boldsymbol{B}$ is zero, because to zeroth order the vector radius of gyration $\boldsymbol{R}$ is necessarily perpendicular to the guiding axis. Then (2.203) leads to (2.198)[80]:

$$\boldsymbol{R} = \frac{m_\alpha}{q_\alpha B^2}(\boldsymbol{w} \wedge \boldsymbol{B}) . \tag{2.198}$$

First order motion: calculation of $\boldsymbol{R}_g$

We have assumed till now that the lines of force are rectilinear. In order to avoid repeating the calculation when tackling point 2) where the lines are curvilinear, we set $\boldsymbol{B} = B\hat{\mathbf{e}}_B$ rather than $\boldsymbol{B} = B\hat{\mathbf{e}}_z$, where $\hat{\mathbf{e}}_B$ is the unit vector tangent to the field line, which takes into account the possible curvature of these lines.
Following Fig. 2.16:

$$\boldsymbol{R}_g = \boldsymbol{r} + \boldsymbol{R} , \tag{2.204}$$

where $\boldsymbol{R}$ describes the guiding centre motion in the frame of the particle, which is itself at position $\boldsymbol{r}$ in the laboratory frame. We can then rewrite $\boldsymbol{R}$ (2.198) in the form:

$$\boldsymbol{R} = \frac{m_\alpha}{q_\alpha B}(\boldsymbol{w} \wedge \hat{\mathbf{e}}_B) . \tag{2.205}$$

[80] In fact, it is sufficient to note that $|\boldsymbol{R}| = m_\alpha w_\perp / q_\alpha B$ ($|R| = r_B$) and that $\boldsymbol{R}$ is perpendicular to $\boldsymbol{w}$ and $\boldsymbol{B}$.

The derivative of (2.204), taking (2.205) into account, gives[81]:

$$\frac{\mathrm{d}\boldsymbol{R}_g}{\mathrm{d}t} = \frac{\mathrm{d}\boldsymbol{r}}{\mathrm{d}t} + \frac{\mathrm{d}\boldsymbol{R}}{\mathrm{d}t} = \boldsymbol{w} - \frac{m_\alpha}{q_\alpha B^2}\frac{\mathrm{d}B}{\mathrm{d}t}(\boldsymbol{w} \wedge \hat{\mathbf{e}}_B) + \frac{m_\alpha}{q_\alpha B}\left(\frac{\mathrm{d}\boldsymbol{w}}{\mathrm{d}t} \wedge \hat{\mathbf{e}}_B\right) + \frac{m_\alpha}{q_\alpha B}\left(\boldsymbol{w} \wedge \frac{\mathrm{d}\hat{\mathbf{e}}_B}{\mathrm{d}t}\right), \tag{2.206}$$

where $\mathrm{d}\hat{\mathbf{e}}_B/\mathrm{d}t = 0$ when we assume that $\boldsymbol{B}$ is directed parallel to the z axis (case 1). In the context of point 2) which follows, where we make the assumption of a weak field curvature, we will neglect the term comprising $\mathrm{d}\hat{\mathbf{e}}_B/\mathrm{d}t$ [82]. We can therefore take $\boldsymbol{B} = \hat{\mathbf{e}}_z B$ and (2.206) reduces to:

$$\frac{\mathrm{d}\boldsymbol{R}_g}{\mathrm{d}t} = \boldsymbol{w} - \frac{m_\alpha}{q_\alpha B^3}\frac{\mathrm{d}B}{\mathrm{d}t}(\boldsymbol{w} \wedge \boldsymbol{B}) + \frac{m_\alpha}{q_\alpha B^2}\left(\frac{\mathrm{d}\boldsymbol{w}}{\mathrm{d}t} \wedge \boldsymbol{B}\right). \tag{2.207}$$

In order to modify the third term on the RHS, we will take the equation of motion $m_\alpha \mathrm{d}\boldsymbol{w}/\mathrm{d}t = q_\alpha(\boldsymbol{w} \wedge \boldsymbol{B})$ and multiply it on the right vectorially by $\boldsymbol{B}$:

$$m_\alpha \frac{\mathrm{d}\boldsymbol{w}}{\mathrm{d}t} \wedge \boldsymbol{B} = q_\alpha(\boldsymbol{w} \wedge \boldsymbol{B}) \wedge \boldsymbol{B}. \tag{2.208}$$

Owing to the properties of the double vectorial product (2.199):

$$(\boldsymbol{w} \wedge \boldsymbol{B}) \wedge \boldsymbol{B} = \boldsymbol{B}(\boldsymbol{B} \cdot \boldsymbol{w}) - \boldsymbol{w}(\boldsymbol{B} \cdot \boldsymbol{B}) \equiv \boldsymbol{B}(B w_\parallel) - \boldsymbol{w}B^2, \tag{2.209}$$

we obtain:

$$m_\alpha \frac{\mathrm{d}\boldsymbol{w}}{\mathrm{d}t} \wedge \boldsymbol{B} = q_\alpha(\boldsymbol{w}_\parallel - \boldsymbol{w})B^2. \tag{2.210}$$

This expression can be substituted in the third term on the RHS of (2.207), which after some reorganisation, becomes:

$$\frac{\mathrm{d}\boldsymbol{R}_g}{\mathrm{d}t} = \boldsymbol{w} + \frac{1}{q_\alpha B^2}\left[q_\alpha(-\boldsymbol{w} + \boldsymbol{w}_\parallel)B^2\right] - \frac{m_\alpha}{q_\alpha B^3}\frac{\mathrm{d}B}{\mathrm{d}t}(\boldsymbol{w} \wedge \boldsymbol{B}). \tag{2.211}$$

After simplification, we find an expression for the (instantaneous) velocity of the guiding centre in the laboratory frame:

$$\frac{\mathrm{d}\boldsymbol{R}_g}{\mathrm{d}t} = \boldsymbol{w}_\parallel - \frac{m_\alpha}{q_\alpha B^3}\frac{\mathrm{d}B}{\mathrm{d}t}(\boldsymbol{w}_\perp \wedge \boldsymbol{B}), \tag{2.212}$$

[81] If $\boldsymbol{B}$ is spatially non-uniform in the laboratory frame, it varies with time in the frame of the particle, as already mentioned.

[82] If we include the term $\mathrm{d}\hat{\mathbf{e}}_B/\mathrm{d}t$, its contribution will be of second order in an expression which is of first order. In effect, $\mathrm{d}\hat{\mathbf{e}}_B/\mathrm{d}t = (\partial\hat{\mathbf{e}}_B/\partial y)\partial y/\partial t$ is a second order term.

where the first term represents the guiding centre velocity along the lines of force of the field $\boldsymbol{B}$ (zeroth order expression) and the second term is that in the direction perpendicular to $\boldsymbol{w}_\perp$ and $\boldsymbol{B}$ (first order expression), a motion that varies with time as a result of the cyclotron trajectory of the particle.

- The average velocity of the guiding centre in the plane perpendicular to $\boldsymbol{w}_\perp$ and to $\boldsymbol{B}$: the gradient of the magnetic field drift velocity
In order to calculate the temporal average of the second term on the RHS of (2.212), we rearrange it in the following form:

$$\begin{aligned} \frac{-m_\alpha}{q_\alpha B^3}\frac{\mathrm{d}B}{\mathrm{d}t}(\boldsymbol{w}\wedge\boldsymbol{B}) &= \frac{-m_\alpha}{q_\alpha B^3}\frac{\partial B}{\partial y}w_y\,(w_x\hat{\mathbf{e}}_x + w_y\hat{\mathbf{e}}_y + w_z\hat{\mathbf{e}}_z)\wedge B_z\hat{\mathbf{e}}_z \\ &= \frac{-m_\alpha}{q_\alpha B^3}\frac{\partial B}{\partial y}w_y(\hat{\mathbf{e}}_x w_y B_z - \hat{\mathbf{e}}_y w_x B_z)\,, \end{aligned} \tag{2.213}$$

where the RHS is now expressed in the laboratory frame. Since the temporal average of $w_x w_y$ is zero[83] and that:

$$\overline{w_y^2} = \frac{1}{2}w_\perp^2\,, \qquad (w_\perp^2 \equiv \overline{w_x^2} + \overline{w_y^2})\,, \tag{2.214}$$

the velocity associated with the average motion of the magnetic field drift finally reduces to:

$$\boldsymbol{w}_{dm} = \frac{-m_\alpha}{q_\alpha B^2}\frac{\partial B}{\partial y}\frac{w_\perp^2}{2}\hat{\mathbf{e}}_x\,. \tag{2.215}$$

This expression can be transformed, since in a direct trihedral coordinate system (contrary to a indirect one) $-\hat{\mathbf{e}}_x = \hat{\mathbf{e}}_z \wedge \hat{\mathbf{e}}_y$, into:

$$\boldsymbol{w}_{dm} = m_\alpha \frac{w_\perp^2}{2}\frac{1}{q_\alpha B^3}(\boldsymbol{B}\wedge\boldsymbol{\nabla}B) \tag{2.216}$$

or, equivalently:

$$\boldsymbol{w}_{dm} = \frac{\mu}{q_\alpha}\frac{(\boldsymbol{B}\wedge\boldsymbol{\nabla}B)}{B^2}\,, \tag{2.217}$$

which is the *magnetic field drift velocity* of a particle in the presence of a gradient in the field perpendicular to $\boldsymbol{B}$ and assumed to have no curvature[84].
The relation (2.217) could have been obtained directly from the general expression giving the drift velocity of charged particles subjected to a magnetic field in the presence of a given force, as is shown in Ap-

[83] Larmor motion: if w_x is proportional to $\sin\omega_c t$ and w_y is proportional to $\cos\omega_c t$, since these two functions are orthogonal, the time integral of $w_x w_y$ over a period is zero.

[84] In fact, this gradient is related to the lines of force because $\beta \simeq 1/\rho$ (XIII.18).

pendix XII. Appendix XIII allows us, in addition, to write (2.217) in the form:

$$\boldsymbol{w}_{dm} = -\frac{1}{\omega_c B}\frac{w_\perp^2}{2}\left(\frac{\boldsymbol{\rho}}{\rho^2} \wedge \boldsymbol{B}\right) , \tag{2.218}$$

where $\boldsymbol{\rho}$ is the radius of curvature (see Fig. XIII.1). It will be useful to compare this expression with that of the curvature drift velocity, which we will now calculate.

2. Accounting for the curvature of the field lines
The magnetic field drift, for which we have just established the equations of motion, cannot exist alone, because the lines of force of $\boldsymbol{B}$ which we have supposed to be rectilinear by setting:

$$\boldsymbol{B} = \hat{\mathbf{e}}_z B_0(1+\beta y) , \tag{2.219}$$

where $\beta \ll 1$, are not really so! In fact, although Maxwell's equation for the divergence of $\boldsymbol{B}$:

$$\boldsymbol{\nabla}\cdot\boldsymbol{B} \equiv \frac{\partial B_x}{\partial x} + \frac{\partial B_y}{\partial y} + \frac{\partial B_z}{\partial z} = 0 \tag{2.220}$$

can be trivially verified, in contrast to Maxwell's equation $\boldsymbol{\nabla}\wedge\boldsymbol{B} = 0$[85], since the curl operates on (2.219). It requires that the field $\boldsymbol{B}$ be in the form:

$$\boldsymbol{B} = \hat{\mathbf{e}}_y(\beta B_0 z) + \hat{\mathbf{e}}_z[B_0(1+\beta y)] , \tag{2.221}$$

as is shown in (XIII.7). Note that the component along y is of first order ($\beta \ll 1$). These field lines, which we find in a toroidal configuration, are schematically drawn in Fig. 2.17: the greater the distance from the origin of the frame, the more the contribution from B_y becomes important.

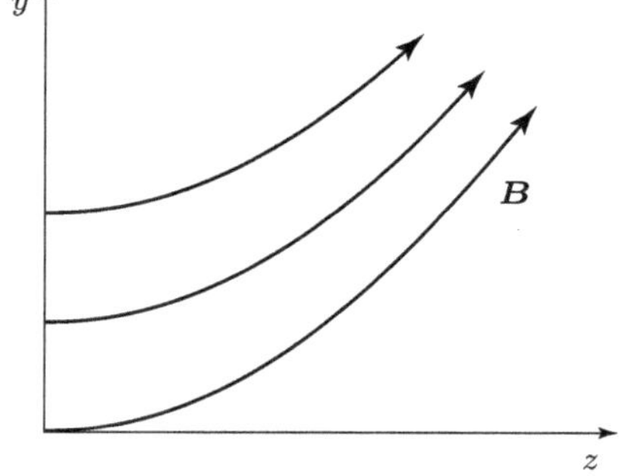

Fig. 2.17 Lines of force $y(z)$ in the presence of a gradient in B in the direction perpendicular to $\boldsymbol{B}$.

[85] In general $\boldsymbol{\nabla}\wedge\boldsymbol{H} = \boldsymbol{J} + \epsilon_0\partial\boldsymbol{E}/\partial t$: however, in the context of the individual trajectories model, we neglect the current associated with the charged particle motion, $\boldsymbol{J} = 0$, as well as the corresponding displacement current $\epsilon_0\partial\boldsymbol{E}/\partial t$. This last term is non zero in the case of a variable electric field $\boldsymbol{E}$ applied from outside.

The expression (2.221) represents a field $\boldsymbol{B}$ with lines of force characterised by a local curvature ρ. Recalling that the *radius of curvature* at a given point A of a curve is the distance between that point and the point of intersection of two normal vectors to the curve situated immediately on either side of A (Fig. XIII.1 of Appendix XIII), one can show that ρ is approximately $1/\beta$ (XIII.18).

- Magnetic curvature drift velocity
  This field curvature is associated with a particular drift motion, perpendicular to the lines of force (hence a velocity perpendicular to $\boldsymbol{B}$, as with the other drift velocities already defined). We will determine the average temporal velocity of this drift called the *magnetic curvature drift*, by resorting to the general expression for the drift of a charged particle subject to a given force $\boldsymbol{F}_D$ in a magnetic field $\boldsymbol{B}$ (Appendix XII).
  For this, we need to know the expression for the force exerted on the particle by the curvature of the lines: during its helical motion around the lines of force, the particle experiences a centrifugal force, for which the corresponding inertia term is of the classical form:

$$\boldsymbol{F}_{cd} = -\frac{m_\alpha w_\parallel^2}{\rho}\,\hat{\mathbf{e}}_y\,, \tag{2.222}$$

  where $\boldsymbol{w}_\parallel$ is the velocity parallel to the line of $\boldsymbol{B}$ at a given point and $\hat{\mathbf{e}}_y$ is the base vector linked to the coordinate system of the particle and directed towards the "instantaneous centre of rotation": we then have $\boldsymbol{\rho} = -\rho\,\hat{\mathbf{e}}_y$. Following (XII.2), the drift velocity in the curved magnetic field is then:

$$\boldsymbol{w}_{dc} = \frac{m_\alpha w_\parallel^2}{q_\alpha \rho^2 B^2}\,\boldsymbol{\rho}\wedge\boldsymbol{B} \tag{2.223}$$

  or equivalently:

$$\boldsymbol{w}_{dc} = -\frac{w_\parallel^2}{\omega_c}\,\frac{\boldsymbol{\rho}\wedge\boldsymbol{B}}{\rho^2 B}\,. \tag{2.224}$$

- Total drift velocity due to the presence of a gradient in B in the direction perpendicular to $\boldsymbol{B}$
  From (2.218) and (2.224), we obtain finally:

$$\boldsymbol{w}_{dm} + \boldsymbol{w}_{dc} = -\underset{\substack{\uparrow\\ \text{Charge}\\ \text{sign}}}{\frac{\boldsymbol{\rho}\wedge\boldsymbol{B}}{B\omega_c\rho^2}}\left[\underset{\substack{\uparrow\\ \text{Magnetic}\\ \text{field drift}}}{\frac{1}{2}w_\perp^2} + \underset{\substack{\uparrow\\ \text{Magnetic}\\ \text{curvature drift}}}{w_\parallel^2}\right]. \tag{2.225}$$

Remark: These two contributions to the drift motion are in the same direction, defined by the vector $-\boldsymbol{\rho}\wedge\boldsymbol{B}$, but whose sense depends on the sign of the charged particle. This drift can therefore create a separation of charges in the plasma, generating an electric field[86]. This effect causes a loss of charged particles in tokomaks, because they are directed to the walls, as we see in the following.

- The evolution of the drift motion tied to the magnetic field in a tokomak
Figure 2.18a is a schematic representation of the configuration of the coils producing the toroidal field in a tokomak: this magnetic field, imposed by the machine, is directed along the z axis. Because the coils are closer towards the central axis of the torus, than at the outer radius, the $\boldsymbol{B}$ field is inhomogeneous as a function of x[87] and, due to this, it acquires a curvature. We will examine the different effects to which the particles are subjected in the presence of this toroidal field by referring to Fig. 2.18b.

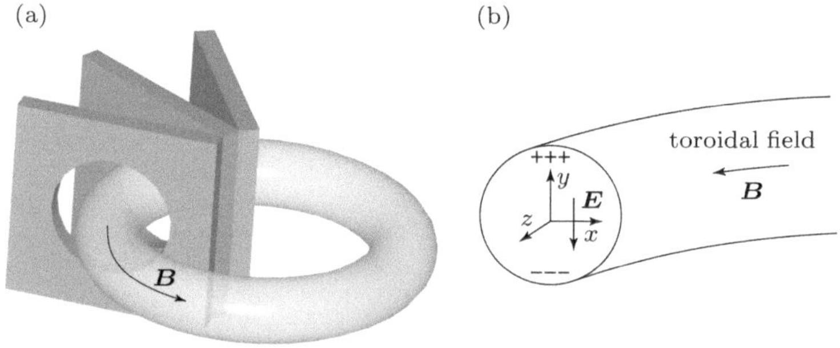

Fig. 2.18 **a** Schematic representation, showing the positioning of some magnetic field coils around a toroidal vessel: because they become closer towards the central axis of the torus, the field $\boldsymbol{B}$ increases along x. **b** Section of the toroidal vessel showing the charge separation created by the particle drifts in the toroidal magnetic field.

- The two magnetic drifts create a separation of charges along y (electrons downwards, ions upwards: the direction of the drift is that of q_α in (2.217)).
- This separation of charges creates a field $\boldsymbol{E}$ (perpendicular to z and x), directed downwards, opposed to the magnetic field drift current, giving rise to a weak current.
- The fields $\boldsymbol{E}$ and $\boldsymbol{B}$ then create an electric field drift, which is oriented according to $\boldsymbol{E} \wedge \boldsymbol{B}$ (crossed-field case, (2.222)). In the electric field drift, positive ions and negative electrons are displaced in the same direction: in the present case, they are directed towards the external wall of the torus (the vector product rule applied to the

[86] Except in structures where the magnetic configuration is closed on itself (a magnetic structure with rotational symmetry, for example).

[87] Note: direction designated by y in the previous discussion.

right trihedral coordinate system yields $-\hat{\mathbf{e}}_y \wedge \hat{\mathbf{e}}_z = -\hat{\mathbf{e}}_x$). These charged particles are then "lost" to the fusion plasma: they recombine at the wall losing their energy, additionally inflicting damage on the wall.

Remark: The plasma particles subjected to a simple toroidal magnetic field do not remain confined, as we have just seen, and a supplementary magnetic field, called a "poloidal" field, is used to reduce the drift effects. This second magnetic field provides a slight azimuthal variation to the toroidal magnetic field line configuration, forming a helix around the minor axis of the torus, in order to prevent particles from traveling to the walls.

Problems

2.1. Consider electrons and ions, of mass m_e and m_i respectively, subject to a constant electric field. Assuming that the average time τ between two collisions is the same for electrons and ions, show that the (average) kinetic energy acquired by an electron in the time τ is m_i/m_e times greater than that acquired by an ion in the same time.

Answer

We know that:

$$F \equiv |q|E = m_e \frac{\mathrm{d}w_e}{\mathrm{d}t} = m_i \frac{\mathrm{d}w_i}{\mathrm{d}t} \,. \tag{2.226}$$

The same force, with opposite sign, acts on the ion and the electron. From this, for a time τ between two collisions:

$$\int_0^\tau F \,\mathrm{d}t = m_e[w_e(\tau) - w_e(0)] \,, \tag{2.227}$$

$$= m_i[w_i(\tau) - w_i(0)] \,, \tag{2.228}$$

and, to simplify the problem, setting $w_e(0) = w_i(0) = 0$, we obtain:

$$w_e = \frac{m_i}{m_e} w_i \,. \tag{2.229}$$

The ratio of the kinetic energy of the electron to that of the ion is then:

$$\frac{\mathcal{E}_{ce}}{\mathcal{E}_{ci}} = \frac{m_e w_e^2/2}{m_i w_i^2/2} = \frac{m_e \left(\frac{m_i}{m_e}\right)^2 w_i^2}{m_i w_i^2} = \frac{m_i}{m_e} \ ! \tag{2.230}$$

2.2. Consider the motion of an electron ($\alpha = e$) or that of an ion ($\alpha = i$) in the plane perpendicular to a magnetic field $\boldsymbol{B} = B\hat{\mathbf{e}}_z$ present in the plasma, as shown in Fig. 2.4.

a) Calculate the direction and amplitude of the magnetic moment μ_α of an electron and an ion in cyclotron rotation about $\boldsymbol{B}$.
b) Calculate the macroscopic magnetisation $\boldsymbol{\mathcal{M}}_\alpha$ (magnetic moment per unit volume, expressed in A/m) induced by the electron ($\alpha = e$) and ion ($\alpha = i$) populations rotating in the field $\boldsymbol{B}$:

$$\mathcal{M}_{z\alpha} = \int \mu_{z\alpha} f_\alpha(w) \, \mathrm{d}\boldsymbol{w} \ . \tag{2.231}$$

Assume that the velocity distributions $f_\alpha(w)$ for the electrons and ions are Maxwellian, with temperatures T_e and T_i, respectively.
c) Calculate the total macroscopic magnetisation $\boldsymbol{\mathcal{M}}$ and discuss the respective contributions of the electron and ion populations to the plasma diamagnetism.
d) Assuming that $n_e = n_i = n$ and $T_e = T_i = T$, deduce the magnetic induction $\boldsymbol{B}$ resulting from both the applied magnetic induction $\boldsymbol{B}_0 = B_0\hat{\mathbf{e}}_z$ and the component $\mu_0\boldsymbol{\mathcal{M}}$ due to the diamagnetism of the plasma. Write the condition for which the intensity of the field $\boldsymbol{B}$ in the plasma becomes equal to half B_0.

Numerical application: $B_0 = 2 \times 10^{-2}$ tesla (200 gauss), $T_e = T_i = 35000$ K.

Answer

a) The cyclotron motion of an electron and an ion in the plane perpendicular to the magnetic induction $\boldsymbol{B}$ can be described by equation (2.68):

$$\boldsymbol{w}_\alpha = \boldsymbol{\omega}_{c\alpha} \wedge \boldsymbol{r}_{B\alpha} \tag{2.232}$$

where:

$$\boldsymbol{\omega}_{c\alpha} = -\frac{q_\alpha \boldsymbol{B}}{m_\alpha} \ . \tag{2.233}$$

For the electrons, $\boldsymbol{\omega}_{ce}$ and $\boldsymbol{B}$ are collinear with the same sign, i.e. the same direction, while for the ions, $\boldsymbol{\omega}_{ci}$ and $\boldsymbol{B}$ have opposite sign. On the other hand, the currents $\boldsymbol{i}_e$ and $\boldsymbol{i}_i$ associated with the cyclotron motion of electrons and ions are in the same direction, because of their opposite charge, as shown in the Fig. 2.4. They induce a magnetic field $\boldsymbol{B}'$ in the

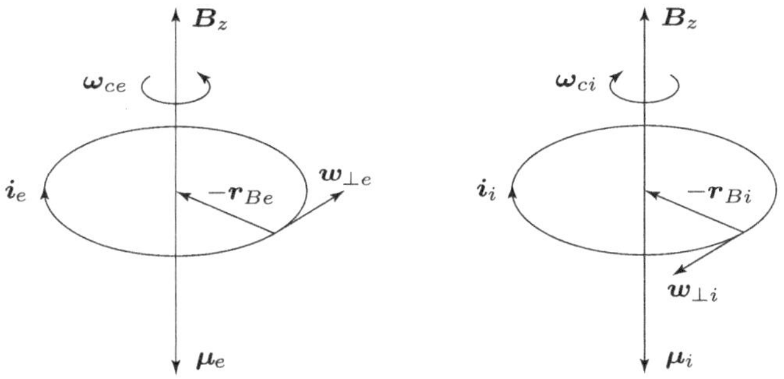

Fig. 2.19 Characteristic parameters of the cyclotron motion of an electron (left) and an ion (right) in a magnetic field.

direction opposite to $\boldsymbol{B}$ (Biot-Savart Law (2.70)); for the same reasons, $\boldsymbol{\mu}_i$ and $\boldsymbol{\mu}_e$ are anti-parallell to $\boldsymbol{B}$, as shown in Fig. 2.19.
The modulus of the magnetic moment μ is defined as the product of a current density i circulating in a closed loop with surface S. In the case of a cyclotron gyration:

$$\mu_{z\alpha} = \pi r_{B\alpha}^2 i_\alpha \,. \tag{2.234}$$

The current induced by the rotational motion is then:

$$i_\alpha = \frac{q_\alpha \omega_{c\alpha}}{2\pi} \,, \tag{2.235}$$

such that (2.148):

$$\mu_{z\alpha} = \frac{m_\alpha w_{\perp\alpha}^2}{2B} \,. \tag{2.236}$$

b) For an ensemble of particles of type α, the average value $\mathcal{M}_z$ of the macroscopic magnetisation (in the hydrodynamic sense) is given by (3.39):

$$\mathcal{M}_{z\alpha} = \int\limits_w \mu_z f_\alpha(w)\, \mathrm{d}\boldsymbol{w} \,, \tag{2.231}$$

i.e. (neglecting the subscript α):

$$\mathcal{M}_z = \frac{1}{B} \int\limits_w \frac{m w_\perp^2}{2} n \left(\frac{m}{2\pi k_B T}\right)^{\frac{3}{2}} \exp\left(-\frac{m w^2}{2 k_B T}\right) \mathrm{d}\boldsymbol{w} \,, \tag{2.237}$$

which can be expanded:

$$\mathcal{M}_z = \frac{nm}{2B}\left(\frac{m}{2\pi k_B T}\right)^{\frac{3}{2}} \left[\int_{-\infty}^{\infty} w_x^2 \exp\left(-\frac{mw_x^2}{2k_BT}\right) \mathrm{d}w_x \right.$$
$$\times \int_{-\infty}^{\infty} \exp\left(-\frac{mw_y^2}{2k_BT}\right) \mathrm{d}w_y \int_{-\infty}^{\infty} \exp\left(-\frac{mw_z^2}{2k_BT}\right) \mathrm{d}w_z$$
$$+ \int_{-\infty}^{\infty} \exp\left(-\frac{mw_x^2}{2k_BT}\right) \mathrm{d}w_x \int_{-\infty}^{\infty} w_y^2 \exp\left(-\frac{mw_y^2}{2k_BT}\right) \mathrm{d}w_y$$
$$\left. \times \int_{-\infty}^{\infty} \exp\left(-\frac{mw_z^2}{2k_BT}\right) \mathrm{d}w_z \right] \tag{2.238}$$

and this reduces to:

$$\mathcal{M}_{z\alpha} = \frac{n_\alpha k_B T_\alpha}{B} \,. \tag{2.239}$$

In vector form, taking account of the sense of the induced magnetisation $\boldsymbol{\mathcal{M}}_\alpha$ with respect to $\boldsymbol{B}$:

$$\boldsymbol{\mathcal{M}}_\alpha = -\frac{n_\alpha k_B T_\alpha}{B^2}\boldsymbol{B} \,. \tag{2.240}$$

c) The total magnetisation is the sum of those induced by the ions and the electrons, i.e.:

$$\boldsymbol{\mathcal{M}} = -\frac{\boldsymbol{B}}{B^2}\left(n_e k_B T_e + n_i k_B T_i\right) . \tag{2.241}$$

If $T_i \ll T_e$, $\mathcal{M}$ is induced solely by the electrons.
If $T_i = T_e$, the contribution from the ions is equal to that from the electrons.

d) For $n_e = n_i = n$ and $T_e = T_i = T$, we have from (2.241):

$$\boldsymbol{\mathcal{M}} = -\frac{2nk_BT}{B^2}\boldsymbol{B} = -\frac{2p}{B^2}\boldsymbol{B} \,, \tag{2.242}$$

where p is the pressure exerted by the charged particles, referred to as the (scalar) kinetic pressure (p. 205).
$\boldsymbol{B}$ is the magnetic induction in the plasma and results from the vector addition of the applied field $\boldsymbol{B}_0$ (which exists in the absence of the plasma) and the field created by the motion of charged particles, i.e. $\mu_0\boldsymbol{\mathcal{M}}$ (the magnetisation $\boldsymbol{\mathcal{M}}$ giving the magnetic field, the corresponding magnetic induction is thus obtained by multiplying the field $\boldsymbol{\mathcal{M}}$ by μ_0, the vacuum magnetic permeability). We then have:

$$\boldsymbol{B} = \boldsymbol{B}_0 + \mu_0 \boldsymbol{\mathcal{M}} = \boldsymbol{B}_0 - \frac{p}{B^2/2\mu_0}\boldsymbol{B} , \tag{2.243}$$

$$\boldsymbol{B}\left(1 + \frac{p}{B^2/2\mu_0}\right) = \boldsymbol{B}_0 . \tag{2.244}$$

The diamagnetism of the plasma causes a reduction of the applied magnetic field, due to the motion of the charged particles in the same field. The diamagnetism can be neglected ($\boldsymbol{B} \simeq \boldsymbol{B}_0$) if:

$$p \ll \frac{B^2}{2\mu_0} , \tag{2.245}$$

i.e. the kinetic pressure p remains much smaller than the *magnetic pressure* $B^2/2\mu_0$.

The magnetic induction in the plasma is half the applied magnetic field ($B = B_0/2$) if:

$$p = \frac{B^2}{2\mu_0} , \tag{2.246}$$

that is:

$$n = \frac{B^2}{2\mu_0 k_B T} . \tag{2.247}$$

Numerical application

From (2.247), we obtain:

$$n = \frac{10^{-4}}{2 \times 4\pi \times 10^{-7} \times 1.38 \times 10^{-23} \times 35000} = 8.24 \times 10^{19}\,\mathrm{m}^{-3} ,$$
$$= 8.24 \times 10^{13}\,\mathrm{cm}^{-3} . \tag{2.248}$$

Remark: Maxwell's Law for the curl of $\boldsymbol{H}$ applied to the field $\boldsymbol{\mathcal{M}}$ leads to $\boldsymbol{J}_{\mathcal{M}} = \boldsymbol{\nabla} \wedge \boldsymbol{\mathcal{M}}$. Since $\boldsymbol{\mathcal{M}}$ is uniform in the plasma:

$$\boldsymbol{\nabla} \wedge \boldsymbol{\mathcal{M}} = 0 , \tag{2.249}$$

thus $\boldsymbol{J}_{\mathcal{M}} = 0$: no macroscopic current is induced. On the other hand, in regions with gradients in $\boldsymbol{\mathcal{M}}$ (boundaries of enclosed plasmas), the diamagnetism of the plasma actually induces magnetisation currents ($\boldsymbol{J}_{\mathcal{M}} \neq 0$).

2.3. Consider a particle of charge q subject to uniform, static magnetic and electric fields which are perpendicular to each other. The particle velocity $\boldsymbol{w}$ can be decomposed according to $\boldsymbol{w} = \boldsymbol{w}_D + \boldsymbol{w}'$, where $\boldsymbol{w}_D$ is the electric field drift. From the equation of motion, show analytically that $\boldsymbol{w}'$ represents the motion the particle would have in the magnetic field alone.

Answer

The equation of motion can be written (2.5):

$$m\frac{\mathrm{d}\boldsymbol{w}}{\mathrm{d}t} = q\left[\boldsymbol{E}_\perp + \boldsymbol{w}\wedge\boldsymbol{B}\right] \tag{2.250}$$

in which we substitute $\boldsymbol{w}_D + \boldsymbol{w}'$ for $\boldsymbol{w}$ knowing that:

$$\boldsymbol{w}_D = \frac{\boldsymbol{E}_\perp\wedge\boldsymbol{B}}{B^2}\,, \tag{IX.2}$$

noting that $\mathrm{d}\boldsymbol{w}_D/\mathrm{d}t = 0$ because $\boldsymbol{E}$ and $\boldsymbol{B}$ are constant in time. We find:

$$m\frac{\mathrm{d}\boldsymbol{w}'}{\mathrm{d}t} = q\left[\boldsymbol{E}_\perp + \left(\frac{\boldsymbol{E}_\perp\wedge\boldsymbol{B}}{B^2}\right)\wedge\boldsymbol{B} + \boldsymbol{w}'\wedge\boldsymbol{B}\right]\,. \tag{2.251}$$

From the double vector product:

$$(\boldsymbol{Q}\wedge\boldsymbol{T})\wedge\boldsymbol{P} = \boldsymbol{T}(\boldsymbol{P}\cdot\boldsymbol{Q}) - \boldsymbol{Q}(\boldsymbol{T}\cdot\boldsymbol{P})\,, \tag{2.252}$$

this becomes:

$$\left(\frac{\boldsymbol{E}\wedge\boldsymbol{B}}{B^2}\right)\wedge\boldsymbol{B} = \frac{\boldsymbol{B}(\boldsymbol{B}\cdot\boldsymbol{E}) - \boldsymbol{E}(B^2)}{B^2} = -\boldsymbol{E}\frac{B^2}{B^2} \tag{2.253}$$

from the assumption that $\boldsymbol{E}$ is perpendicular to $\boldsymbol{B}$.
Finally:

$$\frac{\mathrm{d}\boldsymbol{w}'}{\mathrm{d}t} = q(\boldsymbol{w}'\wedge\boldsymbol{B})\,, \tag{2.254}$$

which is precisely the motion of a particle in a magnetic field alone.

2.4. Consider the motion of a charged particle in a uniform and static magnetic field $\boldsymbol{B}$, and a uniform electric field $\boldsymbol{E}$, directed perpendicular to $\boldsymbol{B}$ and slowly varying in time. The velocity of the particle is denoted by $\boldsymbol{w}$.

a) Show, by expressing the velocity $\boldsymbol{w}$ in terms of the following three velocity components:

$$\boldsymbol{w} = \boldsymbol{w}_D + \boldsymbol{w}' + \boldsymbol{w}_p\,, \tag{2.255}$$

where $\boldsymbol{w}_D$ is the electric field drift velocity and:

$$\boldsymbol{w}_p = \frac{m}{qB^2}\frac{\partial\boldsymbol{E}}{\partial t}\,, \tag{2.256}$$

that $\boldsymbol{w}'$ and $\boldsymbol{w}_p$ then obey the equation of motion:

$$m\dot{\boldsymbol{w}}' + m\dot{\boldsymbol{w}}_p = q(\boldsymbol{w}'\wedge\boldsymbol{B})\,, \tag{2.257}$$

where m is the mass of the particle and q its charge.

b) Consider a periodically varying field $E(t)$, with an angular frequency ω. Show that if the frequency of the field oscillation is small compared to ω_c for the cyclotron gyration, then the $\boldsymbol{w}'$ component describes the cyclotron motion of the particle in the $\boldsymbol{B}$ field alone.

c) Show that there is no net current (ions and electrons) associated with $\boldsymbol{w}_D$ while, on the other hand, $\boldsymbol{w}_p$ leads to a current which is referred to as the polarisation current:

$$\boldsymbol{J}_p = \frac{\rho_m}{B^2}\dot{\boldsymbol{E}} , \tag{2.258}$$

where $\rho_m = (m_e + m_i)n$ is the mass density of the electrons and ions (of masses m_e and m_i respectively) and n, the charged particle density. The velocity $\boldsymbol{w}_p$ is called the polarisation drift velocity.

d) By considering the total charge current (the conduction current and the displacement current $\partial \boldsymbol{D}/\partial t$), show that the relative permittivity of the medium with respect to vacuum is given by:

$$\epsilon_p = 1 + \frac{\rho_m}{\epsilon_0 B^2} . \tag{2.259}$$

To do this, recall from (2.45) that:

$$\boldsymbol{J}_T = \frac{\partial \boldsymbol{D}}{\partial t} + \boldsymbol{J}_c = \frac{\partial \boldsymbol{D}'}{\partial t} , \tag{2.260}$$

where $\boldsymbol{D}' = \epsilon_0 \epsilon_p \boldsymbol{E}$ is the displacement current in the dielectric description (see (2.43)).

Answer

a) Equation (2.257) signifies that the presence of the field $\boldsymbol{E}$ does not qualitatively modify the helical motion (described by $\boldsymbol{w}'$: to be shown in b) of the particle.
We know that the equation of motion is linked to the Lorentz force by:

$$m\dot{\boldsymbol{w}} = q[\boldsymbol{E} + \boldsymbol{w} \wedge \boldsymbol{B}] , \tag{2.261}$$

independent of the form of $\boldsymbol{E}$ and $\boldsymbol{B}$.
In the present case, we make the assumption that the total velocity can be expressed in terms of the three vectors given by (2.255). Developing (2.261) in terms of these different velocities, we obtain:

$$m\dot{\boldsymbol{w}}_D + m\dot{\boldsymbol{w}}' + m\dot{\boldsymbol{w}}_p = q[\boldsymbol{E} + (\boldsymbol{w}_D \wedge \boldsymbol{B}) + (\boldsymbol{w}' \wedge \boldsymbol{B}) + (\boldsymbol{w}_p \wedge \boldsymbol{B})] \tag{2.262}$$

and, replacing $\boldsymbol{w}_p$ on the RHS by equation (2.256) and $\boldsymbol{w}_D$ by its vector form:

$$\boldsymbol{w}_D = \frac{\boldsymbol{E} \wedge \boldsymbol{B}}{B^2} , \tag{2.263}$$

we obtain:

$$m\dot{\boldsymbol{w}}_D + m\dot{\boldsymbol{w}}' + m\dot{\boldsymbol{w}}_p = q\left\{\boldsymbol{E} + \underbrace{\left[\left(\frac{\boldsymbol{E}\wedge\boldsymbol{B}}{B^2}\right)\wedge\boldsymbol{B}\right]}_{\substack{\text{double vector product}\\ -\boldsymbol{E}}} + (\boldsymbol{w}'\wedge\boldsymbol{B}) + \underbrace{\frac{m}{q}\frac{\mathrm{d}}{\mathrm{d}t}\left(\frac{\boldsymbol{E}\wedge\boldsymbol{B}}{B^2}\right)}_{= m\dot{\boldsymbol{w}}_D}\right\} . \tag{2.264}$$

From the double vector product (see problem 2.3), this becomes:

$$m\dot{\boldsymbol{w}}_D + m\dot{\boldsymbol{w}}' + m\dot{\boldsymbol{w}}_p = q\left\{\boldsymbol{E} - \boldsymbol{E} + (\boldsymbol{w}'\wedge\boldsymbol{B})\right\} + m\dot{\boldsymbol{w}}_D \tag{2.265}$$

and finally:

$$m\dot{\boldsymbol{w}}' + m\dot{\boldsymbol{w}}_p = q(\boldsymbol{w}'\wedge\boldsymbol{B}) . \tag{2.257}$$

b) We need to show, starting from (2.256) and (2.257), that $\left|\dot{\boldsymbol{w}}_p/\dot{\boldsymbol{w}}'\right| \ll 1$. Setting $E = E_0 \mathrm{e}^{\mathrm{i}\omega t}$, we can write:

$$\left|\frac{m\dot{\boldsymbol{w}}_p}{m\dot{\boldsymbol{w}}' + m\dot{\boldsymbol{w}}_p}\right| = \left|\frac{m^2\omega^2}{qB^2}E_0\right|\left|\frac{1}{qw'B}\right| = \left|\frac{\omega^2}{\omega_c^2}\right|\left|\frac{E_0}{Bw'}\right| = \left|\frac{\omega^2}{\omega_c^2}\right|\left|\frac{w_D}{w'}\right| , \tag{2.266}$$

which indicates that we require not only $\omega/\omega_c \le 1$ but also, preferably, $w_D \lesssim w'$, which is an acceptable hypothesis.

c) Since the velocity $\boldsymbol{w}_D$ does not depend on the charge of the particles, the corresponding conduction current density is zero, because:

$$\boldsymbol{J}_D = \sum_\alpha n_\alpha q_\alpha \boldsymbol{w}_D = n\boldsymbol{w}_D(e - e) = 0 . \tag{2.267}$$

For the *conduction* current referred to as the *polarisation* current, we have:

$$\boldsymbol{J}_p \equiv \sum_\alpha n_\alpha q_\alpha \boldsymbol{w}_{p\alpha} = \frac{\dot{\boldsymbol{E}}}{B^2}\left[\frac{n_e(-e)m_e}{-e} + \frac{n_i(e)m_i}{e}\right]$$

$$= \frac{\dot{\boldsymbol{E}}}{B^2}\, n(m_i + m_e) , \tag{2.268}$$

$$= \frac{\dot{\boldsymbol{E}}}{B^2}\, \rho_m . \tag{2.269}$$

d) The conduction current $\boldsymbol{J}_c$ reduces to $\boldsymbol{J}_p$, as we have just shown. In addition, because quite generally:

$$\boldsymbol{J}_T \equiv \frac{\partial \boldsymbol{D}}{\partial t} + \boldsymbol{J}_c = \frac{\partial \boldsymbol{D}'}{\partial t} , \tag{2.260}$$

in the present case this reduces to:

$$\boldsymbol{J}_T \equiv \epsilon_0 \dot{\boldsymbol{E}} + \boldsymbol{J}_p = \epsilon_p \epsilon_0 \dot{\boldsymbol{E}} \tag{2.270}$$

and from (2.269):

$$\boldsymbol{J}_T \equiv \epsilon_0 \dot{\boldsymbol{E}} + \frac{\dot{\boldsymbol{E}}}{B^2}\rho_m = \epsilon_p \epsilon_0 \dot{\boldsymbol{E}} , \tag{2.271}$$

such that we obtain, as required:

$$\epsilon_p = 1 + \frac{\rho_m}{B^2 \epsilon_0} . \tag{2.259}$$

2.5. Consider a plasma subject to a high frequency electric field $\boldsymbol{E}_0 \mathrm{e}^{\mathrm{i}\omega t}$, directed arbitrarily with respect to a static magnetic field of intensity B, both fields being spatially uniform. In the framework of the "individual motion of charged particles" description, calculate the conductivity and permittivity tensors for electrons whose motion is associated with the particular solution of the non-resonant equation of motion. Assume $\boldsymbol{B}$ is directed along the Oz axis and express $\boldsymbol{E}_\perp$ in terms of the Cartesian coordinates x and y. The multiplying factor for the tensor $\underline{\boldsymbol{\sigma}}$ should be such that it reduces to a unitary matrix for $B = 0$.

Answer

In order to obtain $\boldsymbol{w}_2$, the particular solution to this problem (see Sect. 2.2.2, p. 123), we used the frame of Fig. 2.10, which led us to the expression:

$$\boldsymbol{w}_2 = \left[a\boldsymbol{E}_{0\parallel} + b\boldsymbol{E}_{0\perp} + c(\boldsymbol{E}_{0\perp} \wedge \boldsymbol{B})\right] \mathrm{e}^{\mathrm{i}\omega t} . \tag{2.127}$$

To transpose this result into Cartesian coordinates (x, y, z), as posed by the question, we write:

$$\boldsymbol{w}_2 = \hat{\mathbf{e}}_z(aE_{0\parallel}) + \hat{\mathbf{e}}_x(bE_{0x}) + \hat{\mathbf{e}}_y(bE_{0y}) + c(\hat{\mathbf{e}}_x E_{0x} + \hat{\mathbf{e}}_y E_{0y}) \wedge \hat{\mathbf{e}}_z B , \tag{2.272}$$

where we have set $\boldsymbol{E}_{0\perp} = \hat{\mathbf{e}}_x E_{0x} + \hat{\mathbf{e}}_y E_{0y}$ and have canceled, for simplicity, the dependence on $\mathrm{e}^{\mathrm{i}\omega t}$. After regrouping the terms along the three base vectors, we find:

$$\boldsymbol{w}_2 = \hat{\mathbf{e}}_x[bE_{0x} + cE_{0y}B] + \hat{\mathbf{e}}_y[bE_{0y} - cE_{0x}B] + \hat{\mathbf{e}}_z[aE_z] . \tag{2.273}$$

Including the coefficients a, b and c for the non-resonance solution from Sect. 2.2.2 (Eq. (2.135)), we obtain:

$$
\boldsymbol{w}_2 = \hat{\mathbf{e}}_x \left[\frac{\mathrm{i}q}{m_\alpha} \frac{\omega}{\omega_c^2 - \omega^2} E_{0x} + \frac{q^2}{m_\alpha^2} \frac{B}{\omega_c^2 - \omega^2} E_{0y} \right]
+ \hat{\mathbf{e}}_y \left[\frac{\mathrm{i}q}{m_\alpha} \frac{\omega}{\omega_c^2 - \omega^2} E_{0y} - \frac{q^2}{m_\alpha^2} \frac{B}{\omega_c^2 - \omega^2} E_{0x} \right] + \hat{\mathbf{e}}_z \left[-\frac{\mathrm{i}q}{\omega m_\alpha} E_z \right] . \quad (2.274)
$$

To develop the electrical conductivity tensor, recall that $J^i = \sigma^{ij} E_j$ (2.121). By definition, the current density $J^i = nqw^i$ and we have, after introducing the factor $-\mathrm{i}nq^2/m_\alpha \omega$:

$$
\boldsymbol{J} = -\frac{\mathrm{i}nq^2}{m_\alpha \omega} \left\{ \hat{\mathbf{e}}_x \left[\left(-\frac{\omega^2}{\omega_c^2 - \omega^2} E_{0x} \right) + \frac{\mathrm{i}q}{m_\alpha} \frac{\omega B}{\omega_c^2 - \omega^2} E_{0y} \right] + \hat{\mathbf{e}}_y \left[-\frac{\omega^2}{\omega_c^2 - \omega^2} E_{0y} - \frac{\mathrm{i}qB\omega}{m_\alpha(\omega_c^2 - \omega^2)} E_{0x} \right] + \hat{\mathbf{e}}_z E_z \right\} . \quad (2.275)
$$

There are two components of $\underline{\boldsymbol{\sigma}}$ along $\hat{\mathbf{e}}_x$, and recalling that $qB/m_\alpha = -\omega_c$, we find:

$$
\sigma_{xx} = \frac{\omega^2}{\omega^2 - \omega_c^2} , \qquad \sigma_{xy} = \frac{\mathrm{i}\omega_c \omega}{\omega^2 - \omega_c^2} , \quad (2.276)
$$

then, along $\hat{\mathbf{e}}_y$:

$$
\sigma_{yy} = \frac{\omega^2}{\omega^2 - \omega_c^2} , \qquad \sigma_{yx} = -\frac{\mathrm{i}\omega_c \omega}{\omega^2 - \omega_c^2} , \quad (2.277)
$$

and, finally, along $\hat{\mathbf{e}}_z$:

$$
\sigma_{zz} = 1 . \quad (2.278)
$$

The tensor $\underline{\boldsymbol{\sigma}}$, represented as a 3×3 matrix, has the value:

$$
\underline{\boldsymbol{\sigma}} = -\frac{\mathrm{i}nq^2}{m_\alpha \omega} \begin{pmatrix} \dfrac{\omega^2}{\omega^2 - \omega_c^2} & \dfrac{\mathrm{i}\omega_c \omega}{\omega^2 - \omega_c^2} & 0 \\ -\dfrac{\mathrm{i}\omega_c \omega}{\omega^2 - \omega_c^2} & \dfrac{\omega^2}{\omega^2 - \omega_c^2} & 0 \\ 0 & 0 & 1 \end{pmatrix} \quad (2.279)
$$

and we can verify that if $B = 0$ ($\omega_c = 0$), the matrix becomes unitary: the plasma is no longer anisotropic.

For the relative permittivity tensor $\underline{\boldsymbol{\epsilon}}_p$, we have for $E_0 \mathrm{e}^{\mathrm{i}\omega t}$:

$$
\underline{\boldsymbol{\epsilon}}_p = \underline{\boldsymbol{I}} + \frac{\underline{\boldsymbol{\sigma}}}{\mathrm{i}\omega \epsilon_0} , \quad (2.125)
$$

such that:

$$\underline{\epsilon}_p = \begin{pmatrix} 1 - \dfrac{nq^2}{m\epsilon_0\omega^2}\dfrac{\omega^2}{\omega^2-\omega_c^2} & -\dfrac{nq^2}{m\epsilon_0\omega^2}\dfrac{\mathrm{i}\omega_c\omega}{\omega^2-\omega_c^2} & 0 \\ \dfrac{nq^2}{m\epsilon_0\omega^2}\dfrac{\mathrm{i}\omega_c\omega}{\omega^2-\omega_c^2} & 1 - \dfrac{nq^2}{m\epsilon_0\omega^2}\dfrac{\omega^2}{\omega^2-\omega_c^2} & 0 \\ 0 & 0 & 1 - \dfrac{nq^2}{m\epsilon_0\omega^2} \end{pmatrix}, \quad (2.280)$$

from which, finally:

$$\underline{\epsilon}_p = \begin{pmatrix} 1 - \dfrac{\omega_{pe}^2}{\omega^2-\omega_c^2} & -\dfrac{\mathrm{i}\omega_{pe}^2}{\omega^2-\omega_c^2}\dfrac{\omega_c}{\omega} & 0 \\ \dfrac{\mathrm{i}\omega_{pe}^2}{\omega^2-\omega_c^2}\dfrac{\omega_c}{\omega} & 1 - \dfrac{\omega_{pe}^2}{\omega^2-\omega_c^2} & 0 \\ 0 & 0 & 1 - \dfrac{\omega_{pe}^2}{\omega^2} \end{pmatrix}. \quad (2.281)$$

2.6. Consider a uniform, alternating electric field of the form $\boldsymbol{E}_0\mathrm{e}^{-\mathrm{i}\omega t}$, together with a uniform, static magnetic field $\boldsymbol{B}$ along the z axis (entering into the page). We wish to study the phenomenon of cyclotron resonance using the coordinate frame rotating in the plane perpendicular to $\boldsymbol{B}$. Expressing the field $\boldsymbol{E}$ in the laboratory Cartesian frame in the form:

$$\boldsymbol{E} = \hat{\mathbf{e}}_x E_x + \hat{\mathbf{e}}_y E_y + \hat{\mathbf{e}}_z E_z \ , \quad (2.282)$$

the same field in the rotating coordinates frame is written:

$$\begin{aligned} \boldsymbol{E} &= \hat{\mathbf{e}}_+ E_+ + \hat{\mathbf{e}}_- E_- + \hat{\mathbf{e}}_z E_z \\ &= \frac{(\hat{\mathbf{e}}_x + \mathrm{i}\hat{\mathbf{e}}_y)}{\sqrt{2}} E_+ + \frac{(\hat{\mathbf{e}}_x - \mathrm{i}\hat{\mathbf{e}}_y)}{\sqrt{2}} E_- + \hat{\mathbf{e}}_z E_z \ . \end{aligned} \quad (2.283)$$

a) Express the components E_+ and E_- in terms of the components E_x and E_y. Determine which of the two unit vectors, $\hat{\mathbf{e}}_+$ or $\hat{\mathbf{e}}_-$, rotate in the same direction as the electrons during their cyclotron motion around $\boldsymbol{B}$.
b) The conductivity tensor, expressed in the laboratory frame, for $\omega \neq \omega_c$, and $E_0\mathrm{e}^{-\mathrm{i}\omega t}$, has the following elements (exercise 2.5):

$$\underline{\sigma} = \sigma_0 \begin{pmatrix} \dfrac{\omega^2}{\omega^2-\omega_c^2} & \dfrac{-\mathrm{i}\omega\omega_c}{\omega^2-\omega_c^2} & 0 \\ \dfrac{\mathrm{i}\omega\omega_c}{\omega^2-\omega_c^2} & \dfrac{\omega^2}{\omega^2-\omega_c^2} & 0 \\ 0 & 0 & 1 \end{pmatrix}. \quad (2.284)$$

Show, by calculating the corresponding terms σ_+ and σ_-, that this tensor is diagonalised in the rotating frame.

c) Show that electron cyclotron resonance leads to an increase in velocity of the particles as a function of time, according to the relation:

$$\boldsymbol{w}_2 = \frac{q}{m}\boldsymbol{E}_+ t \, . \tag{2.285}$$

In other words, the electrons in their own frame "see" a continuous electric field ($\omega = 0$) which accelerates them continuously between two collisions. For this, develop the relation (2.146) in the laboratory frame.

Answer

a) We can develop (2.283) by regrouping the terms along the unit vectors in the laboratory frame:

$$\boldsymbol{E} = \frac{1}{\sqrt{2}}\hat{\mathbf{e}}_x[E_+ + E_-] + \frac{\mathrm{i}}{\sqrt{2}}\hat{\mathbf{e}}_y[E_+ - E_-] + \hat{\mathbf{e}}_z E_z \, , \tag{2.286}$$

which must be equal to the same vector $\boldsymbol{E}$ expressed in the laboratory frame:

$$\boldsymbol{E} = \hat{\mathbf{e}}_x E_x + \hat{\mathbf{e}}_y E_y + \hat{\mathbf{e}}_z E_z \, , \tag{2.287}$$

from which:

$$E_x = \frac{1}{\sqrt{2}}[E_+ + E_-] \, , \tag{2.288}$$

$$E_y = \frac{\mathrm{i}}{\sqrt{2}}[E_+ - E_-] \, . \tag{2.289}$$

By combining (2.288) and (2.289), we obtain the components of the field $\boldsymbol{E}$ in the rotating frame:

$$E_+ = \frac{E_x - \mathrm{i}E_y}{\sqrt{2}} \tag{2.290}$$

and, similarly:

$$E_- = \frac{E_x + \mathrm{i}E_y}{\sqrt{2}} \, . \tag{2.291}$$

The components E_+ and E_-, in terms of the components E_x and E_y, thus correspond to the concept of a rotating field: the superposition of two oscillating fields with the same frequency, perpendicular to each other and in phase quadrature.

Since the rotating field $\boldsymbol{E}_+$ can be written:

$$\boldsymbol{E}_+ = \frac{\hat{\mathbf{e}}_x + \mathrm{i}\hat{\mathbf{e}}_y}{\sqrt{2}} E_+ \mathrm{e}^{-\mathrm{i}\omega t} \, , \tag{2.292}$$

taking the real part, we obtain:

$$\boldsymbol{E}_+ = \frac{E_+}{\sqrt{2}} \left(\hat{\mathbf{e}}_x \cos\omega t + \hat{\mathbf{e}}_y \sin\omega t \right) \ , \tag{2.293}$$

from which it is easy to verify, in the xOy plane, that the vector $\hat{\mathbf{e}}_+$ (and thus the field $\boldsymbol{E}_+$) rotates in the clockwise direction, thus, according to our convention (field $\boldsymbol{B}$ entering the page and $\omega_c > 0$), in the same direction as the electrons, as shown in the figure. Note that the intensity in expression (2.293) originating from (2.290), is constant in the rotating frame.
Make sure that the orientation of the axes x and y is such that the field along z enters into the page (right-handed frame).

b) Equation (2.284) gives us the components of the tensor in the laboratory frame. Note that $\sigma_{xx} = \sigma_{yy}$ and $\sigma_{xy} = -\sigma_{yx}$. Thus, by expanding the current density $\boldsymbol{J} = \underline{\boldsymbol{\sigma}} \cdot \boldsymbol{E}$ in the laboratory frame, we can write:

$$\underline{\boldsymbol{\sigma}} \cdot \boldsymbol{E} = \sigma_{xx}\hat{\mathbf{e}}_x E_x + \sigma_{xy}\hat{\mathbf{e}}_x E_y - \sigma_{xy}\hat{\mathbf{e}}_y E_x + \sigma_{xx}\hat{\mathbf{e}}_y E_y + \sigma_{zz}\hat{\mathbf{e}}_z E_z \ . \tag{2.294}$$

Replacing the components of the fields in the laboratory frame with those of the rotating frame ((2.288) and (2.289)), we have:

$$\begin{aligned} \underline{\boldsymbol{\sigma}} \cdot \boldsymbol{E} = & \ \sigma_{xx}\hat{\mathbf{e}}_x \frac{1}{\sqrt{2}}[E_+ + E_-] + \sigma_{xy}\hat{\mathbf{e}}_x \frac{\mathrm{i}}{\sqrt{2}}[E_+ - E_-] \\ & - \sigma_{xy}\hat{\mathbf{e}}_y \frac{1}{\sqrt{2}}[E_+ + E_-] + \sigma_{xx}\hat{\mathbf{e}}_y \frac{\mathrm{i}}{\sqrt{2}}[E_+ - E_-] + \sigma_{zz}\hat{\mathbf{e}}_z E_z \ . \end{aligned} \tag{2.295}$$

Regrouping the terms in E_+ and those in E_-:

$$\frac{E_+}{\sqrt{2}} \left[\hat{\mathbf{e}}_x\, \sigma_{xx} + \mathrm{i}\hat{\mathbf{e}}_x\, \sigma_{xy} - \hat{\mathbf{e}}_y\, \sigma_{xy} + \mathrm{i}\hat{\mathbf{e}}_y\, \sigma_{xx} \right] \ , \tag{2.296}$$

$$\frac{E_-}{\sqrt{2}} \left[\hat{\mathbf{e}}_x\, \sigma_{xx} - \mathrm{i}\hat{\mathbf{e}}_x\, \sigma_{xy} - \hat{\mathbf{e}}_y\, \sigma_{xy} - \mathrm{i}\hat{\mathbf{e}}_y\, \sigma_{xx} \right] \ , \tag{2.297}$$

and introducing the base vectors $\hat{\mathbf{e}}_+$ and $\hat{\mathbf{e}}_-$, we find:

$$\begin{aligned} & E_+ \left[\frac{\hat{\mathbf{e}}_x + \mathrm{i}\hat{\mathbf{e}}_y}{\sqrt{2}}\, \sigma_{xx} + \mathrm{i}\frac{\hat{\mathbf{e}}_x + \mathrm{i}\hat{\mathbf{e}}_y}{\sqrt{2}}\, \sigma_{xy} \right] + \\ & E_- \left[\frac{\hat{\mathbf{e}}_x - \mathrm{i}\hat{\mathbf{e}}_y}{\sqrt{2}}\, \sigma_{xx} - \mathrm{i}\frac{\hat{\mathbf{e}}_x - \mathrm{i}\hat{\mathbf{e}}_y}{\sqrt{2}}\, \sigma_{xy} \right] \ . \end{aligned} \tag{2.298}$$

Finally, we obtain:

$$\underline{\boldsymbol{\sigma}} \cdot \boldsymbol{E} = E_+\hat{\mathbf{e}}_+\sigma_{xx} + \mathrm{i}E_+\hat{\mathbf{e}}_+\sigma_{xy} + E_-\hat{\mathbf{e}}_-\sigma_{xx} - \mathrm{i}E_-\hat{\mathbf{e}}_-\sigma_{xy} + \sigma_{zz}\hat{\mathbf{e}}_z E_z \ . \tag{2.299}$$

We can now introduce the elements of the tensor $\underline{\boldsymbol{\sigma}}$ in the rotating frame:

$$\underline{\boldsymbol{\sigma}} \cdot \boldsymbol{E} = \Big[E_+\hat{\mathbf{e}}_+\underbrace{(\sigma_{xx} + \mathrm{i}\sigma_{xy})}_{\sigma_+} + E_-\hat{\mathbf{e}}_-\underbrace{(\sigma_{xx} - \mathrm{i}\sigma_{xy})}_{\sigma_-}\Big] + \sigma_{zz}\hat{\mathbf{e}}_z E_z \ , \qquad (2.300)$$

which is an expression showing that, in the rotating frame, the tensor has been diagonalised (there are no mixed components $E_+\hat{\mathbf{e}}_-$ or $E_-\hat{\mathbf{e}}_+$). The matrix can now be represented by:

$$\underline{\boldsymbol{\sigma}} = \sigma_0 \begin{pmatrix} \dfrac{\omega}{\omega - \omega_c} & 0 & 0 \\ 0 & \dfrac{\omega}{\omega + \omega_c} & 0 \\ 0 & 0 & 1 \end{pmatrix} \ , \qquad (2.301)$$

where the base vectors are $\hat{\mathbf{e}}_+$, $\hat{\mathbf{e}}_-$ and $\hat{\mathbf{e}}_z$, respectively.

We can thus verify that the electron resonance is indeed directed along $\hat{\mathbf{e}}_+$ ($\omega = \omega_c$).

c) Consider (2.146):

$$\boldsymbol{w}_2 = \left[-\frac{\mathrm{i}q_\alpha}{m_\alpha\omega}\boldsymbol{E}_{0\|} + \frac{q_\alpha}{2m_\alpha\omega}(\omega t - \mathrm{i})\boldsymbol{E}_{0\perp} - \frac{\mathrm{i}q_\alpha^2 t}{2\omega m_\alpha^2}(\boldsymbol{E}_{0\perp} \wedge \boldsymbol{B})\right] \mathrm{e}^{\mathrm{i}\omega t} \ , \qquad (2.302)$$

where we have replaced the dependence $\mathrm{e}^{\mathrm{i}\omega t}$ of the field $\boldsymbol{E}$ by $\mathrm{e}^{-\mathrm{i}\omega t}$, as indicated in the introduction.

We will replace i by $-\mathrm{i}$ in (2.302). Since we are only interested in the resonance phenomenon, we will ignore the term in the $\hat{\mathbf{e}}_z$ direction and the second term in $\boldsymbol{E}_{0\perp}$, which is time independent (which rapidly becomes negligible). Writing this in Cartesian coordinates:

$$\boldsymbol{E}_{0\perp} = \hat{\mathbf{e}}_x E_{0x} + \hat{\mathbf{e}}_y E_{0y} \ , \qquad (2.303)$$

this becomes:

$$\boldsymbol{w}_{20\perp} = \frac{qt}{2m}\left(\hat{\mathbf{e}}_x E_{0x} + \hat{\mathbf{e}}_y E_{0y}\right) + \frac{\mathrm{i}q^2 Bt}{2\omega m^2}\left(\hat{\mathbf{e}}_x E_{0y} - \hat{\mathbf{e}}_y E_{0x}\right) \ . \qquad (2.304)$$

Knowing that at resonance, $-qB/m = \omega_c$ and $\omega = \omega_c$, from (2.304):

$$\boldsymbol{w}_{20\perp} = \frac{qt}{2m}\left[(\hat{\mathbf{e}}_x E_{0x} + \hat{\mathbf{e}}_y E_{0y}) - \mathrm{i}(\hat{\mathbf{e}}_x E_{0y} - \hat{\mathbf{e}}_y E_{0x})\right] \ , \qquad (2.305)$$

$$\boldsymbol{w}_{20\perp} = \frac{qt}{2m}\left[\hat{\mathbf{e}}_x(E_{0x} - \mathrm{i}E_{0y}) + \hat{\mathbf{e}}_y(E_{0y} + \mathrm{i}E_{0x})\right] \ , \qquad (2.306)$$

$$\boldsymbol{w}_{20\perp} = \frac{qt}{m}\left[\left(\frac{\hat{\mathbf{e}}_x + \mathrm{i}\hat{\mathbf{e}}_y}{2}\right)(E_{0x} - \mathrm{i}E_{0y})\right] \ , \qquad (2.307)$$

such that, from (2.283) and (2.290), and multiplying by $\mathrm{e}^{-\mathrm{i}\omega t}$:

$$\boldsymbol{w}_{2\perp} = \frac{qt}{m}E_+\left(\frac{\hat{\mathbf{e}}_x + \mathrm{i}\hat{\mathbf{e}}_y}{\sqrt{2}}\right)\mathrm{e}^{-\mathrm{i}\omega t} = \frac{qt}{m}\hat{\mathbf{e}}_+ E_+ \mathrm{e}^{-\mathrm{i}\omega t} = \frac{qt}{m}\boldsymbol{E}_+ \ . \qquad (2.308)$$

2.7. Consider a homogeneous, static magnetic field $\boldsymbol{B} = \hat{\mathbf{e}}_z B_0$ and a homogeneous alternating electric field $\boldsymbol{E} = \hat{\mathbf{e}}_x E_0 \cos \omega t$ ($\hat{\mathbf{e}}_x$, $\hat{\mathbf{e}}_y$ and $\hat{\mathbf{e}}_z$ are the unit vectors along the Cartesian axes x, y and z).

a) Show that at cyclotron resonance, the contribution of this effect to the velocity of particles of mass m is given by:

$$\boldsymbol{w} = \frac{q}{2m} E_0 t \left[\cos(\omega t)\hat{\mathbf{e}}_x + \sin(\omega t)\hat{\mathbf{e}}_y\right] + \frac{q}{2m\omega} E_0 \sin(\omega t)\hat{\mathbf{e}}_x \,. \tag{2.309}$$

b) Write the form of this motion at resonance explicitly. What does it represent?

c) In an alternating electric field $\boldsymbol{E}$ directed instead along $\hat{\mathbf{e}}_y$:

$$\boldsymbol{E} = E_0 \sin(\omega t)\hat{\mathbf{e}}_y \,, \tag{2.310}$$

show that the contribution to the particle velocity at cyclotron resonance can be written in the form:

$$\boldsymbol{w}' = \frac{q}{2m} E_0 \left(t - \frac{\pi}{2\omega}\right) \left[\sin(\omega t)\hat{\mathbf{e}}_y + \cos(\omega t)\hat{\mathbf{e}}_x\right] - \frac{q}{2m\omega} E_0 \cos(\omega t)\hat{\mathbf{e}}_y \,. \tag{2.311}$$

d) A rotating electric field is applied in the xOy plane with an amplitude such that $E_x = E_y = E_0$. Following the chosen direction of rotation, we have:

$$\boldsymbol{E}_+ = E_0 \left[\cos(\omega t)\hat{\mathbf{e}}_x + \sin(\omega t)\hat{\mathbf{e}}_y\right] \tag{2.312}$$

or:

$$\boldsymbol{E}_- = E_0 \left[\cos(\omega t)\hat{\mathbf{e}}_x - \sin(\omega t)\hat{\mathbf{e}}_y\right] \,. \tag{2.313}$$

Based on the expressions for $\boldsymbol{w}$ and $\boldsymbol{w}'$, calculate the resultant velocity for a particle at cyclotron resonance $\omega_c > 0$ in a field rotating to the right (E_+) and then a field rotating to the left (E_-). What can you conclude?

Answer

a) In the presence of a magnetic field:

$$\boldsymbol{B} = B_0 \hat{\mathbf{e}}_z \tag{2.314}$$

and in a periodic electric field perpendicular to it:

$$\boldsymbol{E} = E_0 \mathrm{e}^{\mathrm{i}\omega t} \hat{\mathbf{e}}_y \,, \tag{2.315}$$

the particular solution (2.146) of the equation of motion at $\omega = \omega_c$ represents the effect of the cyclotron resonance on the particle velocity. We will ignore the contribution to the velocity in the direction parallel to the field $\boldsymbol{B}$, since it is not affected by the field $\boldsymbol{E}$, because it is perpendicular to $\boldsymbol{B}$.

In the xOy plane (2.142) we have:

$$\boldsymbol{w}_2 = \left[\frac{q}{2m\omega}(\omega t - \mathrm{i})E_0\hat{\mathbf{e}}_x - \frac{\mathrm{i}q^2t}{2\omega m^2}(E_0\hat{\mathbf{e}}_x \wedge B_0\hat{\mathbf{e}}_z)\right]\mathrm{e}^{\mathrm{i}\omega t} . \tag{2.316}$$

We can then take the real part of this expression (knowing that $\mathrm{e}^{\mathrm{i}\omega t} = \cos(\omega t) + \mathrm{i}\sin(\omega t)$), from which:

$$\boldsymbol{w}_2 = \frac{q}{2m\omega}\omega t E_0\cos(\omega t)\hat{\mathbf{e}}_x + \frac{q}{2m\omega}E_0\sin(\omega t)\hat{\mathbf{e}}_x + \frac{q^2t}{2\omega m^2}E_0\sin(\omega t)(\hat{\mathbf{e}}_x \wedge \hat{\mathbf{e}}_z)B_0 . \tag{2.317}$$

Since $\hat{\mathbf{e}}_x \wedge \hat{\mathbf{e}}_z = -\hat{\mathbf{e}}_y$ (right-handed frame) and $\omega = \omega_c \equiv -qB_0/m$, this becomes:

$$\boldsymbol{w}_2 = \frac{q}{2m}E_0t\left[\cos(\omega t)\hat{\mathbf{e}}_x - \frac{qm}{(-1)qB_0m}\sin(\omega t)B_0\hat{\mathbf{e}}_y\right] + \frac{q}{2m\omega}E_0\sin(\omega t)\hat{\mathbf{e}}_x . \tag{2.318}$$

By setting $\boldsymbol{w}_2 = \boldsymbol{w}$, we obtain the relation from statement a):

$$\boldsymbol{w} = \frac{q}{2m}E_0t[\cos(\omega t)\hat{\mathbf{e}}_x + \sin(\omega t)\hat{\mathbf{e}}_y] + \frac{q}{2m\omega}E_0\sin(\omega t)\hat{\mathbf{e}}_x . \tag{2.309}$$

b) The third term on the RHS of (2.309) describes a periodic motion in the direction of the field $\boldsymbol{E}$ (this is normal: it would also exist in the absence of $\boldsymbol{B}$) while the first and the second conjugate terms represent a periodic rotational motion, of frequency ω_c, whose amplitude continuously increases with time: in other words, the particle describes a spiral. The modulus of the corresponding velocity of this motion is $w_0 = qE_0t/2m$, because the contribution to the periodic motion along $\hat{\mathbf{e}}_x$ rapidly becomes negligible. This increase in amplitude comes precisely from the resonance between ω and ω_c, called cyclotron resonance.
c) We use equation (2.317), in which the direction of the field $\boldsymbol{E}$ is along $\hat{\mathbf{e}}_y$ rather than $\hat{\mathbf{e}}_x$. Since $\hat{\mathbf{e}}_y \wedge \hat{\mathbf{e}}_z = \hat{\mathbf{e}}_x$, we have:

$$\begin{aligned}\boldsymbol{w}' &= \frac{qt}{2m}E_0\cos(\omega t)\hat{\mathbf{e}}_y + \frac{q}{2m\omega}E_0\sin(\omega t)\hat{\mathbf{e}}_y + \frac{q^2t}{2\omega m^2}E_0B_0\sin(\omega t)\hat{\mathbf{e}}_x \\ &= \frac{q}{2m}E_0t\left[\cos(\omega t)\hat{\mathbf{e}}_y - \sin(\omega t)\hat{\mathbf{e}}_x\right] + \frac{q}{2\omega m}E_0\sin(\omega t)\hat{\mathbf{e}}_y .\end{aligned} \tag{2.319}$$

We now change the time origin by replacing t by $t - \pi/2\omega$: as $\cos(\omega t - \pi/2) = \sin(\omega t)$ and $\sin(\omega t - \pi/2) = -\cos(\omega t)$, we obtain from (2.319):

$$\boldsymbol{w}' = \frac{q}{2m} E_0 \left(t - \frac{\pi}{2\omega}\right) [\sin(\omega t)\hat{\mathbf{e}}_y + \cos(\omega t)\hat{\mathbf{e}}_x]$$
$$- \frac{q}{2m\omega} E_0 \cos(\omega t)\hat{\mathbf{e}}_y \ . \tag{2.311}$$

d) To get the resulting particle velocity in the field E_+ (defined by (2.312)), we simply add the velocities $\boldsymbol{w} + \boldsymbol{w}'$ from equations (2.309) and (2.311):

$$\boldsymbol{w}_{\mathrm{TOT}+} = \frac{q}{2m} E_0 t [2\cos(\omega t)\hat{\mathbf{e}}_x + 2\sin(\omega t)\hat{\mathbf{e}}_y] \tag{2.320}$$
$$- \frac{qE_0\pi}{4m\omega}[\sin(\omega t)\hat{\mathbf{e}}_y + \cos(\omega t)\hat{\mathbf{e}}_x] + \frac{qE_0}{2m\omega}[\sin(\omega t)\hat{\mathbf{e}}_x - \cos(\omega t)\hat{\mathbf{e}}_y] \ .$$

To obtain the resulting particle velocity in the field E_-, we add the velocity $\boldsymbol{w}$ with $\boldsymbol{w}''$, where $\boldsymbol{w}'' = -\boldsymbol{w}$, since then $\boldsymbol{E} = -E_0 \sin(\omega t)\hat{\mathbf{e}}_y$. This yields:

$$\boldsymbol{w}_{\mathrm{TOT}-} \equiv \boldsymbol{w} - \boldsymbol{w}'$$
$$= \frac{q}{2m} E_0 t \, [\, \overbrace{\cos(\omega t)\hat{\mathbf{e}}_x + \sin(\omega t)\hat{\mathbf{e}}_y - \sin(\omega t)\hat{\mathbf{e}}_y - \cos(\omega t)\hat{\mathbf{e}}_x}^{=0} \,]$$
$$+ \frac{q}{4m}\frac{\pi}{\omega} E_0 [\sin(\omega t)\hat{\mathbf{e}}_y + \cos(\omega t)\hat{\mathbf{e}}_x]$$
$$+ \frac{q}{2m\omega} E_0 [\sin(\omega t)\hat{\mathbf{e}}_x + \cos(\omega t)\hat{\mathbf{e}}_y] \ . \tag{2.321}$$

We can conclude that if the field turns in the positive direction of ω_c (i.e. if the particle velocity is not in the direction of the rotating field), the velocity of the particle does not increase linearly in time and, therefore, there is no cyclotron resonance.

2.8. Consider a magnetic field of the form:

$$\boldsymbol{B} = B_0(1 - \epsilon \cos kx)\hat{\mathbf{e}}_x \ , \tag{2.322}$$

where ϵ is a parameter smaller than unity and k is a constant. This field is used to axially confine charged particles at each end of a linear machine whose centre is at $x = 0$.

a) Find the expression for $\boldsymbol{w}_\parallel$ as a function of $\boldsymbol{w}_\parallel(0)$, $\boldsymbol{w}_\perp(0)$, ϵ and k.
b) Show that particles will be effectively trapped if:

$$w_\parallel^2(0) \leq w_\perp^2(0)\frac{2\epsilon}{1-\epsilon} \ . \tag{2.323}$$

c) Assuming an isotropic velocity distribution at $x = 0$, calculate the number of trapped particles as a fraction of the total number of particles.

Answer

a) This is a case in which the magnetic field has a (weak) non uniformity in its own direction, which corresponds to the situation treated in Sect. 2.2.3 (page 138); note that we are using here the x axis as the direction of $\boldsymbol{B}$ rather than z. We know that the expression:

$$\boldsymbol{B} = B_0(1 - \epsilon \cos kx)\hat{\mathbf{e}}_x \tag{2.324}$$

is not complete, because the component of $\boldsymbol{B}$ (of order 1), required to satisfy Maxwell's equation $\boldsymbol{\nabla} \cdot \boldsymbol{B} = 0$, is missing. Nevertheless, this correction does not appear in the calculation of the $\boldsymbol{w}_{\parallel}$ component. Finally, note that the minimum value of $|\boldsymbol{B}|$, $|\boldsymbol{B}| = B_0(1 - \epsilon)$ is found at $x = 0$ and thus corresponds to the region situated between the mirrors: correctly speaking, there is no region of uniform field in this machine, only two mirrors at each end with a minimum in the magnetic field in between.
By transposing equation (2.176) along $\hat{\mathbf{e}}_x$ and setting $B_{\parallel} = B(x = 0)$, we have:

$$\boldsymbol{w}_{\parallel}(t) = \boldsymbol{w}_{\parallel}(0) - \frac{\hat{\mathbf{e}}_x}{2} r_B^2 \omega_c^2 \frac{1}{B(x=0)} \left(\frac{\partial B_x}{\partial x} \right)_{y=z=0} t \, , \tag{2.325}$$

where ω_c corresponds to the value at $B(x = 0)$. Since $\partial B_x / \partial x = B_0 \epsilon k \sin kx$ and $r_B^2 \omega_c^2 = w_{\perp}^2(0)$, we find:

$$\boldsymbol{w}_{\parallel}(t) = \boldsymbol{w}_{\parallel}(0) - \frac{\hat{\mathbf{e}}_x}{2} w_{\perp}^2(0) \frac{\epsilon k \sin kx}{1 - \epsilon} t \tag{2.326}$$

(we could neglect the parameter ϵ compared to 1 in the denominator).

b) We have shown in Sect. 2.2.3 (p. 138) that the particles coming from the central region of the machine ($x = 0$) are reflected by the magnetic mirror if the angle α_0 of their vector velocity with respect to the axis (of components $w_{0\parallel} = w_0 \cos \alpha_0$ and $w_{0\perp} = w_0 \sin \alpha_0$) in the section of uniform field (the region of the weakest field between the two mirrors) has a value greater than α_{0m}, defined by (2.189):

$$\sin \alpha_{0m} = \sqrt{\frac{B(x=0)}{B_{\max}}} \, . \tag{2.327}$$

$B_{\max}$ is reached for $\cos kx = -1$, from which $B_{\max} = 1 + \epsilon$ and $B(x = 0) = 1 - \epsilon$, such that (2.327) gives:

$$\sin^2 \alpha_{0m} = \frac{1 - \epsilon}{1 + \epsilon} \, . \tag{2.328}$$

Noting that the ratio of the velocities $w_{\perp}(x = 0)/w(x = 0)$ corresponds to $\sin \alpha_{0m}$, we have:

$$\frac{w_\perp^2(0)}{w^2(0)} = \frac{1-\epsilon}{1+\epsilon} , \tag{2.329}$$

such that:

$$w_\perp^2(0)(1+\epsilon) = (w_\perp^2(0) + w_\parallel^2(0))(1-\epsilon) , \tag{2.330}$$

$$w_\perp^2(0)[(1+\epsilon) + (\epsilon - 1)] = w_\parallel^2(0)(1-\epsilon) , \tag{2.331}$$

and finally:

$$w_\parallel^2(0) = w_\perp^2(0)\frac{2\epsilon}{1-\epsilon} . \tag{2.332}$$

The condition for reflection is defined by the inequality:

$$w_\parallel^2(0) \leq w_\perp^2(0)\frac{2\epsilon}{1-\epsilon} . \tag{2.333}$$

c) The reflection coefficient C_r of particles in a magnetic mirror (2.196) leads us to:

$$C_r \equiv \frac{\Gamma_r}{\Gamma_{\text{inc}}} = 1 - \frac{1}{\mathcal{R}} = 1 - \frac{1}{B_{\max}/B(x=0)} = 1 - \frac{1-\epsilon}{1+\epsilon} = \frac{2\epsilon}{1+\epsilon} . \tag{2.334}$$

2.9. Consider a magnetic field directed **principally** along z but subject to a slight curvature, represented by the term $\partial B_x/\partial z$ (we suppose that the curvature is in two dimensions only, in the plane xz). The origin of the Cartesian frame is chosen such that the field $\boldsymbol{B}$ is directed along the z axis while, on each side of the origin, there is a contribution from the x and z components of the field, as shown in the figure. The radius of curvature ρ of this field line is assumed to be much greater than the particle Larmor radius (of charge q and mass m).

a) Show that, in the immediate region of the origin, $\boldsymbol{B}$ is described by:

$$\boldsymbol{B} = B_0\left(\hat{\mathbf{e}}_x\frac{z}{\rho} + \hat{\mathbf{e}}_z\right) , \tag{2.335}$$

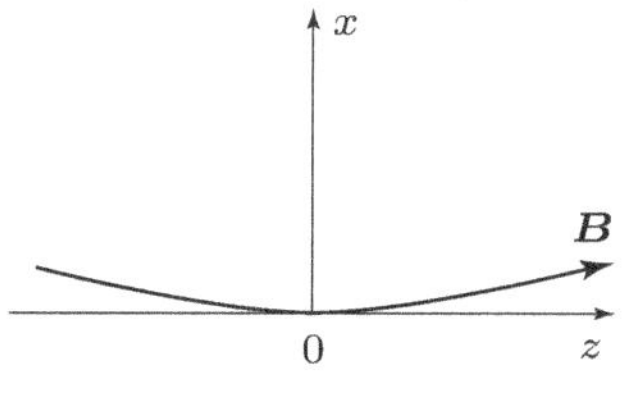

where B_0 is the intensity of the field at $z = 0$ and $\hat{\mathbf{e}}_x$, $\hat{\mathbf{e}}_z$ are the unit base vectors in the Cartesian frame (x, y, z); $\rho^{-1} = \mathrm{d}^2x/\mathrm{d}z^2$ in the case that $\mathrm{d}x/\mathrm{d}z$ is not too large.

b) Assume $\boldsymbol{w}$ is the particle velocity. Using the field given by (2.335), express the components of $\dot{\boldsymbol{w}}$ to first order, in Cartesian coordinates.

c) Determine, to order zero, the three components of velocity and position, using the following initial conditions:

$$x^0 = y^0 = z^0 = 0 , \tag{2.336}$$

$$\boldsymbol{w}^0 = w_{\perp 0}\hat{\mathbf{e}}_x + w_{z0}\hat{\mathbf{e}}_z , \tag{2.337}$$

(the zero superscript in A^0 signifies that the quantity A is expressed to order zero).

d) Show that the calculation to first order leads to:

$$w_x = \left(\frac{w_{z0}^2}{\rho}\right) t + w_{\perp 0} \cos \omega_c t \; , \tag{2.338}$$

$$w_y = -\left(\frac{w_{z0}^2}{\rho\omega_c}\right) + w_{\perp 0} \sin \omega_c t \; , \tag{2.339}$$

where the constants of integration are fixed such that if we set $\rho^{-1} = 0$, we will recover the zeroth order solution. In this first order calculation, we have replaced the values of w_z and z, which appear in the expressions obtained for w_x and w_y in b), by their values approaching $z = 0$, i.e. to order zero, namely w_{z0} and $w_{z0}t$ respectively.

e) Finally, show that the position of the guiding centre satisfies the expression:

$$x = \left(\frac{1}{2\rho}\right) z^2 \; . \tag{2.340}$$

Answer

a) The magnetic field is directed, at the origin of the system, along the z axis and is affected by a (symmetric) curvature of the lines of force in the xOz plane. Equation (2.335) in the terms of the problem suggests that the B_x component is a correction term for B_z in the neighbourhood of $z = 0$. A similar question was treated in Sect. 2.2.3, p. 147 and p. 152 (2.221) with the inhomogeneity along $\hat{\mathbf{e}}_y$ instead of $\hat{\mathbf{e}}_x$. Achieving this transposition yields:

$$\boldsymbol{B} = \hat{\mathbf{e}}_x(\beta B_0 z) + \hat{\mathbf{e}}_z[B_0(1 + \beta x)] \; . \tag{2.341}$$

The principal component B_z is subject to a first order correction, which we can neglect in comparison to B_0, of order zero; the correction along the x axis introduces a (small) term β, where $\beta \simeq \rho^{-1}$ (XIII.18): equation (2.335) of the question is therefore demonstrated.

We can also treat the problem from the beginning, without recourse to the results of Sect. 2.2.3 as we have just done. Since B_x is a first order correction, a development in series, limited to first order, is justified:

$$B_x = \underbrace{B_x(z=0)}_{=0} + \frac{\partial B_x}{\partial x} x + \frac{\partial B_x}{\partial y} y + \frac{\partial B_x}{\partial z} z \simeq \frac{\partial B_x}{\partial z} z \; , \tag{2.342}$$

the B_x component depending only on z (see figure). Since $\boldsymbol{B} = \hat{\mathbf{e}}_x B_x + \hat{\mathbf{e}}_z B_z$ and $B_z = B_0$, we obtain:

$$\boldsymbol{B} = \hat{\mathbf{e}}_x \frac{\partial B_x}{\partial z} z + \hat{\mathbf{e}}_z B_0 \,. \tag{2.343}$$

To calculate $\partial B_x/\partial z$, we consider a magnetic field line whose slope $\mathrm{d}x/\mathrm{d}z$ is, as a general rule, equal to the ratio B_x/B_z of the local magnetic field components:

$$\frac{\mathrm{d}x}{\mathrm{d}z} = \frac{B_x}{B_z} \,. \tag{2.344}$$

From (2.344), we have:

$$\frac{\mathrm{d}^2 x}{\mathrm{d}z^2} = \frac{\partial B_x}{\partial z}\frac{1}{B_z} - \frac{\partial B_z}{\partial z}\frac{B_x}{B_z^2} \,, \tag{2.345}$$

where the second term on the RHS is negligible compared to the first term because it is of order 2: from (2.341) and (2.342), $\partial B_x/\partial z$ and B_x are both non zero only to order 1.
By assumption, $\mathrm{d}x/\mathrm{d}z$ is not too large in the neighbourhood of the origin, then (XIII.2):

$$\frac{\mathrm{d}^2 x}{\mathrm{d}z^2} \simeq \frac{1}{\rho} \,. \tag{2.346}$$

Finally, from (2.345) and (2.346):

$$\frac{\partial B_x}{\partial z} = \frac{B_0}{\rho} \,, \tag{2.347}$$

from which, following (2.343), we have thus demonstrated the validity of (2.335).

b) We simply return to the general equations shown in Sect. 2.1, which we can write, taking $-qB_0/m = \omega_c$ and setting $E = 0$ (Eqs. (2.6)–(2.8)) in the form:

$$\dot{w}_x = \frac{q}{m}[B_z w_y - B_y w_z] = \frac{q}{m}[B_0 w_y] = -\omega_c w_y \,, \tag{2.348}$$

$$\begin{aligned}\dot{w}_y &= \frac{q}{m}[B_x w_z - B_z w_x] = \frac{q}{m}\left[\frac{B_0 z}{\rho} w_z - B_0 w_x\right] \\ &= -\omega_c \left[\underline{\frac{z}{\rho} w_z} - w_x\right] \,,\end{aligned} \tag{2.349}$$

$$\dot{w}_z = \frac{q}{m}[B_y w_x - B_x w_y] = -\frac{q}{m}\left[\frac{B_0 z}{\rho} w_y\right] = \omega_c \left[\underline{\frac{z}{\rho} w_y}\right] \,, \tag{2.350}$$

where the underlined terms are of first order, because they tend to zero if the radius of curvature tends to infinity (rectilinear lines of force).

c) To obtain the expression for the components of $\dot{\boldsymbol{w}}$ to order zero, we only need to set $\rho \to \infty$ in (2.348)–(2.350). We then get:

$$\dot{w}_x = -\omega_c w_y \ , \tag{2.351}$$

$$\dot{w}_y = \omega_c w_x \ , \tag{2.352}$$

$$\dot{w}_z = 0 \ . \tag{2.353}$$

We can start by calculating the expressions for w_x and x. To do this, we differentiate (2.351):

$$\dot{w}_x = -\omega_c \dot{w}_y \tag{2.354}$$

and substituting $\dot{w}_y$ from (2.352) gives:

$$\dot{w}_x = -\omega_c(\omega_c w_x) \ , \tag{2.355}$$

from which:

$$\dot{w}_x + \omega_c^2 w_x = 0 \ , \tag{2.356}$$

for which the solution is clearly:

$$w_x = A_1 \cos \omega_c t + A_2 \sin \omega_c t \ , \tag{2.357}$$

where A_1 and A_2 are constants, depending on the initial conditions (equations (2.336) and (2.337)). This leads to:

$$w_x = w_{\perp 0} \cos \omega_c t \tag{2.358}$$

and:

$$x = \frac{w_{\perp 0}}{\omega_c} \sin \omega_c t \ . \tag{2.359}$$

To calculate the expressions for w_y and y, we will take (2.352) and substitute therein the value of w_x from (2.358). After integration, and application of initial conditions, we obtain:

$$w_y = w_{\perp 0} \sin \omega_c t \tag{2.360}$$

and:

$$y = \frac{w_{\perp 0}}{\omega_c}(1 - \cos \omega_c t) \ . \tag{2.361}$$

As for w_z and z, we deduce from (2.353):

$$w_z = w_{z0} \quad \text{and} \quad z = w_{z0} t \ . \tag{2.362}$$

d) The system (2.348)–(2.350) can be resolved quite easily provided that, in the terms in $1/\rho$ (terms of first order), the expressions for velocity and position along $\hat{\mathbf{e}}_z$ are those of (2.362), which are of order zero. Note that, as a result, these terms remain of order one. Then from (2.349), we write down:

$$\dot{w}_y = -\omega_c \left[\frac{w_{z0}^2 t}{\rho} - w_x \right] \ . \tag{2.363}$$

To obtain w_x, we start by differentiating (2.348):

$$\dot{w}_x = -\omega_c \dot{w}_y \ , \tag{2.364}$$

which we can render homogeneous in w_x by substituting the expression for $\dot{w}_y$ from (2.363):

$$\dot{w}_x = \omega_c^2 \left[\frac{w_{z0}^2 t}{\rho} - w_x \right] \ , \tag{2.365}$$

from which:

$$\dot{w}_x + \omega_c^2 w_x = \frac{\omega_c^2 w_{z0}^2 t}{\rho} \ . \tag{2.366}$$

The solution is the general solution with no RHS, namely (2.358), to which must be added a particular solution (of first order): $w_x^{(1)} = w_{z0}^2 t/\rho$ (following (2.366)), from which, in total:

$$w_x = w_{\perp 0} \cos \omega_c t + \frac{w_{z0}^2 t}{\rho} \ . \tag{2.367}$$

For w_y, differentiate (2.363) and we have:

$$\dot{w}_y = -\omega_c \left[\frac{w_{z0}^2}{\rho} - \dot{w}_x \right] \ , \tag{2.368}$$

an expression in which we can replace $\dot{w}_x$ by its expression in (2.348), such that:

$$\dot{w}_y = -\omega_c \left[\frac{w_{z0}^2}{\rho} + \omega_c w_y \right] \ , \tag{2.369}$$

from which:

$$\dot{w}_y + \omega_c^2 w_y = -\frac{\omega_c}{\rho} w_{z0}^2 \ , \tag{2.370}$$

whose solution, following the method of (2.367), is:

$$w_y = w_{\perp 0} \sin \omega_c t - \frac{1}{\rho \omega_c} w_{z0}^2 \ . \tag{2.371}$$

e) To calculate the guiding centre velocity, we simply ignore the sinusoidal terms in (2.367) and (2.371), such that:

$$w_x = \frac{w_{z0}^2 t}{\rho} \ , \tag{2.372}$$

$$w_y = -\frac{w_{z0}^2}{\rho \omega_c} \ , \tag{2.373}$$

and, clearly ($\boldsymbol{E} = 0$):

$$w_z = w_{z0} \ . \tag{2.374}$$

The position of the guiding centre is then described by:

$$x = \frac{w_{z0}^2 t^2}{2\rho} , \tag{2.375}$$

$$y = -\frac{w_{z0}^2 t}{\rho \omega_c} , \tag{2.376}$$

$$z = w_{z0} t . \tag{2.377}$$

Combining (2.375) and (2.377) for x and z, we obtain:

$$x = \frac{z^2}{2\rho} . \tag{2.378}$$

This guiding centre motion is due to the magnetic curvature drift (Sect. 2.2.3, p. 147). To obtain the contribution from the magnetic drift, as was shown in that section, we need to include the correction terms (to first order) in B_z, which we have neglected in our calculation so far (see (2.363)).

Remark: We can also obtain (2.378) by recalling that the motion of the guiding centre due to the curvature drift is effected along the lines of force $\boldsymbol{B}$. It is sufficient to use the parameterisation of the lines of force calculated in Appendix XIII. From equation (XIII.7), we have $\mathrm{d}y/\mathrm{d}z = z/\rho$ which, adapted to the direction of curvature in the present exercise, becomes $dx/dz = z/\rho$, from which $x = z^2/2\rho$, since the constant of integration is zero, because at $z = 0$, $B_x = 0$.

2.10. Consider an axial, linear magnetic confinement machine whose magnetic field in the homogeneous region has a magnitude B_0. This machine is fitted with a magnetic mirror situated at $z \geq 0$ such that the value of the magnetic field is:

$$\boldsymbol{B} = \hat{\mathbf{e}}_z B_0 \left[1 + \left(\frac{z}{a} \right)^2 \right] , \tag{2.379}$$

where a is a constant. Find an expression giving the position z, in the mirror zone, at which a particle is reflected if its vector velocity makes an angle α_0 with the direction of the magnetic field in the homogeneous region.

Answer

The position at which a particle is reflected in the mirror zone is independent of the magnitude of the velocity, but depends on the angle α_0 (in the homogeneous field region) of its vector velocity with respect to the axis of the machine. More exactly, we have seen that:

$$\sin\alpha = \sin\alpha_0 \sqrt{\frac{B(z)}{B_0}}\,, \tag{2.187}$$

where the angle α is the angle of its vector velocity with respect to the axis of the machine in the mirror region at the position z, where the magnetic field is $B(z)$ (see Fig. 2.15). The value of z at which the particle is reflected is simply obtained by setting $\sin\alpha = 1$ in the preceding equation. Since:

$$B(z) = B_0\left[1 + \left(\frac{z}{a}\right)^2\right]\,, \tag{2.379}$$

we then obtain:

$$1 = \sin\alpha_0 \sqrt{\frac{B_0\left[1 + \left(\frac{z}{a}\right)^2\right]}{B_0}} \tag{2.380}$$

from which:

$$\left(\frac{1}{\sin\alpha_0}\right)^2 = 1 + \left(\frac{z}{a}\right)^2\,, \tag{2.381}$$

i.e.:

$$z = a\left[\left(\frac{1}{\sin\alpha_0}\right)^2 - 1\right]^{\frac{1}{2}} = \pm a\cot\alpha_0\,. \tag{2.382}$$

Remark: Although the field $\boldsymbol{B}$ has components B_x and B_y, which are necessary to satisfy $\boldsymbol{\nabla}\cdot\boldsymbol{B} = 0$, the important component for the mirror effect is B_z (remember that in the adiabatic approximation, only F_z enters into the conservation of μ_z).

2.11. Consider the motion of an electron in a uniform magnetic field $\boldsymbol{B}$ directed along the z axis and symmetric about this axis. This magnetic field varies slowly as a function of time, such that $B_z = B_0(1 - \alpha t)$, where the intensity B_0 is constant and α is a very small parameter.

a) Verify that the field $\boldsymbol{B}$ satisfies Maxwell's equations
b) Show that, in a Cartesian frame, the equation of motion of the electron can be written in the form:

$$\dot{w}_x = -\omega_{ce}\left[w_y(1 - \alpha t) - \frac{\alpha y}{2}\right]\,, \tag{2.383}$$

$$\dot{w}_y = \omega_{ce}\left[w_x(1 - \alpha t) - \frac{\alpha x}{2}\right]\,, \tag{2.384}$$

$$\dot{w}_z = 0\,, \tag{2.385}$$

where $\boldsymbol{w} = (w_x, w_y, w_z)$ is the velocity of the electron, and ω_{ce} its cyclotron frequency in the field B_0.
In this calculation, why is it important that α be small?

Answer

a) Conformity of the field $\boldsymbol{B}$ with Maxwell's equations.
Due to the fact that the magnetic field varies with time, it creates an associated electric field $\boldsymbol{E}$, described by Maxwell's equation:

$$\boldsymbol{\nabla} \wedge \boldsymbol{E} = -\frac{\partial \boldsymbol{B}}{\partial t} , \tag{2.386}$$

and this field $\boldsymbol{E}$ is perpendicular to $\boldsymbol{B}$. Since $\boldsymbol{B} = \hat{\mathbf{e}}_z B_0(1-\alpha t)$, expanding equation (2.386) gives:

$$\hat{\mathbf{e}}_x \left(-\frac{\partial E_y}{\partial z}\right) + \hat{\mathbf{e}}_y \left(\frac{\partial E_x}{\partial z}\right) + \hat{\mathbf{e}}_z \left(\frac{\partial E_y}{\partial x} - \frac{\partial E_x}{\partial y}\right) = \alpha B_0 \hat{\mathbf{e}}_z . \tag{2.387}$$

This requires that $\partial E_y/\partial z = \partial E_x/\partial z = 0$ and $\partial E_y/\partial x - \partial E_x/\partial y = \alpha B_0$, i.e. that E_x and E_y are independent of z. Under these conditions, due to the axial symmetry with respect to z, we can infer that the rotational term along the z axis can be written:

$$E_x = -\frac{y}{2}(\alpha B_0) , \tag{2.388}$$

$$E_y = \frac{x}{2}(\alpha B_0) . \tag{2.389}$$

In fact, the vector $\boldsymbol{E}$ describes a circle in the $x\mathrm{O}y$ plane.
For the equation $\boldsymbol{\nabla} \cdot \boldsymbol{D} = \rho$, where $\rho = 0$ in the framework of individual trajectories, following (2.388) and (2.389), we have $\partial E_x/\partial x + \partial E_y/\partial y = 0$, and this equation is verified.
The verification of $\boldsymbol{\nabla} \cdot \boldsymbol{B} = 0$ is trivial, since $\boldsymbol{B}$ is independent of position.
Finally, for $\boldsymbol{\nabla} \wedge \boldsymbol{H} = \boldsymbol{J} + \partial \boldsymbol{D}/\partial t$, where $\boldsymbol{J} = 0$ (individual trajectory), $\boldsymbol{\nabla} \wedge \boldsymbol{H} = 0$ because $\boldsymbol{H}$ is uniform, and the term $\partial \boldsymbol{D}/\partial t = \epsilon_0 \partial \boldsymbol{E}/\partial t = 0$.

b) Equation of motion. Following (2.6)–(2.8), we have:

$$m_e \frac{\mathrm{d}^2 x}{\mathrm{d}t^2} = q_e \left[E_x + B_z \frac{\mathrm{d}y}{\mathrm{d}t}\right] = -q_e \frac{\alpha B_0}{2} y + q_e B_0 (1 - \alpha t) \dot{y} , \tag{2.390}$$

$$m_e \frac{\mathrm{d}^2 y}{\mathrm{d}t^2} = q_e \left[E_y - B_z \frac{\mathrm{d}x}{\mathrm{d}t}\right] = q_e \frac{\alpha B_0}{2} x - q_e B_0 (1 - \alpha t) \dot{x} , \tag{2.391}$$

$$m_e \frac{\mathrm{d}^2 z}{\mathrm{d}t^2} = 0 . \tag{2.392}$$

Since the motion is uniform along z, setting $\omega_{ce} = -q_e B/m$ and $w_x \equiv \mathrm{d}x/\mathrm{d}t$, $w_y \equiv \mathrm{d}y/\mathrm{d}t$, we have then verified (2.383) and (2.384):

$$\dot{w}_x = -\omega_{ce}\left[(1-\alpha t)w_y - \frac{\alpha y}{2}\right] , \quad (2.393)$$

$$\dot{w}_y = \omega_{ce}\left[(1-\alpha t)w_x - \frac{\alpha x}{2}\right] , \quad (2.394)$$

$$\dot{w}_z = 0 . \quad (2.395)$$

The parameter α must be chosen to be small in order to keep a positive intensity of $\boldsymbol{B}$ with time ($\alpha t < 1$).

2.12. Consider the motion of an electron near the origin of a given frame. The electron is subject to a magnetic field which is constant in time, but slightly inhomogeneous along the field lines. Its equation is given by:

$$\boldsymbol{B} = \hat{\mathbf{e}}_z B_0(1+\alpha z) , \quad (2.396)$$

where α is sufficiently small, such that $\alpha z \ll 1$. Assume that the field is axially symmetric about the z axis. The initial conditions are as follows: $x(0) = -x_0$, $y(0) = 0$, $z(0) = 0$, $w_x(0) = 0$, $w_y(0) = w_{\perp 0}$, $w_z(0) = w_{z0}$.

a) Show that the x, y and z components of the equation of motion due to the Lorentz force are described by the following equations:

$$\ddot{x} = -\omega_{ce}\left[\dot{y} + \alpha\left(z\dot{y} + \frac{1}{2}y\dot{z}\right)\right] , \quad (2.397)$$

$$\ddot{y} = \omega_{ce}\left[\dot{x} + \alpha\left(z\dot{x} + \frac{1}{2}x\dot{z}\right)\right] , \quad (2.398)$$

$$\ddot{z} = -\omega_{ce}\left(\frac{\alpha}{2}\right)[x\dot{y} - y\dot{x}] . \quad (2.399)$$

b) Supposing that the initial velocity is given by:

$$\boldsymbol{w}_0 = w_{\perp 0}\hat{\mathbf{e}}_y + w_{z0}\hat{\mathbf{e}}_z , \quad (2.400)$$

show that:

$$\dot{z} \equiv w_z = -\frac{1}{2}\alpha w_{\perp 0}^2 t + w_{z0} . \quad (2.401)$$

Answer

a) The field we are considering has a slight inhomogeneity in its own direction. Under these conditions, and due to the axial symmetry of $\boldsymbol{B}$, we know (Sect. 2.2.3, p. 138) there is, in fact, a component B_r (of order 1 with respect to the component B_z, of order zero) which has been ignored in (2.396).
Including an expression for this in Cartesian coordinates (2.163), the complete expression for the field $\boldsymbol{B}$ is now:

$$\boldsymbol{B} = -\frac{1}{2}x\left(\frac{\partial B_z}{\partial z}\right)_{x=y=0}\hat{\mathbf{e}}_x - \frac{1}{2}y\left(\frac{\partial B_z}{\partial z}\right)_{0,0}\hat{\mathbf{e}}_y + B_0(1+\alpha z)\hat{\mathbf{e}}_z \ , \quad (2.402)$$

where $\partial B_z/\partial z$ is calculated from (2.396), and finally:

$$\boldsymbol{B} = -\frac{1}{2}xB_0\,\alpha\hat{\mathbf{e}}_x - \frac{1}{2}yB_0\,\alpha\hat{\mathbf{e}}_y + B_0(1+\alpha z)\hat{\mathbf{e}}_z \ . \quad (2.403)$$

Developing the equation of motion along the three axes of the Cartesian frame, and setting $E = 0$, we have:

along x:
$$\dot{x} = \frac{q_e}{m_e}(B_z\dot{y} - B_y\dot{z}) = \frac{q_e}{m_e}\left[B_0(1+\alpha z)\dot{y} + \frac{B_0}{2}\alpha y\dot{z}\right]$$
$$= -\omega_{ce}\left[\dot{y} + \alpha\left(z\dot{y} + \frac{y\dot{z}}{2}\right)\right] , \quad (2.404)$$

along y:
$$\dot{y} = \omega_{ce}\left[\dot{x} + \alpha\left(z\dot{x} + \frac{x\dot{z}}{2}\right)\right] , \quad (2.405)$$

and along z:
$$\dot{z} = -\omega_{ce}\frac{\alpha}{2}(x\dot{y} - y\dot{x}) \ . \quad (2.406)$$

b) To calculate the velocity along the guiding axis, we can observe, following (2.406), that it is of order 1, due to the presence of the small parameter α (assuming $\alpha z \ll 1$), which agrees with the assumption of the guiding centre approximation. We therefore simply need to replace the positions and velocities of the cyclotron motion appearing in (2.406) by their expansion, limited to order zero.
The equations describing the zeroth order motion in the perpendicular plane are obtained by setting $\alpha = 0$ in (2.404) and (2.405):

$$\dot{x} = -\omega_{ce}\dot{y} \ , \quad (2.407)$$
$$\dot{y} = \omega_{ce}\dot{x} \ . \quad (2.408)$$

Now we must resolve the system of 2 equation in 2 unknowns.
To calculate the motion along y, we perform a first integration of (2.407) over time:

$$\dot{x} = -\omega_{ce}y + C_1 \ , \quad (2.409)$$

where the constant $C = 0$, because $\dot{x}(0) = 0$ and $y(0) = 0$, such that:

$$\dot{x} = -\omega_{ce}y \ . \quad (2.410)$$

Substituting this equation in (2.408), we obtain:

$$\dot{y} + \omega_{ce}^2 y = 0 \ , \quad (2.411)$$

for which the solution is:

$$y = A_1 \cos \omega_{ce} t + A_2 \sin \omega_{ce} t \ . \tag{2.412}$$

To determine the constants A_1 and A_2, we note that:

$$y(0) \equiv 0 = A_1 \ , \tag{2.413}$$

such that:

$$w_y = A_2 \omega_{ce} \cos \omega_{ce} t \ , \tag{2.414}$$

with the initial condition $w_y(0) = w_{\perp 0}$.
Finally:

$$w_y = w_{\perp 0} \cos \omega_{ce} t \tag{2.415}$$

and:

$$y = \frac{w_{\perp 0}}{\omega_{ce}} \sin \omega_{ce} t \ . \tag{2.416}$$

To calculate the motion along the x axis, we continue in an analogous fashion. Integrating (2.408), we obtain:

$$\dot{y} = \omega_{ce} x + C_2 \ , \tag{2.417}$$

where, because of the initial conditions, the constant of integration $C_2 = w_{\perp 0} - \omega_{ce} x_0$.
Substituting (2.417) in (2.407):

$$\dot{x} + \omega_{ce}^2 x = -\omega_{ce} w_{\perp 0} + \omega_{ce}^2 x_0 \ , \tag{2.418}$$

whose complete solution has the form:

$$x = A_1 \cos \omega_{ce} t + A_2 \sin \omega_{ce} t - \frac{w_{\perp 0}}{\omega_{ce}} - x_0 \ . \tag{2.419}$$

Using the initial conditions $x(0) = -x_0$ and $w_x(0) = 0$, we find successively $A_1 = w_{\perp 0}/\omega_{ce}$, from which:

$$x = \frac{w_{\perp 0}}{\omega_{ce}} \cos \omega_{ce} t \tag{2.420}$$

and $A_2 = 0$, from which:

$$w_x = -w_{\perp 0} \sin \omega_{ce} t \ . \tag{2.421}$$

To calculate the motion along z, we substitute the values for the motion along x and y in (2.406):

$$\dot{z} = -\frac{\omega_{ce}\alpha}{2} \left[\frac{w_{\perp 0}^2}{\omega_{ce}} \cos^2 \omega_{ce} t + \frac{w_{\perp 0}^2}{\omega_{ce}} \sin^2 \omega_{ce} t \right] = -\frac{\alpha}{2} w_{\perp 0}^2 \ , \tag{2.422}$$

which, on integration gives:

$$\dot{z} = -\frac{\alpha}{2} w_{\perp 0}^2 t + C_3 \ , \tag{2.423}$$

where $\dot{z}(0) \equiv w_{z0} = C_3$, which indeed leads us to (2.401):

$$\dot{z} = -\frac{\alpha}{2} w_{\perp 0}^2 t + w_{z0} \ .$$

Remark: The choice of $x(0) = -x_0$ rather than $x(0) = 0$ as initial condition enables us to obtain simpler expressions for the components of velocity and position!

2.13. In the context of the individual trajectories model, consider an applied magnetic field:

$$\boldsymbol{B} = B_0 h(t) \hat{\mathbf{e}}_z \ , \tag{2.424}$$

where B_0 is a constant and $h(t)$ is a slowly varying function of time.

a) Verify that Maxwell's equations are satisfied if the field $\boldsymbol{E}$ induced by $\mathrm{d}\boldsymbol{B}/\mathrm{d}t$ is expressed by:

$$\boldsymbol{E} = \frac{1}{2} B_0 \dot{h} (y \hat{\mathbf{e}}_x - x \hat{\mathbf{e}}_y) \ , \tag{2.425}$$

where $\dot{h} = \mathrm{d}h(t)/\mathrm{d}t$. Specify the restrictions, if necessary.

b) Using the values of the fields $\boldsymbol{E}$ and $\boldsymbol{B}$, show that:

$$\frac{\mathrm{d}}{\mathrm{d}t}\left(\frac{1}{2} m w_\perp^2\right) = -\frac{1}{2} m \omega_c \dot{h} (y w_x - x w_y) \ , \tag{2.426}$$

where $\boldsymbol{w}_\perp = w_x \hat{\mathbf{e}}_x + w_y \hat{\mathbf{e}}_y$, and ω_c is the cyclotron frequency of the charged particle.

c) Find the solutions for x, y, w_x, w_y, to zeroth order (initial conditions $x = y = z = 0$; $w_x(0) = w_{x0}$, $w_y(0) = w_{y0}$, $w_z(0) = w_{z0}$).

d) Show that the quantity $m w_\perp^2 / 2B$ remains constant to order zero in h.

Answer

a) We want to check whether the field $\boldsymbol{E}$ induced by the time variation of $\boldsymbol{B}$ and the field $\boldsymbol{B}$ itself obey Maxwell's four equations:

1. $$\boldsymbol{\nabla} \wedge \boldsymbol{E} = -\frac{\partial \boldsymbol{B}}{\partial t} \ . \tag{2.427}$$

The calculation of the LHS gives:

$$\nabla \wedge \boldsymbol{E} \equiv \nabla \wedge \left[\frac{1}{2} B_0 \dot{h}(y\hat{\mathbf{e}}_x - x\hat{\mathbf{e}}_y)\right] = \frac{1}{2} B_0 \dot{h} \begin{vmatrix} \hat{\mathbf{e}}_x & \hat{\mathbf{e}}_y & \hat{\mathbf{e}}_z \\ \dfrac{\partial}{\partial x} & \dfrac{\partial}{\partial y} & \dfrac{\partial}{\partial z} \\ y & -x & 0 \end{vmatrix} ,$$

$$\equiv \frac{1}{2} B_0 \dot{h} \hat{\mathbf{e}}_z (-1 - 1) = -B_0 \dot{h} \hat{\mathbf{e}}_z , \tag{2.428}$$

but:

$$- B_0 \dot{h} \hat{\mathbf{e}}_z \equiv -\frac{\partial \boldsymbol{B}}{\partial t} , \tag{2.429}$$

which corresponds to the RHS of (2.427): this equation is therefore satisfied.

2. $$\nabla \cdot \epsilon_0 \boldsymbol{E} = 0 . \tag{2.430}$$

From Poisson's equation $\nabla \cdot \boldsymbol{D} \equiv \nabla \cdot \epsilon_0 \boldsymbol{E} = \rho$, where $\rho = 0$ in the individual trajectories model; in effect, the assumption in this description is that there are no charges to induce the field $\boldsymbol{E}$. Equation (2.430) is effectively verified since:

$$\nabla \cdot \epsilon_0 \boldsymbol{E} \equiv \nabla \cdot \left[\epsilon_0 B_0 \dot{h}(y\hat{\mathbf{e}}_x - x\hat{\mathbf{e}}_y)\right] \tag{2.431}$$

$$= \epsilon_0 B_0 \dot{h} \left[\frac{\partial}{\partial x}(y) + \frac{\partial}{\partial y}(-x) + \frac{\partial}{\partial z}(0)\right] \equiv 0 .$$

3. $$\nabla \wedge \boldsymbol{B} = \epsilon_0 \mu_0 \frac{\partial \boldsymbol{E}}{\partial t} + \mu_0 \boldsymbol{J} . \tag{2.432}$$

where $\boldsymbol{J}$, the conduction current, is zero in the framework of individual trajectories. Expanding the LHS of this equation gives:

$$\nabla \wedge \boldsymbol{B} \equiv \begin{vmatrix} \hat{\mathbf{e}}_x & \hat{\mathbf{e}}_y & \hat{\mathbf{e}}_z \\ \dfrac{\partial}{\partial x} & \dfrac{\partial}{\partial y} & \dfrac{\partial}{\partial z} \\ 0 & 0 & B_0 h(t) \end{vmatrix} \tag{2.433}$$

$$\equiv \frac{\partial}{\partial y}(B_0 h(t))\hat{\mathbf{e}}_x - \frac{\partial}{\partial x}(B_0 h(t))\hat{\mathbf{e}}_y \equiv 0 ,$$

since $B_0 h(t)$ is independent of position.
It remains to show that the RHS, $\partial \boldsymbol{E}/\partial t$, is also zero. From (2.425), we have:

$$\begin{aligned}\frac{\partial \boldsymbol{E}}{\partial t} &= \frac{\partial}{\partial t}\left[\frac{1}{2}B_0\dot{h}(y\hat{\mathbf{e}}_x - x\hat{\mathbf{e}}_y)\right] \\ &= \frac{B_0}{2}\left[\dot{h}(w_y\hat{\mathbf{e}}_x - w_x\hat{\mathbf{e}}_y) + \ddot{h}(y\hat{\mathbf{e}}_x - x\hat{\mathbf{e}}_y)\right] . \end{aligned} \tag{2.434}$$

This expression is zero to zeroth order ($\dot{h} \simeq \ddot{h} \simeq 0$), but not to first order.

Remark: We cannot set a priori, $\partial \boldsymbol{E}/\partial t = 0$ because this requires us, in the present case, to neglect the field $\boldsymbol{E}$ induced by $\partial \boldsymbol{B}/\partial t$. On the other hand, the field $\boldsymbol{E}$ induced by the particle motion ($\boldsymbol{J} = \sigma \boldsymbol{E}$) is effectively zero in the framework of individual particles since $J = 0$.

4. $$\boldsymbol{\nabla} \cdot \boldsymbol{B} = 0 . \tag{2.435}$$

This equation is trivially verified because B_0 and $h(t)$ are independent of position. In effect:

$$\frac{\partial}{\partial x}(0) + \frac{\partial}{\partial y}(0) + \frac{\partial}{\partial z}(B_0 h(t)) = 0 . \tag{2.436}$$

The four Maxwell equations are satisfied, but only to order zero ($\dot{h} = \ddot{h} = 0$) for $\boldsymbol{\nabla} \wedge \boldsymbol{B} = \partial \boldsymbol{D}/\partial t$.

b) We know that only the electric field affects the work done (Sect. 2.1); in the present case, the actual electric field is that induced by the variation of $\boldsymbol{B}$. The component of the kinetic energy perpendicular to the field $\boldsymbol{B}$ is given by:

$$\frac{\mathrm{d}}{\mathrm{d}t}\left(\frac{1}{2}mw_\perp^2\right) = q\boldsymbol{E} \cdot \boldsymbol{w}_\perp , \tag{2.437}$$

but this is also the total work effected because $\boldsymbol{E}$ is entirely in the plane perpendicular to $\boldsymbol{B}$.
We then have:

$$\begin{aligned}\frac{\mathrm{d}}{\mathrm{d}t}\left(\frac{1}{2}mw_\perp^2\right) &= q\frac{1}{2}B_0\dot{h}(y\hat{\mathbf{e}}_x - x\hat{\mathbf{e}}_y) \cdot (w_x\hat{\mathbf{e}}_x + w_y\hat{\mathbf{e}}_y) , \\ &= \left(\frac{qB_0}{m}\right)\frac{m\dot{h}}{2}(yw_x - xw_y) \end{aligned} \tag{2.438}$$

and, noting $\omega_c = -qB_0/m$, we then retrieve:

$$\frac{\mathrm{d}}{\mathrm{d}t}\left(\frac{1}{2}mw_\perp^2\right) = -\frac{1}{2}m\omega_c\dot{h}(yw_x - xw_y) . \tag{2.426}$$

c) The equations of motion in Cartesian coordinates (Sect. 2.1) are:

$$\dot{w}_x = \frac{q}{m}\left[\frac{1}{2}B_0\dot{h}y + B_0h(t)w_y\right] , \tag{2.439}$$

$$\dot{w}_y = \frac{q}{m}\left[-\frac{1}{2}B_0\dot{h}x - B_0h(t)w_x\right] , \tag{2.440}$$

$$\dot{w}_z = \frac{q}{m}[0] . \tag{2.441}$$

To integrate these equations, we will use the guiding centre approximation, and consider that there are two timescales, that of the cyclotron motion, and that of the much slower motion due to $h(t)$: thus $h(t)$ will be a constant to zeroth order and we will then put $\dot{h} = 0$ in Eqs. (2.439) and (2.440). Setting $-qB_0h/m = \Omega$, we can, by identification, use the solutions for a constant field B, taken from Sect. 2.2.2, p. 113.
However, in this case, $\boldsymbol{B}$ is along $\hat{\mathbf{e}}_x$ while here it needs to be along $\hat{\mathbf{e}}_z$: exchanging $x \leftrightarrow z$ and recalling that, to maintain a right-handed frame, we must have $\hat{\mathbf{e}}_x \wedge \hat{\mathbf{e}}_y = \hat{\mathbf{e}}_z$, it is necessary to replace z by $-x$ in such a permutation.
Finally:

$$x = -\frac{w_{x0}}{\Omega}\sin\Omega t + \frac{w_{y0}}{\Omega}\cos\Omega t - \frac{w_{y0}}{\Omega} , \tag{2.442}$$

$$y = \frac{w_{x0}}{\Omega}\cos\Omega t + \frac{w_{y0}}{\Omega}\sin\Omega t - \frac{w_{x0}}{\Omega} , \tag{2.443}$$

$$z = w_{z0}t , \tag{2.444}$$

$$w_x = -w_{x0}\cos\Omega t - w_{y0}\sin\Omega t , \tag{2.445}$$

$$w_y = -w_{x0}\sin\Omega t + w_{y0}\cos\Omega t , \tag{2.446}$$

$$w_z = w_{z0} . \tag{2.447}$$

We can easily verify that the initial conditions are respected: we recover $x = y = z = 0$ at $t = 0$ and $w_x(0) = w_{x0}$, $w_y(0) = w_{y0}$, $w_z(0) = w_{z0}$.

Another solution

The initial conditions are the same except for $w_x(0) = 0$. The expressions are then much simpler:

$$x = \frac{w_{y0}}{\Omega}\cos\Omega t - \frac{w_{y0}}{\Omega} , \tag{2.448}$$

$$y = \frac{w_{y0}}{\Omega}\sin\Omega t , \tag{2.449}$$

$$z = w_{z0}t , \tag{2.450}$$

$$w_x = -w_{y0} \sin \Omega t \ , \tag{2.451}$$

$$w_y = w_{y0} \cos \Omega t \ , \tag{2.452}$$

$$w_z = w_{z0} \ . \tag{2.453}$$

d) We need to verify that the first adiabatic invariant is, in effect, constant to zeroth order, in the present configuration.
We will consider the temporal variation of the magnetic moment:

$$\frac{\mathrm{d}}{\mathrm{d}t}\mu \equiv \frac{\mathrm{d}}{\mathrm{d}t}\left(\frac{mw_\perp^2}{2B}\right) = \frac{1}{B}\frac{\mathrm{d}}{\mathrm{d}t}\left(\frac{mw_\perp^2}{2}\right) - \frac{1}{B^2}\left(\frac{\mathrm{d}B}{\mathrm{d}t}\right)\left(\frac{mw_\perp^2}{2}\right) . \tag{2.454}$$

The derivative appearing in the first term on the RHS is given by (2.426), while the second term is calculated by making use of the zeroth order velocities ((2.445) and (2.446)), conforming to the concept of adiabatic invariance:

$$\frac{\mathrm{d}}{\mathrm{d}t}\mu = \frac{1}{B}\frac{\mathrm{d}}{\mathrm{d}t}\left[-\frac{1}{2}m\omega_c\dot{h}(yw_x - xw_y)\right] - \frac{1}{B_0^2 h^2}B_0\dot{h}\left[\frac{1}{2}m(w_{x0}^2 + w_{y0}^2)\right] . \tag{2.455}$$

It is clear that the RHS, due to the presence of $\dot{h}$, is of first order, therefore zero to zeroth order, which leads to the fact that μ is indeed a constant to order zero.

2.14. Consider a static magnetic field (toroidal field):

$$\boldsymbol{B} = \hat{\mathbf{e}}_x B_0 \alpha z + \hat{\mathbf{e}}_z B_0 (1 + \alpha x) \ , \tag{2.456}$$

where α is a constant much less than unity.

a) Express the equation of motion of an individual particle in Cartesian coordinates. Underline the terms which are linked to the curvature of the lines of force of the magnetic field $\boldsymbol{B}$.
b) Find the solutions of the motion to zeroth order, knowing that the initial conditions are:

$$\begin{aligned} w_x(0) &= 0 & x(0) &= r_B \\ w_y(0) &= w_{y0} & y(0) &= 0 \\ w_z(0) &= w_{z0} & z(0) &= 0 \end{aligned} \tag{2.457}$$

with $r_B = w_{y0}/\omega_c$, where ω_c is the cyclotron frequency and r_B, the cyclotron gyro-radius.
c) Then show that to order one, the following equation is obtained for w_x.

$$\ddot{w}_x + \omega_c^2 w_x \simeq \alpha\omega_c^2 \left(\frac{3}{2} r_B w_{y0} \sin 2\omega_c t + w_{z0}^2 t\right) . \tag{2.458}$$

d) Find the solution for w_x from (2.458).

Answer

a) This problem corresponds to the case studied in Sect. 2.2.3, p. 152. The component of $\boldsymbol{B}$ in the direction of $\hat{\mathbf{e}}_x$ is that responsible for the curvature of the field lines of $\boldsymbol{B}$. The field already satisfies Maxwell's equations, since $\boldsymbol{\nabla} \cdot \boldsymbol{B} = 0$ (Sect. 2.2.3, p. 138) and clearly from (2.457), $\boldsymbol{\nabla} \wedge \boldsymbol{B} = 0$.
The equation of motion in Cartesian coordinates, with $E = 0$, $B_y = 0$ and $\omega_c = -qB/m$, is obtained from (2.6)–(2.8):

$$\dot{w}_x = \frac{q}{m}\left[B_z w_y\right] = \frac{q}{m}\left[B_0(1+\alpha x) w_y\right] = -\omega_c w_y (1+\alpha x) , \tag{2.459}$$

$$\begin{aligned} \dot{w}_y &= \frac{q}{m}\left[B_x w_z - B_z w_x\right] = \frac{q}{m}\left[B_0 \alpha z w_z - B_0(1+\alpha x) w_x\right] \\ &= \omega_c \left[w_x + \alpha(x w_x - \underline{z w_z})\right] , \end{aligned} \tag{2.460}$$

$$\dot{w}_z = \frac{q}{m}\left[-B_x w_y\right] = -\frac{q}{m}\left[B_0 \alpha z w_y\right] = \underline{\omega_c \alpha z w_y} , \tag{2.461}$$

where the first order quantities underlined are related to the curvature of the field (contribution from the B_x component).

b) In order to reduce (2.459)–(2.461) to order zero, we only need to set $\alpha = 0$:

$$\dot{w}_x = -\omega_c w_y , \tag{2.462}$$

$$\dot{w}_y = \omega_c w_x , \tag{2.463}$$

$$\dot{w}_z = 0 . \tag{2.464}$$

To calculate the w_x component, we start by differentiating equation (7), to introduce $\dot{w}_y$:

$$\dot{w}_x = -\omega_c \dot{w}_y , \tag{2.465}$$

then use (2.463) to obtain:

$$\dot{w}_x + \omega_c^2 w_x = 0 , \tag{2.466}$$

which has the (harmonic oscillator) solution:

$$w_x = A_1 \sin \omega_c t + A_2 \cos \omega_c t , \tag{2.467}$$

where the constants A_1 and A_2 must be determined from the initial conditions. Since $w_x(0) = 0$, $A_2 = 0$. For A_1, we have, by integrating (2.467):

$$x = -\frac{A_1}{\omega_c}\cos\omega_c t \tag{2.468}$$

and since $x(0) = r_B$, $A_1 = -r_B\omega_c$:

$$w_x = -r_B\omega_c \sin\omega_c t \ , \tag{2.469}$$

$$x = r_B \cos\omega_c t \ . \tag{2.470}$$

w_y, it is found by the same method:

$$w_y = A_1 \sin\omega_c t + A_2 \cos\omega_c t \ , \tag{2.471}$$

which leads to:

$$w_y = w_{y0}\cos\omega_c t \tag{2.472}$$

and:

$$y = r_B \sin\omega_c t \ . \tag{2.473}$$

Finally, for the w_z component, since $\dot{w}_z = 0$ (2.461), we obtain:

$$w_z = w_{z0} \tag{2.474}$$

and:

$$z = w_{z0}t \ . \tag{2.475}$$

c) By substituting the values of the zero order components of $\boldsymbol{w}$ and $\boldsymbol{r}$ in the terms involving α ((2.459)–(2.461)), we find:

$$\dot{w}_x = -\omega_c\left[w_y + \alpha w_{y0} r_B \cos^2\omega_c t\right] \ , \tag{2.476}$$

$$\dot{w}_y = \omega_c\left[w_x + \alpha(-r_B^2\omega_c \sin\omega_c t\cos\omega_c t - w_{z0}^2 t)\right] \ , \tag{2.477}$$

$$\dot{w}_z = \alpha\omega_c w_{y0} w_{z0} t\cos\omega_c t \ . \tag{2.478}$$

To obtain an homogeneous equation for w_x, we proceed in the same fashion as b), differentiating (2.476) with respect to time t, then replacing $\dot{w}_y$ by its value taken from (2.477):

$$\ddot{w}_x = -\omega_c\left[\dot{w}_y - 2\alpha w_{y0} r_B\omega_c \cos\omega_c t\sin\omega_c t\right] \ ,$$

$$\ddot{w}_x = -\omega_c\left\{\omega_c\left[w_x + \alpha\left(-\frac{r_B w_{y0}}{2}\sin 2\omega_c t - w_{z0}^2 t\right)\right]\right\}$$
$$+\omega_c^2\alpha w_{y0} r_B \sin 2\omega_c t \ ,$$

i.e.:

$$\ddot{w}_x + \omega_c^2 w_x = \alpha\omega_c^2\left[\frac{3}{2} r_B w_{y0}\sin 2\omega_c t + w_{z0}^2 t\right] \ . \tag{2.479}$$

d) The solution of the differential equation (2.479) is the sum of the general solution without the RHS:

$$w_x = A_1 \cos \omega_c t + A_2 \sin \omega_c t \tag{2.480}$$

and a particular solution with the RHS (not obvious!):

$$w_x = -\frac{1}{2} \alpha r_B w_{y0} \sin 2\omega_c t + \alpha w_{z0}^2 t \,. \tag{2.481}$$

We can verify that (2.481) is indeed a particular solution of (2.479). From (2.479) and (2.480),

$$\begin{aligned} \dot{w}_x + \omega_c^2 w_x &\equiv 2\alpha \omega_c^2 r_B w_{y0} \sin 2\omega_c t - \frac{\omega_c^2}{2} \alpha r_B w_{y0} \sin 2\omega_c t + \omega_c^2 \alpha w_{z0}^2 t \\ &\equiv \frac{3}{2} \alpha \omega_c^2 r_B w_{y0} \sin 2\omega_c t + \omega_c^2 \alpha w_{z0}^2 t \,, \end{aligned}$$

which corresponds exactly to the RHS of (2.479).
We can fix the constants A_1 and A_2 in (2.480) by the values they have at $t = 0$. Since $w_x(0) = 0$, then $A_1 = 0$. To obtain A_2, we integrate the complete solution to obtain x:

$$x = -\frac{A_2}{\omega_c} \cos \omega_c t + \frac{1}{2} \frac{\alpha r_B w_{y0}}{2\omega_c} \cos 2\omega_c t + \frac{\alpha w_{z0}^2 t^2}{2} \,, \tag{2.482}$$

from which:

$$x(0) = -\frac{A_2}{\omega_c} + \frac{1}{4} \frac{\alpha r_B w_{y0}}{\omega_c} = r_B \tag{2.483}$$

and, thus, since $r_B = \dfrac{w_{0\perp}}{\omega_c} = \dfrac{w_{0y}}{\omega_c}$:

$$A_2 = \left(\frac{1}{4} \frac{\alpha r_B w_{y0}}{\omega_c} - r_B \right) \omega_c = \frac{1}{4} \alpha r_B w_{y0} - w_{y0} = w_{y0} \left(\frac{\alpha r_B}{4} - 1 \right) \,, \tag{2.484}$$

i.e.:

$$w_x = w_{y0} \left(\frac{\alpha r_B}{4} - 1 \right) \sin \omega_c t - \frac{1}{2} \alpha r_B w_{y0} \sin 2\omega_c t + \alpha w_{z0}^2 t \,. \tag{2.485}$$

2.15. Consider a linear magnetic confinement machine, limited at each end by magnetic mirrors. We can arrange for the two mirrors to be displaced with respect to each other, each being given a velocity v_M in the laboratory frame. We will consider a particle of charge q and mass m which, in the homogeneous magnetic field region of the machine, is characterised initially by a velocity $\boldsymbol{w}_0$, such that $w_{0\perp} = w_{0\parallel}$, and by its kinetic energy $\mathcal{E}_i$.

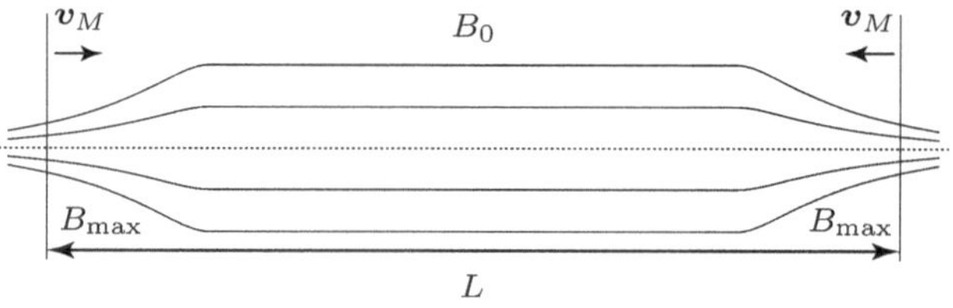

a) Show that at each reflection by the mirrors, the magnitude of the parallel velocity of the particle is increased by $2v_M$.
b) Explain why the particle will eventually leave the mirror.
c) Calculate the energy $\mathcal{E}_p$ of the particle when it leaves the mirror: express $\mathcal{E}_p$ as a function of the initial energy $\mathcal{E}_i$ and the mirror ratio $\mathcal{R} = B_{max}/B_0$.
d) Find the expression giving the number of reflections n that the particle experiences before leaving the system, as a function of v_M, $\mathcal{E}_i$, $\mathcal{R}$, and m.

Answer

a) In the laboratory frame, the particle with velocity $\boldsymbol{w}_{0\|}$ is directed towards the mirror M, which is itself in motion, with a velocity $\boldsymbol{v}_M$, in the direction of the incident particle, as illustrated in the figure. The velocity of the particle along the z axis, in this frame, is:

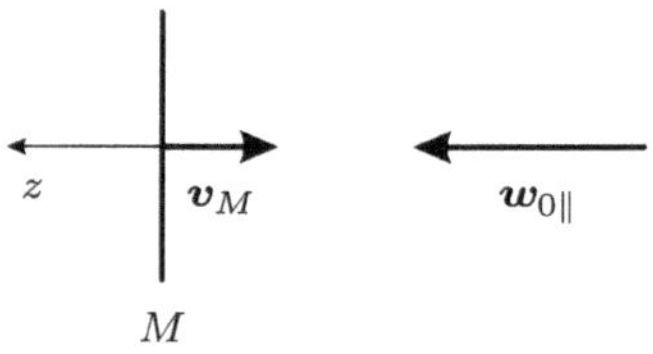

$$\boldsymbol{w}_z = \boldsymbol{w}_{0\|} \ , \tag{2.486}$$

taking the positive sign as the direction to the mirror.
An observer tied to the mirror sees the particle coming towards him with an increased velocity:

$$\boldsymbol{w}_{zM} = \boldsymbol{w}_{0\|} - \boldsymbol{v}_M \ , \tag{2.487}$$

since $\boldsymbol{v}_M$ is negative.
In the same frame, after reflection, the particle returns with the opposite velocity:

$$\boldsymbol{w}'_{zM} = -(\boldsymbol{w}_{0\|} - \boldsymbol{v}_M) \ . \tag{2.488}$$

Returning to the laboratory frame, we must, this time, add $\boldsymbol{v}_M$ to the particle velocity (inverse operation as for (2.487):

$$\boldsymbol{w}'_z = -(\boldsymbol{w}_{0\|} - \boldsymbol{v}_M) + \boldsymbol{v}_M = -(\boldsymbol{w}_{0\|} - 2\boldsymbol{v}_M) \ . \tag{2.489}$$

The speed of the particle after reflection is thus augmented by $|2\boldsymbol{v}_M|$ (compare (2.486) and (2.489)).

Another solution for a)

We can consider the reflection of the particle on the mirror as a head-on elastic collision between a light particle of mass m and a moving wall, i.e. a heavy particle with a quasi-infinite mass M ($m \ll M$).
The relative velocity $\boldsymbol{w}_{mM}$ of the two particles, independent of the frame, is expressed by:

$$\boldsymbol{w}_{mM} = \boldsymbol{w}_{\|0} - \boldsymbol{v}_M \ . \tag{1.70}$$

After reflection, considering an elastic collision, the relative velocity is opposite and with the same modulus (after (1.79)):

$$\boldsymbol{w}'_{mM} = -\boldsymbol{w}_{mM} = -(\boldsymbol{w}_{\|0} - \boldsymbol{v}_M) \ . \tag{2.490}$$

In the laboratory frame, the velocity $\boldsymbol{w}'_{\|0}$ of the light particle after reflection becomes (1.74):

$$\boldsymbol{w}'_{\|0} = \boldsymbol{w}'_{CM} + \left(\frac{M}{m+M}\right) \boldsymbol{w}'_{mM} \ , \tag{2.491}$$

where $\boldsymbol{w}'_{CM}$ is the velocity of the center of mass after reflection. Then, from (1.69) and (1.70):

$$\boldsymbol{w}'_{CM} = \boldsymbol{w}_{CM} = \frac{m\boldsymbol{w}_{\|0} + M\boldsymbol{v}_M}{m+M} \ . \tag{2.492}$$

From (2.491) and (2.492), we obtain:

$$\boldsymbol{w}'_{\|0} = \frac{m\boldsymbol{w}_{\|0} + M\boldsymbol{v}_M + M\boldsymbol{v}_M - M\boldsymbol{w}_{\|0}}{m+M} \ , \tag{2.493}$$

which, assuming $M \gg m$, gives:

$$\boldsymbol{w}'_{\|0} = -\boldsymbol{w}_{\|0} + 2\boldsymbol{v}_M \ . \tag{2.494}$$

b) We have just shown in a) that $\boldsymbol{w}_{\|}$, the component of the particle velocity parallel to the field $\boldsymbol{B}$, is increased at each reflection, while the component $\boldsymbol{w}_{\perp}$ remains unchanged, with $\boldsymbol{w}_{\perp}^{(n)} = \boldsymbol{w}_{\perp}^{(0)}$ after a number of reflections n. As a result, assuming that in the uniform region before the first reflection we have:

$$w_{0\perp}^{(0)} = w_0 \sin\alpha_0 \ , \tag{2.495}$$

then after the first reflection, this becomes:

$$w_{0\perp}^{(1)} \equiv w_0^{(1)} \sin\alpha_0^{(1)} = w_0^{(0)} \sin\alpha_0^{(0)} \ . \tag{2.496}$$

Since w_0, the modulus of $\boldsymbol{w}$, increases, the angle $\alpha_0^{(1)}$ must decrease, for $w_{0\perp}$ to remain constant: after n reflections, the angle $\alpha_0^{(n)}$ will be smaller than α_{0m} and, following (2.189), the particle will find itself in the loss cone.

c) By assumption, initially:

$$\mathcal{E}_i = \frac{1}{2}m(w_{0\perp}^2 + w_{0\parallel}^2) = m w_{0\parallel}^2 \tag{2.497}$$

and, as discussed in b), the particle will leave the machine when α_0 is sufficiently small, i.e. when:

$$\frac{w_{0\perp}}{w_0^{(n)}} \leq \sin\alpha_{0m} = \frac{1}{\sqrt{\mathcal{R}}} \, . \tag{2.498}$$

Setting the equality in (2.498), where w_0 is increased at each reflection, such that after n reflections:

$$w_0^{(n)} = \left[w_{0\perp}^2 + (w_{0\parallel} + 2n\,v_M)^2\right]^{\frac{1}{2}} \, . \tag{2.499}$$

Owing to our initial assumption $w_{0\parallel} = w_{0\perp}$, from (2.498) we have:

$$w_0^{(n)} = w_{0\perp}\sqrt{\mathcal{R}} = w_{0\parallel}\sqrt{\mathcal{R}} \, , \tag{2.500}$$

such that, by multiplying the square of equation (2.499) by $m/2$ and substituting (2.500), we have:

$$\mathcal{E}_p \equiv \left[\frac{m w_{0\perp}^2}{2} + \frac{m w_{0\parallel}^2}{2} + 2mn\,w_{0\parallel}v_M + 2mn^2 v_M^2\right] = \frac{1}{2} m w_{0\parallel}^2 \mathcal{R} \, , \tag{2.501}$$

thus:

$$\mathcal{E}_p = \mathcal{E}_i \frac{\mathcal{R}}{2} \, .$$

d) We can calculate n from its quadratic expression in (2.501):

$$n^2 + \frac{w_{0\parallel}}{v_M} n - \mathcal{E}_i \left(\frac{\mathcal{R}}{2} - 1\right) \frac{1}{2 m v_M^2} = 0 \, , \tag{2.502}$$

from which we obtain:

$$n = -\frac{w_{0\parallel}}{2v_M} + \frac{1}{2}\sqrt{\left(\frac{w_{0\parallel}}{v_M}\right)^2 + 2\mathcal{E}_i \left(\frac{\mathcal{R}}{2} - 1\right) \frac{1}{m v_M^2}} \, , \tag{2.503}$$

because the solution with a negative sign in front of the square root would give a negative value for n. Continuing the development further, we have:

$$n = -\frac{w_{0\parallel}}{2v_M} + \frac{1}{2}\sqrt{\left(\frac{w_{0\parallel}}{v_M}\right)^2 + 2\left(\frac{w_{0\parallel}}{v_M}\right)^2\left(\frac{\mathcal{R}}{2} - 1\right)}\,, \tag{2.504}$$

$$n = -\frac{w_{0\parallel}}{2v_M} + \frac{w_{0\parallel}}{2v_M}\sqrt{1 + \mathcal{R} - 2}\,, \tag{2.505}$$

$$n = \frac{w_{0\parallel}}{2v_M}\left[(\mathcal{R}-1)^{\frac{1}{2}} - 1\right] = \left(\frac{\mathcal{E}_i}{m}\right)^{\frac{1}{2}}\frac{1}{2v_M}\left[(\mathcal{R}-1)^{\frac{1}{2}} - 1\right]\,, \tag{2.506}$$

of which we will take the nearest upper integer value!

Another solution for d)

We have:

$$w_{f\parallel} = w_{0\parallel} + 2nv_M\,, \tag{2.507}$$

where $w_{f\parallel}$ is the parallel velocity of the particle on leaving the machine, defined by:

$$\sin\alpha_{0m} \geq \frac{w_{0\perp}}{(w_{f\parallel}^2 + w_{0\perp}^2)^{\frac{1}{2}}}\,. \tag{2.508}$$

Considering equality (2.508), we have:

$$\sin\alpha_{0m}^2 = \frac{w_{0\perp}^2}{w_{f\parallel}^2 + w_{0\perp}^2} = \frac{1}{\mathcal{R}}\,, \tag{2.509}$$

which becomes:

$$w_{0\perp}^2(\mathcal{R}-1) = w_{f\parallel}^2\,. \tag{2.510}$$

From (2.507), we can then write (with, by assumption, $w_{0\parallel} = w_{0\perp}$ in the present case):

$$n = \frac{w_{f\parallel} - w_{0\perp}}{2v_M} = \frac{w_{0\parallel}[(\mathcal{R}-1)^{\frac{1}{2}} - 1]}{2v_M}$$

and, using (2.497):

$$n = \left(\frac{\mathcal{E}_i}{m}\right)^{\frac{1}{2}}\frac{1}{2v_M}\left[(\mathcal{R}-1)^{\frac{1}{2}} - 1\right]\,. \tag{2.506}$$

2.16. A magnetic mirror is situated at each end of a machine, and its axis is along the z axis. The magnetic configuration of these mirrors is symmetric with respect to the plane $z = 0$. The planes of the mirrors are situated at z_M and $-z_M$.
The magnetic field has been constructed such that, along the z axis, we have:

$$B(z) = B_0\left[1 + \left(\frac{z}{a}\right)^2\right]\,, \tag{2.511}$$

where a is a constant.

a) Show that the period of oscillation of a particle of mass m between the two mirrors is given by:

$$\mathcal{T}_p = 2\pi a \left[\frac{m}{2\mu B_0}\right]^{\frac{1}{2}} , \tag{2.512}$$

where μ is the orbital magnetic moment. Assume the adiabatic approximation is valid along the whole trajectory.

b) Calculate the particle velocity along z (the particle is assumed to remain within the machine) and show that:

$$w_z = \left(\frac{2B_0\mu}{m}\right)^{\frac{1}{2}} \frac{z_M}{a} \cos\left(\frac{2\pi t}{\mathcal{T}_p}\right) \tag{2.513}$$

and finally:

$$w_z = \left(\frac{2\mu}{m}\right)^{\frac{1}{2}} [B(z_M) - B(z)]^{\frac{1}{2}} . \tag{2.514}$$

Answer

a) To the extent that the magnetic field is only slightly non uniform (adiabatic condition) in its own direction, we can write (2.177):

$$m\dot{\boldsymbol{w}} = \boldsymbol{\mu} \cdot \boldsymbol{\nabla} \boldsymbol{B} \tag{2.515}$$

because $\boldsymbol{\mu} = -\mu \hat{\mathbf{e}}_z$ (p. 136). Along z, in the present case:

$$m\dot{w}_z = -\mu \frac{\partial B_z}{\partial z} = -\frac{2\mu B_0 z}{a^2} , \tag{2.516}$$

where equivalently:

$$\frac{\mathrm{d}^2 z}{\mathrm{d}t^2} = -\frac{2\mu B_0}{ma^2} z , \tag{2.517}$$

which has a permissible solution, for $z = 0$ at $t = 0$:

$$z = z_M \sin\left(\frac{2\mu B_0}{ma^2}\right)^{\frac{1}{2}} t , \tag{2.518}$$

from which the period:

$$\mathcal{T}_p = 2\pi a \sqrt{\frac{m}{2\mu B_0}} . \tag{2.519}$$

b) At $z = \pm z_M$, due to the assumption that the particle does not leave the mirror, we must have $w_\parallel(z_M) = 0$, that is:

$$\frac{1}{2} m w_\parallel^2(z_M) = 0 . \tag{2.520}$$

The conservation of kinetic energy gives:

$$\frac{1}{2}mw_{\parallel}^2(z) + \frac{1}{2}mw_{\perp}^2(z) = \frac{1}{2}mw_{\perp}^2(z_M) . \tag{2.521}$$

Knowing that μ is a constant of the motion, we have:

$$\mu = \frac{1}{2}\frac{mw_{\perp}^2(z)}{B(z)} = \text{constant} , \tag{2.522}$$

such that, from (2.521) and (2.522):

$$\frac{1}{2}mw_{\parallel}^2(z) = \mu[B(z_M) - B(z)] , \tag{2.523}$$

from which:

$$w_{\parallel}(z) = \sqrt{\frac{2\mu}{m}[B(z_M) - B(z)]} . \tag{2.524}$$

If we now replace $B(z_M)$ and $B(z)$ by their respective values:

$$w_{\parallel}(z) = \sqrt{\frac{2\mu B_0}{m}}\left[1 + \left(\frac{z_M}{a}\right)^2 - 1 - \left(\frac{z}{a}\right)^2\right]^{\frac{1}{2}} , \tag{2.525}$$

$$= \sqrt{\frac{2\mu B_0}{m}}\frac{z_M}{a}\left[1 - \left(\frac{z}{z_M}\right)^2\right]^{\frac{1}{2}} , \tag{2.526}$$

then, from (2.518) with (2.519):

$$w_{\parallel}(z) = \sqrt{\frac{2\mu B_0}{m}}\frac{z_M}{a}\left[1 - \sin^2\left(\frac{2\pi t}{\mathcal{T}_p}\right)\right]^{\frac{1}{2}}$$

$$= \sqrt{\frac{2\mu B_0}{m}}\frac{z_M}{a}\cos\left(\frac{2\pi t}{\mathcal{T}_p}\right) . \tag{2.527}$$

2.17. The Hamilton-Jacobi formalism of classical mechanics allows us to introduce the concept of an adiabatic invariant I as the average of an action integral (LHS of (2.528) below). In the case where the particle is subject to a periodic motion, this integral takes a constant value:

$$\frac{1}{2\pi}\oint \mathrm{pdq} = I , \tag{2.528}$$

where q is a generalised canonical coordinate[88] (for example a position variable) and p is the conjugate canonical moment (for example, the momentum

[88] Do not confuse the variables p and q of the Hamilton-Jacobi formalism with pressure and charge.

corresponding to the position q). Given the kinetic energy $\mathcal{E}_c$ of the system, the value of p is obtained from the equation:

$$p = \frac{\partial \mathcal{E}_c}{\partial \dot{q}} \; . \tag{2.529}$$

Consider a linear discharge, confined by a static magnetic field $\boldsymbol{B}$, directed axially, and terminated at its two ends by magnetic mirrors.

a) Show that the action integral taken over the cyclotron motion of a charged particle leads, within a constant coefficient, to a constant value of the orbital magnetic motion μ of the particle. Specify the assumptions used for your calculation.
b) Show that the charged particle oscillates between the mirrors with a period $\mathcal{T}$, given by:

$$\mathcal{T} = \oint \frac{\mathrm{d}z}{\left[\left(\frac{2}{m}\right)(\mathcal{E}_c - \mu B)\right]^{\frac{1}{2}}} \; . \tag{2.530}$$

Specify your assumptions.
c) Calculate the second invariant I_2, linked to the axial motion, to show that:

$$I_2 = \frac{1}{2\pi} \oint \left[2m(\mathcal{E}_c - V)\right]^{\frac{1}{2}} \mathrm{d}z \; , \tag{2.531}$$

where we have introduced the pseudo potential $V \equiv \mu B$, to give the integral a more general aspect.
d) Consider the case where the discharge is confined by a magnetic field:

$$\boldsymbol{B} = B_0 \left[1 + \left(\frac{z}{a}\right)^2\right] \hat{\mathbf{e}}_z \; , \tag{2.532}$$

where a is a constant such that $\partial B/\partial z$ varies very slowly along z: calculate the period of oscillation of a particle in such a linear machine.

Answer

a) Since this concerns cyclotron motion, we will describe the system in cylindrical coordinates (r, φ, z). We then have $x = r_B \cos\varphi$, $y = r_B \sin\varphi$, where r_B is the Larmor radius, which we assume to be constant because B varies slowly along the axis.
Since:

$$\mathcal{E}_c = \frac{1}{2} m(\dot{x}^2 + \dot{y}^2 + \dot{z}^2) \; , \tag{2.533}$$

this can be written in cylindrical coordinates as:

$$\mathcal{E}_c = \frac{1}{2} m(r_B^2 \dot{\varphi}^2 + \dot{z}^2) \; . \tag{2.534}$$

Because $\dot{x}^2 + \dot{y}^2 = (-r_B \sin\varphi\,\dot{\varphi})^2 + (r_B \cos\varphi\,\dot{\varphi})^2$ and setting $q = \varphi$, we obtain for p_φ:

$$p_\varphi = \frac{\partial \mathcal{E}_c}{\partial \dot{\varphi}} = m r_B^2 \dot{\varphi} \,, \tag{2.535}$$

where $\dot{\varphi} \equiv \mathrm{d}\varphi/\mathrm{d}t = \omega_c$, and then:

$$I_1 \equiv \frac{1}{2\pi} \oint p_\varphi \mathrm{d}\varphi = \frac{1}{2\pi} \int\limits_{\substack{\text{cyclotron}\\ \text{period}}} m r_B^2 \dot{\varphi}\, \mathrm{d}\varphi = \frac{1}{2\pi} m r_B^2 \omega_c \int \mathrm{d}\varphi = \frac{m w_\perp^2}{\omega_c} \,, \tag{2.536}$$

recalling that $\omega_c = w_\perp / r_B$. Knowing that $\mu \equiv m w_\perp^2/2B$ and $\omega_c = |q|B/m$, we obtain:

$$I_1 = \left(\frac{2m}{|q|}\right) \mu \,. \tag{2.537}$$

b) The period of oscillation $\mathcal{T}$ along the axis of the discharge is found by integrating the closed (cyclic) motion along z:

$$\mathcal{T} = \oint \frac{\mathrm{d}z}{w_\parallel} = \int \frac{\mathrm{d}z}{\dot{z}} \,. \tag{2.538}$$

In addition, we have from (2.534):

$$\mathcal{E}_c = \frac{1}{2} m r_B^2 \omega_c^2 + \frac{1}{2} m \dot{z}^2 = \frac{1}{2} m w_\perp^2 + \frac{1}{2} m \dot{z}^2 = \mu B + \frac{1}{2} m \dot{z}^2 \,, \tag{2.539}$$

which allows us to calculate the axial velocity $\dot{z}$:

$$\frac{\mathrm{d}z}{\mathrm{d}t} = \dot{z} = \left[\frac{2}{m}(\mathcal{E}_c - \mu B)\right]^{\frac{1}{2}} \,, \tag{2.540}$$

thus, over a cycle back and forth, the period of oscillation of the particle is:

$$\mathcal{T} \equiv \int_0^{\mathcal{T}} \mathrm{d}t = \oint \frac{\mathrm{d}z}{\left[\frac{2}{m}(\mathcal{E}_c - \mu B)\right]^{\frac{1}{2}}} \,. \tag{2.541}$$

c) Taking advantage of the definition (2.528), we can introduce the adiabatic invariant linked to the axial motion, by setting $q = z$ and $p = p_z$:

$$I_2 = \frac{1}{2\pi} \oint p_z \, \mathrm{d}z \,. \tag{2.542}$$

Since $p_z \equiv m\dot{z}$, we obtain:

$$I_2 = \frac{1}{2\pi} \oint m\dot{z} \, \mathrm{d}z \,, \tag{2.543}$$

which can be evaluated from (2.540), giving:

$$I_2 = \frac{1}{2\pi} \oint m \left[\frac{2}{m} (\mathcal{E}_c - \mu B) \right]^{\frac{1}{2}} \mathrm{d}z \,,$$
$$= \frac{1}{2\pi} \oint [2m(\mathcal{E}_c - \mu B)]^{\frac{1}{2}} \mathrm{d}z \,. \tag{2.544}$$

d) We know from (2.177) that:

$$F_z = -\mu \frac{\partial B_z}{\partial z} \,, \tag{2.545}$$

such that, in the present case:

$$F_z = -\mu \frac{B_0 2z}{a^2} \tag{2.546}$$

and, since $F_z = m\dot{z}$, this leads to:

$$\dot{z} + \frac{\mu}{m} \frac{B_0 2}{a^2} z = 0 \,, \tag{2.547}$$

which is the equation of periodic motion. Its oscillation frequency ω is given by:

$$\omega = \left(\frac{2B_0\mu}{m} \right)^{\frac{1}{2}} \frac{1}{a} \tag{2.548}$$

and since $\omega = 2\pi/\mathcal{T}$:

$$\mathcal{T} = 2\pi a \left(\frac{m}{2B_0\mu} \right)^{\frac{1}{2}} \,. \tag{2.549}$$

See also problem 2.16.

2.18. Calculate the current density of ions and electrons in the ionosphere, due to the gravitational gradient in the terrestrial magnetic field at an altitude of 300 km. Assume that the magnetic induction vector $\boldsymbol{B}$ is perpendicular to the terrestrial gravitational field.
The general relation for the gravitational force exerted on a mass m due to a mass M, when the masses are separated by a distance r, is:

$$\boldsymbol{F}_g = -\frac{\boldsymbol{G}Mm}{r^2} \,, \tag{2.550}$$

where $\boldsymbol{G}$ is the universal gravitational constant. By definition, at the surface of the earth, of mass M, a mass m is subject to a force:

$$\boldsymbol{F}_{gt} = -\frac{\boldsymbol{G}Mm}{R^2} = -\boldsymbol{g}_0 m \,, \tag{2.551}$$

where R is the radius of the earth (the mass M is localised at the centre of the earth!).
Numerically, consider the O^+ ions to have a density, $1.8\times10^{12}\,\mathrm{m}^{-3}$, equal to that of the electrons; $|\boldsymbol{B}| = 10^{-4}$ tesla (1 gauss). The mass m_e of the electrons is 9.11×10^{-28} g and that of O^+, is $m_i = m_e \times 1836 \times 16$. The earth radius is $4\times10^7/2\pi$ metres and g_0, the gravitational field at the surface of the earth, is $9.8\,\mathrm{m\,s}^{-2}$.

Answer

a) The expression for the drift velocity in the gravitational field
It has been shown (Appendix XII) that the drift velocity of a charged particle in a magnetic field $\boldsymbol{B}$ due to an arbitrary force $\boldsymbol{F}_D$ is given by:

$$\boldsymbol{w}_D = \frac{\boldsymbol{F}_D \wedge \boldsymbol{B}}{qB^2}, \tag{2.552}$$

where $\boldsymbol{F}_D$ is the gravitational force, in the present case.
For the electrons $\boldsymbol{F}_{De} = -\boldsymbol{G}Mm_e/r^2$ and for the ions $\boldsymbol{F}_{Di} = -\boldsymbol{G}Mm_i/r^2$.
As can be seen from (2.552), the two types of particles, because of the opposite sign of their charge, drift in opposite directions: there will be a net current.

b) Calculation of the gravitational force at an altitude of 300 km with respect to the surface of the earth
At a given altitude z with respect to the surface of the earth such that $z \ll R$ (which is the case here: $(300/40000)2\pi \simeq 0.05$), we can develop (2.550) to first order with respect to the surface of the earth ($z = 0$):

$$\boldsymbol{F}_g = -\frac{\boldsymbol{G}Mm}{(R+z)^2} = -\frac{\boldsymbol{G}Mm}{R^2\left(1+\dfrac{z}{R}\right)^2} \simeq -\frac{\boldsymbol{G}Mm}{R^2}\left(1-\frac{2z}{R}\right). \tag{2.553}$$

Since, at the surface of the earth:

$$\boldsymbol{F}_{gt} = -\frac{\boldsymbol{G}Mm}{R^2} = -m\boldsymbol{g}_0, \tag{2.551}$$

then from (2.551) and (2.553):

$$F_g = -mg_0\left(1-\frac{2z}{R}\right) \tag{2.554}$$

and finally, because $\boldsymbol{B} \perp \boldsymbol{F}_g$, the expression of drift velocity ((2.551) and (2.552)) is:

$$w_D = \frac{mg_0\left(1-\dfrac{2z}{R}\right)}{qB}. \tag{2.555}$$

c) The current density of the gravitational drift is due to the contribution of the ions and the electrons which, moving in opposite directions (contrary to the electric field drift in electric and magnetic fields), creates a net current:

$$\boldsymbol{J} = ne\boldsymbol{v}_{Di} - ne\boldsymbol{v}_{De} \ , \tag{2.556}$$

where $\boldsymbol{v}_{Di}$ and $\boldsymbol{v}_{De}$ are the drift velocities of the ions and electrons respectively. This leads to:

$$J = ne\left(1 - \frac{2z}{R}\right)\left[\frac{m_i g_0}{eB} - \frac{m_e g_0}{-eB}\right] \ , \tag{2.557}$$

$$J = \frac{n g_0}{B}\left(1 - \frac{2z}{R}\right)(m_i + m_e) \tag{2.558}$$

and, because $m_i \gg m_e$:

$$J \simeq \frac{n g_0}{B}\left(1 - \frac{2z}{R}\right) m_i \ . \tag{2.559}$$

This drift is perpendicular to the direction of the earth radius and to B.

d) Numerical application:

$$J \simeq \frac{1.8 \times 10^{12} \times 9.8 \left(1 - \dfrac{2 \times 300 \times 10^3 \times 2\pi}{4 \times 10^7}\right)}{10^{-4}} 1837 \times 16 \times 9.11 \times 10^{-31}$$
$$= 4.3 \times 10^{-9}\,\mathrm{A\,m^{-2}} \simeq 4\,\mathrm{nA\,m^{-2}}\,!$$

Chapter 3
Hydrodynamic Description of a Plasma

In Chapter 2, we analysed the motion of an isolated charged particle, subjected to externally applied electric ($\boldsymbol{E}$) and magnetic ($\boldsymbol{B}$) fields, ignoring the interaction with other particles. In this chapter, we will introduce a model that considers an ensemble of particles; the motion of these charged particles self consistently produces magnetic and electric fields, referred to as induced fields, which are superimposed on the external fields applied to the plasma. A further difference from Chap. 2 is that we will also account for collisions. This model is *hydrodynamic* in nature, such that the parameters describing the plasma (density, particle diffusion, fluid velocity $\boldsymbol{v}$, temperature, kinetic pressure...) are average values taken over a velocity distribution in a volume element. These values are said to be *macroscopic*.

The hydrodynamic model concept: assumption of a continuous medium

In this description, we follow the motion of small volume elements in the plasma, without taking account of the microscopic phenomena taking place therein. This assumes:

1. That there are enough particles in the volume for the fluctuations around the average values to be negligible, leading to well centred average values. In the same way, we assume that the effect of electric and magnetic microfields produced by the charged particles in the volume element considered is taken account of on the macroscopic scale by average fields assumed to be acting globally on this same volume element.
2. That the volume elements are sufficiently small to provide an accurate spatial description.

These conditions are generally realised in laboratory plasmas: for example, a cube of gas with sides of length 10 μm at atmospheric pressure contains 2.7×10^{10} molecules.

M. Moisan, J. Pelletier, *Physics of Collisional Plasmas*,
DOI 10.1007/978-94-007-4558-2_3,

This macroscopic description of a plasma is analogous to that of ordinary fluids; the particles in the same volume element move together[89], forming a *continuous medium* although it is discontinuous on the molecular scale.

Relationship between the hydrodynamic model and the kinetic description of a plasma

The hydrodynamic description of a plasma introduces macroscopic quantities such as the temperature, the pressure, the mobility of charged particles, the diffusion coefficient of different types of particles, etc: these values are averages, calculated from the distribution function $f(\boldsymbol{r}, \boldsymbol{w}, t)$ of the microscopic velocities $\boldsymbol{w}$ of the particles (Sect. 3.3). This distribution function is obtained in the framework of the kinetic theory of gases.

The hydrodynamic model enables us to describe all the physical phenomena taking place in the plasma in a relatively complete way, by performing calculations which are much simpler than those from kinetic theory, which are much more difficult and complex to derive and interpret.

A plasma is not an ordinary fluid

The charged particles in a plasma constitute one or more fluids, whose motion results in a current density $\boldsymbol{J}$: this motion involves the coupling of the $\boldsymbol{E}$ and $\boldsymbol{B}$ external fields with the particles, through the terms $\boldsymbol{J} \wedge \boldsymbol{B}$ in the equation of motion and $\boldsymbol{J} \cdot \boldsymbol{E}$ in the energy balance equation[90]. The study of fluid conductors subjected to electromagnetic fields has given birth to a branch of plasma physics called magneto-hydrodynamics (MHD).

The term MHD is also used to designate a particular technology, related to this area of plasma physics: this term can also be used to refer to ion propulsion or MHD electricity converters, for example.

Cold plasma and warm plasma: two levels of approximation of the hydrodynamic model

In the framework of the hydrodynamic model, if we neglect all thermal motion of the particles ($T = 0$), we obtain what is conveniently referred to as a *cold plasma*; if, on the other hand, we take account of the random motion of the

---

[89] In fact, particles pass from one volume element to another, but the average number of particles in each element remains almost constant, or varies slowly.

[90] The term $\boldsymbol{J} \wedge \boldsymbol{B}$, the magnetic part of the Lorentz force $\boldsymbol{F} = q(\boldsymbol{E} + \boldsymbol{v} \wedge \boldsymbol{B})$, is also referred to as the Laplace force. The term $\boldsymbol{J} \cdot \boldsymbol{E}$ describes the transfer of energy from the field $\boldsymbol{E}$ to the charged particles, and is related to Ohm's Law ($P = VI$). From the fact that $\boldsymbol{J} = \sigma \boldsymbol{E}$, we obtain $\boldsymbol{J} \cdot \boldsymbol{E} = \sigma E^2$; this term is referred to as the Joule *heating effect.*

particles, but assume an isotropic scalar pressure[91], $p_\alpha = n_\alpha k_B T_\alpha$, we call this the *warm plasma* approximation.

Description of the plasma by one, two or many fluids

- In the case of a *weak interaction* between the different particle species (electrons, ions, neutrals), each type of particle is generally characterised by a different temperature T_α. We can thus describe the motion of electrons, ions and neutrals separately, knowing however that these fluids are partially coupled by their collisional interaction terms. Moreover, since the neutrals do not react with the $\boldsymbol{E}$ and $\boldsymbol{B}$ fields, we usually do not include their equation of motion (for example, in the case of propagation of a wave), but we take account of their influence through the collision terms in the fluid equations of the charged particles; we can thus reduce the model to two fluids (electrons and ions).
  A particularly interesting case of weak coupling is that in which we limit ourselves to a single fluid, that of the electrons, neglecting the motion of the ions, which are much heavier. In this case, the ions are assumed to form a stationary and continuous background, ensuring macroscopic neutrality and influencing the collisional interaction term of the electron fluid. This is the *Lorentz plasma* model, used in particular to describe plasmas produced by HF fields.
- In the case of a *strong interaction* between the different types of particles, as is the case in a plasma in LTE, the different fluids are so well coupled that they can be considered as a single fluid possessing, a single and equal temperature, that is sufficient to describe the plasma.

3.1 Fundamental aspects of the Boltzmann equation

Before giving a detailed description of the hydrodynamic approach, we will discuss some rudimentary kinetic theory.

3.1.1 Formal derivation of the Boltzmann equation

The Boltzmann equation is derived rigorously from Liouville's theorem (see for example Delcroix and Bers, Sect. 10.2). We can, however, obtain this equation simply, but in a purely formal way, by initially assuming there are no collisions between the particles, and in a further calculation, including the

[91] When there is no isotropy, it is convenient to use the (hydrodynamic) kinetic pressure tensor, $\underline{\boldsymbol{\Psi}}$,which is a 2nd order tensor (Sects. 3.3 and 3.5).

effect of collisions. Consider the particles of a particular type that, at time t, occupy a volume element $\mathrm{d}\boldsymbol{r}\,\mathrm{d}\boldsymbol{w}$ in *phase space*, centred at the values $\boldsymbol{r}$, $\boldsymbol{w}$. By definition, the number of particles is given by:

$$f(\boldsymbol{r},\boldsymbol{w},t)\,\mathrm{d}\boldsymbol{r}\,\mathrm{d}\boldsymbol{w}\;, \tag{3.1}$$

where $f(\boldsymbol{r},\boldsymbol{w},t)$ is the *velocity distribution function* of the particles. In the absence of collisions, the flow in phase space is that of an incompressible fluid (Liouville's theorem), such that at a later time $t+\mathrm{d}t$, the same particles of the volume element $\mathrm{d}\boldsymbol{r}\,\mathrm{d}\boldsymbol{w}$ will, under the influence of the forces present, be situated at the point $\boldsymbol{r}+\boldsymbol{w}\,\mathrm{d}t$, $\boldsymbol{w}+(\mathrm{d}\boldsymbol{w}/\mathrm{d}t)\mathrm{d}t$ in phase space, where $\mathrm{d}\boldsymbol{w}/\mathrm{d}t=\boldsymbol{F}/m$ represents the acceleration produced by the fields imposed externally on the plasma and those induced by the micro-fields. Expanding the distribution function at the new point with respect to the initial point, using a first order Taylor expansion, we obtain[92]:

$$f(\boldsymbol{r}+\boldsymbol{w}\mathrm{d}t,\,\boldsymbol{w}+\frac{\boldsymbol{F}}{m}\mathrm{d}t,\,t+\mathrm{d}t)\approx f(\boldsymbol{r},\boldsymbol{w},t)+\sum_{i=1}^{3}\frac{\partial f}{\partial x_i}w_i\,\mathrm{d}t+\sum_{i=1}^{3}\frac{\partial f}{\partial w_i}\frac{F_i}{m}\mathrm{d}t+\frac{\partial f}{\partial t}\mathrm{d}t \tag{3.2}$$

and we can then express the total variation, $\mathrm{d}f/\mathrm{d}t$, of f between the two points, in the time step $\mathrm{d}t$, in the form:

$$\frac{\mathrm{d}f}{\mathrm{d}t}\approx\sum_{i=1}^{3}w_i\frac{\partial f}{\partial x_i}+\sum_{i=1}^{3}\frac{F_i}{m}\frac{\partial f}{\partial w_i}+\frac{\partial f}{\partial t}=\boldsymbol{w}\cdot\boldsymbol{\nabla}_r f+\frac{\boldsymbol{F}}{m}\cdot\boldsymbol{\nabla}_w f+\frac{\partial f}{\partial t}\;, \tag{3.3}$$

where $\boldsymbol{\nabla}_r$ and $\boldsymbol{\nabla}_w$ are the differential operators in spatial coordinate space and velocity space respectively.

In the absence of collisions, the number of particles contained in a volume element in phase space is conserved and the distribution function is therefore not modified, such that $\mathrm{d}f/\mathrm{d}t=0$, from which:

$$\frac{\partial f}{\partial t}+\boldsymbol{w}\cdot\boldsymbol{\nabla}_r f+\frac{\boldsymbol{F}}{m}\cdot\boldsymbol{\nabla}_w f=0\;. \tag{3.4}$$

This is the Vlasov equation (or the collisionless Boltzmann equation), an equation that is particularly useful for providing a simple treatment of wave propagation in a plasma: the field of the wave acts on the charged particles in the plasma, and these, in their turn, modify the field of the wave. This effect is self consistently accounted for by the force term $\boldsymbol{F}/m$. We can set $\partial f/\partial t=0$ if we wish to study the stationary solution, which however is not applicable in the case of an alternating electric field, such as HF discharges.

[92] Recall that $f(x+\Delta x)\simeq f(x)+(\partial f/\partial x)\Delta x$, where Δx is very small (this development is limited to first order).

For the second step of our formal derivation, we will consider the influence of binary collisions. Without further justification, therefore, we will suppose that, for species α:

$$\frac{\mathrm{d}f}{\mathrm{d}t} \equiv \frac{\partial f}{\partial t} + \boldsymbol{w} \cdot \boldsymbol{\nabla}_r f + \frac{\boldsymbol{F}}{m} \cdot \boldsymbol{\nabla}_w f = \left(\frac{\partial f}{\partial t}\right)_{\text{coll.}} , \tag{3.5}$$

where $(\partial f/\partial t)_{\text{coll.}}$ is the collision (or interaction) term in the Boltzmann equation: it expresses the variation of the number of particles in the volume element of phase space, centred about $\boldsymbol{r}, \boldsymbol{w}$, due to elastic and inelastic collisions. To avoid any confusion in notation, we will replace $(\partial f/\partial t)_{\text{coll.}}$ by $S(f)$, where S denotes the global *collision operator*:

$$\frac{\partial f}{\partial t} + \boldsymbol{w} \cdot \boldsymbol{\nabla}_r f + \frac{\boldsymbol{F}}{m} \cdot \boldsymbol{\nabla}_w f = S(f) . \tag{3.6}$$

Equation (3.6) is the *Boltzmann equation*. It describes the evolution of the particle distribution function in phase space under the influence, on the one hand, of the gradients affecting this distribution, and on the other hand, the forces and the presence of collisions.

In the case of elastic collisions, the collision operator can be expressed in the form of an integral (assuming binary collisions, as well as weak correlations):

$$S(f_\alpha) = \sum_{\beta \neq \alpha} S(f_\alpha)_\beta = \sum_{\beta \neq \alpha} \int_{\boldsymbol{w}_\beta} \int_{\Omega} (f'_\alpha f'_\beta - f_\alpha f_\beta) \, |\boldsymbol{w}_\alpha - \boldsymbol{w}_\beta| \, \hat{\sigma}_{\alpha\beta}(\Omega) \, \mathrm{d}\Omega \, \mathrm{d}\boldsymbol{w}_\beta \tag{3.7}$$

where f'_α, f'_β and f_α, f_β are the distribution functions after and before collisions, respectively; $\boldsymbol{w}_\beta$ is the velocity of the target molecule before the collision, $\boldsymbol{w}_\alpha$, that of the incident molecule; $\hat{\sigma}_{\alpha\beta}(\Omega)$ is the microscopic differential cross-section (Sect. 1.7.3) for elastic collisions and $\mathrm{d}\Omega = \sin\theta \, \mathrm{d}\theta \mathrm{d}\varphi$ is the solid angle element. Note that these collisions are characterised by the usual parameters, in particular the scattering angle θ. The integral (3.7) is the *elastic collision integral* for species α[93]. Since (3.6) includes the collision term in the form of (3.7), is also referred to as *Boltzmann's integro-differential equation.*

Remarks:

1. In (3.7), the assumption of weak correlation enables us to replace the double distribution functions $f_{\alpha\beta}$ and $f'_{\alpha\beta}$ by the product of the single functions $f_\alpha f_\beta$ and $f'_\alpha f'_\beta$ (see Sect. 3.2 for details).
2. It is interesting to compare the composition of the term (3.7) with that of the binary collision frequency $\langle \nu_{\alpha\beta} \rangle$ (1.139).

[93] For a demonstration, see Golant et al. (Sect. 6.2).

3. The Boltzmann equation applies to neutral gases whose density is not too large, and to weakly ionised plasmas that are essentially governed by short-range binary interactions between neutrals and charged particles (ions and electrons).
   If the assumption of weak correlation is not valid, the Lenard-Balescu equation, in which the distribution function $f_{\alpha\beta}$ is decomposed into a correlated part and an uncorrelated part, enables us to take account of both collective and individual phenomena simultaneously. Finally, the Fokker-Planck equation can be considered as an extension of the Boltzmann equation for plasmas in which the long range Coulomb interactions (but screened at the Debye length) predominate.

3.1.2 Approximation to the Boltzmann elastic collision term: relaxation of the distribution function towards an isotropic state

In a plasma subjected to a force $\boldsymbol{F}$ resulting from a field $\boldsymbol{E}$, particles are accelerated along $\boldsymbol{F}$. Elastic collisions will, however, tend to reduce this directed velocity; in other words, they reduce the anisotropy in the velocity distribution function $f(\boldsymbol{r}, \boldsymbol{w}, t)$.

To express this physical mechanism, we introduce the collision operator:

$$S(f) \equiv -\nu(w)[f(\boldsymbol{r}, \boldsymbol{w}, t) - f_0(\boldsymbol{r}, w)] \,, \tag{3.8}$$

which, in this form, is called the *relaxation operator to the isotropic distribution function* $f_0(\boldsymbol{r}, w)$, where w is a scalar velocity; f_0 represents the velocity distribution function in the absence of the force $\boldsymbol{F}$ and $f(\boldsymbol{r}, \boldsymbol{w}, t)$ describes the distribution function at time t, in the presence of $\boldsymbol{F}$ or shortly after $\boldsymbol{F}$ has been suppressed, $\nu(w)$ being the microscopic collision frequency.

- Evolution of the time dependent function $f(\boldsymbol{w}, t)$ towards the isotropic function $f_0(w)$
  To simplify this calculation, we will assume that the plasma is spatially uniform, such that $\boldsymbol{\nabla}_r f = 0$ and $f(\boldsymbol{r}, \boldsymbol{w}, t) = f(\boldsymbol{w}, t)$. We will then follow the development for $t \geq 0$ when $\boldsymbol{F}$ is zero after $t = 0$. The Boltzmann equation then reduces to:

$$\frac{\partial f}{\partial t} = -\nu[f(\boldsymbol{w}, t) - f_0(w)] \,. \tag{3.9}$$

  Since $\partial f_0(w)/\partial t = 0$, (3.9) is equivalent to:

$$\frac{\partial}{\partial t}\left[f(\boldsymbol{w}, t) - f_0(w)\right] = -\nu \left[f(\boldsymbol{w}, t) - f_0(w)\right] \,, \tag{3.10}$$

which has the solution:

$$f(\boldsymbol{w},t) - f_0(w) = [f(\boldsymbol{w},0) - f_0(w)]\exp(-\nu t)\ . \tag{3.11}$$

The difference $f(\boldsymbol{w},t) - f_0(w)$, which appears in the collisional term (3.8), decreases exponentially, such that the function $f(\boldsymbol{w},t)$ tends towards the isotropic function $f_0(w)$, with a time constant equal to $1/\nu$.

- The collisional relaxation operator applied to a steady state Boltzmann equation ($\partial f/\partial t = 0$)
  We have, from (3.6) and (3.8):

$$\boldsymbol{w}\cdot\boldsymbol{\nabla}_r f + \frac{\boldsymbol{F}}{m}\cdot\boldsymbol{\nabla}_w f = -\nu[f(\boldsymbol{r},\boldsymbol{w},t) - f_0(\boldsymbol{r},w)]\ , \tag{3.12}$$

and this leads to the following expression:

$$f(\boldsymbol{r},\boldsymbol{w},t) - f_0(\boldsymbol{r},w) = -\frac{1}{\nu}\left[\boldsymbol{w}\cdot\boldsymbol{\nabla}_r f + \frac{\boldsymbol{F}}{m}\cdot\boldsymbol{\nabla}_w f\right]\ , \tag{3.13}$$

which shows that relaxation to isotropy occurs if the RHS of (3.13) tends to zero, i.e. if the value of $\boldsymbol{F}$ is not too high and collisions are sufficiently numerous.

Solution of (3.13) by an iterative method

In the case of weak anisotropy, (3.13) can be resolved by an iterative method, where the zeroth order approximation of the distribution function is given by $f^{(0)}(\boldsymbol{r},\boldsymbol{w},t) = f_0(\boldsymbol{r},w)$. The first order approximation for the distribution function is obtained by substituting $f_0(\boldsymbol{r},w)$ for $f(\boldsymbol{r},\boldsymbol{w},t)$ in the terms $\boldsymbol{\nabla}_r f$ and $\boldsymbol{\nabla}_w f$ in (3.13):

$$f^{(1)}(\boldsymbol{r},\boldsymbol{w},t) = f_0(\boldsymbol{r},w) - \frac{1}{\nu}\left[\boldsymbol{w}\cdot\boldsymbol{\nabla}_r f_0 + \frac{\boldsymbol{F}}{m}\cdot\boldsymbol{\nabla}_w f_0\right]\ . \tag{3.14}$$

The 2nd order approximation is:

$$f^{(2)}(\boldsymbol{r},\boldsymbol{w},t) = f_0(\boldsymbol{r},w) - \frac{1}{\nu}\left[\boldsymbol{w}\cdot\boldsymbol{\nabla}_r f^{(1)} + \frac{\boldsymbol{F}}{m}\cdot\boldsymbol{\nabla}_w f^{(1)}\right] \tag{3.15}$$

$$\begin{aligned} &= f_0(\boldsymbol{r},w) \\ &\quad -\frac{1}{\nu}\left[\boldsymbol{w}\cdot\boldsymbol{\nabla}_r\left(f_0(\boldsymbol{r},w) - \frac{1}{\nu}\left\{\boldsymbol{w}\cdot\boldsymbol{\nabla}_r f_0 + \frac{\boldsymbol{F}}{m}\cdot\boldsymbol{\nabla}_w f_0\right\}\right)\right. \\ &\quad \left.+\frac{\boldsymbol{F}}{m}\cdot\boldsymbol{\nabla}_w\left(f_0(\boldsymbol{r},w) - \frac{1}{\nu}\left\{\boldsymbol{w}\cdot\boldsymbol{\nabla}_r f_0 + \frac{\boldsymbol{F}}{m}\cdot\boldsymbol{\nabla}_w f_0\right\}\right)\right]\ , \end{aligned} \tag{3.16}$$

and so on until the kth order approximation.

Remark: It can be shown that, in the present case, the collision frequency $\nu(w)$ is the momentum transfer frequency.

3.1.3 Two classical methods to find an analytic solution to the Boltzmann equation

Chapman Enskog

We assume that the function we are seeking is not far from a Maxwell-Boltzmann distribution $f_M(\boldsymbol{r}, w, t)$; the difference is characterised by a parameter $\eta \ll 1$, by setting:

$$f(\boldsymbol{r}, \boldsymbol{w}, t) = f_M(\boldsymbol{r}, w, t) + \eta f_1(\boldsymbol{r}, \boldsymbol{w}, t) + \eta^2 f_2(\boldsymbol{r}, \boldsymbol{w}, t) + \cdots . \tag{3.17}$$

This form of solution enables us to treat the deviations from isotropy with respect to f_M. However, in all cases, it is necessary that $f(\boldsymbol{r}, \boldsymbol{w}, t)$ is not very far from a Maxwell-Boltzmann distribution.

Development in spherical harmonics in velocity space[94]

The presence of an electric field renders the distribution f anisotropic. The degree of anisotropy remains weak as long as the directed velocity created by the electric field is small with respect to the mean velocity of thermal motion. The method proposed by W.P. Allis consists of developing $f(\boldsymbol{r}, \boldsymbol{w}, t)$ in spherical harmonics (Legendre polynomials, see Appendix XIV), such that:

$$f(\boldsymbol{r}, \boldsymbol{w}, t) = f_0(\boldsymbol{r}, w, t) + f_1(\boldsymbol{r}, w, t)\cos\theta + f_2(\boldsymbol{r}, w, t)\frac{3\cos^2\theta - 1}{2} + \cdots , \tag{3.18}$$

where f_0 is isotropic, but not necessarily Maxwellian (Golant et al. Sect. 5.2). The angle θ is that of the spherical coordinate system, where z is in the direction of the anisotropy: thus, we write $w_z = w\cos\theta$ (see the example in Sect. 3.4). The development (3.18) assumes symmetry in φ (otherwise, see Delcroix and Bers Sect. 12.3).

---

[94] We obtain spherical harmonics $Y(\theta, \varphi)$ in the form $Y(\theta, \varphi) = \Theta(\theta)\Phi(\varphi)$, when we apply a separation of variables to the Schrödinger equation for a one-electron atom.

3.2 Velocity distribution functions and the concept of correlation between particles

We wish to define a distribution function $f(\boldsymbol{r}, \boldsymbol{w}, t)$ that is valid at each point in phase space and which characterises each point in that space individually, i.e. without there being any relation between that point and the other points in this space. Such a function is said to be a *single-point distribution function*, in contrast to a two-point distribution function, a three-point distribution function... functions, which introduce a dependence between pairs of points, triplets of points... of this space. We will initially consider this question, starting with the phase-space joint probability of an ensemble of completely correlated particles, and then examine how and under what conditions, it is possible to partially or fully neglect these correlations. Correlations are due to the interactions between particles; in general, all particles have an influence on each of the others, so there is total correlation.

3.2.1 *Probability density of finding a particle in phase space*

- The probability that, at the same time, (distinguishable) particle 1 is at $\boldsymbol{r}_1, \boldsymbol{w}_1$ [95], particle 2 is at $\boldsymbol{r}_2, \boldsymbol{w}_2$, particle 3 at is at $\boldsymbol{r}_3, \boldsymbol{w}_3 \cdots$, is expressed symbolically by:

$$\mathcal{D}(\boldsymbol{r}_1, \boldsymbol{w}_1, \cdots, \boldsymbol{r}_N, \boldsymbol{w}_N)\, \mathrm{d}\boldsymbol{r}_1\, \mathrm{d}\boldsymbol{w}_1 \cdots \mathrm{d}\boldsymbol{r}_N\, \mathrm{d}\boldsymbol{w}_N\,, \tag{3.19}$$

  where $\mathcal{D}$ designates the *probability density of presence* in the $6N$ dimensional phase space, and N is the total number of particles. The probability density $\mathcal{D}$ obeys the Liouville equation; it contains all the information necessary to completely characterise the system of N particles.
- The probability of finding the N particles in a given state in phase space is obtained by integrating the probability density $\mathcal{D}$ over the ensemble of volume elements in phase space:

$$\int \mathcal{D}(\boldsymbol{r}_1, \boldsymbol{w}_1, \cdots, \boldsymbol{r}_N, \boldsymbol{w}_N)\, \mathrm{d}\boldsymbol{r}_1\, \mathrm{d}\boldsymbol{w}_1\, \mathrm{d}\boldsymbol{r}_2\, \mathrm{d}\boldsymbol{w}_2 \cdots \mathrm{d}\boldsymbol{r}_N\, \mathrm{d}\boldsymbol{w}_N = 1\,. \tag{3.20}$$

  This integral is unity because, statistically, it includes a summation over all the possibilities, and is therefore certain.
- The probability of finding (distinguishable) particle 1 at $\boldsymbol{r}_1, \boldsymbol{w}_1$, independently of the positions and velocities of the other particles, is found by integrating (3.19) over all the volume elements except $\mathrm{d}\boldsymbol{r}_1 \mathrm{d}\boldsymbol{w}_1$:

---

[95] More exactly, in a volume element $\mathrm{d}\boldsymbol{r}_1 \mathrm{d}\boldsymbol{w}_1$ in phase space centred at $\boldsymbol{r}_1, \boldsymbol{w}_1 \ldots$

$$\left[\int \mathcal{D}(\boldsymbol{r}_1, \boldsymbol{w}_1, \cdots, \boldsymbol{r}_N, \boldsymbol{w}_N)\, \mathrm{d}\boldsymbol{r}_2\, \mathrm{d}\boldsymbol{w}_2 \cdots \mathrm{d}\boldsymbol{r}_N\, \mathrm{d}\boldsymbol{w}_N\right] \mathrm{d}\boldsymbol{r}_1\, \mathrm{d}\boldsymbol{w}_1 \,. \tag{3.21}$$

Clearly, the result of this integration then depends only on $\boldsymbol{r}_1$, $\boldsymbol{w}_1$. Since, in reality, the particles are indistinguishable, expression (3.21) represents the probability of finding any of the N particles at $\boldsymbol{r}_1, \boldsymbol{w}_1$ in $\mathrm{d}\boldsymbol{r}_1 \mathrm{d}\boldsymbol{w}_1$ at time t.

- The probable number $[\mathrm{d}N]$ of (indistinguishable) particles found at $\boldsymbol{r}_1, \boldsymbol{w}_1$ is thus given by the probability of finding a particular particle at that position in phase space (3.21), multiplied by the total number, N, of particles in the system which can be written:

$$[\mathrm{d}N] = N\left[\int \mathcal{D}(\boldsymbol{r}_1, \boldsymbol{w}_1, \cdots, \boldsymbol{r}_N, \boldsymbol{w}_N)\, \mathrm{d}\boldsymbol{r}_2\, \mathrm{d}\boldsymbol{w}_2 \cdots \mathrm{d}\boldsymbol{r}_N\, \mathrm{d}\boldsymbol{w}_N\right] \mathrm{d}\boldsymbol{r}_1\, \mathrm{d}\boldsymbol{w}_1 \,. \tag{3.22}$$

Remark: The actual number $\mathrm{d}N$ of particles located in a volume element $\mathrm{d}\boldsymbol{r}_1$, $\mathrm{d}\boldsymbol{w}_1$ centred at $\boldsymbol{r}_1, \boldsymbol{w}_1$, can be replaced by the probable number of particles $[\mathrm{d}N]$, if the number of particles is sufficiently large for statistical fluctuations to be negligible.

3.2.2 *Single-point distribution function (the case of correlated particles)*

The actual number of particles in the volume element $\mathrm{d}\boldsymbol{r}\, \mathrm{d}\boldsymbol{w}$ centred at $\boldsymbol{r}, \boldsymbol{w}$ is, by definition, given by:

$$\mathrm{d}N = f_1(\boldsymbol{r}, \boldsymbol{w}, t)\, \mathrm{d}\boldsymbol{r}\, \mathrm{d}\boldsymbol{w} \,, \tag{3.23}$$

where f_1 is referred to as the single-point (in phase space) velocity distribution function. If, as we have noted, the number of particles is very large, $[\mathrm{d}N] = \mathrm{d}N$, so from (3.22) and (3.23):

$$f_1(\boldsymbol{r}_1, \boldsymbol{w}_1, t)\, \mathrm{d}\boldsymbol{r}_1\, \mathrm{d}\boldsymbol{w}_1 = N\left[\int \mathcal{D}(\boldsymbol{r}_1, \boldsymbol{w}_1, \cdots, \boldsymbol{r}_N, \boldsymbol{w}_N)\, \mathrm{d}\boldsymbol{r}_2\, \mathrm{d}\boldsymbol{w}_2 \cdots \mathrm{d}\boldsymbol{r}_N\, \mathrm{d}\boldsymbol{w}_N\right] \mathrm{d}\boldsymbol{r}_1\, \mathrm{d}\boldsymbol{w}_1 \,, \tag{3.24}$$

which is the expression for the *single-point distribution function* obtained for complete correlation of particles. Because the particles are indistinguishable (a quantum property), we will remove the "labels" from them. The function enables us to calculate the number of particles present at a given point in phase space independently of particles at another point, or other points in this space.

The integration of (3.23) over $d\boldsymbol{r}_1\, d\boldsymbol{w}_1$ leads, by definition, to:

$$\int f_1(\boldsymbol{r}, \boldsymbol{w}, t)\, d\boldsymbol{r}\, d\boldsymbol{w} = N\,. \tag{3.25}$$

3.2.3 Single-point distribution function (uncorrelated particles)

Contrary to the previous case, in complete absence of correlations (permanent state of *molecular chaos*), the total probability density $\mathcal{D}$ is the product of individual probability densities $\mathcal{D}_i$ ($i = 1, 2, 3 \ldots N$), and we can write:

$$\mathcal{D} = \mathcal{D}_1 \mathcal{D}_2 \mathcal{D}_3 \ldots \mathcal{D}_N\,, \tag{3.26}$$

Suppose, once again, that the particles are indistinguishable, in which case the individual probability densities are all equal, so that:

$$\mathcal{D} = (\mathcal{D}_0)^N\,. \tag{3.27}$$

Inserting (3.27) into (3.20), the decomposition of $\mathcal{D}$ leads to:

$$\left[\int \mathcal{D}_0\, d\boldsymbol{r}\, d\boldsymbol{w}\right]^N = 1\,,$$

such that:

$$\int \mathcal{D}_0\, d\boldsymbol{r}\, d\boldsymbol{w} = 1\,. \tag{3.28}$$

In this case, from (3.24) and (3.28):

$$f_1(\boldsymbol{r}, \boldsymbol{w}, t) = N\mathcal{D}_0 \left[\int \mathcal{D}_0\, d\boldsymbol{r}\, d\boldsymbol{w}\right]^{N-1} = N\mathcal{D}_0\,. \tag{3.29}$$

To calculate $f_1(\boldsymbol{r}, \boldsymbol{w}, t)$, we will thus use either (3.24) for completely correlated particles, or (3.29) for completely uncorrelated particles.

3.2.4 Two-point distribution function (correlated particles)

The two-point distribution function $f_{12}(\boldsymbol{r}_1, \boldsymbol{w}_1, \boldsymbol{r}_2, \boldsymbol{w}_2, t)$, by definition, considers pairs of particles which are found simultaneously, at time t, at two distinct points $\boldsymbol{r}_1$, $\boldsymbol{w}_1$ and $\boldsymbol{r}_2$, $\boldsymbol{w}_2$ in phase space. The probability that particle 1 is at $\boldsymbol{r}_1$, $\boldsymbol{w}_1$, while particle 2 is at $\boldsymbol{r}_2$, $\boldsymbol{w}_2$ (at the same time t), according to the formalism developed above, is given by:

$$\left[\int \mathcal{D}(\boldsymbol{r}_1, \boldsymbol{w}_1, \cdots, \boldsymbol{r}_N, \boldsymbol{w}_N)\, \mathrm{d}\boldsymbol{r}_3 \mathrm{d}\boldsymbol{w}_3 \cdots \mathrm{d}\boldsymbol{r}_N \mathrm{d}\boldsymbol{w}_N\right] \mathrm{d}\boldsymbol{r}_1\, \mathrm{d}\boldsymbol{w}_1\, \mathrm{d}\boldsymbol{r}_2\, \mathrm{d}\boldsymbol{w}_2\ . \quad (3.30)$$

Since the particles are indistinguishable, the total number of pairs of particles which can occupy two given points in space is found, analogously with (3.22), by multiplying (3.30) by the total number of ordered pairs that it is possible to form with N particles, namely $N(N-1)$:

$$f_{12}(\boldsymbol{r}_1, \boldsymbol{w}_1, \boldsymbol{r}_2, \boldsymbol{w}_2, t)\, \mathrm{d}\boldsymbol{r}_1 \mathrm{d}\boldsymbol{w}_1\, \mathrm{d}\boldsymbol{r}_2\, \mathrm{d}\boldsymbol{w}_2 = \quad (3.31)$$
$$N(N-1)\left[\int \mathcal{D}(\boldsymbol{r}_1, \boldsymbol{w}_1, \cdots, \boldsymbol{r}_N, \boldsymbol{w}_N)\, \mathrm{d}\boldsymbol{r}_3 \cdots \mathrm{d}\boldsymbol{w}_N\right] \mathrm{d}\boldsymbol{r}_1\, \mathrm{d}\boldsymbol{w}_1\, \mathrm{d}\boldsymbol{r}_2\, \mathrm{d}\boldsymbol{w}_2\ ,$$

where, from the normalisation (3.20):

$$\int f_{12}\, \mathrm{d}\boldsymbol{r}_1\, \mathrm{d}\boldsymbol{w}_1\, \mathrm{d}\boldsymbol{r}_2\, \mathrm{d}\boldsymbol{w}_2 = N(N-1) \quad (3.32)$$

giving the total number of (ordered) pairs of particles that can be formed[96].

The two-point function is used in particular used to describe binary collisions: in this case, the pairs of particles travelling with velocities $\boldsymbol{w}_\alpha$ and $\boldsymbol{w}_\beta$ before collision enter into the collisional term through the double function $f_{\alpha\beta}$ (see Remark 1 following (3.7)). The recourse to a pair of ordered particles is arbitrary, but reasonable in a number of problems where the two points play different roles, for example, due to different physical environments (presence of forces, spatial inhomogeneity).

3.2.5 Two-point distribution function (uncorrelated particles)

If we completely neglect the correlation between particles, we can, from (3.27), express the two-point function in the form:

$$f_{12} = N(N-1)\mathcal{D}_0^2\left[\int \mathcal{D}_0(\boldsymbol{r}, \boldsymbol{w})\, \mathrm{d}\boldsymbol{r}\, \mathrm{d}\boldsymbol{w}\right]^{N-2} = (N\mathcal{D}_0)(N\mathcal{D}_0)\frac{N-1}{N}\ , \quad (3.33)$$

which, from (3.29) for f_1, gives $f_{12} = f_1(\boldsymbol{r}, \boldsymbol{w}, t) f_2(\boldsymbol{r}, \boldsymbol{w}, t)(N-1)/N$, such that for very large values of N:

$$f_{12} \approx f_1(\boldsymbol{r}, \boldsymbol{w}, t) f_2(\boldsymbol{r}, \boldsymbol{w}, t)\ . \quad (3.34)$$

---

[96] For example, for $N = 3$, there are 6 possible ordered pairs; 12, 21, 13, 31, 23 and 32.

Thus, in the case of uncorrelated particles, the two-point function, which leads to the number of pairs of particles for which one is situated at $\boldsymbol{r}_1, \boldsymbol{w}_1$ while the other is at $\boldsymbol{r}_2, \boldsymbol{w}_2$, is simply expressed as the product of two single-point (in phase space) functions, as expected.

3.2.6 N-point distribution functions

From the generalisation of (3.31), we know how to write these functions, so in the case of correlated particles:

$$f_{12\cdots N}(\boldsymbol{r}_1, \boldsymbol{w}_1, \cdots \boldsymbol{r}_N,\ \boldsymbol{w}_N) = N!\,\mathcal{D}(\boldsymbol{r}_1, \boldsymbol{w}_1, \cdots \boldsymbol{r}_N,\ \boldsymbol{w}_N) \tag{3.35}$$

and in the complete absence of correlation:

$$f_{12\cdots N} \approx f_1 f_2 \cdots f_N\,. \tag{3.36}$$

In the following sections, we will mainly consider single-point distribution functions. We will, however, encounter two-point functions in the binary collision integral of the hydrodynamic equation for momentum transfer. More generally, the multiple-point distribution functions appear in the kinetic BBGKY hierarchy (Sect. 3.6).

3.3 Distribution functions and hydrodynamic quantities

The single-point velocity distribution function enables us to calculate the *mean value* of certain molecular properties, also referred to as corpuscular or microscopic properties, for each position $\boldsymbol{r}$ and time t. As a result, these are called *hydrodynamic* (or macroscopic) *quantities*. For any particular molecular property $\Upsilon(\boldsymbol{r}, \boldsymbol{w}, t)$ (Υ is the Greek capital letter upsilon), the most general definition of the mean value, denoted by $\langle \Upsilon(\boldsymbol{r}, t)\rangle$, is given by the expression:

$$\langle \Upsilon(\boldsymbol{r}, t)\rangle = \frac{\displaystyle\int_{\boldsymbol{w}} \Upsilon(\boldsymbol{r}, \boldsymbol{w}, t) f(\boldsymbol{r}, \boldsymbol{w}, t)\, \mathrm{d}\boldsymbol{w}}{\displaystyle\int_{\boldsymbol{w}} f(\boldsymbol{r}, \boldsymbol{w}, t)\, \mathrm{d}\boldsymbol{w}} \tag{3.37}$$

in which the denominator represents the particle density per unit volume[97]:

[97] Consider, in this regard, that the term $\left[\int_{\boldsymbol{w}} f(\boldsymbol{r}, \boldsymbol{w}, t)\, \mathrm{d}\boldsymbol{w}\right] \mathrm{d}\boldsymbol{r}$ represents the number of particles, for all velocities, in the volume element $\mathrm{d}\boldsymbol{r}$ in position space.

$$n(\boldsymbol{r},t) = \int\limits_{\boldsymbol{w}} f(\boldsymbol{r},\boldsymbol{w},t)\,\mathrm{d}\boldsymbol{w}\ . \tag{3.38}$$

Equation (3.38) is the *normalisation condition* on $f(\boldsymbol{r},\boldsymbol{w},t)$, the un-separated distribution function in $\boldsymbol{r}$ and $\boldsymbol{w}$.
We then find:

$$\langle \Upsilon(\boldsymbol{r},t)\rangle \equiv \frac{\int\limits_{\boldsymbol{w}} \Upsilon(\boldsymbol{r},\boldsymbol{w},t) f(\boldsymbol{r},\boldsymbol{w},t)\,\mathrm{d}\boldsymbol{w}}{n(\boldsymbol{r},t)}\ . \tag{3.39}$$

Definition of some typical hydrodynamic quantities

The *mean velocity*:

$$\boldsymbol{v}(\boldsymbol{r},t) = \frac{1}{n(\boldsymbol{r},t)} \int\limits_{\boldsymbol{w}} \boldsymbol{w} f(\boldsymbol{r},\boldsymbol{w},t)\,\mathrm{d}\boldsymbol{w}\ , \tag{3.40}$$

the *mean kinetic energy*:

$$\bar{\mathcal{E}}_{\mathrm{cin}} = \frac{m_\alpha}{2n(\boldsymbol{r},t)} \int\limits_{\boldsymbol{w}} w^2 f(\boldsymbol{r},\boldsymbol{w},t)\,\mathrm{d}\boldsymbol{w} \tag{3.41}$$

and the *kinetic pressure tensor*[98]:

$$\underline{\boldsymbol{\Psi}}(\boldsymbol{r},\boldsymbol{v},t) = m_\alpha \int\limits_{\boldsymbol{w}} (\boldsymbol{w}-\boldsymbol{v})\otimes(\boldsymbol{w}-\boldsymbol{v}) f(\boldsymbol{r},\boldsymbol{w},t)\,\mathrm{d}\boldsymbol{w}\ , \tag{3.42}$$

where the operator $\otimes$ represents the tensorial product of two vectors.

Hydrodynamic quantities in the particular case where the distribution function $f(r, w, t)$ is separable

The *distribution function* f is *separable* if we can write:

$$f(\boldsymbol{r},\boldsymbol{w},t) = n(\boldsymbol{r},t) h(\boldsymbol{w},t) \tag{3.43}$$

or, when time-independent:

$$f(\boldsymbol{r},\boldsymbol{w}) = n(\boldsymbol{r}) h(\boldsymbol{w})\ . \tag{3.44}$$

[98] The significance of this parameter is discussed further in Sect. 3.5. Note that $(\boldsymbol{w}-\boldsymbol{v})\otimes(\boldsymbol{w}-\boldsymbol{v})$ represents a tensorial product which results in a 2nd order tensor (see Appendix VII for tensor notation). Note also that the density $n(\boldsymbol{r},t)$ does not appear explicitly in the definition of $\underline{\boldsymbol{\Psi}}(\boldsymbol{r},t)$ (3.42).

In this case, the normalisation condition on the distribution function depending only on velocity (cf. (3.38) for comparison) is by definition:

$$\int_{\boldsymbol{w}} h(\boldsymbol{w})\,\mathrm{d}\boldsymbol{w} = 1\ . \tag{3.45}$$

The mean values (3.40), (3.41) and (3.42) then take the form:

$$\boldsymbol{v} = \int_{\boldsymbol{w}} \boldsymbol{w} h(\boldsymbol{w})\,\mathrm{d}\boldsymbol{w}\ , \tag{3.46}$$

$$\bar{\mathcal{E}}_{\mathrm{cin}} = \frac{m_\alpha}{2}\int_{\boldsymbol{w}} w^2 h(\boldsymbol{w})\,\mathrm{d}\boldsymbol{w}\ , \tag{3.47}$$

$$\underline{\underline{\boldsymbol{\Psi}}} = m_\alpha n(\boldsymbol{r})\int_{\boldsymbol{w}} (\boldsymbol{w}-\boldsymbol{v})\otimes(\boldsymbol{w}-\boldsymbol{v})h(\boldsymbol{w})\,\mathrm{d}\boldsymbol{w}\ . \tag{3.48}$$

Remarks:

1. In the following, we will use the notation f to denote the velocity distribution function, whether it is separable or not: if the argument of f does not contain the position vector $\boldsymbol{r}$, the function is assumed to be separable, i.e. $f(\boldsymbol{w}) \equiv h(\boldsymbol{w})$.
2. A sufficient condition for the function f to be separable is that the plasma has uniform density.

Calculation of a hydrodynamic quantity from a distribution function expanded in spherical harmonics (separable function)

The normalisation condition (3.45), expanded in spherical coordinates (3.18), leads to:

$$\int_{\boldsymbol{w}} f(\boldsymbol{w},t)\,\mathrm{d}\boldsymbol{w} = 4\pi\int_{w} f_0(w,t)\,w^2\,\mathrm{d}w = 1\ . \tag{3.49}$$

The RHS only contains the isotropic contribution from the spherical harmonic expansion (3.18), as we will show. The second term of the expansion (3.18) is proportional to $\cos\theta$, and since the volume element $\mathrm{d}\boldsymbol{w} \equiv \mathrm{d}^3w = w^2\sin\theta\,\mathrm{d}\theta\,\mathrm{d}\varphi\,\mathrm{d}w$ contains the term $\sin\theta$, the integral of $\sin\theta\cos\theta\mathrm{d}\theta$ over θ from 0 to π, of odd parity, is zero! Writing $\cos\theta = \tau$ and noting that $\sin\theta\,\mathrm{d}\theta = -\mathrm{d}(\cos\theta)$, the integral of $\cos\theta\sin\theta\,\mathrm{d}\theta$ over θ from 0 to π is $(\tau^2/2)\big|_{-1}^{1}$. The contribution from the third term in the expansion (3.18) is proportional to $3\cos^2\theta - 1$: since these two terms have the same parity in τ, they cancel after evaluation over $(1,-1)$, and this is true for all higher order terms of the expansion.

Applied example: calculation of the mean velocity in the direction of anisotropy z in velocity space. The coordinate system in this space is $w_x = w\sin\theta\cos\varphi$, $w_y = w\sin\theta\sin\varphi$ and $w_z = w\cos\theta$. Since symmetry in φ has been assumed in the expansion (3.18), we have (recalling that f is a separable distribution function in the present case!):

$$v_z = -\int_{\theta=0}^{\pi}\int_{w=0}^{\infty} 2\pi(w\cos\theta)f(\boldsymbol{w},t)w^2\,\mathrm{d}(\cos\theta)\,\mathrm{d}w\;. \tag{3.50}$$

Replacing $f(\boldsymbol{w},t)$ by its expansion (3.18), and once again choosing $\tau = \cos\theta$ as variable of integration, we find:

$$v_z = \int_0^\infty \underbrace{\left[\frac{1}{2}\tau^2\right]_{-1}^{1}}_{0} 2\pi w^3 f_0(w,t)\mathrm{d}w + \int_0^\infty \underbrace{\left[\frac{\tau^3}{3}\right]_{-1}^{1}}_{2/3} 2\pi w^3 f_1(w,t)\mathrm{d}w + \cdots\;. \tag{3.51}$$

The isotropic terms and the terms of order higher than 1 do not contribute to v, and we obtain:

$$v_z = \frac{4\pi}{3}\int_0^\infty w^3 f_1(w,t)\,\mathrm{d}w\;. \tag{3.52}$$

3.4 Kinetic and hydrodynamic conductivities of electrons in a plasma in the presence of a HF electromagnetic field

The electrical conductivity is an essential parameter in the description of a plasma, because it enables us to establish a link between the motion of the charged particles and the electric field, or fields, with which they interact, including both externally applied fields and those created by the charged particles themselves. The electrical conductivity will be found in both the kinetic model and the hydrodynamic model.

We will start by deriving the expression for the conductivity of electrons in its kinetic form, which requires the electron velocity distribution function. We will use this expression to deduce the hydrodynamic conductivity, in which the relationship with the kinetic conductivity introduces the concept of effective collision frequency. These various expressions for the conductivity will be developed in the presence of a high frequency (HF) field, as a preview to the treatment of HF discharges in Chap. 4.

3.4.1 Kinetic form of the electrical conductivity due to electrons in an HF field

Solution to the Boltzmann equation with a weakly anisotropic velocity

Consider a uniform plasma[99] subjected to a small amplitude HF electromagnetic field (i.e. the field does not introduce non-linear effects or contribute to the plasma ionisation), directed along z. The Boltzmann equation for electrons is then written in terms of the distribution function $f(\boldsymbol{w}, t)$:

$$\frac{\partial f}{\partial t} - \frac{eE_0}{m_e} \mathrm{e}^{\mathrm{i}\omega t} \frac{\partial f}{\partial w_z} = S(f) \,. \tag{3.53}$$

In (3.53), we have neglected the effect of the field $\boldsymbol{H}$ of the electromagnetic wave on the electrons and we have assumed that the wavelength is much larger than the plasma dimensions: this is the *electrostatic approximation*, allowing us to neglect the term $\mathrm{e}^{-\mathrm{i}\beta z}$ in the phase term $\mathrm{e}^{-i(\beta z - \omega t)}$ of the HF field. In addition, we will make the approximation that, in the absence of the field $\boldsymbol{E}$, the particle distribution function, created by a mechanism other than that of the HF field, $f_0(w)$, is an isotropic function, but not necessarily Maxwellian.

To resolve (3.53), including the anisotropy due to $\boldsymbol{E}$, assumed to be small, we expand the velocity distribution function in spherical harmonics. Limiting the expansion to first order, we have:

$$f(\boldsymbol{w}, t) = f_0(w) + \frac{w_z}{w} f_1(w) \mathrm{e}^{\mathrm{i}\omega t} \tag{3.54}$$

because $\cos\theta = w_z/w$. Note that the dependence of the function $f(\boldsymbol{w}, t)$ on t, explicitly expressed in the second term of the expansion, reflects the periodic variation of the HF field. In accord with our assumption of a small amplitude field, hence of weak isotropy, we can express the collision operator in the form of a collisional relaxation term (3.8), associated with the microscopic collision frequency $\nu(w)$. To first order in the spherical harmonics expansion of $f(\boldsymbol{w}, t)$, we find from (3.8):

$$S(f) = -\nu(w) \left[\frac{w_z}{w} f_1(w) \mathrm{e}^{\mathrm{i}\omega t} \right] \,. \tag{3.55}$$

We wish to obtain the expression for the function $f_1(w)$, which characterises the departure from isotropy induced by the field $\boldsymbol{E}$ (3.53). We must first show that:

$$\frac{\partial f}{\partial w_z} = \frac{w_z}{w} \frac{\partial f}{\partial w} \,. \tag{3.56}$$

---

[99] One can treat the case of an inhomogeneous plasma in the same way, provided that the function f is separable.

This comes from:

$$\frac{\partial}{\partial w_z} = \frac{\partial}{\partial w}\frac{\partial w}{\partial w_z} \text{ and } w = (w_x^2 + w_y^2 + w_z^2)^{\frac{1}{2}}, \text{ thus } \frac{\partial w}{\partial w_z} = \frac{1}{2}\frac{2w_z}{w} = \frac{w_z}{w} \,.$$

Importing (3.54) and (3.55) in (3.53) and taking account of (3.56), and canceling the factor $e^{i\omega t}$ appearing on all terms, we obtain:

$$\frac{w_z}{w} f_1(w) i\omega - \frac{eE_0}{m_e}\frac{w_z}{w}\frac{\partial}{\partial w}\left[f_0(w) + \frac{w_z}{w} f_1(w) e^{i\omega t}\right] = -\nu(w)\frac{w_z}{w} f_1(w) \,, \quad (3.57)$$

where the second order term $\partial/\partial w\,[f_1(w)w_z/w]$, is neglected compared to the other terms, which are all of first order. Regrouping, we have:

$$f_1(w)\Big[\nu(w) + i\omega\Big] = \frac{eE_0}{m_e}\frac{\partial f_0}{\partial w} \,, \quad (3.58)$$

from which:

$$f_1(w) = \frac{eE_0/m_e}{\nu(w) + i\omega}\frac{\partial f_0}{\partial w} \,. \quad (3.59)$$

Electron conductivity in the HF field

- Current density of electrons due to the field $\boldsymbol{E}$
  Since $f(\boldsymbol{w}, t)$ is a separable distribution function, we have:

$$J \equiv -nev_z = -ne \int_{\boldsymbol{w}} w_z f(\boldsymbol{w}, t)\, d\boldsymbol{w} \,. \quad (3.60)$$

  Using (3.52), which gives v_z specifically for a separable function f, it follows that:

$$J = -ne\frac{4\pi}{3}\int_0^\infty w^3 f_1(w, t)\, dw \,, \quad (3.61)$$

  where in the present case, $f_1(w, t) = f_1(w) e^{i\omega t}$.
  Finally, including (3.59), we arrive at:

$$\boldsymbol{J} = -\frac{4\pi}{3}\frac{ne^2}{m_e}\boldsymbol{E}_0 e^{i\omega t}\int_0^\infty \frac{1}{\nu(w) + i\omega}\frac{\partial f_0}{\partial w} w^3\, dw \,. \quad (3.62)$$

- Expression for the conductivity
  Setting $\boldsymbol{J} = \sigma\boldsymbol{E}$, we obtain, by identification from (3.62):

$$\sigma = -\frac{4\pi}{3}\frac{ne^2}{m_e}\int_0^\infty \frac{1}{\nu(w)+\mathrm{i}\omega}\frac{\partial f_0}{\partial w}w^3\,\mathrm{d}w\,, \tag{3.63}$$

referred to as the *Boltzmann conductivity* or *kinetic conductivity*. After integration by parts, we find the following equivalent form:

$$\sigma = \frac{4\pi}{3}\frac{ne^2}{m_e}\int_0^\infty \frac{\partial}{\partial w}\left[\frac{w^3}{\nu(w)+\mathrm{i}\omega}\right] f_0(w)\,\mathrm{d}w \tag{3.64}$$

because $f_0(\infty) = 0$ and the contribution from the limit $w = 0$ is zero.

- Dielectric permittivity of the plasma (relative to that of vacuum)
  From (2.40), we know that for $E = E_0\mathrm{e}^{\mathrm{i}\omega t}$:

$$\epsilon_p = 1 - \frac{\mathrm{i}\sigma}{\omega\epsilon_0}\,, \tag{3.65}$$

(Note: $\epsilon_p = 1 + \mathrm{i}\sigma/\omega\epsilon_0$ if $E = E_0\mathrm{e}^{-\mathrm{i}\omega t}$) then from (3.63):

$$\epsilon_p = 1 + \frac{\omega_{pe}^2}{\omega}\frac{4\pi}{3}\int_0^\infty \frac{1}{\omega - \mathrm{i}\nu(w)}\frac{\partial f_0(w)}{\partial w}w^3\,\mathrm{d}w\,. \tag{3.66}$$

3.4.2 Hydrodynamic form of the electrical conductivity due to electrons in an HF field

Noteworthy expressions for the electrical conductivity

Collision frequency, independent of the velocity w[100]

In this case, $\nu(w) = \nu$, and from (3.64):

$$\sigma = \frac{4}{3}\frac{\pi ne^2}{m_e(\nu+\mathrm{i}\omega)}\int_0^\infty f_0(w)\,3w^2\,\mathrm{d}w = \frac{ne^2}{m_e(\nu+\mathrm{i}\omega)}\underbrace{\left[4\pi\int_0^\infty f_0(w)w^2\,\mathrm{d}w\right]}_{=1\ (\text{see } (3.49))} \tag{3.67}$$

from which:

$$\sigma = \frac{ne^2}{m_e(\nu+\mathrm{i}\omega)}\,, \tag{3.68}$$

which is the (hydrodynamic) *Lorentz conductivity*.

[100] Since $\nu(w) = \hat{\sigma}(w)wN$, this is only true if $\nu(w)$ does not depend on w, for example, if $\hat{\sigma}(w) \approx w^{-1}$.

Remark: We have already encountered this expression for the conductivity σ in Chap. 2 (2.39), by considering the electrons to be moving in a viscous hydrodynamic medium (viscosity term $-\nu \boldsymbol{v}$ in the equation of motion (2.27)). We can see that the true condition allowing us to obtain (3.68) is $\nu(w) =$ constant, $\nu(w)$ being the electron-neutral *microscopic collision frequency* for momentum transfer. Recall that, under the same conditions, the permittivity of the plasma is given by (2.41):

$$\epsilon_p = 1 - \frac{\omega_{pe}^2}{\omega(\omega - \mathrm{i}\nu)} . \tag{3.69}$$

Maxwell velocity distribution function

We have (Appendix I):

$$f_0(w) = \left(\frac{m_e}{2\pi k_B T_e}\right)^{3/2} \exp - \left(\frac{m_e w^2}{2k_B T_e}\right) , \tag{I.1}$$

and thus:

$$\sigma = \frac{8}{3\pi^{1/2}} \frac{ne^2}{m_e} \int_0^\infty \frac{1}{\nu(u_e) + \mathrm{i}\omega} u_e^4 \exp(-u_e^2) \, \mathrm{d}u_e , \tag{3.70}$$

where $u_e \equiv w/(2\pi k_B T_e/m_e)^{1/2}$. We can then calculate σ, provided $\nu(u_e)$ is known.

Expression for the effective collision frequency

We can preserve the Lorentz form of the conductivity (3.68), even in the case where ν depends on w, by substituting an effective frequency ν_{eff} for ν in (3.63), which we will now define. This approximation leads to relatively simple expressions in two limit cases:

1. The field $\boldsymbol{E}$ is constant ($\omega = 0$) or the angular frequency ω is such that $\omega \ll \nu$. In this limit, the continuous current (CC) limit, the Lorentz conductivity takes the (purely real) form:

$$\sigma = \frac{ne^2}{m_e \nu} , \tag{3.71}$$

while the Boltzmann conductivity (3.63), in the same limit, is written:

$$\sigma = -\frac{4\pi}{3} \frac{ne^2}{m_e} \int \frac{1}{\nu(w)} \frac{\partial f_0(w)}{\partial w} w^3 \, \mathrm{d}w . \tag{3.72}$$

Equation (3.72) can be expressed in the form (3.71) provided we can introduce an effective collision frequency $\nu_{\mathrm{eff(cc)}}$, such that:

$$\frac{1}{\nu_{\text{eff(cc)}}} \equiv -\frac{4\pi}{3}\int \frac{1}{\nu(w)}\frac{\partial f_0(w)}{\partial w} w^3 \, \mathrm{d}w \, . \tag{3.73}$$

2. The field $\boldsymbol{E}$ varies periodically, but rapidly enough that $\omega \gg \nu$, called the high frequency (HF) approximation. In this case, the Lorentz conductivity (3.68) can be written:

$$\sigma = \frac{ne^2}{m_e\,\omega\left(\dfrac{\nu}{\omega}+\mathrm{i}\right)} = \frac{ne^2}{m_e\,\omega}\left(\frac{\nu}{\omega}-\mathrm{i}\right)\frac{1}{1+\left(\dfrac{\nu}{\omega}\right)^2} \, , \tag{3.74}$$

which leads, in the present limit case, to:

$$\sigma \approx \frac{ne^2}{m_e}\left(\frac{\nu}{\omega^2}-\frac{\mathrm{i}}{\omega}\right) . \tag{3.75}$$

Furthermore, the Boltzmann conductivity (3.63), expressed with the help of the function f_0, in the same limit gives:

$$\sigma = -\frac{4\pi}{3}\frac{ne^2}{m_e}\int\left[\frac{\nu(w)}{\omega^2}-\frac{\mathrm{i}}{\omega}\right]\frac{\partial f_0}{\partial w}w^3\,\mathrm{d}w \, , \tag{3.76}$$

such that, in comparing (3.76) and (3.75), leads us to define:

$$\nu_{\text{eff(HF)}} = -\frac{4\pi}{3}\int \nu(w)\frac{\partial f_0}{\partial w}w^3\,\mathrm{d}w \, . \tag{3.77}$$

This finally enables us to express the conductivity, in this limit, in its Lorentzian form as:

$$\sigma = \frac{ne^2}{m_e}\left(\frac{\nu_{\text{eff(HF)}}}{\omega^2}-\frac{\mathrm{i}}{\omega}\right) . \tag{3.78}$$

To show that the term:

$$-\frac{4}{3}\pi\int\left(\frac{-\mathrm{i}}{\omega}\right)\frac{\partial f_0}{\partial w}w^3\,\mathrm{d}w \tag{3.79}$$

in (3.76) actually reduces to $-\mathrm{i}/\omega$ in (3.75), it is sufficient to realise that the term:

$$-\frac{4\pi}{3}\int\frac{\partial f_0}{\partial w}w^3\,\mathrm{d}w \tag{3.80}$$

is unity, which we show by integrating the expression by parts and then applying the normalisation condition (3.49):

$$-\frac{4\pi}{3}\int\frac{\partial f_0}{\partial w}w^3\,\mathrm{d}w = \frac{4\pi}{3}3\int f_0 w^2\,\mathrm{d}w = 4\pi\int f_0 w^2\,\mathrm{d}w = 1 \, . \tag{3.81}$$

3.5 Transport equations

In the hydrodynamic model, for each microscopic variable ($\Upsilon(\boldsymbol{r},\boldsymbol{w},t)=1$, $m\boldsymbol{w}$, $mw^2/2$, $m(\boldsymbol{w}-\boldsymbol{v})\otimes(\boldsymbol{w}-\boldsymbol{v}),\cdots$), there is a corresponding macroscopic flux due to gradients in these quantities in phase space, which is described by a set of so called transport equations (hydrodynamic equations).

To obtain these equations, we multiply the Boltzmann equation (3.6) for the single-point distribution function $f_\alpha(\boldsymbol{r},\boldsymbol{w},t)$, by the variable Υ, then integrate over all velocities:

$$\int_{\boldsymbol{w}} \Upsilon\frac{\partial f_\alpha}{\partial t}\mathrm{d}\boldsymbol{w}+\int_{\boldsymbol{w}} \Upsilon\boldsymbol{w}\cdot\boldsymbol{\nabla}_r f_\alpha\mathrm{d}\boldsymbol{w}+\int_{\boldsymbol{w}} \Upsilon\frac{\boldsymbol{F}}{m_\alpha}\cdot\boldsymbol{\nabla}_w f_\alpha\mathrm{d}\boldsymbol{w}=\int_{\boldsymbol{w}} \Upsilon S(f_\alpha)\mathrm{d}\boldsymbol{w}\ . \tag{3.82}$$

In the following, for reasons of simplicity, we will ignore the subscript α of the species of particles considered, the subscript r attached to the differential operator in the spatial coordinates, and the symbol $\otimes$ for the tensorial product ($\boldsymbol{w}\otimes\boldsymbol{w}\equiv\boldsymbol{w}\boldsymbol{w}$). Examination of the different terms of the LHS of (3.82) enables us to obtain mean values (3.39) of a given microscopic variable Υ explicitly, as we will now show.

The time dependent term (1st term)

This can be written in the form:

$$\int_{\boldsymbol{w}} \Upsilon\frac{\partial f}{\partial t}\,\mathrm{d}\boldsymbol{w}=\frac{\partial}{\partial t}\int_{\boldsymbol{w}} \Upsilon f\,\mathrm{d}\boldsymbol{w}-\int_{\boldsymbol{w}} f\frac{\partial\Upsilon}{\partial t}\,\mathrm{d}\boldsymbol{w}\ , \tag{3.83}$$

such that:

$$\int_{\boldsymbol{w}} \Upsilon\frac{\partial f}{\partial t}\,\mathrm{d}\boldsymbol{w}=\frac{\partial}{\partial t}\left[n\langle\Upsilon\rangle\right]-n\langle\frac{\partial\Upsilon}{\partial t}\rangle\ , \tag{3.84}$$

where the brackets $\langle\ \rangle$ designate an average taken over the (un-separated) distribution function.

The term including the spatial gradient of *f* (2nd term)

Knowing that [101]:

$$\boldsymbol{\nabla}\cdot\int_{\boldsymbol{w}} \boldsymbol{w}\Upsilon f\,\mathrm{d}\boldsymbol{w}=\int_{\boldsymbol{w}} \Upsilon\boldsymbol{w}\cdot\boldsymbol{\nabla} f\,\mathrm{d}\boldsymbol{w}+\int_{\boldsymbol{w}} f\boldsymbol{w}\cdot\boldsymbol{\nabla}\Upsilon\,\mathrm{d}\boldsymbol{w}\ , \tag{3.85}$$

we can write:

[101] See Appendices VII and VIII for details.

$$\int_{\boldsymbol{w}} \Upsilon \boldsymbol{w} \cdot \boldsymbol{\nabla} f \, \mathrm{d}\boldsymbol{w} = \boldsymbol{\nabla} \cdot n\langle \boldsymbol{w}\Upsilon \rangle - n\langle \boldsymbol{w} \cdot \boldsymbol{\nabla}\Upsilon \rangle . \tag{3.86}$$

The term including the gradient of f in velocity space (3rd term)

$$\begin{aligned}\int_{\boldsymbol{w}} \Upsilon \frac{\boldsymbol{F}}{m} \cdot \boldsymbol{\nabla}_w f \, \mathrm{d}\boldsymbol{w} \equiv & \int_{w_x}\int_{w_y}\int_{w_z} \Upsilon \frac{F_x}{m}\frac{\partial f}{\partial w_x} \mathrm{d}w_x \, \mathrm{d}w_y \, \mathrm{d}w_z \\ & + \int_{w_x}\int_{w_y}\int_{w_z} \Upsilon \frac{F_y}{m}\frac{\partial f}{\partial w_y} \mathrm{d}w_x \, \mathrm{d}w_y \, \mathrm{d}w_z \\ & + \int_{w_x}\int_{w_y}\int_{w_z} \Upsilon \frac{F_z}{m}\frac{\partial f}{\partial w_z} \mathrm{d}w_x \, \mathrm{d}w_y \, \mathrm{d}w_z .\end{aligned} \tag{3.87}$$

Integrating, for instance, the first term on the RHS by parts along w_x, we obtain:

$$\begin{aligned}&\int_{w_x}\int_{w_y}\int_{w_z} \Upsilon \frac{F_x}{m}\frac{\partial f}{\partial w_x} \mathrm{d}w_x \mathrm{d}w_y \mathrm{d}w_z = \\ &\int_{w_y}\int_{w_z} \mathrm{d}w_y \mathrm{d}w_z \left\{ \left[\Upsilon \frac{F_x}{m} f \right]_{w_x=-\infty}^{+\infty} - \int f \frac{\partial}{\partial w_x}\left(\Upsilon \frac{F_x}{m} \right) \mathrm{d}w_x \right\} ,\end{aligned} \tag{3.88}$$

where the first term on the RHS of (3.88) is zero because $f(\pm\infty) = 0$. The second term is easily calculated if we suppose that:

$$\frac{\partial F_x}{\partial w_x} = \frac{\partial F_y}{\partial w_y} = \frac{\partial F_z}{\partial w_z} = 0 . \tag{3.89}$$

This condition is satisfied for the two types of force that we will encounter:

- the force due to an electric field $\boldsymbol{E}$. This force, which acts on the charged particles, is independent of their velocities.
- the force due to a magnetic field $\boldsymbol{B}$. The component of this force in a given direction depends only on the components of the velocity in the other two directions.

According to (3.89), since F_x is a constant with respect to w_x, we can exclude it from the derivative in (3.88), and the term containing the force $\boldsymbol{F}$ (3.87) can be written:

$$\int_{\boldsymbol{w}} \Upsilon \frac{\boldsymbol{F}}{m} \cdot \boldsymbol{\nabla}_w f \, \mathrm{d}\boldsymbol{w} = -n\langle \frac{\boldsymbol{F}}{m} \cdot \boldsymbol{\nabla}_w \Upsilon \rangle . \tag{3.90}$$

By substituting (3.84), (3.86) and (3.90) in (3.82), we obtain the evolution of the macroscopic parameters of the microscopic property Υ:

$$\frac{\partial}{\partial t}[n\langle\Upsilon\rangle] - n\langle\frac{\partial\Upsilon}{\partial t}\rangle + \boldsymbol{\nabla}\cdot[n\langle\boldsymbol{w}\Upsilon\rangle] - n\langle\boldsymbol{w}\cdot\boldsymbol{\nabla}\Upsilon\rangle - n\langle\frac{\boldsymbol{F}}{m}\cdot\boldsymbol{\nabla}_w\Upsilon\rangle$$
$$= \int\limits_{\boldsymbol{w}} \Upsilon S(f)\,\mathrm{d}\boldsymbol{w}\,. \quad (3.91)$$

We will now use this relationship to obtain the different hydrodynamic moments.

3.5.1 The continuity equation (1st hydrodynamic moment, of zero order in w)

This equation describes *particle transport* (their flux), taking account of the various influences affecting their motion (field of force $\boldsymbol{F}$ and collisions). It corresponds to the microscopic variable:

$$\Upsilon = 1\,, \quad (3.92)$$

so that:

$$\frac{\partial\Upsilon}{\partial t} = 0, \quad \boldsymbol{\nabla}\Upsilon = 0, \quad \boldsymbol{\nabla}_w\Upsilon = 0\,. \quad (3.93)$$

Equation (3.91) then reduces to:

$$\frac{\partial n}{\partial t} + \boldsymbol{\nabla}\cdot n\boldsymbol{v} = \int\limits_{\boldsymbol{w}} S(f)\,\mathrm{d}\boldsymbol{w}\,. \quad (3.94)$$

This scalar (zero order tensor) equation is called the *equation for conservation of particles* or the *continuity equation*[102].

Collisional term: assumptions required

The term $S(f)\mathrm{d}\boldsymbol{w}$ represents the net number of particles entering (or leaving, if negative) the interval $\boldsymbol{w}$, $\boldsymbol{w}+\mathrm{d}\boldsymbol{w}$ in velocity space as a result of collisions[103]. In the case of elastic collisions, there is neither creation nor disappearance of particles in the plasma volume. These collisions only modify the velocity distribution of the particles, whose total number does not change locally, and the integral over all velocities is then necessarily zero.

[102] By multiplying (3.94) by m_α, the mass of the species α, or by q_α, the charge of species α, we obtain the law of conservation of mass or the law of conservation of electric charge respectively, the latter being $\partial\rho/\partial t + \boldsymbol{\nabla}\cdot\boldsymbol{J} = q_\alpha\int_{\boldsymbol{w}} S(f)\,\mathrm{d}\boldsymbol{w}$.

[103] Writing the collision operator $S(f)$ in the form $(\partial f/\partial t)_{\mathrm{col.}}$ has the advantage that it describes the variation in f as a function of time, resulting from collisions (see (3.5)).

Thus:

$$\frac{\partial n}{\partial t} + \boldsymbol{\nabla} \cdot (n\boldsymbol{v}) = 0 \,. \tag{3.95}$$

However, in general, creation of charged particles occurs in the discharge volume (for example, by electron-neutral collisions) and destruction of these particles takes place, either by volume recombination or by recombination as a result of diffusion of these particles to the wall (Sect. 1.8). When there is volume recombination, the integral of the collision term takes the following complete form:

$$\int\limits_{\boldsymbol{w}} S(f)\, \mathrm{d}\boldsymbol{w} = (\bar{\nu}_i - \bar{\nu}_r)n \,, \tag{3.96}$$

where $\bar{\nu}_i$ is the mean ionisation frequency and $\bar{\nu}_r$, that of volume recombination. In the case where losses only occur due to volume recombination, in the stationary state ($\partial n/\partial t = 0$), we have $\bar{\nu}_i = \bar{\nu}_r$, and the term $\boldsymbol{\nabla} \cdot n\boldsymbol{v}$ in (3.94) becomes zero. On the other hand, if diffusion is the predominant mechanism for the loss of charged particles, the ionisation term dominates the volume recombination term and the integral (3.96) is non zero (for a more detailed discussion, see Delcroix and Bers, Appendix A9-1).

We are now in a position to interpret (3.94) in more depth: the variation of the number of particles of species α in a volume V as a function of time is equal to the net number of these particles resulting from creation and volume loss, less the flux of this species leaving the volume V by diffusion. Equation (3.94) can thus be written in integral form:

$$\int\limits_V \frac{\partial n}{\partial t}\, \mathrm{d}V = \int\limits_V \int\limits_{\boldsymbol{w}} S(f)\, \mathrm{d}\boldsymbol{w}\, \mathrm{d}V - \int\limits_V \boldsymbol{\nabla} \cdot n\boldsymbol{v}\, \mathrm{d}V \,, \tag{3.97}$$

and applying the Ostrogradski theorem:

$$\int\limits_V \frac{\partial n}{\partial t}\, \mathrm{d}V = \int\limits_V \int\limits_{\boldsymbol{w}} S(f)\, \mathrm{d}\boldsymbol{w}\, \mathrm{d}V - \int\limits_{S=\partial V} n\boldsymbol{v} \cdot \mathrm{d}\boldsymbol{S} \,, \tag{3.98}$$

where S is the limiting surface of the volume V, and $\mathrm{d}\boldsymbol{S}$ is a surface element normal to the surface S and directed outwards from the volume.

3.5.2 The momentum transport equation (2nd hydrodynamic moment, 1st order in w)

This moment corresponds to the microscopic variable:

$$\boldsymbol{\Upsilon} = m\boldsymbol{w} \,, \tag{3.99}$$

a vector which satisfies:

$$\frac{\partial m\boldsymbol{w}}{\partial t} = 0 \,, \quad \boldsymbol{\nabla}_r m\boldsymbol{w} = 0 \,, \quad \boldsymbol{\nabla}_w m\boldsymbol{w} = m\underline{\boldsymbol{I}} \,, \tag{3.100}$$

as well as:

$$\frac{\partial \boldsymbol{\Upsilon}}{\partial t} = \boldsymbol{0} \,, \quad \boldsymbol{\nabla}\boldsymbol{\Upsilon} = \underline{\boldsymbol{0}} \,, \quad \boldsymbol{\nabla}_w \boldsymbol{\Upsilon} = m\underline{\boldsymbol{I}} \,, \tag{3.101}$$

where $\underline{\boldsymbol{I}}$ is the second order unit tensor; this tensor has components δ_{ij},where δ_{ij} is the Kronecker delta ($\delta_{ij} = 1$ if $i = j$, $\delta_{ij} = 0$ if $i \neq j$). Equation (3.91) can then be written:

$$\frac{\partial}{\partial t}\left[nm\langle \boldsymbol{w}\rangle\right] + \boldsymbol{\nabla}\cdot\left[nm\langle \boldsymbol{w}\boldsymbol{w}\rangle\right] - n\langle \boldsymbol{F}\cdot\underline{\boldsymbol{I}}\rangle = \int_{\boldsymbol{w}} m\boldsymbol{w}S(f)\,\mathrm{d}\boldsymbol{w} \,, \tag{3.102}$$

which is a vector equation (1st order tensor).

To evaluate the dyad $\langle \boldsymbol{w}\boldsymbol{w}\rangle$ (2nd order tensor), set:

$$\boldsymbol{w} = \boldsymbol{v} + \boldsymbol{u} \,, \tag{3.103}$$

where $\boldsymbol{u}$ is the velocity of a particle relative to the mean velocity $\langle \boldsymbol{w}\rangle = \boldsymbol{v}$ of the ensemble of particles. The velocity $\boldsymbol{u}$ thus has a mean value of zero ($\langle \boldsymbol{u}\rangle = 0$). In statistical terminology, $\boldsymbol{u}$ is a centred quantity with respect to its mean value. We can use (3.103) to obtain:

$$nm\langle \boldsymbol{w}\boldsymbol{w}\rangle = nm\langle \boldsymbol{u}\boldsymbol{u} + 2\boldsymbol{u}\boldsymbol{v} + \boldsymbol{v}\boldsymbol{v}\rangle \,. \tag{3.104}$$

Since $\langle \boldsymbol{u}\boldsymbol{v}\rangle = \langle \boldsymbol{u}\rangle \boldsymbol{v} = 0$, and $\langle \boldsymbol{v}\boldsymbol{v}\rangle = \boldsymbol{v}\boldsymbol{v}$, (3.104) can be written:

$$nm\langle \boldsymbol{w}\boldsymbol{w}\rangle = \underline{\boldsymbol{\Psi}} + nm\boldsymbol{v}\boldsymbol{v} \,, \tag{3.105}$$

where, from (3.39) and (3.42):

$$\underline{\boldsymbol{\Psi}} = nm\langle \boldsymbol{u}\boldsymbol{u}\rangle \tag{3.106}$$

is the kinetic pressure tensor[104] (of order 2).

Equation (3.102) thus takes the form:

$$\frac{\partial}{\partial t}(nm\boldsymbol{v}) + \boldsymbol{\nabla}\cdot\underline{\boldsymbol{\Psi}} + \boldsymbol{\nabla}\cdot(nm\boldsymbol{v}\boldsymbol{v}) - n\langle \boldsymbol{F}\rangle = \int_{\boldsymbol{w}} m\boldsymbol{w}S(f)\,\mathrm{d}\boldsymbol{w} \,. \tag{3.107}$$

However, knowing that:

$$\boldsymbol{\nabla}\cdot(n\boldsymbol{v}\boldsymbol{v}) = (n\boldsymbol{v}\cdot\boldsymbol{\nabla})\boldsymbol{v} + \boldsymbol{v}(\boldsymbol{\nabla}\cdot n\boldsymbol{v}) \,, \tag{3.108}$$

[104] The term $(nm)\langle \boldsymbol{w}\boldsymbol{w}\rangle$ represents a *total* "agitation" density (due to n), while $(nm)\boldsymbol{v}\boldsymbol{v}$ is a convective (directed) "agitation" density and $(nm)\langle \boldsymbol{u}\boldsymbol{u}\rangle$ is a purely thermal (random in direction) "agitation" density. *Agitation* is a quantity with dimensions of energy, whose tensorial nature takes account of the anisotropies of the medium.

we obtain from (3.107):

$$\frac{\partial}{\partial t}(nm\boldsymbol{v})+\boldsymbol{\nabla}\cdot\underline{\boldsymbol{\Psi}}+nm(\boldsymbol{v}\cdot\boldsymbol{\nabla})\boldsymbol{v}+m\boldsymbol{v}(\boldsymbol{\nabla}\cdot n\boldsymbol{v})-n\langle\boldsymbol{F}\rangle=\int\limits_{\boldsymbol{w}} m\boldsymbol{w}S(f)\mathrm{d}\boldsymbol{w}\ . \tag{3.109}$$

From the continuity equation (3.95) (the particular case where the collision term is zero), and noting that $\partial(nm\boldsymbol{v})/\partial t = m\boldsymbol{v}\partial n/\partial t + nm\partial\boldsymbol{v}/\partial t$, (3.109) leads to the following well known expression:

$$nm\left(\frac{\partial}{\partial t}+\boldsymbol{v}\cdot\boldsymbol{\nabla}\right)\boldsymbol{v}+\boldsymbol{\nabla}\cdot\underline{\boldsymbol{\Psi}}-n\langle\boldsymbol{F}\rangle=\int\limits_{\boldsymbol{w}} m\boldsymbol{w}S(f)\,\mathrm{d}\boldsymbol{w}\ . \tag{3.110}$$

We will now successively examine the different terms of this equation, to clarify their physical significance and give more details on some of them:

1. The *convective term* $\boldsymbol{v}\cdot\boldsymbol{\nabla}\boldsymbol{v}$ is non linear in $\boldsymbol{v}$, which complicates the solution of (3.110). Fortunately, its contribution is often negligible in comparison to the other terms: this term obviously becomes important when $\boldsymbol{v}$ or its gradient is large.
2. The kinetic pressure tensor appears above in its completely general form (the dyad $\langle\boldsymbol{uu}\rangle$ being anisotropic, see (3.114), further in the text). The term $\boldsymbol{\nabla}\cdot\underline{\boldsymbol{\Psi}}$ appears as a force per unit volume (it has the same dimensions as the term $n\langle\boldsymbol{F}\rangle$), referred to as the *kinetic pressure force*.
   To clarify the significance of $\underline{\boldsymbol{\Psi}}$, consider the corresponding total force acting on the given volume V. The Ostrogradski relation (a particular case of the Stokes-Cartan theorem) enables us to write:

$$\int\limits_{V}\boldsymbol{\nabla}\cdot\underline{\boldsymbol{\Psi}}\,\mathrm{d}V=\int\limits_{S=\partial V}\underline{\boldsymbol{\Psi}}\cdot\mathrm{d}\boldsymbol{S}\ , \tag{3.111}$$

   introducing a force $\underline{\boldsymbol{\Psi}}\cdot\mathrm{d}\boldsymbol{S}$, exerted on a surface element on the boundary of the volume. Then, noting that $\underline{\boldsymbol{\Psi}}\cdot\mathrm{d}\boldsymbol{S}=\underline{\boldsymbol{\Psi}}\cdot\hat{\mathbf{e}}_s\,\mathrm{d}S$, where $\hat{\mathbf{e}}_s$ is a unit vector perpendicular to the surface element $\mathrm{d}S$, we can conclude that $\underline{\boldsymbol{\Psi}}\cdot\hat{\mathbf{e}}_s$ is a force [105] per unit area, i.e. a pressure!
   Still with the aim of clarifying the meaning of $\underline{\boldsymbol{\Psi}}$ (3.106), we will now consider the dyad $\langle\boldsymbol{uu}\rangle$. Since this is a 2nd order tensor, we can represent it by the matrix:

$$\langle\boldsymbol{uu}\rangle=\begin{pmatrix}\langle u_x^2\rangle & \langle u_xu_y\rangle & \langle u_xu_z\rangle\\ \langle u_yu_x\rangle & \langle u_y^2\rangle & \langle u_yu_z\rangle\\ \langle u_zu_x\rangle & \langle u_zu_y\rangle & \langle u_z^2\rangle\end{pmatrix}\ . \tag{3.112}$$

[105] Since $\underline{\Psi}$ is a second order tensor, the contracted product $\underline{\Psi}\cdot\hat{\mathbf{e}}_s$ is therefore a vector, as it must be the case for a force.

The non diagonal terms are zero since $\int_{w_i} u_i f \, \mathrm{d}w_i = 0 \; (i = x, y, z)$ [106], implying for example:

$$\langle u_x u_y \rangle = \frac{1}{n} \int\limits_{w_z} \int\limits_{w_x} u_x \left[\int\limits_{w_y} u_y f \, \mathrm{d}w_y \right] \mathrm{d}w_x \, \mathrm{d}w_z = 0 \, . \tag{3.113}$$

Hence the tensor reduces to a diagonal matrix:

$$\langle \boldsymbol{uu} \rangle = \begin{pmatrix} \langle u_x^2 \rangle & 0 & 0 \\ 0 & \langle u_y^2 \rangle & 0 \\ 0 & 0 & \langle u_z^2 \rangle \end{pmatrix} . \tag{3.114}$$

Particular case: the distribution of velocities $\boldsymbol{u}$ is isotropic ($u_x = u_y = u_z$). Making the further assumption of a Maxwell-Boltzmann distribution, we obtain:

$$\langle u_x^2 \rangle = \langle u_y^2 \rangle = \langle u_z^2 \rangle = \frac{\langle u^2 \rangle}{3} = \frac{k_B T}{m} \, . \tag{3.115}$$

With these two assumptions, and taking account of (3.106), we can write:

$$\boldsymbol{\nabla} \cdot \underline{\boldsymbol{\Psi}} \equiv \boldsymbol{\nabla} \cdot \left[(n k_B T) \, \underline{\boldsymbol{I}} \, \right] , \tag{3.116}$$

where $\underline{\boldsymbol{I}}$ is the 2nd order unit tensor. Therefore, introducing the partial pressure $p_\alpha = n_\alpha k_B T_\alpha$ associated with the particles of species α, we finally obtain:

$$\boldsymbol{\nabla} \cdot \underline{\boldsymbol{\Psi}} = \boldsymbol{\nabla} p_\alpha \, . \tag{3.117}$$

With these assumptions, the momentum transport equation (3.110), for particles of species α, takes the form:

$$n_\alpha m_\alpha \left(\frac{\partial}{\partial t} + \boldsymbol{v}_\alpha \cdot \boldsymbol{\nabla} \right) \boldsymbol{v}_\alpha + \boldsymbol{\nabla} p_\alpha - n_\alpha q_\alpha \left[\boldsymbol{E} + \boldsymbol{v}_\alpha \wedge \boldsymbol{B} \right] = \int\limits_{\boldsymbol{w}_\alpha} m_\alpha \boldsymbol{w}_\alpha S(f_\alpha) \, \mathrm{d}\boldsymbol{w}_\alpha \, , \tag{3.118}$$

where, in general, the fields $\boldsymbol{E}$ and $\boldsymbol{B}$ denote both the externally applied fields and the (macroscopic) induced fields; the collision operator $S(f_\alpha)_\beta$ is defined by (3.7).

[106] In fact, since $\boldsymbol{u}$ is, by definition, a centred velocity with mean value zero, we have

$$\iint\limits_{w_y w_z} \left[\int\limits_{w_x} u_x f(\boldsymbol{w}) \, \mathrm{d}w_x \right] \mathrm{d}w_y \, \mathrm{d}w_z = n \langle u_x \rangle = 0 \, ,$$

and the same for $\langle u_y \rangle$ and $\langle u_z \rangle$, then more generally $\int_{w_i} u_i f(\boldsymbol{w}) \, \mathrm{d}w_i = 0$.

3. The collision term appearing on the RHS of (3.110), the second hydrodynamic moment, represents the total momentum "gained" or "lost" by particles α, as a result of elastic and inelastic interactions, with *other* types of particles only: collisions between particles of the same species can neither lead to a *net* gain nor *net* loss of momentum! The collision term can then be written formally as:

$$\boldsymbol{\mathcal{P}}_\alpha = \sum_{\beta \neq \alpha} \boldsymbol{\mathcal{P}}_{\alpha\beta} , \tag{3.119}$$

where:

$$\boldsymbol{\mathcal{P}}_{\alpha\beta} = \int_{\boldsymbol{w}_\alpha} m_\alpha \boldsymbol{w}_\alpha S(f_\alpha)_\beta \, \mathrm{d}\boldsymbol{w}_\alpha . \tag{3.120}$$

In order to obtain an expression describing the net transfer of momentum from one group of particles to another (for example, from electrons to neutrals), we will first take a phenomenological approach (with an approximate expression for $\boldsymbol{\mathcal{P}}_{\alpha\beta}$) then, in a second step, we will calculate its exact value.

- Approximate expression for $\boldsymbol{\mathcal{P}}_{\alpha\beta}$ for elastic collisions
We have already shown (Sect. 1.7.2) that the momentum $\Delta \boldsymbol{p}_{\alpha\beta}$ transferred from one particle to another during a collision depends on the relative velocities of the two particles before collision:

$$\Delta \boldsymbol{p}_{\alpha\beta} = -\frac{m_\alpha m_\beta}{m_\alpha + m_\beta}(1 - \cos\theta)(\boldsymbol{w}_\alpha - \boldsymbol{w}_\beta) . \tag{1.91}$$

To characterise the net transfer of momentum per unit volume, $\Delta\boldsymbol{\mathcal{P}}_{\alpha\beta}$, from particles of type α, density n_α, to particles of type β ($\beta \neq \alpha$), density n_β, over their entire velocity distribution, we will make use of the mean velocities $\boldsymbol{v}_\alpha$ and $\boldsymbol{v}_\beta$ rather than the integrals over the velocities $\boldsymbol{w}_\alpha$ and $\boldsymbol{w}_\beta$ (this assumes that the microscopic collision frequencies $\nu_{\alpha\beta}$ are independent of the relative velocity $w_{\alpha\beta}$). Then, the momentum $\boldsymbol{\mathcal{P}}_\alpha$ lost by particles α to the benefit of particles β, per **unit time**, per **unit volume**, is of order:

number of momentum transfers per second, per volume unit ↓

$$\boldsymbol{\mathcal{P}}_\alpha = \frac{\Delta\boldsymbol{\mathcal{P}}_{\alpha\beta}}{\Delta t} = \sum_{\beta\neq\alpha} \boldsymbol{\mathcal{P}}_{\alpha\beta} \approx -\sum_{\beta\neq\alpha} \frac{m_\alpha m_\beta}{m_\alpha + m_\beta} \overbrace{\nu_{\alpha\beta} n_\alpha}(\boldsymbol{v}_\alpha - \boldsymbol{v}_\beta) . \tag{3.121}$$

↑ decrease of average momentum to the benefit of particle β

↑ number of momentum transfer per second per particle α

Supposing that the particles of type α are electrons:

$$\boldsymbol{\mathcal{P}}_{\alpha\beta} \simeq -\nu_{\alpha\beta} n_\alpha m_\alpha (\boldsymbol{v}_\alpha - \boldsymbol{v}_\beta) , \tag{3.122}$$

because the reduced mass $\mu_{\alpha\beta} \equiv m_\alpha m_\beta/(m_\alpha + m_\beta) \approx m_\alpha$.
This relatively simple relation, obtained for the case of elastic collisions, can also be applied to inelastic collisions, to the extent that $\nu_{\alpha\beta}$ is the sum of the elastic and inelastic collision frequencies expressed in an appropriate fashion, and that the velocity distribution is isotropic (Golant et al., Sect. 6.3).

- Exact expression for $\boldsymbol{\mathcal{P}}_{\alpha\beta}$ for elastic collisions
The rigorous expression for $\boldsymbol{\mathcal{P}}_{\alpha\beta}$ is given by Golant et al., Sect. 6.3:

$$\boldsymbol{\mathcal{P}}_{\alpha\beta} = \frac{m_\alpha m_\beta}{m_\alpha + m_\beta} \int\limits_{w_\alpha}\int\limits_{w_\beta} (\boldsymbol{w}_\beta - \boldsymbol{w}_\alpha) \frac{\nu_{\alpha\beta}}{n_\beta} f_\alpha(\boldsymbol{w}_\alpha) f_\beta(\boldsymbol{w}_\beta)\, \mathrm{d}\boldsymbol{w}_\alpha\, \mathrm{d}\boldsymbol{w}_\beta , \tag{3.123}$$

where the functions $f_\alpha(\boldsymbol{w}_\alpha)$ and $f_\beta(\boldsymbol{w}_\beta)$ are the single-point, unseparated, velocity distribution functions for particles α and β (the dependence on $\boldsymbol{r}$ has been omitted for convenience of notation). The (microscopic) collision frequency of particles α on particles β, $\nu_{\alpha\beta}$ is written (from (1.140)):

$$\nu_{\alpha\beta} = n_\beta w_{\alpha\beta} \int_0^\pi 2\pi\hat{\sigma}(\theta)(1-\cos\theta)\sin\theta\, \mathrm{d}\theta , \tag{3.124}$$

where $w_{\alpha\beta}$ is the modulus of the relative velocity, $|\boldsymbol{w}_\alpha - \boldsymbol{w}_\beta|$, of the particles α and β, and where the integral corresponds to the total microscopic momentum transfer cross-section (1.111).
Remember that $\nu_{\alpha\beta}$ is different from $\nu_{\beta\alpha}$ because:

$$\nu_{\beta\alpha} = n_\alpha w_{\alpha\beta} \int_0^\pi 2\pi\hat{\sigma}(\theta)(1-\cos\theta)\sin\theta\, \mathrm{d}\theta , \tag{3.125}$$

so that, combining (3.124) and (3.125), we obtain:

$$\frac{\nu_{\alpha\beta}}{n_\beta} = \frac{\nu_{\beta\alpha}}{n_\alpha} . \tag{3.126}$$

This results in:

$$\boldsymbol{\mathcal{P}}_{\alpha\beta} = -\boldsymbol{\mathcal{P}}_{\beta\alpha} \tag{3.127}$$

and, in particular:

$$\boldsymbol{\mathcal{P}}_{\alpha\alpha} = -\boldsymbol{\mathcal{P}}_{\alpha\alpha} = 0 . \tag{3.128}$$

There is therefore conservation of global momentum during elastic collisions between particles ($\boldsymbol{\mathcal{P}}_\alpha = 0$).

In the general case, $\nu_{\alpha\beta}$ depends on the modulus of the relative velocities $w_{\alpha\beta}$, and the calculation of the integral (3.123) is not straight forward. On the other hand, if one can make the approximation that $\nu_{\alpha\beta}$ has no such dependence[107], Eq. (3.123) is simple to integrate and we obtain:

$$\boldsymbol{\mathcal{P}}_{\alpha\beta} = -\mu_{\alpha\beta} n_\alpha \nu_{\alpha\beta} (\boldsymbol{v}_\alpha - \boldsymbol{v}_\beta) , \tag{3.129}$$

which is identical to (3.121).

Remark: In the case of inelastic collisions, the calculation is much more complex, because the variation in kinetic energy of the particle α is equal to the sum of the variation in both internal (potential) energy and kinetic energy of the particle β. However, the problem is considerably simplified if the relative kinetic energy $\mu_{\alpha\beta} w_{\alpha\beta}^2/2$ is entirely converted to potential energy $\Delta\mathcal{E}$ in the course of the collision[108], i.e. from (1.79):

$$\mu_{\alpha\beta} w_{\alpha\beta}^{'2} = 0 . \tag{3.130}$$

The variation in momentum $\Delta\boldsymbol{p}_{\alpha\beta}$ of the particle α (1.81) can thus be simplified by taking account of (1.82):

$$\Delta\boldsymbol{p}_{\alpha\beta} \equiv -\mu_{\alpha\beta}(\boldsymbol{w}_{\alpha\beta} - \boldsymbol{w}'_{\alpha\beta}) = -\mu_{\alpha\beta}\boldsymbol{w}_{\alpha\beta} . \tag{3.131}$$

Under these conditions, the inelastic collision term, after integration, can be written:

$$\boldsymbol{\mathcal{P}}_{\alpha\beta} = -\mu_{\alpha\beta} n_\alpha \nu_{\alpha\beta} (\boldsymbol{v}_\alpha - \boldsymbol{v}_\beta) , \tag{3.132}$$

an identical equation to that for elastic collisions (3.129).

4. We can now reunite the different terms of the momentum equation of order 1 in $\boldsymbol{w}$. To do this, we set:

$$\frac{\mathrm{d}}{\mathrm{d}t} \equiv \frac{\partial}{\partial t} + \boldsymbol{v}\cdot\boldsymbol{\nabla} , \tag{3.133}$$

where $\mathrm{d}/\mathrm{d}t$ is the *total derivative* operator[109]. It is used when the observer follows the motion of a volume element (Lagrange *description)*. On the other hand, the RHS of (3.133) can be considered as describing the motion of a volume element in the laboratory frame: this motion depends, on

[107] The effective collision frequency $\nu_{\alpha\beta}$ is independent of the relative velocity if the momentum transfer collision cross-section $\hat{\sigma}_{\alpha\beta}$ is inversely proportional to the relative velocity. One can show that this case corresponds to an interaction potential in $1/r^4$, that is approximately valid for collisions between charged and neutral species.

[108] With this assumption, the collision is by definition head-on because $\boldsymbol{w}'_{\alpha\beta} = 0$ (3.130) and $\boldsymbol{w}_{\alpha\beta}$, $\boldsymbol{w}_{\alpha 0}$ and $\boldsymbol{w}_{\beta 0}$ are collinear (1.73).

[109] Note that, in the stationary regime, the total derivative is non zero because the convective term $\boldsymbol{v}\cdot\boldsymbol{\nabla}$ remains: only the term $\partial/\partial t$ cancels.

the one hand, on the time variation of the local velocity, and on the other hand, on the ensemble motion (convection) of the fluid (Euler description). Finally, by substituting (3.129) and (3.133) in (3.118), we obtain the well known form[110]:

$$m_\alpha \frac{\mathrm{d}}{\mathrm{d}t}\boldsymbol{v}_\alpha = q_\alpha\left[\boldsymbol{E} + \boldsymbol{v}_\alpha \wedge \boldsymbol{B}\right] - \frac{1}{n_\alpha}\boldsymbol{\nabla} p_\alpha - \sum_{\beta \neq \alpha} \mu_{\alpha\beta}\nu_{\alpha\beta}[\boldsymbol{v}_\alpha - \boldsymbol{v}_\beta] \ . \quad (3.134)$$

This is the 2nd moment equation (first order moment in $\boldsymbol{w}$) or the momentum transport equation for particles of species α, having an isotropic distribution function, and colliding elastically with particles β that are different from α. This equation is also called the Langevin equation.
Equation (3.134) determines the acceleration of the fluid under the influence of different forces, including electric and magnetic forces, the pressure gradient and collisional viscosity.

Remark: It is interesting to compare (3.134) with the hydrodynamic Navier-Stokes *equation* for the transfer of momentum in an incompressible (in this case, ρ_M, the mass per unit volume is constant, and the continuity equation leads to $\boldsymbol{\nabla}\cdot\boldsymbol{v} = 0$) and viscous fluid. Under these conditions, the equation can be expressed as (see Landau and Lifschitz):

$$\underbrace{\rho_M\left[\frac{\partial \boldsymbol{v}}{\partial t} + (\boldsymbol{v}\cdot\boldsymbol{\nabla})\boldsymbol{v}\right]}_{\substack{\uparrow \\ \text{Force per volume unit}}} = -\boldsymbol{\nabla} p + \underbrace{\eta_v \Delta \boldsymbol{v}}_{\substack{\uparrow \\ \text{Interaction term} \\ \text{(viscosity)}}} + n\boldsymbol{F} \ , \quad (3.135)$$

where η_v is the coefficient of fluid viscosity.

3.5.3 Moment equations of second order in w

We can distinguish two cases, in which the second order moment is either w^2 or $\boldsymbol{ww}$.

Transport equation for kinetic energy

This equation is also called the *energy balance equation.* This moment corresponds to the microscopic variable:

$$\Upsilon = \frac{1}{2} m_\alpha w^2 \ , \quad (3.136)$$

110 Remember that Eq. (3.118), taken from (3.110), is obtained from (3.95), the continuity equation without a RHS term.

which leads to[111]:

$$\frac{\partial \Upsilon}{\partial t} = 0 \, , \quad \boldsymbol{\nabla} \Upsilon = \boldsymbol{0} \, , \quad \boldsymbol{\nabla}_w \Upsilon = m_\alpha \boldsymbol{w} \, . \tag{3.137}$$

Equation (3.91) can then be written:

$$\frac{\partial}{\partial t}\left(\frac{1}{2} n_\alpha m_\alpha \langle w^2 \rangle\right) + \boldsymbol{\nabla} \cdot \left[\frac{1}{2} n_\alpha m_\alpha \langle \boldsymbol{w} w^2 \rangle\right] - n_\alpha \langle \boldsymbol{F} \cdot \boldsymbol{w} \rangle = \int \frac{1}{2} m_\alpha w^2 S(f) \, \mathrm{d}\boldsymbol{w} \, , \tag{3.138}$$

which is a scalar equation (zero order tensor).

Noting that $\langle w^2 \rangle = \langle u^2 \rangle + v^2$ and assuming the velocities are isotropic, with $\langle u^2 \rangle = 3k_B T_\alpha / m_\alpha$, we obtain, for the species α:

$$\underbrace{\frac{\partial}{\partial t}\left(\frac{1}{2} n_\alpha m_\alpha v_\alpha^2\right)}_{\substack{\text{kinetic energy} \\ \text{directed} \\ \text{(convection)}}} + \underbrace{\frac{\partial}{\partial t}\left(\frac{3}{2} n_\alpha k_B T_\alpha\right)}_{\substack{\text{kinetic energy} \\ \text{random}}} = -\boldsymbol{\nabla} \cdot \boldsymbol{q}_\alpha + n_\alpha q_\alpha \boldsymbol{E} \cdot \boldsymbol{v}_\alpha + \mathcal{R}_\alpha \, , \tag{3.139}$$

where:

$$\boldsymbol{q}_\alpha = \frac{n_\alpha m_\alpha}{2} \langle \boldsymbol{w}_\alpha w_\alpha^2 \rangle \tag{3.140}$$

is the vector-flux[112] of the total kinetic energy of the particles of species α, and where the term:

$$\mathcal{R}_\alpha = \sum_{\beta \neq \alpha} \mathcal{R}_{\alpha\beta} \tag{3.141}$$

represents the total energy "gained" or "lost" by particles α, following elastic and inelastic collisions with **other** types of particles only: collisions between particles of the same species cannot lead to either a loss or gain of kinetic energy.

The variation of the density of the total kinetic energy of the fluid of particles α (LHS of (3.139)) occurs due to three mechanisms, represented by the three terms on the RHS: 1st term: transport of kinetic energy from one point to another in the plasma due to a spatial gradient[113]; 2nd term: deposition of energy (heating) in the plasma by the current of particles moving in the field $\boldsymbol{E}$ (Ohm's law): 3rd term: variation of kinetic energy of the particles α due to their collisions with other types of particles.

---

111 Note that (3.136) can also be written in the form $\Upsilon = (m_\alpha \boldsymbol{w} \cdot \boldsymbol{w})/2$, and therefore $\boldsymbol{\nabla}_w \Upsilon = m_\alpha \boldsymbol{w}$: the gradient of a scalar must be a vector.

112 In general, the vector $\overline{n\boldsymbol{w}\Upsilon(w)}$ is the *vector flux* of the molecular property $\Upsilon(w)$.

113 For example, in the case of a temperature gradient, the heat flux can be expressed by $\boldsymbol{q}_\alpha = -\lambda_\alpha \boldsymbol{\nabla} T_\alpha$, where λ_α is the *thermal conductivity* of species α.

The collision term $\mathcal{R}_{\alpha\beta}$, resulting from the elastic collision of particles α with particles β, can be written in the same way as for the first order moment, if the collision frequency $\nu_{\alpha\beta}$ is independent of the velocity (Golant et al. Sect. 6.4):

$$\mathcal{R}_{\alpha\beta} = -\frac{2m_\alpha m_\beta}{(m_\alpha + m_\beta)^2} n_\alpha \nu_{\alpha\beta} \left[\frac{m_\alpha \langle w_\alpha^2 \rangle}{2} - \frac{m_\beta \langle w_\beta^2 \rangle}{2} + \frac{m_\beta - m_\alpha}{2} (\boldsymbol{v}_\alpha \cdot \boldsymbol{v}_\beta) \right] . \tag{3.142}$$

Knowing that:

$$m_\alpha \langle w_\alpha^2 \rangle = m_\alpha \left[\langle u_\alpha^2 \rangle + v_\alpha^2 \right] = 3k_B T_\alpha + m_\alpha v_\alpha^2 , \tag{3.143}$$

and using the energy transfer coefficient δ (1.103), Eq. (3.142) can now be written:

$$\mathcal{R}_{\alpha\beta} = -\delta n_\alpha \nu_{\alpha\beta} \left[\frac{m_\alpha v_\alpha^2}{2} - \frac{m_\beta v_\beta^2}{2} + \frac{3}{2} k_B (T_\alpha - T_\beta) + \frac{(m_\beta - m_\alpha)}{2} (\boldsymbol{v}_\alpha \cdot \boldsymbol{v}_\beta) \right] . \tag{3.144}$$

It should be noted that $\mathcal{R}_{\alpha\alpha} = 0$. Although this can be easily verified in (3.144), it is not necessary to assume that $\nu_{\alpha\beta}$ is independent of the velocity to obtain this result, because the kinetic energy is a collisional invariant.
In many cases, the gas has no directed velocity ($\boldsymbol{v}_\alpha = \boldsymbol{v}_\beta = 0$), or all the particles have the same velocity ($\boldsymbol{v}_\alpha = \boldsymbol{v}_\beta$). In these cases, the terms containing $\boldsymbol{v}_\alpha$ and $\boldsymbol{v}_\beta$ disappear and we are left with:

$$\mathcal{R}_{\alpha\beta} = -\delta n_\alpha \nu_{\alpha\beta} \left(\frac{3}{2} k_B T_\alpha - \frac{3}{2} k_B T_\beta \right) , \tag{3.145}$$

which clearly illustrates the exchange of kinetic energy during elastic collisions.
Remark: In the case of inelastic collisions, the calculation is much more complex, because the change in kinetic energy of particle α is equal to the sum of the change in internal energy and in kinetic energy of particle β. However, the problem is considerably simplified if we suppose that the change in kinetic energy of particle β is negligible in comparison to the change in internal energy, which is generally the case for collisions with electrons (1.80). With this assumption, for a collision frequency $\nu_{\alpha\beta}$, the collision term $\mathcal{R}_{\alpha\beta}$ resulting from inelastic collisions of particles α (electrons) with particles β is then:

$$\mathcal{R}_{\alpha\beta} = -n_\alpha \nu_{\alpha\beta} \mathcal{E}_k , \tag{3.146}$$

where $\mathcal{E}_k$ represents the energy threshold of the inelastic collision considered.

The transport equation for the kinetic pressure tensor $\underline{\boldsymbol{\Psi}}$

This tensor $\underline{\boldsymbol{\Psi}}$ leads to the true second order moment.

The kinetic pressure tensor $\underline{\boldsymbol{\Psi}}$ (3.42) corresponds to the microscopic property:

$$\underline{\boldsymbol{\Upsilon}} = m_\alpha(\boldsymbol{w}-\boldsymbol{v})(\boldsymbol{w}-\boldsymbol{v}) \tag{3.147}$$

in which $\boldsymbol{v} = \boldsymbol{v}(\boldsymbol{r},\, t)$. Writing $\boldsymbol{w} = \boldsymbol{v} + \boldsymbol{u}$, we obtain:

$$\frac{\partial \underline{\boldsymbol{\Upsilon}}}{\partial t} = -m_\alpha \left(\boldsymbol{u}\frac{\partial \boldsymbol{v}}{\partial t} + \frac{\partial \boldsymbol{v}}{\partial t}\boldsymbol{u} \right) , \tag{3.148}$$

$$\boldsymbol{\nabla}\underline{\boldsymbol{\Upsilon}} = -m_\alpha \boldsymbol{\nabla}\boldsymbol{v}\boldsymbol{u} - m_\alpha \boldsymbol{\nabla}\boldsymbol{u}\boldsymbol{v} = -m_\alpha \boldsymbol{\nabla}\boldsymbol{v}\boldsymbol{u} - (m_\alpha \boldsymbol{\nabla}\boldsymbol{v}\boldsymbol{u})^T , \tag{3.149}$$

$$\boldsymbol{\nabla}_w \underline{\boldsymbol{\Upsilon}} = m_\alpha(\boldsymbol{u}\underline{\boldsymbol{I}} + \underline{\boldsymbol{I}}\boldsymbol{u}) , \tag{3.150}$$

where $\underline{\boldsymbol{I}}$ is the second order unit tensor and the suffix T indicates the transpose of the matrix representing the tensor[114]. All the terms are 2nd order tensors. Equation (3.91) can then be written:

$$\begin{aligned} &\frac{\partial}{\partial t}\left[n_\alpha m_\alpha \langle (\boldsymbol{w}-\boldsymbol{v})(\boldsymbol{w}-\boldsymbol{v})\rangle\right] + n_\alpha m_\alpha \langle \boldsymbol{u}\frac{\partial \boldsymbol{v}}{\partial t} + \frac{\partial \boldsymbol{v}}{\partial t}\boldsymbol{u}\rangle \\ &\quad + \boldsymbol{\nabla}\cdot\left[n_\alpha m_\alpha \langle \boldsymbol{w}(\boldsymbol{w}-\boldsymbol{v})(\boldsymbol{w}-\boldsymbol{v})\rangle\right] + n_\alpha m_\alpha \langle \boldsymbol{w}\cdot\boldsymbol{\nabla}\boldsymbol{v}\boldsymbol{u} + \boldsymbol{w}\cdot(\boldsymbol{\nabla}\boldsymbol{v}\boldsymbol{u})^T\rangle \\ &\quad - n_\alpha \langle \boldsymbol{F}\cdot(\boldsymbol{u}\underline{\boldsymbol{I}} + \underline{\boldsymbol{I}}\boldsymbol{u})\rangle = \int_{\boldsymbol{w}} m_\alpha(\boldsymbol{w}-\boldsymbol{v})(\boldsymbol{w}-\boldsymbol{v})S(f)\,\mathrm{d}\boldsymbol{w} , \end{aligned} \tag{3.151}$$

which is a 2nd order tensorial equation.

Rewriting (3.151), again with $\boldsymbol{w} = \boldsymbol{v}+\boldsymbol{u}$, and suppressing the terms whose mean value is zero, we obtain:

$$\begin{aligned} &\frac{\partial}{\partial t} n_\alpha m_\alpha \langle \boldsymbol{u}\boldsymbol{u}\rangle + \boldsymbol{\nabla}\cdot n_\alpha m_\alpha \langle \boldsymbol{v}\boldsymbol{u}\boldsymbol{u} + \boldsymbol{u}\boldsymbol{u}\boldsymbol{u}\rangle + n_\alpha m_\alpha \langle \boldsymbol{u}\cdot\boldsymbol{\nabla}\boldsymbol{v}\boldsymbol{u} + (\boldsymbol{u}\cdot\boldsymbol{\nabla}\boldsymbol{v}\boldsymbol{u})^T\rangle \\ &\qquad - n_\alpha \langle \boldsymbol{F}\cdot(\boldsymbol{u}\underline{\boldsymbol{I}} + \underline{\boldsymbol{I}}\boldsymbol{u})\rangle = \underline{\boldsymbol{\mathcal{R}}}_\alpha , \end{aligned} \tag{3.152}$$

where:

$$\underline{\boldsymbol{\mathcal{R}}}_\alpha = \sum_{\beta\neq\alpha} \underline{\boldsymbol{\mathcal{R}}}_{\alpha\beta} \tag{3.153}$$

is the collision term (Appendix XV). Expanding (3.152), we obtain:

---

[114] The tensor $\underline{\boldsymbol{A}}^T$, with elements α_{ij}^T is the *transpose of the tensor* of the same order $\underline{\boldsymbol{A}}$, with elements α_{ij} such that $\alpha_{ij}^T = \alpha_{ji}$.

Hence: $\boldsymbol{\nabla}\boldsymbol{\Upsilon} = -m_\alpha(\boldsymbol{\nabla}\boldsymbol{v}\boldsymbol{u} + \boldsymbol{\nabla}\boldsymbol{u}\boldsymbol{v}) = -m_\alpha \boldsymbol{\nabla}\boldsymbol{v}\boldsymbol{u} - (m_\alpha \boldsymbol{\nabla}\boldsymbol{v}\boldsymbol{u})^T$.

$$\frac{\partial}{\partial t}\langle n_\alpha m_\alpha \boldsymbol{uu}\rangle + (\boldsymbol{\nabla}\cdot\boldsymbol{v})\langle n_\alpha m_\alpha \boldsymbol{uu}\rangle + (\boldsymbol{v}\cdot\boldsymbol{\nabla})\langle n_\alpha m_\alpha \boldsymbol{uu}\rangle$$
$$+\boldsymbol{\nabla}\cdot n_\alpha m_\alpha\langle\boldsymbol{uuu}\rangle + n_\alpha m_\alpha\langle\boldsymbol{uu}\rangle\cdot\boldsymbol{\nabla v} + (n_\alpha m_\alpha\langle\boldsymbol{uu}\rangle\cdot\boldsymbol{\nabla v})^T$$
$$- n_\alpha\langle\boldsymbol{F}\cdot(\boldsymbol{u}\underline{\boldsymbol{I}} + \underline{\boldsymbol{I}}\boldsymbol{u})\rangle = \underline{\boldsymbol{\mathcal{R}}}_\alpha\ . \tag{3.154}$$

It should be noted that the last term of the LHS of (3.154) is zero if the force $\boldsymbol{F}$ is independent of velocity (e.g. in the case $\boldsymbol{F} = q_\alpha \boldsymbol{E}$): $\boldsymbol{F}$ can then be removed from the expression between brackets, which is then zero.

We can then transform (3.154) to the form:

$$\left[\frac{\partial}{\partial t} + \boldsymbol{v}\cdot\boldsymbol{\nabla} + (\boldsymbol{\nabla}\cdot\boldsymbol{v})\right]\underline{\boldsymbol{\Psi}} + \boldsymbol{\nabla}\cdot\underline{\underline{\boldsymbol{Q}}} + \underline{\boldsymbol{\Psi}}\cdot\boldsymbol{\nabla v} + (\underline{\boldsymbol{\Psi}}\cdot\boldsymbol{\nabla v})^T - \underline{\boldsymbol{M}} = \underline{\boldsymbol{\mathcal{R}}}_\alpha\ , \tag{3.155}$$

where $\underline{\boldsymbol{\Psi}}$ is the kinetic pressure tensor (3.106) and $\underline{\boldsymbol{M}}$, which is also a 2nd order tensor, results from the action of an external magnetic field (Appendix XV); on the other hand, $\underline{\underline{\boldsymbol{Q}}}$ is a 3rd order tensor, defined by:

$$\underline{\underline{\boldsymbol{Q}}} = m_\alpha \int_w (\boldsymbol{w}-\boldsymbol{v})(\boldsymbol{w}-\boldsymbol{v})(\boldsymbol{w}-\boldsymbol{v}) f(\boldsymbol{r},\boldsymbol{w},t)\,\mathrm{d}\boldsymbol{w}\ , \tag{3.156}$$

known as the *thermal energy flux tensor*: this is a third order, centred moment of velocities with respect to the mean velocity of the distribution function $f(\boldsymbol{r},\boldsymbol{w},t)$.

We would like to rearrange the first term of (3.155). To do this, we note that the continuity equation (3.28) with RHS zero can be expanded as:

$$\frac{\partial n}{\partial t} + (\boldsymbol{v}\cdot\boldsymbol{\nabla})n + n\boldsymbol{\nabla}\cdot\boldsymbol{v} = 0\ , \tag{3.157}$$

from which, including the total derivative (3.133):

$$\boldsymbol{\nabla}\cdot\boldsymbol{v} = -\frac{1}{n}\frac{\mathrm{d}n}{\mathrm{d}t}\ . \tag{3.158}$$

We can then transform the first three terms of (3.155), to obtain:

$$\left(\frac{\partial}{\partial t} + \boldsymbol{v}\cdot\boldsymbol{\nabla} + \boldsymbol{\nabla}\cdot\boldsymbol{v}\right)\underline{\boldsymbol{\Psi}} \equiv \frac{\mathrm{d}}{\mathrm{d}t}\underline{\boldsymbol{\Psi}} - \frac{\underline{\boldsymbol{\Psi}}}{n}\frac{\mathrm{d}n}{\mathrm{d}t} \equiv n\frac{\mathrm{d}}{\mathrm{d}t}\left(\frac{\underline{\boldsymbol{\Psi}}}{n}\right)\ . \tag{3.159}$$

This equation then reduces to:

$$n\frac{\mathrm{d}}{\mathrm{d}t}\left(\frac{\underline{\boldsymbol{\Psi}}}{n}\right) + \boldsymbol{\nabla}\cdot\underline{\underline{\boldsymbol{Q}}} + (\underline{\boldsymbol{\Psi}}\cdot\boldsymbol{\nabla})\,\boldsymbol{v} + [(\underline{\boldsymbol{\Psi}}\cdot\boldsymbol{\nabla})\,\boldsymbol{v}]^T - \underline{\boldsymbol{M}} = \underline{\boldsymbol{\mathcal{R}}}_\alpha\ . \tag{3.160}$$

Remark: The kinetic pressure tensor:

$$\underline{\underline{\Psi}} = m_\alpha \int\limits_w (\boldsymbol{w} - \boldsymbol{v})(\boldsymbol{w} - \boldsymbol{v}) f(\boldsymbol{r}, \boldsymbol{w}, t)\, \mathrm{d}\boldsymbol{w}$$

appears as a centred moment (with respect to $\boldsymbol{v}$) of order 2 and is related to a *variance*[115] calculated with respect to an average value which, in the present case, is the velocity $\boldsymbol{v}$.

We can see this by noting that:

$$\underline{\underline{\Psi}} = m_\alpha \int\limits_w \boldsymbol{w}\boldsymbol{w} f\, \mathrm{d}\boldsymbol{w} - m_\alpha \int\limits_w \boldsymbol{v}\boldsymbol{v} f\, \mathrm{d}\boldsymbol{w} = n m_\alpha \langle \boldsymbol{w}\boldsymbol{w} \rangle - n m_\alpha \boldsymbol{v}\boldsymbol{v} \,, \tag{3.161}$$

which is of the form $E[X^2] - E[X]^2$, where $E[X]$ is the *mathematical expectation* of the variable X. In summary, we have discussed four aspects of the kinetic pressure tensor:

- $\boldsymbol{\nabla} \cdot \underline{\underline{\Psi}}$ represents a force per unit volume.
- $\underline{\underline{\Psi}} \cdot \hat{\mathbf{e}}_s$ ($\underline{\underline{\Psi}}$ projected normal to a unit surface) is a force per unit area, the kinetic pressure. It is the generalisation of the scalar pressure in an anisotropic gas.
- $\underline{\underline{\Psi}}$ is the second order, centred moment of the velocity distribution function with respect to the average value $\boldsymbol{v}$: it is related to the variance of the microscopic velocities.
- $\underline{\underline{\Psi}}$ has the dimensions of momentum flux: in fact, while $n\boldsymbol{v}$ is a particle flux, $n\boldsymbol{v}(m\boldsymbol{v})$ in (3.161) represents a momentum flux.

3.5.4 Higher order moment equations

We can write the transport equation for the thermal energy flux $\underline{\underline{Q}}$ (3.156), yielding a 3rd order moment in $\boldsymbol{w}$, and so on for the higher moments, which leads to an infinite number of hydrodynamic equations.

Remark: In general, we require a set of hydrodynamic equations for each type of particle. However, in some cases, the single fluid of electrons is adequate for our purposes (Sect. 3.7).

---

115 The variance D of a random variable X is expressed as a function of the mathematical expectation E according to:

$$D[X] \equiv E[(X - m)^2] = E[X^2] - E[X]^2$$

where $E(X) = m$. This characterises the importance of the deviation of the ensemble of values of the distribution, with respect to their average value m.

3.6 Closure of the transport equations

The transport equations of the parameters $n, m\boldsymbol{v}, \underline{\boldsymbol{\Psi}}, \underline{\underline{\boldsymbol{Q}}} \cdots$ provide a good description of the evolution of a plasma at the macroscopic level, but unfortunately they constitute an **indeterminate system**:

- the equation for the conservation of the number of particles n contains $\boldsymbol{v}$,
- the equation describing the evolution of $\boldsymbol{v}$ introduces a second order tensor $\underline{\boldsymbol{\Psi}}$,
- the equation describing the evolution of the kinetic pressure tensor $\underline{\boldsymbol{\Psi}}$ introduces a third order tensor $\underline{\underline{\boldsymbol{Q}}}$,
- and so on...

In summary, the evolution of a given variable is always dependent on a further variable whose tensorial order is one order higher. Such a system is called a *hierarchy*. In general, we use only the first 2 or 3 transport equations. In order to break this hierarchical dependence, it is necessary to make a simplifying assumption on the highest order tensor appearing in the highest order equation we wish to consider. This procedure is commonly referred to as *closure of the transport equations* (see the following examples for methods of closure).

Remark: An analogous problem, but different in its physical meaning, is found in kinetic theory. The integration of the Liouville equation $\mathrm{d}\mathcal{D}/\mathrm{d}t = 0$ ($\mathcal{D}$ is the probability density defined in Sect. 3.3) over all positions $\boldsymbol{r}_i, \boldsymbol{w}_i$ in phase space except $\boldsymbol{r}_1, \boldsymbol{w}_1$, leads to:

$$\frac{\partial f_1}{\partial t} + \boldsymbol{w} \cdot \boldsymbol{\nabla}_r f_1 + \frac{\boldsymbol{F}}{m} \cdot \boldsymbol{\nabla}_w f_1 = S_1(f_{12}) \,. \tag{3.162}$$

This equation (the Boltzmann equation), which describes the evolution of the single-point function f_1 (3.24), introduces a binary interaction term between particles in the form of the two-point function f_{12} (3.30). Integrating the Liouville equation in the same way, but this time over all variables except $\boldsymbol{r}_1, \boldsymbol{w}_1$ and $\boldsymbol{r}_2, \boldsymbol{w}_2$, we obtain:

$$\frac{\partial f_{12}}{\partial t} + \boldsymbol{w} \cdot \boldsymbol{\nabla}_r f_{12} + \frac{\boldsymbol{F}}{m} \cdot \boldsymbol{\nabla}_w f_{12} = S_{12}(f_{123}) \,, \tag{3.163}$$

where the term $S(f_{123})$ represents the ternary interactions between particles, and so on for f_{123},f_{1234}... Once again, we have an indeterminate system of equations, called the BBGKY[116] hierarchy. In order to use (3.162) independently of (3.163), we set, as conditions of closure, the assumption of weak binary correlations between particles (3.34), such that $S(f_{12}) \simeq S(f_1 f_2)$.

[116] The names of the physicists Born, Bogolioubov, Green, Kirkwood, Yvon are in alphabetic order, and apparently the exact inverse of the historical order.

Note that the BBGKY hierarchy is constructed from the collision operator, while the hierarchy of hydrodynamic equations has a completely different origin. It results from the existence of gradients in real space, and appears in the form of a divergence of a tensor of one order higher than the order of the hydrodynamic equation considered. Note, furthermore, that the set of hydrodynamic equations we have just developed proceed from the calculation of mean values, taken uniquely over the Boltzmann equation, the very first equation of the BBGKY hierarchy.

Method of closure

We can limit the hydrodynamic description to k equations by "simplifying" the tensor $\underline{\underline{\boldsymbol{\Upsilon}}}$ of order $k+1$ that generally appears, as we have just indicated, in the form $\boldsymbol{\nabla}\cdot\underline{\underline{\boldsymbol{\Upsilon}}}$: this is usually achieved by assuming that $\underline{\underline{\boldsymbol{\Upsilon}}}$ can be replaced by a tensor quantity of one order lower. Among the simplifying assumptions for *closure of the hydrodynamic equations*, we will consider two that are most commonly used:

1. *Cold plasma.* We completely neglect thermal motion, and assume that $T = 0$, which allows us to write $\underline{\underline{\boldsymbol{\Psi}}} = \underline{\mathbf{0}}$ in the equation for momentum transport. The hydrodynamic equations for n and $\boldsymbol{v}$ then form a two-equation determinate system, to which we can add Maxwell's equations. This approximation is particularly applicable to:

   - the description of the properties of an electron beam
   - wave phenomena in plasmas (for a phase velocity much greater than the mean velocity of the thermal motion of the particles).

2. *Warm plasma.* This approximation, which is less restrictive than the preceding one, has the advantage of accounting for the thermal motion, which is assumed to obey a Maxwell-Boltzmann law. This approximation therefore enables us to describe a larger number of observed phenomena, being particularly applicable to the following cases:

   - neutral gases: since the electric field $\boldsymbol{E}$ cannot transfer energy into the system and set particles in motion, it is thus essential to include their thermal motion,
   - plasmas for which the "cold plasma" approximation is regarded as too crude, when it comes, for instance, to describing the propagation of waves with low phase velocity.

   In the warm plasma approximation, various assumptions are possible, the main ones being:

   a. *Isothermal approximation.* In this case, we assume that the kinetic pressure tensor is no longer zero but has the form $\underline{\underline{\boldsymbol{\Psi}}} = nk_B\underline{\boldsymbol{T}}$, a simplified form of this tensor, where the temperature values T_{ij} of the tensor $\underline{\boldsymbol{T}}$ are independent of the spatial coordinates (T_{ij} = constant). Under this assumption, two situations can be envisaged:

- isotropic plasma. The kinetic pressure tensor then reduces to the sole isotropic and scalar pressure $p = nk_BT$. Then, according to (3.116) and (3.117), we can substitute $\boldsymbol{\nabla} p$ for $\boldsymbol{\nabla} \cdot \underline{\boldsymbol{\Psi}}$, both a tensorial term also of order one (a vector) in which, from our assumptions we can deduce:

$$\boldsymbol{\nabla} k_B T = \mathbf{0} \ . \tag{3.164}$$

  As in the case of the cold plasma approximation, the closure of the system of equations is achieved for the first order moment in $\boldsymbol{w}$, i.e. we are left with only two hydrodynamic equations for the fluid considered.

- anisotropic plasma. The kinetic pressure tensor, following (3.116), takes the form $\underline{\boldsymbol{\Psi}} = nk_B\underline{\boldsymbol{T}}$, where this time the components T_{ii} are different for different values of i. Since by hypothesis, the components of $\underline{\boldsymbol{T}}$ do not depend on position, $\boldsymbol{\nabla} \cdot k_B\underline{\boldsymbol{T}} = \mathbf{0}$. The closure of the equation of moment of order 1 in $\boldsymbol{w}$ is written as:

$$\boldsymbol{\nabla} \cdot \underline{\boldsymbol{\Psi}} = \boldsymbol{\nabla} \cdot nk_B\underline{\boldsymbol{T}} = k_B\underline{\boldsymbol{T}} \cdot \boldsymbol{\nabla} n + n(\boldsymbol{\nabla} \cdot k_B\underline{\boldsymbol{T}}) = k_B\underline{\boldsymbol{T}} \cdot \boldsymbol{\nabla} n \ . \tag{3.165}$$

b. *Adiabatic approximation*[117]. The perturbing system (for example a wave) does not have the time to exchange energy with its environment. To determine the equation of state $p(n)$, we need to consider the first three hydrodynamic equations. The closure condition is applied to the transport equation for $\underline{\boldsymbol{\Psi}}$ (moment in $\underline{\boldsymbol{\Psi}}$) by setting $\underline{\boldsymbol{\mathcal{R}}} = \underline{\mathbf{0}}$ and requiring $\boldsymbol{\nabla} \cdot \underline{\underline{\boldsymbol{Q}}} = \underline{\mathbf{0}}$: there cannot be transfer of energy between particles nor transport of thermal energy by the particles because the compression is adiabatic. If, under these conditions, we require that $\underline{\boldsymbol{\Psi}}$ reduces to the scalar kinetic pressure p, we can show (Appendix XVI) that the transport equation for the kinetic pressure tensor leads to the following purely scalar equation:

$$n\frac{\mathrm{d}}{\mathrm{d}t}\frac{3}{2}\frac{p}{n} + p\boldsymbol{\nabla} \cdot \boldsymbol{v} = 0 \ . \tag{3.166}$$

Then, taking account of the expression $\boldsymbol{\nabla} \cdot \boldsymbol{v} = -\dfrac{1}{n}\dfrac{\mathrm{d}n}{\mathrm{d}t}$ (3.138), Eq. (3.166) finally reduces to:

$$n\frac{\mathrm{d}}{\mathrm{d}t}\left(\frac{3}{2}\frac{p}{n}\right) = \frac{p}{n}\frac{\mathrm{d}n}{\mathrm{d}t} \tag{3.167}$$

[117] A change of state of a system is *adiabatic* if there is neither a gain, nor a loss of thermal energy of the system. Two situations are possible: 1) the system is isolated; 2) the process considered (for example, compression of the plasma exerted by a wave) is so rapid that there is no time to transfer energy by conduction.

for which the solution is:

$$pn^{-\gamma} = \text{constant} , \tag{3.168}$$

where $\gamma = 5/3$ (perfect gas). This is the *adiabatic relation.* In the case where the fluid considered is isotropic (for example, electrons in the absence of a magnetic field $\boldsymbol{B}$) and resembling a perfect gas with density n, the adiabatic constant γ (ratio of specific heats c_p/c_v) is effectively 5/3.
Other values for γ are possible. For instance, for a (linear) one dimensional flow in the presence of a field $\boldsymbol{B}_0$ (the medium is thus anisotropic), we use $\gamma = 3$ for compression parallel to $\boldsymbol{B}_0$ and $\gamma = 1$ for compression perpendicular to $\boldsymbol{B}_0$. More generally, if the molecules have $\bar{\delta}$ degrees of freedom (vibration, rotation, translation), then $\gamma = 1 + 2/\bar{\delta}$. Thus, $\bar{\delta} = 2$ in the case of azimuthal symmetry and $\bar{\delta} = 3$ for a three dimensional, spherically symmetric compression, for which, $\gamma = 5/3$. The case $\gamma = 5/3$ is also referred to as the *Euler* or *scalar approximation* (because the compression is spherically symmetric).
In summary, the assumption of an adiabatic gas flow is used, for example, in situations where the particles take part in the propagation of a sound wave or when the fluid flows extremely fast. Assuming a perfect gas where $p = nk_BT$, the adiabatic equation can also be expressed as:

$$Tn^{1-\gamma} = \text{constant} , \tag{3.169}$$

illustrating that in this case, since n is spatially varying, T also varies spatially.

3.7 The Lorentz electron plasma model

This model can be regarded as a first application of the hydrodynamic equations and closure methods to the case of an electron fluid. Consider a plasma composed of electrons, ions and neutral atoms. We will treat the case in which the degree of ionisation is weak ($n_e \ll n_0$): the electron-electron, ion-ion and electron-ion collisions can then be neglected in comparison to the electron-neutral collisions, which are much more numerous and therefore are the dominant mechanism in the momentum exchange through collisions. From this fact, the energetic exchanges between the electron fluid and that of the ions (but not the interaction for the space charge electric field) are negligible, leading to $T_i < T_e$: we can thus consider that we are dealing with an electron gas and an ion gas which are quasi independent of each other.

Furthermore, because $T_e > T_i$ (or more commonly $T_e \gg T_i$), and the mass of the electrons is much smaller than the mass of the ions and neutrals, we can assume the ions and neutrals to be at rest compared to the motion of electrons. The situation finally reduces to one in which we can consider **a single fluid**, that of electrons, which moves in contact with a continuous fluid of ions and neutral atoms at rest, providing an effective viscosity to the electron motion. Furthermore, in addition to the "viscosity" just invoked, the interaction between electrons and ions also manifests itself in creating a space-charge electric field (Poisson's equation) affecting the motion of electrons.

The Lorentz equation for plasma electrons

If we neglect the convection term in the Langevin equation (3.134)[118], and since $v_e \gg v_i$ and v_n in the collisional term, this equation can be simplified to give:

$$m_e \frac{\partial \boldsymbol{v}_e}{\partial t} = q_e[\boldsymbol{E} + \boldsymbol{v}_e \wedge \boldsymbol{B}] - \frac{1}{n_e}\boldsymbol{\nabla} p_e - m_e \boldsymbol{v}_e(\nu_{en} + \nu_{ei}) \,, \tag{3.170}$$

where we set $\nu_{en} + \nu_{ei} \simeq \nu$ ($\nu_{ei} \ll \nu_{en}$). Assuming that $p_e = n_e k_B T_e$, with T_e independent of position, (3.170) reduces to:

$$m_e \frac{\partial \boldsymbol{v}_e}{\partial t} = q_e[\boldsymbol{E} + \boldsymbol{v}_e \wedge \boldsymbol{B}] - k_B T_e \frac{\boldsymbol{\nabla} n_e}{n_e} - m_e \nu \boldsymbol{v}_e \,. \tag{3.171}$$

The gradient $\boldsymbol{\nabla} p_e$ expresses the spatial evolution of the thermal pressure due to the electrons (isothermal assumption, Sect. 3.6): it is not related to a fluid compression.

A particular case of the Lorentz plasma

Cold electron plasma ($T_e = 0$) subjected to a periodic field $\boldsymbol{E} = \boldsymbol{E}_0 \mathrm{e}^{\mathrm{i}\omega t}$.

We will consider such a plasma, in one dimension and without a magnetic field. Then, relation (3.171) becomes:

$$m_e \frac{\mathrm{d}^2 x}{\mathrm{d}t^2} = -eE_0 \mathrm{e}^{\mathrm{i}\omega t} - m_e \nu \frac{\mathrm{d}x}{\mathrm{d}t} \,, \tag{3.172}$$

which is an equation we have proposed earlier, without proof (see Sect. 1.7.9, Eq. (1.147)) and we are now in a position to better understand its physical content. Recall that in a cold plasma, the movement of particles is uniquely created by the field $E_0 \mathrm{e}^{\mathrm{i}\omega t}$, so $v(t) = v_0 \mathrm{e}^{\mathrm{i}\omega t}$, then from (3.172):

$$m_e \mathrm{i}\omega v_0 \mathrm{e}^{\mathrm{i}\omega t} = -eE_0 \mathrm{e}^{\mathrm{i}\omega t} - m_e \nu v_0 \mathrm{e}^{\mathrm{i}\omega t} \tag{3.173}$$

and finally:

[118] We consider that $v < v_{th}$.

$$v_0 = \frac{-eE_0}{m_e(\nu + \mathrm{i}\omega)} , \tag{3.174}$$

so that the expression for the electron current density is:

$$J \equiv nqv_0 = \frac{ne^2}{m_e(\nu + \mathrm{i}\omega)} E_0 , \tag{3.175}$$

from which we obtain the expression for the scalar conductivity:

$$\sigma = \frac{ne^2}{m_e(\nu + \mathrm{i}\omega)} . \tag{3.176}$$

This result has been obtained previously, by integrating the velocity distribution function, assuming that $\nu(w) =$ constant. This expression for σ was then termed the Lorentz *conductivity* (3.68): we now see the origin of this name.

3.8 Diffusion and mobility of charged particles

3.8.1 The concepts of diffusion and mobility

The mobility and diffusion of charged particles are two hydrodynamic quantities, intrinsically linked to the presence of collisions in a plasma.

Diffusion

Diffusion stems from the kinetic pressure gradient in the Langevin equation (3.134). In the case where $\boldsymbol{\nabla} p \equiv \boldsymbol{\nabla}(nk_BT)$, diffusion results from either a gradient in particle density, or a gradient in their mean energy (temperature), or both. We will consider the two cases successively:

The case of a particle density gradient

Collisions between particles at any given point are random, and in the absence of strong fields, isotropic: there is then an equal probability for the directions of the scattered particles, after a sufficient number of collisions. We will consider, in one dimension, two points A and B in space, such that the density n of the gas at A is greater than that at B. Due to the equi-partition of collisions at each point in space, the flux in both possible directions at the point A is greater than that in both directions at point B. This means that the gas flux from A towards B is greater than that moving from B to A. There is therefore a net flux $n\boldsymbol{v}$ from A to B, where the mean velocity $\boldsymbol{v}$ is that of the fluid circulating from A to B.

The case of a temperature gradient

In this case, there is a flux of energy from the region of high temperature to that at low temperature, the energy transported by the high energy particles being larger! This energy flux is associated with the flux of particles.

Mobility

This parameter characterises the mean progression, or *drift*, of charged particles of a given type subjected to an electric field $\boldsymbol{E}$, when the motion is retarded by collisions.

Combined diffusion and mobility

In the case of a plasma that is both inhomogeneous and subjected to an (external or induced) field $\boldsymbol{E}$, there will be a combined diffusion and drift of the charged particles of type α, characterised by a total directed (mean) velocity $\boldsymbol{v}_\alpha$. We will study these two phenomena, with the help of the Langevin equation (3.134), considering a stationary state and neglecting the convective term $\boldsymbol{v}_\alpha \cdot \boldsymbol{\nabla} \boldsymbol{v}_\alpha$. In the case where there is no temperature gradient, this equation reduces to:

$$\boldsymbol{v}_\alpha = \frac{1}{m_\alpha \nu_{\alpha\beta}} \left\{ q_\alpha (\boldsymbol{E} + \boldsymbol{v}_\alpha \wedge \boldsymbol{B}) - k_B T_\alpha \frac{\boldsymbol{\nabla} n_\alpha}{n_\alpha} \right\} . \tag{3.177}$$

3.8.2 Solution of the Langevin equation with zero total derivative (dv/dt = 0)

We can rewrite (3.177) in a form such that the LHS is homogeneous in $\boldsymbol{v}$:

$$m_\alpha \nu_{\alpha\beta} \boldsymbol{v}_\alpha - q_\alpha (\boldsymbol{v}_\alpha \wedge \boldsymbol{B}) = q_\alpha \boldsymbol{E} - k_B T_\alpha \frac{\boldsymbol{\nabla} n_\alpha}{n_\alpha} . \tag{3.178}$$

In the following, to simplify the notation, we will suppress the indices α and β.

The general solution to this equation without the RHS is $\boldsymbol{v} = 0$, because $\boldsymbol{v}$ and $\boldsymbol{v} \wedge \boldsymbol{B}$ are orthogonal. The particular solution of the equation including the RHS will be the sum of two solutions obtained separately:

1. that with $\boldsymbol{\nabla} n = 0$ (drift velocity only),
2. that with $q\boldsymbol{E} = 0$ for $\boldsymbol{\nabla} n \neq 0$ (diffusion velocity only).

The proposed method of solution implies that the two following conditions are satisfied:

1. The convective term $(\boldsymbol{v} \cdot \boldsymbol{\nabla})\boldsymbol{v}$ appearing in the Langevin equation (3.135) can be assumed negligible, which is the case if the total directed velocity $\boldsymbol{v}$ has a small absolute value. This condition ensures that the sum of the drift velocity and the diffusion velocity is equal to the combined velocity of these two phenomena.
2. In order for the scalar pressure approximation in the Langevin equation to be valid, the velocity distribution must be isotropic, which implies that the diffusion and drift velocities are small compared to v_{th}.

Expression for the drift velocity

Setting $\boldsymbol{\nabla} n = 0$ in (3.177) in order to eliminate the contribution due to diffusion, and choosing $\hat{\mathbf{e}}_z$ to be along $\boldsymbol{B}$, we have:

$$v_x = \frac{q}{m\nu}[E_x + v_y B_z] \,, \tag{3.179}$$

$$v_y = \frac{q}{m\nu}[E_y - v_x B_z] \,, \tag{3.180}$$

$$v_z = \frac{q}{m\nu} E_z \,. \tag{3.181}$$

We will first define the mobility in the absence of $\boldsymbol{B}$:

$$\mu \equiv \frac{q}{m\nu} \,, \tag{3.182}$$

which is the mobility of a charged particle in a constant field $\boldsymbol{E}$ (note that μ is completely determined, if we know ν). This mobility enables us to rewrite (3.181) in the form:

$$v_z = \mu E_z \,, \tag{3.183}$$

which defines the mean velocity or *drift velocity* of one type of particle subject to an electric field $\boldsymbol{E}$ in a collisional plasma (remember that, by convention, and in contrast to ions, the electrons will drift in the opposite direction to that of the field $\boldsymbol{E}$: the value of μ_e is negative). The mobility thus defined is called the *linear mobility*, to signify that μ does not depend upon E_z. In the presence of a magnetic field $\boldsymbol{B}$, we find, for v_x and v_y respectively:

$$v_x = \frac{\mu_{\parallel}\nu^2}{\nu^2 + \omega_c^2} E_x + \frac{\mu_{\parallel}\nu\omega_c}{\nu^2 + \omega_c^2} E_y \,, \tag{3.184}$$

$$v_y = -\frac{\mu_{\parallel}\nu\omega_c}{\nu^2 + \omega_c^2} E_x + \frac{\mu_{\parallel}\nu^2}{\nu^2 + \omega_c^2} E_y \,, \tag{3.185}$$

where $\omega_c = -qB/m$ is the cyclotron frequency (Sect. 2.2.2). We can conclude that the drift velocities along $\hat{\mathbf{e}}_x$ and $\hat{\mathbf{e}}_y$ becomes weaker as the magnetic field increases ($\nu \ll \omega_c$).

The mobility tensor

The previous results can be expressed in tensor notation, by writing:

$$\mu_\perp = \frac{\mu_\| \nu^2}{\nu^2 + \omega_c^2} , \qquad \mu_H = \frac{\mu_\| \nu \omega_c}{\nu^2 + \omega_c^2} , \tag{3.186}$$

where $\mu_\perp$ and μ_H are the mobility perpendicular to the magnetic field and the Hall mobility respectively. These two coefficients allow us to define the 2nd order mobility tensor, $\underline{\boldsymbol{\mu}}$, by the general relation:

$$\boldsymbol{v} = \underline{\boldsymbol{\mu}} \cdot \boldsymbol{E} , \quad \text{or} \quad v_i = \sum_i \mu_{ij} E_j \tag{3.187}$$

with, as its representative matrix:

$$\underline{\boldsymbol{\mu}} = \begin{pmatrix} \mu_\perp & \mu_H & 0 \\ -\mu_H & \mu_\perp & 0 \\ 0 & 0 & \mu_\| \end{pmatrix} . \tag{3.188}$$

We can also relate it to the electrical conductivity tensor. Knowing that $\boldsymbol{J} = nq\boldsymbol{v}$, where $\boldsymbol{v}$ is the drift velocity of the particles in the field $\boldsymbol{E}$, we can write from (3.187):

$$\boldsymbol{J} = nq\underline{\boldsymbol{\mu}} \cdot \boldsymbol{E} \tag{3.189}$$

and since $\boldsymbol{J} = \underline{\boldsymbol{\sigma}} \cdot \boldsymbol{E}$ (Sect. 2.2.2), the relation between the conductivity and mobility tensors is then given by:

$$\underline{\boldsymbol{\sigma}} = nq\underline{\boldsymbol{\mu}} . \tag{3.190}$$

Remarks:

1. The expressions for the mobility and conductivity of electrons parallel to a magnetic field $\boldsymbol{B}$ (or in the absence of a magnetic field) are:

$$\mu_e = -\frac{e}{m_e \nu} , \tag{3.191}$$

$$\sigma_e = \frac{ne^2}{m_e \nu} , \tag{3.192}$$

where ν is the average electron-neutral collision frequency for momentum transfer. If the assumptions in the Lorentz model are valid, this electrical conductivity describes the electrical conductivity of the ensemble of particles in the plasma.

2. Ion mobility. We can define the ion mobility analogously to that for electrons:

$$\mu_i = \frac{e}{m_i \nu_{in}} \,, \tag{3.193}$$

where ν_{in} is the ion-neutral collision frequency. Note that the mobility of a positive ion is positive, whilst the mobility of an electron is negative; note also that some authors define mobility to be always positive.

3. Note that if $\omega_c \to 0$, the mobility tensor reduces to the scalar mobility coefficient, as expected.

4. Reduced mobility. For a given type of particle, $\mu_\parallel$ varies only with ν. Then, for a given electron (ion) temperature (mean energy), the value of ν depends only on N, the number of atomic targets per unit volume, the neutral atoms, in the present case. Therefore, it is practical to record the values of μ at one reference pressure and one reference temperature of the neutral atoms: these references have been fixed at $p_A = 760$ torr and 0°Celsius. The corresponding mobility is referred to as the *reduced mobility*, μ_{e0}: the neutral density in this case refers to the Lochsmidt *number*, $N_L = 2.69 \times 10^{19}$ atoms cm^{-3}. The mobility at any particular pressure p (torr) and temperature T_C (degrees Celsius) with respect to the reference pressure p_A and temperature 0°C, can be written:

$$\mu_e = \mu_{e0} \frac{N_L}{N_p} = \frac{\mu_{e0}\, p_A}{273\, p} (273 + T_c) = \frac{\mu_{e0}\,(273 + T_c)}{273\, p'} \,, \tag{3.194}$$

where N_p is the density of neutrals at the (dimensionless) "pressure" $p' = p/p_A$ and at the temperature T_c (°C).

5. Mobility in a periodic electric field. Recalling that the Lorentz conductivity in a periodic electric field is given by:

$$\sigma_e = \frac{n_e e^2}{m_e(\nu + \mathrm{i}\omega)} \tag{3.176}$$

(see (3.176)), we can define the corresponding electron mobility, knowing that $\sigma = nq\mu$ (3.190), such that:

$$\mu_e = -\frac{e}{m_e(\nu + \mathrm{i}\omega)} \,. \tag{3.195}$$

Setting $\omega = 0$, we obtain the expression for the mobility in a continuous electric field $\boldsymbol{E}$ (3.182).

Expression for the diffusion velocity

By setting the field $\boldsymbol{E}$ to zero in (3.177), we obtain the diffusion velocity in the direction of the magnetic field:

$$v_z = -\frac{k_B T_\alpha}{m_\alpha \nu_{\alpha\beta}} \frac{1}{n_\alpha} \frac{\partial n_\alpha}{\partial z} , \tag{3.196}$$

which enables us to define the diffusion coefficient parallel to the direction of $\boldsymbol{B}$ (or in the absence of a magnetic field) for particles of type α:

$$D_\parallel = \frac{k_B T_\alpha}{m_\alpha \nu_{\alpha\beta}} . \tag{3.197}$$

For the components of $\boldsymbol{v}$ in the plane perpendicular to $\boldsymbol{B}$, we obtain, from (3.177) (where the subscript α is retained solely for the temperature):

$$v_x = \frac{q}{m\nu}(v_y B_z) - \frac{k_B T_\alpha}{m\nu} \frac{1}{n} \frac{\partial n}{\partial x} , \tag{3.198}$$

$$v_y = -\frac{q}{m\nu}(v_x B_z) - \frac{k_B T_\alpha}{m\nu} \frac{1}{n} \frac{\partial n}{\partial y} , \tag{3.199}$$

and, as in the case of mobility, we introduce the two coefficients of diffusion relative to the plane perpendicular to $\boldsymbol{B}$:

$$D_\perp = \frac{D_\parallel \nu^2}{\nu^2 + \omega_c^2} , \qquad D_H = \frac{D_\parallel \nu \omega_c}{\nu^2 + \omega_c^2} , \tag{3.200}$$

which enables us to rewrite the three components of the diffusion velocity as:

$$v_x = -D_\perp \frac{1}{n}\frac{\partial n}{\partial x} - D_H \frac{1}{n}\frac{\partial n}{\partial y} , \tag{3.201}$$

$$v_y = D_H \frac{1}{n}\frac{\partial n}{\partial x} - D_\perp \frac{1}{n}\frac{\partial n}{\partial y} , \tag{3.202}$$

$$v_z = -D_\parallel \frac{1}{n}\frac{\partial n}{\partial z} , \tag{3.203}$$

whence finally the representative matrix of the *diffusion tensor*:

$$\underline{\boldsymbol{D}} = \begin{pmatrix} D_\perp & D_H & 0 \\ -D_H & D_\perp & 0 \\ 0 & 0 & D_\parallel \end{pmatrix} = \begin{pmatrix} \dfrac{D_\parallel \nu^2}{\nu^2+\omega_c^2} & \dfrac{D_\parallel \nu\omega_c}{\nu^2+\omega_c^2} & 0 \\ -\dfrac{D_\parallel \nu\omega_c}{\nu^2+\omega_c^2} & \dfrac{D_\parallel \nu^2}{\nu^2+\omega_c^2} & 0 \\ 0 & 0 & \dfrac{k_B T_\alpha}{m\nu_{\alpha\beta}} \end{pmatrix} \tag{3.204}$$

and the *diffusion velocity vector*:

$$\boldsymbol{v} = -\underline{\boldsymbol{D}} \cdot \frac{\boldsymbol{\nabla} n}{n} \qquad \left(v_i = -\sum_j D_{ij} \frac{\partial_j n}{n} \right) \tag{3.205}$$

resulting from the plasma density gradient.

Remarks:

1. We have just defined in (3.205) the *free diffusion tensor* $\underline{\boldsymbol{D}}$, so named because the electron (ion) diffusion occurs completely independently of all (collective) interactions with ions (electrons): in this case (low plasma density), the space charge field between the different species of charged particles is not large enough for any important coupling of the two different species to occur.
2. Note that as $\omega_c \to 0$, we recover the scalar diffusion coefficient, as expected.
3. We can associate a particle flux to the density gradient, by noting that:

$$\boldsymbol{\Gamma} \equiv n\boldsymbol{v} = -\underline{\boldsymbol{D}} \cdot \boldsymbol{\nabla} n \; . \tag{3.206}$$

   $\boldsymbol{\Gamma}$ is also called the *particle current* because, when multiplied by the charge e, it becomes a current density.
4. In the general case, where D depends on position and given any microscopic velocity distribution function, it can be shown that (Delcroix, Sect. 13.2):

$$\boldsymbol{\Gamma} = -\frac{1}{3} \boldsymbol{\nabla} \left[n \langle w^2 / \nu(w) \rangle \right] \; , \tag{3.207}$$

   where the brackets denote an average taken over the distribution function. From this expression, we extract the most general form of the diffusion coefficient in the absence of a magnetic field $\boldsymbol{B}$ and for a velocity distribution function that is independent of position:

$$D = \frac{1}{3} \langle w^2 / \nu(w) \rangle \; . \tag{3.208}$$

   Moreover, if ν is independent of w, and the distribution function $f(\boldsymbol{r}, \boldsymbol{w})$ is Maxwellian, then:

$$D = \frac{1}{3} \left\{ \frac{1}{2} m \langle w^2 \rangle \right\} \frac{2}{m\nu} = \frac{1}{3} \left\{ \frac{3}{2} k_B T_\alpha \right\} \frac{2}{m\nu} = \frac{k_B T_\alpha}{m\nu} \; , \tag{3.209}$$

   in agreement with (3.197).

Total particle current and total current density (complete solution)

If the above mentioned assumptions on the vector $\boldsymbol{v}$ are valid (velocity $\boldsymbol{v}$ sufficiently weak for the convective term to be neglected, absence of temperature gradient, stationary state) and provided that the electron density is sufficiently small, such that there is no coupling with the ions through the space charge field, the solutions obtained from (3.187) and (3.205) can be added to obtain the total flux of charged particles of one given species, and the corresponding current density:

$$\boldsymbol{\Gamma} = -\underline{\boldsymbol{D}} \cdot \boldsymbol{\nabla} n + n\underline{\boldsymbol{\mu}} \cdot \boldsymbol{E} , \tag{3.210}$$

$$\boldsymbol{J} = -q\underline{\boldsymbol{D}} \cdot \boldsymbol{\nabla} n + \underline{\boldsymbol{\sigma}} \cdot \boldsymbol{E} . \tag{3.211}$$

Remark: The ratio of the coefficients D_α and μ_α leads to:

$$\frac{D_\alpha}{\mu_\alpha} = \left(\frac{k_B T_\alpha}{m_\alpha \nu_{\alpha\beta}}\right)\left(\frac{m_\alpha \nu_{\alpha\beta}}{q_\alpha}\right) = \frac{k_B T_\alpha}{q_\alpha} , \tag{3.212}$$

referred to as the *Einstein relation.* It is usual to set $D_e/|\mu_e| \equiv u_k$, where u_k is the *characteristic electron energy* (expressed in eV).

The effect of a magnetic field on the diffusion coefficient: confinement of charged particles

- Spatial orientation of the confinement with respect to $\boldsymbol{B}$
  The field $\boldsymbol{B} = \hat{\mathbf{e}}_z B_0$ only affects the mobility and diffusion in the plane perpendicular to it (B does not appear in the expressions for $\mu_\parallel$ and $D_\parallel$). This effect depends on the ratio ν/ω_c. For the simple case where $\boldsymbol{E} = 0$, with $\partial n/\partial y = 0$ in the equation for v_x (3.201); the expression for Γ_x takes the form:

$$\Gamma_x = -D_\perp \frac{\partial n}{\partial x} = \frac{-D_\parallel \nu^2}{\nu^2 + \omega_c^2} \frac{\partial n}{\partial x} = \frac{-D_\parallel}{1 + (\omega_c/\nu)^2} \frac{\partial n}{\partial x} , \tag{3.213}$$

  which clearly shows that, when $\nu \lesssim \omega_c$, the particle flux in the direction perpendicular to $\boldsymbol{B}$ is smaller than the flux Γ_z in the direction of $\boldsymbol{B}$; in the case where $\nu \ll \omega_c$, the components $D_\perp \approx (\nu/\omega_c)^2 D_\parallel$ and $D_H \approx (\nu/\omega_c) D_\parallel$ are even more strongly reduced.

  Remark: The reduction in the motion perpendicular to $\boldsymbol{B}$ in the case of individual trajectories ($\nu = 0$, $E_\perp = 0$, $\boldsymbol{B} = \hat{\mathbf{e}}_z B_0$, p. 117, remark 1) is due to the fact that the gyration radius of the particle becomes smaller as B_0

increases. When there is particle diffusion, the cyclotron gyration can be seen as retarding the diffusion perpendicular to $\boldsymbol{B}$.

- Diffusion losses in a long plasma column, subjected to a stationary, axially directed magnetic field $\boldsymbol{B}$
  In a plasma column, with radius small compared to its length, the diffusion losses of the charged particles (by recombination at the walls) takes place mainly in the radial direction. A magnetic field, directed axially, can however greatly reduce this diffusion.
  Consider the vector:

$$\boldsymbol{\nabla} n = \left(\frac{\partial n}{\partial r}, \frac{1}{r}\frac{\partial n}{\partial \varphi}, \frac{\partial n}{\partial z} \right)$$

in cylindrical coordinates. The plasma column being cylindrically symmetric, $\partial n/\partial \varphi = 0$; in addition $\partial n/\partial z \approx 0$ if the column is very long with respect to its radius[119]. The radial flux then reduces to[120]:

$$\Gamma_r \hat{\mathbf{e}}_r = -\underline{\boldsymbol{D}}\cdot(\boldsymbol{\nabla} n)_r = -\underline{\boldsymbol{D}}\cdot\hat{\mathbf{e}}_r \frac{\partial n}{\partial r} = -\hat{\mathbf{e}}_r D_\perp \frac{\partial n}{\partial r} = -\hat{\mathbf{e}}_r \frac{D_\| \nu^2}{\nu^2 + \omega_c^2}\frac{\partial n}{\partial r} \qquad (3.214)$$

and for $\nu \ll \omega_c$,

$$\Gamma \approx -\frac{k_B T_\alpha}{m\nu}\frac{\nu^2}{q^2 B^2} m^2 \frac{\partial n}{\partial r} = -\left(\frac{k_B T_\alpha m \nu}{q^2 B^2}\right)\frac{\partial n}{\partial r} \propto \frac{1}{B^2}, \qquad (3.215)$$

which shows that the magnetic field strongly reduces diffusion losses.
Remark: The diffusion of charged particles perpendicular to $\boldsymbol{B}$ often gives rise to a much larger transport of particles than that predicted by the hydrodynamic model: this type of diffusion is referred to as *anomalous diffusion* (with respect to the hydrodynamic model). See further in Sect. 3.11.

3.9 Normal modes of diffusion and spatial density distribution of charged particles

The ultimate goal of the present section is to show how the diffusion of charged particles to the walls (where the ions recombine with the electrons) determines $n(\boldsymbol{r})$, the spatial distribution of charged particles.

---

[119] The axial density profile $n(z)/n(0)$ for a direct current (DC) discharge has only a weak dependence on the length of the column, such that, in this case $\partial n/\partial z$ is small.

[120] Since only one of the three components of $\boldsymbol{\nabla} n$ is non zero, in this case $\partial n/\partial r$, we can conclude that only the coefficient $D_\perp$ plays a role in the radial flux causing diffusion losses; the azimuthal diffusion flux within the plasma, $\Gamma_\varphi = D_H \partial n/\partial r$, does not contribute to these losses. Note that the (Hall) diffusion coefficient in this relation is $-D_H$ (see (3.204)).

This being obviously a problem of particle transport, it is natural to try solving it by considering the continuity equation, written this time with a non zero source term $\mathcal{S}$[121]:

$$\frac{\partial n}{\partial t} + \boldsymbol{\nabla} \cdot (n\boldsymbol{v}) = \mathcal{S} \, , \tag{3.216}$$

where $\partial n/\partial t$ describes the particle density variation as a function of time and $\boldsymbol{\nabla} \cdot (n\boldsymbol{v})$ represents the variation of the particle flux as a function of position. This flux results from the diffusion and drift of the charged particles. The term $\mathcal{S}$ takes account of the variations in particle density as a result of **volume** collisions, principally ionisation and recombination.

In the absence of an electric field (applied or from space charge)[122], the particle flux can be written $\boldsymbol{\Gamma} \equiv n\boldsymbol{v} = -D\boldsymbol{\nabla} n$. Equation (3.216), in steady state, is then simply:

$$\boldsymbol{\nabla} \cdot (-D\boldsymbol{\nabla} n) = \mathcal{S} \, . \tag{3.217}$$

In the case where the diffusion losses are greater than those due to volume recombination, the collisional interaction term (see Sect. 1.8, and Eq. (3.96) in Sect. 3.5) reduces to that for volume ionisation. If, in addition, the coefficient D is independent of position (isothermal approximation), we finally obtain the relation:

$$- D\nabla^2 n = \nu_i n \, , \tag{3.218}$$

where ν_i is the mean ionisation frequency for the gas considered (Sect. 1.8.3). This equation expresses the fact that the particles, created in the volume, are dragged towards the walls under the influence of their gradient density. We can rewrite (3.218) in the form of an eigen-value equation:

$$\nabla^2 n = - \left(\frac{\nu_i}{D}\right) n \, . \tag{3.219}$$

The results that we will now obtain are of major importance in the establishment of the maintenance conditions of the discharge, particularly as to the value of the $\boldsymbol{E}$ field intensity within the discharge as opposed to the externally applied field. In Chap. 4 (Sect. 4.2) we will establish the fact that this intensity, in the case of a plasma subject to diffusion, is independent of that of the externally applied electric field.

To completely understand the solution to (3.218), we need to introduce the concept of normal modes of diffusion, which is more readily demonstrated in the case of a time varying post-discharge (also called time-afterglow): it refers to a plasma in which, at $t = 0$, we delete the source term, for example the HF field.

---

[121] The term $\mathcal{S}$ results from the integration of the collision operator $\mathcal{S}(f)$ over $\boldsymbol{w}_\alpha$.

[122] If there is no coupling between the electrons and ions through the space charge electric field $\boldsymbol{E}_D$, D is the free diffusion coefficient. If coupling is important, the particle flux, which includes a contribution from the drift of the particles in the field $\boldsymbol{E}$, takes an identical form, but the value of the diffusion coefficient must be modified (Sect. 3.10).

3.9.1 Concept of normal modes of diffusion: study of a time varying post-discharge

We will examine the evolution of the spatial electron density distribution, as a function of time, in a post-discharge in the diffusion regime. Equation (3.216) then reduces to:

$$\frac{\partial n}{\partial t} + \boldsymbol{\nabla} \cdot (n\boldsymbol{v}) = 0 \,. \tag{3.220}$$

Assuming that the mean particle energy is spatially uniform (isothermal approximation) and setting $\boldsymbol{\Gamma} \equiv n\boldsymbol{v} = -D\boldsymbol{\nabla} n$, (3.220) becomes:

$$\frac{\partial n}{\partial t} - D\nabla^2 n = 0 \,. \tag{3.221}$$

To solve this equation, we need to know how the plasma density decreases as a function of time in the diffusion regime: it can be shown that, with certain exceptions, the density at any given point decreases exponentially, i.e.:

$$n(\boldsymbol{r}, t) = n(\boldsymbol{r}, t = 0) \exp(-\nu_D t) \text{ for } t \geq 0 \,, \tag{3.222}$$

where ν_D is the *characteristic frequency for diffusion loss* ($\tau_D = \nu_D^{-1}$ is the characteristic time for the decrease of plasma density by diffusion). Equation (3.221) then becomes:

$$\nabla^2 n = -\left(\frac{\nu_D}{D}\right) n \,, \tag{3.223}$$

where $n = n(\boldsymbol{r}, t)$, and D is the diffusion coefficient for the particles considered[123]. The second order differential equation (3.223) has eigenvalues of the form $\nabla^2 n = -\lambda_p n$ (*characteristic equation*). In the case of a cylindrical column of internal radius R, the boundary conditions are $(\mathrm{d}n/\mathrm{d}r)_{r=0} = 0$ and $n(r = R) = 0$, where this latter condition must be considered as an approximation[124].

---

[123] Note that (3.223) is not limited to charged particles, but can be applied to all the species in the plasma created in the volume and lost to the walls, such as excited neutral species, as well as atoms (O, N, H. . .), molecules or radicals.

[124] The value $n(r = R)$ is always much smaller than $n(r = 0)$, because in the diffusion regime, the walls constitute a region where the charged particles as well as the neutral species (some excited) are lost: by neutralisation for charged particles and through de-excitation, recombination or absorption for neutral species. However, if there is a “reflection” of part of the particle flux at the walls, the condition $n(r = R) = 0$ is not valid. This is the case for rare gases atoms and gaseous molecules (O_2, N_2, H_2. . .) in their ground state, for which the coefficient of reflection is 100% (no losses at the wall, $n(R) = n(0)$). For excited or dissociated neutral species, the de-excitation or recombination may not be complete at the walls, and one part of the species flux is thus reflected. In this case, it is necessary to think of the loss of species in terms of flux rather than density.

To simplify the solution of the characteristic equation, set:

$$\frac{1}{\Lambda^2} \equiv \frac{\nu_D}{D} \,. \tag{3.224}$$

We will show later that Λ, the characteristic diffusion length, depends only on the geometric configuration and plasma dimensions. Consequently, an increase in D translates into a proportional increase in ν_D.

Planar configuration

This situation corresponds to the case of a plasma confined between two parallel plates (conductors or dielectrics), separated by a distance L (x axis), but extending to infinity in the other dimensions (y and z axes), as illustrated in Fig. 3.1.

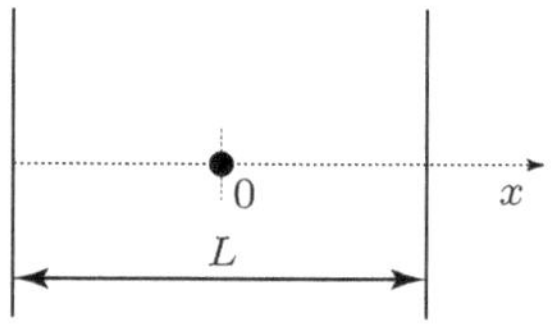

Fig. 3.1 One dimensional representation of a plasma confined between two parallel plates, extending to infinity and separated by a distance L.

The solution of (3.223) is then:

$$n(x,t) = \sum_{k=1}^{\infty} n_k(t=0)\exp(-\nu_{Dk}t)\cos(x/\Lambda_k) \,, \tag{3.225}$$

where:

$$\frac{1}{\Lambda_k^2} \equiv \frac{\nu_{\mathrm{D}k}}{D} \,, \tag{3.226}$$

and including the general condition that the density is zero at the wall:

$$\frac{L}{2\Lambda_k} = (2k-1)\frac{\pi}{2} \,, \tag{3.227}$$

where k is always positive.

Remarks:

1. The solution $\sin(x/\Lambda_k)$ is not acceptable, because it is not symmetric with respect to the axis ($x = 0$), as is required by the diffusion process.
2. Considered one by one, only the fundamental mode ($k = 1$) in (3.225) has a physical significance, because the higher modes result in negative densities along x. However, their sum according to (3.225) constitutes a physical solution (without negative values).

Characteristic frequency ν_{Dk} of the different modes: relative values compared to that of the fundamental mode

We wish to evaluate the relative contributions to electron density of the various diffusion modes appearing in the sum (3.225); to do this, we extract from Eqs. (3.226) and (3.227) for $k = 1$, $\nu_{D1} = D\pi^2/L^2$; for $k = 2$, $\nu_{D2} = 9\, D\pi^2/L^2$; for $k = 3$, $\nu_{D3} = 25\, D\pi^2/L^2$, such that $\nu_{D2}/\nu_{D1} = 9$, $\nu_{D3}/\nu_{D1} = 25$, etc. Then, it is clear that the fundamental mode decays the least rapidly in (3.225): no matter what the initial spatial density distribution was (obtained, for example, as a result of a laser pulse focussed at a point inside the discharge vessel), after a certain time, it will take the shape of the fundamental mode, with the other terms in (3.225) making no significant contribution, as is illustrated in Fig. 3.2.

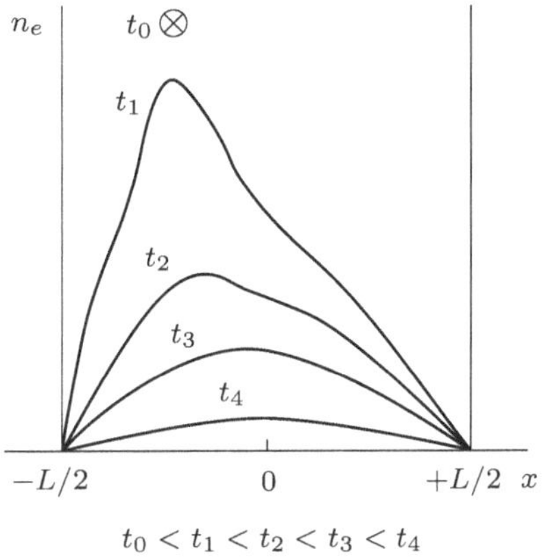

Fig. 3.2 Approximate time evolution of the electron density, in the diffusion regime, of a plasma created locally ($\otimes$) at time t_0.

Diffusion length

For the fundamental mode ($k = 1$), we have $\dfrac{L}{2\Lambda} = \dfrac{\pi}{2}$, from which:

$$\Lambda = L/\pi \, , \tag{3.228}$$

where Λ appears as a characteristic diffusion length in the planar configuration.

Cylindrical configuration

We will consider a cylindrical vessel, closed at each end by planar surfaces (Fig. 3.3).

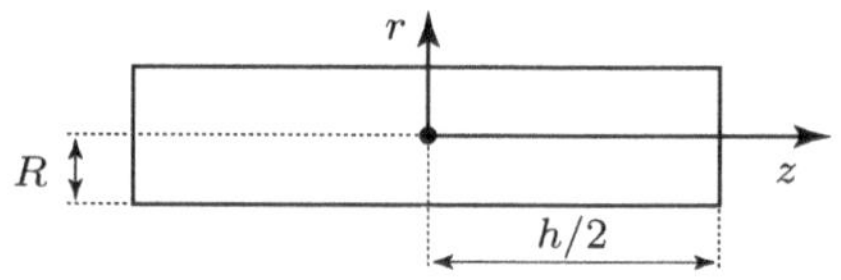

Fig. 3.3 Two dimensional representation of a cylindrical plasma vessel, with its axis z, radius R and length h.

For the fundamental mode, the eigenvalue solution to (3.223) gives the particle density at the point r, z:

$$n(r,z) = n_0 \cos(az) J_0(br) , \tag{3.229}$$

where J_0 is the zero order Bessel function of the first kind (Fig. 3.4): the coefficients a and b obey the relation:

$$a^2 + b^2 = \frac{\nu_{D1}}{D} . \tag{3.230}$$

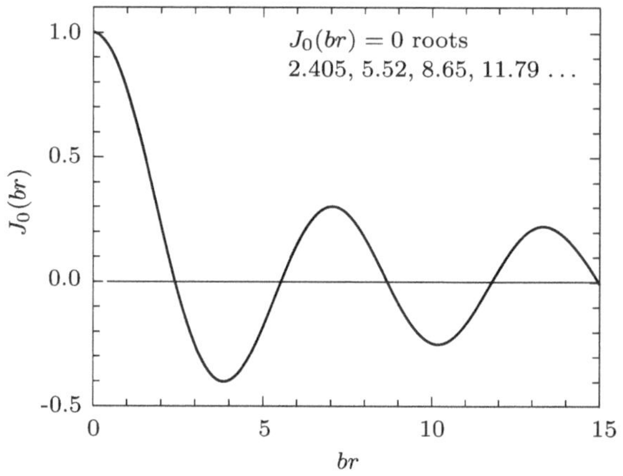

Fig. 3.4 Plot of the zero order Bessel function of the first kind.

Diffusion length

By requiring that the density of the particles be zero at the walls (at $r = R$ and $z = \pm h/2$), we find $a = \pi/h$ and $b = 2.405/R$, from which:

$$\frac{1}{\Lambda^2} = a^2 + b^2 = \left(\frac{\pi}{h}\right)^2 + \left(\frac{2.405}{R}\right)^2 . \tag{3.231}$$

In a *long cylindrical column*, by definition $h \gg R$, from which:

$$\Lambda = \frac{R}{2.405} . \tag{3.232}$$

The diffusion is thus principally in the radial direction, such that:

$$n(r, z) \simeq n(r) = n_0 J_0(2.405\, r/R) \ . \tag{3.233}$$

3.9.2 *Spatial distribution of charged particle density in the stationary diffusion regime*

The source term (ionisation)

In the case of a weakly ionised plasma ($n_e \ll N$, where N is the density of neutrals), the electron-neutral collisions dominate, but they are not sufficient to equalise the mean energy of the different particles; the mean energy of electrons in electric discharges is, in fact, greater than that of the ions and neutrals (Sect. 1.4.3). Moreover, the mean energies of the heavy particles, in this case, remain well below the ionisation threshold: only electrons are able to ionise through impact collisions with atoms. In the following, we assume that the atoms are initially in their ground state, before being ionised by a **single** electron collision: we will neglect multi-step ionisation, where atoms in excited states serve as a relay to ionisation (Appendix VI). This assumes that the intermediate states are not highly populated.

The "source" term S of the continuity equation (3.216) represents a number of particles per unit volume per second. Since we assume the plasma is in the diffusion regime, the source term includes ionisation only. Because ionisation occurs through a single electron impact on a neutral atom in the ground state, we can simply write the "source" term of the continuity equation as $\mathcal{S} = \mathcal{S}_i \equiv \nu_i n$ where $\nu_i = \mathcal{N}_0 \langle \hat{\sigma}_{ti}(w_e) w_e \rangle$ (1.152).

The charged particle balance equation

We know, from the conditions just described, that the conservation equation for particles in a diffusion plasma can be written:

$$-D\nabla^2 n = \nu_i n \ , \tag{3.218}$$

where the diffusion losses of charged particles are exactly balanced by the volume ionisation.

Very often, we can assume that the frequency ν_i is independent of position. However, in plasmas produced by HF fields, ν_i is a function of the spatial variation of the field $\boldsymbol{E}$, but in most cases where diffusion clearly dominates volume recombination, the effect of this dependence on the particle distribution function is minimal due to the global nature (leveling effect), as opposed to the local nature, of the diffusion mechanism.

Note that (3.218) has the same form as (3.223), describing the spatial electron density distribution subject to diffusion in a time dependent post-discharge: it is therefore an eigenvalue equation (the boundary conditions are the same as in Sect. 3.9.1). Thus, by analogy with Sect. 3.9.1, for a long cylindrical column (3.232), we have:

$$n_i(r) \approx n_e(r) = n(0) J_0(2.405\, r/R) \,, \tag{3.234}$$

which describes the radial density distribution corresponding to the fundamental diffusion mode[125]. Following the solution to (3.223), we set this time:

$$\frac{\nu_i}{D} = \frac{1}{\Lambda^2} \,, \tag{3.235}$$

which can be rewritten in the form:

$$\nu_i = \frac{D}{\Lambda^2} \tag{3.236}$$

or again, from (3.132):

$$\nu_i = \nu_D \,, \tag{3.237}$$

and we obtain the equilibrium equation, in a steady-state discharge, between the number of charged particles created and the number of those escaping by diffusion. This equation is also referred to as the *charged particle balance equation.*

Remarks:

1. In the case where the charged particles disappear (in part or in total) by volume recombination, we must add a term to the ionisation term $\mathcal{S}_i$. For example, in the case of atomic ions of the same charge, this would be:

$$\mathcal{S}_r = -\alpha_{ar} n^3 \,, \tag{3.238}$$

   where α_{ar} is the three-body volume-recombination coefficient.
   A more general form of the particle equilibrium equation can thus be written:

$$-D\nabla^2 n = \nu_i n - \alpha_{ar} n^3 \,, \tag{3.239}$$

   where α_{ar} and ν_i can be spatially dependent.
   In a time dependent post-discharge, where volume recombination exceeds the loss of charged particles by diffusion (see (3.222)), in the specific case of a three-body recombination (3.238), the decrease in n is given by:

---

125 This solution once again assumes that the contribution from higher modes is not important: they disappear once the transitory breakdown regime of the discharge terminates, analogous to what happens in a post-discharge.

$$\frac{1}{n^2(t)} = \frac{1}{n^2(0)} + 2\alpha_{ar} t \; . \qquad (3.240)$$

It is then straightforward, in principle, to distinguish a volume recombination regime from a diffusion regime by examining the time evolution of the post-discharge, unless the charged particle losses for both mechanisms are comparable in importance.

2. Density conditions at the walls
As we have already indicated[124], the simplifying condition $n(r = R) = 0$ at the walls is not always valid. In fact, at low-pressures (typically below 100 mtorr (13.3 Pa) in argon), there is a plasma-wall interface in the form of a narrow ion sheath, corresponding to a few λ_{De} (Sect. 3.14). In this case, the losses at the walls are expressed in terms of the flux traversing this ion sheath. At higher pressures, the plasma can often be considered as extending completely to the walls (negligible sheath), and the assumption $n(r = R) = 0$ is then acceptable.

3.10 The ambipolar diffusion regime

The assumption of a free diffusion of electrons, i.e. independent of ions, implied in Sect. 3.8, is only valid if the charged particle density is low. In reality, the electrons have a diffusion coefficient which is much larger than that of the ions (compare $D_e \propto T_e/m_e$ with $D_i \propto T_i/m_i$), which implies that their escape to the walls is more rapid[126]. This leads to a departure from neutrality (charge separation) and therefore to the creation of an electric field E_D (the space charge field). This electric field acts to slow the diffusion of electrons and accelerate the diffusion of ions. Above a certain plasma density (of order 10^8 cm^{-3}), the space-charge field intensity becomes important to the point that the ions and electrons diffuse with a common velocity, the charge separation and, thus, the actual electric field intensity adjusting accordingly: this is *ambipolar diffusion* (in its pure or perfect form), characterised by a single diffusion coefficient D_a for the two species of particle, for which we will now derive the expression.

---

126 To see this, write the diffusion flux of the two types of particles (not accounting for the space charge), in the x direction:

$$\Gamma_{ex} \equiv n_e v_{ex} = -D_e \partial n_e/\partial x \quad \text{and} \quad \Gamma_{ix} \equiv n_i v_{ix} = -D_i \partial n_i/\partial x \; .$$

Since $m_e \ll M$ and $T_e \geq T_i$, although $\nu_{en} \gtrsim \nu_{in}$, one has $D_e \gg D_i$. Knowing that $n_e \simeq n_i$, we obtain $\partial n_e/\partial x \approx \partial n_i/\partial x$, and, finally $v_{ex} \gg v_{ix}$, hence there is a much larger electron flux.

3.10.1 Assumptions required for a completely analytic description of the ambipolar diffusion regime

1. We set the flux of electrons and ions to be equal:

$$\boldsymbol{\Gamma}_e = \boldsymbol{\Gamma}_i = \boldsymbol{\Gamma} \,. \tag{3.241}$$

This is only a first order approximation because, rigorously, we should rather write:

$$\boldsymbol{\nabla} \cdot \boldsymbol{\Gamma}_e = \boldsymbol{\nabla} \cdot \boldsymbol{\Gamma}_i \,. \tag{3.242}$$

In fact, the equation of continuity in the stationary state gives, for all regimes:

$$\boldsymbol{\nabla} \cdot \boldsymbol{\Gamma}_e = \mathcal{S}_e \,, \tag{3.243}$$
$$\boldsymbol{\nabla} \cdot \boldsymbol{\Gamma}_i = \mathcal{S}_i \,, \tag{3.244}$$

where $\mathcal{S}_e = \mathcal{S}_i$, because the charged particles appear and disappear as electron-ion pairs[127].
Equation (3.241) is therefore only an approximation (except in one dimension); it is called the *congruence approximation.*

2. We assume *spatial independence of the diffusion coefficient*:

$$D_k(\boldsymbol{r}) = \text{ constant,} \tag{3.245}$$

where $k = e$, i or a (a standing for ambipolar).

3. We assume that in general:

$$\frac{n_i(r)}{n_e(r)} = C \,, \tag{3.246}$$

where C is a spatially independent parameter, whose value depends on the space charge. In the presence of a particle flux, in the stationary state, the intensity of the space-charge electric field E_D grows, starting from zero for free diffusion, gradually increasing into the ambipolar regime then decreasing again, without ever being completely zero at the perfect ambipolar diffusion limit (see below (3.270)). The *proportionality condition* (3.246) will enable us to treat the transition between free diffusion and perfect ambipolar diffusion. We can verify the validity of this relation in two extreme cases of the transition regime:

- Low plasma density: free diffusion regime ($E_D = 0$)
The two species of particle diffuse freely, thus a priori $C \neq 1$. To determine the value of C, we use the exact relation (3.242), where $\boldsymbol{\Gamma}_\alpha = -D_\alpha \boldsymbol{\nabla} n_\alpha$, and D_α is spatially independent. Since $\mathcal{S}_e = \mathcal{S}_i$, this

[127] Assuming that there are only positive ions in the discharge, and that each of them carries only a single positive charge.

leads to:

$$-D_e\boldsymbol{\nabla}\cdot\boldsymbol{\nabla}n_e = -D_i\boldsymbol{\nabla}\cdot\boldsymbol{\nabla}n_i \ , \tag{3.247}$$

or equivalently:

$$\nabla^2 n_e = +\frac{D_i}{D_e}\nabla^2 n_i \ , \tag{3.248}$$

such that, from (3.246):

$$C = D_e/D_i \ , \tag{3.249}$$

which quantifies (3.246) by giving the parameter C a precise value. In the free diffusion regime, the ion and electron diffusion fluxes are therefore equal provided that $n_i/n_e = D_e/D_i$.

- High plasma density (perfect ambipolar diffusion regime: no charge separation)
  In this case, $C = 1$ ($n_e(\boldsymbol{r}) = n_i(\boldsymbol{r}) = n(\boldsymbol{r})$) and from the congruence approximation (3.241):

$$\boldsymbol{v}_e(\boldsymbol{r}) = \boldsymbol{v}_i(\boldsymbol{r}) = \boldsymbol{v}(\boldsymbol{r}) \ , \tag{3.250}$$

  which requires that the diffusion is governed by a single diffusion coefficient D_a, common to both species, as will be shown in the following section.

3.10.2 *Equations governing the ambipolar diffusion regime and the transition from the free diffusion to the ambipolar regime*

In the absence of external fields, the fluxes resulting from diffusion, as well as that from the drift due to the space charge electric field $\boldsymbol{E}_D$, are given by (3.210):

$$\boldsymbol{\Gamma}_i = -D_i\boldsymbol{\nabla}n_i + \mu_i n_i \boldsymbol{E}_D \ , \tag{3.251}$$

$$\boldsymbol{\Gamma}_e = -D_e\boldsymbol{\nabla}n_e + \mu_e n_e \boldsymbol{E}_D \ , \tag{3.252}$$

where the field $\boldsymbol{E}_D$ is related to the departure from neutrality through Poisson's equation (1.1):

$$\boldsymbol{\nabla}\cdot\boldsymbol{E}_D = -(n_e - n_i)e/\epsilon_0 \ . \tag{3.253}$$

Perfect ambipolar regime ($n_e(\boldsymbol{r}) = n_i(\boldsymbol{r})$)

If we multiply (3.252) by $\mu_i n_i$ and (3.251) by $\mu_e n_e$ and take the difference (3.252) - (3.251), including the congruence approximation (3.241)

($\boldsymbol{\Gamma}_e = \boldsymbol{\Gamma}_i = \boldsymbol{\Gamma}$), we obtain:

$$(\mu_i n_i - \mu_e n_e)\boldsymbol{\Gamma} = - D_e \mu_i n_i \boldsymbol{\nabla} n_e + \mu_e \mu_i n_e n_i \boldsymbol{E}_D + D_i \mu_e n_e \boldsymbol{\nabla} n_i - \mu_e \mu_i n_e n_i \boldsymbol{E}_D \; . \tag{3.254}$$

In the present case, $n_e = n_i = n$, so:

$$\boldsymbol{\Gamma} = \frac{-(D_e\mu_i - D_i\mu_e)\boldsymbol{\nabla} n}{\mu_i - \mu_e} \; , \tag{3.255}$$

which leads to the definition of the *ambipolar diffusion coefficient* D_a [128]:

$$D_a \equiv \frac{D_e\mu_i - D_i\mu_e}{\mu_i - \mu_e} \; , \tag{3.256}$$

so the electron and ion fluxes ((3.251) and (3.252)) can both be represented by the same single expression:

$$\boldsymbol{\Gamma} = -D_a \boldsymbol{\nabla} n \; . \tag{3.257}$$

Note, in (3.256), that D_a is positive because the value of μ_e is negative.

By analogy with the expression obtained for free diffusion (3.235), in the stationary state, the creation-loss equilibrium condition for particles in the ambipolar diffusion regime is:

$$\nu_i = \frac{D_a}{\Lambda^2} \; . \tag{3.258}$$

For a long, cylindrical column, the expression for the radial distribution of the electron and ion densities following the development of (3.233) is thus:

$$n(r) = n(0)J_0(2.405\, r/R) \; . \tag{3.259}$$

The transition region between the ambipolar ($n_e = n_i$) and free diffusion ($n_i = Cn_e$) regimes

We start from (3.254), where the congruence approximation ($\boldsymbol{\Gamma}_e = \boldsymbol{\Gamma}_i$) is used, but this time we consider that $n_e \neq n_i$ and $\boldsymbol{v}_e \neq \boldsymbol{v}_i$. The proportionality condition (3.246) can be expressed in the form:

$$n_i \boldsymbol{\nabla} n_e = n_e \boldsymbol{\nabla} n_i \; , \tag{3.260}$$

and then, from (3.254), we can extract:

---

128 The expression for D_a remains the same if an external electric field $\boldsymbol{E}_{\text{ext}}$ is added to $\boldsymbol{E}_D$ (even if their orientation is different): $\boldsymbol{E}_{\text{ext}}$ can be eliminated in equation (3.254) in the same way as $\boldsymbol{E}_D$.

$$\boldsymbol{\Gamma} = -\left[\frac{(D_e\mu_i - D_i\mu_e)n_i}{\mu_i n_i - \mu_e n_e}\right]\boldsymbol{\nabla} n_e \equiv -D_s\boldsymbol{\nabla} n_e\ , \tag{3.261}$$

where D_s is the *effective diffusion coefficient*. Although the expression for D_s contains n_e and n_i, which are functions of position, D_s is not, because we have assumed that the constant C is spatially independent.

We require an expression for D_s as a function of D_a, in which the difference $n_i - n_e$ (determining the intensity of the field $\boldsymbol{E}_D$ coupling the particles with each other) appears explicitly. Recall the expressions for the total electrical conductivity:

$$\sigma = (\mu_i n_i - \mu_e n_e)e \tag{3.262}$$

(because μ_e is negative, $\sigma > 0$, as it should be) and the total charge density:

$$\rho = (n_i - n_e)e\ . \tag{3.263}$$

We can then write:

$$\begin{aligned} 1 - \frac{\mu_e\rho}{\sigma} &\equiv 1 - \frac{\mu_e(n_i - n_e)e}{(\mu_i n_i - \mu_e n_e)e} \equiv \frac{\mu_i n_i - \mu_e n_e - \mu_e n_i + \mu_e n_e}{\mu_i n_i - \mu_e n_e} \\ &= \frac{(\mu_i - \mu_e)n_i}{\mu_i n_i - \mu_e n_e} \end{aligned} \tag{3.264}$$

and, further, that:

$$\begin{aligned} D_a\left(1 - \frac{\mu_e\rho}{\sigma}\right) &= \left(\frac{D_e\mu_i - D_i\mu_e}{\mu_i - \mu_e}\right)\frac{(\mu_i - \mu_e)n_i}{\mu_i n_i - \mu_e n_e} \\ &= \frac{(D_e\mu_i - D_i\mu_e)n_i}{\mu_i n_i - \mu_e n_e} \equiv D_s\ . \end{aligned} \tag{3.265}$$

The expression:

$$D_s = D_a\left(1 - \frac{\mu_e\rho}{\sigma}\right) \tag{3.266}$$

shows that D_s depends explicitly on the electrical conductivity and the deviation from charge neutrality. Since the coefficient μ_e is negative, D_a is the minimum value of $D_s(\rho \simeq 0)$.

3.10.3 The value of the space-charge electric field intensity

From (3.210), we know that the expression for the electron particle flux is:

$$\boldsymbol{\Gamma}_e = -D_e\boldsymbol{\nabla} n_e + \mu_e n_e\boldsymbol{E}_D\ . \tag{3.267}$$

Moreover, the effective diffusion coefficient D_s enables us to write the ion and electron fluxes (3.251) and (3.252) in the same form $\boldsymbol{\Gamma} = -D_s \boldsymbol{\nabla} n_e$. From (3.267) and (3.261), we then have:

$$\boldsymbol{E}_D = \frac{(D_e - D_s)}{\mu_e} \frac{\boldsymbol{\nabla} n_e}{n_e} . \tag{3.268}$$

We can study this equation in two interesting limits:

- $\boldsymbol{E}_D = 0$ (strictly zero)
  This case corresponds to free diffusion, as shown in (3.268), because then $D_s = D_e$ (we can find the same result by comparing (3.267) with $\boldsymbol{E}_D = 0$ and (3.261))
- Perfect ambipolar diffusion field $\boldsymbol{E}_{Da}$
  Setting $D_s = D_a$ in (3.268) gives:

$$\begin{aligned}\boldsymbol{E}_{Da} &= \frac{1}{\mu_e} \left[D_e - \frac{(D_e \mu_i - D_i \mu_e)}{\mu_i - \mu_e} \right] \frac{\boldsymbol{\nabla} n_e}{n_e} \\ &= \frac{1}{\mu_e} \left[\frac{D_e \mu_i - D_e \mu_e - D_e \mu_i + D_i \mu_e}{\mu_i - \mu_e} \right] \frac{\boldsymbol{\nabla} n_e}{n_e}\end{aligned} \tag{3.269}$$

  and finally:

$$\boldsymbol{E}_{Da} = -\frac{(D_e - D_i)}{\mu_i - \mu_e} \frac{\boldsymbol{\nabla} n_e}{n_e} . \tag{3.270}$$

  This field can become very weak, but never exactly zero, in the ambipolar diffusion regime. More exactly, this means that we can set the strict equality $n_e = n_i$ in order to calculate the ambipolar diffusion coefficient, but not in Poisson's equation. In the latter case, ρ must be different from zero, even if only slightly, for the space charge to exist. Note also that the space charge electric field on the axis of a plasma column must be zero, because of cylindrical symmetry[129] and thus the charge density ρ_0 on the axis must also be zero.
  Since, in general, $D_e \gg D_i$ and $|\mu_e| \gg \mu_i$, a commonly used approximation for the expression (3.270) for $\boldsymbol{E}_{Da}$ is:

$$\boldsymbol{E}_{Da} \approx \frac{D_e}{\mu_e} \frac{\boldsymbol{\nabla} n_e}{n_e} \equiv u_k \frac{\boldsymbol{\nabla} n_e}{n_e} , \tag{3.271}$$

  where u_k is the characteristic energy of the electrons (3.212).

---

[129] There is no reason for the field $\boldsymbol{E}_D$ on the axis to be pointing in any specific radial direction. As a matter of fact, we use the condition $\boldsymbol{\nabla} n_e = 0$ as a boundary condition on the axis.

3.10.4 *The expression for the charge density ρ_0 on the axis: limits to the validity of the analytic calculation*

The charge density ρ can be obtained explicitly from Poisson's equation $\boldsymbol{\nabla} \cdot \boldsymbol{E}_D = \rho/\epsilon_0$, where the expression for $\boldsymbol{E}_D$ is from its general relation (3.268):

$$\begin{aligned}\rho &= -\frac{\epsilon_0(D_s - D_e)}{\mu_e}\boldsymbol{\nabla} \cdot \left[\frac{\boldsymbol{\nabla} n_e}{n_e}\right] \\ &= \frac{\epsilon_0(D_s - D_e)}{\mu_e}\left[\left(\frac{\boldsymbol{\nabla} n_e}{n_e}\right)^2 - \frac{\nabla^2 n_e}{n_e}\right] . \qquad (3.272)\end{aligned}$$

The value ρ_0 of ρ on the axis is obtained by noting that the plasma density is symmetric with respect to this axis, so that $\boldsymbol{\nabla} n_e = 0$[130]; furthermore, since $\nabla^2 n_e/n_e = -1/\Lambda^2$ (Sect. 3.9.2), one gets from (3.272):

$$\rho_0 = \frac{\epsilon_0(D_s - D_e)}{\mu_e \Lambda^2} . \qquad (3.273)$$

In the case of free diffusion ($D_s = D_e$), we do have $\rho_0 = 0$ (negligible space charge). Knowing that $D_e \geq D_s \geq D_a$[131] (and $\mu_e < 0$, according to our convention), ρ_0 increases from 0 to $\rho_{0\max}$ as D_s decreases from D_e to D_a while we should have $\rho_0 = 0$ on the axis, whatever the diffusion regime: the field $\boldsymbol{E}$ on the axis is always zero due to cylindrical symmetry. The analytic calculation is therefore only approximate, which is not surprising, given the rather restrictive assumptions (notably $n_i = Cn_e$) that we have used.

The more exact description of the regime is better treated numerically. In order to compare analytic and numerical calculations, we will express the effective diffusion coefficient as a function of the conductivity on the axis, σ_0. To do so, we substitute (3.273) into (3.266), which gives:

$$D_s = D_a \left[\frac{D_e + \Lambda^2\sigma_0/\epsilon_0}{D_a + \Lambda^2\sigma_0/\epsilon_0}\right] . \qquad (3.274)$$

Assuming that the conductivity σ_0 is primarily due to the electrons, and introducing the electron Debye length (1.42), we obtain:

---

[130] Note that we set $\boldsymbol{\nabla} n_e = 0$ in (3.272) only in the RHS, i.e. once the divergence has been expanded: the fact that the first derivative of n_e is zero at a point does not require that the second derivative also be zero at this point (consider $y = x^2$ at $x = 0$). In the present case $\nabla^2 n_e < 0$ and n_e passes through a maximum on the axis.

[131] And $D_i \leq D_a$, because in the ambipolar diffusion regime, the space charge field accelerates the flux of ions. To see this, use (3.256) for example, with $|\mu_e| \gg |\mu_i|$.

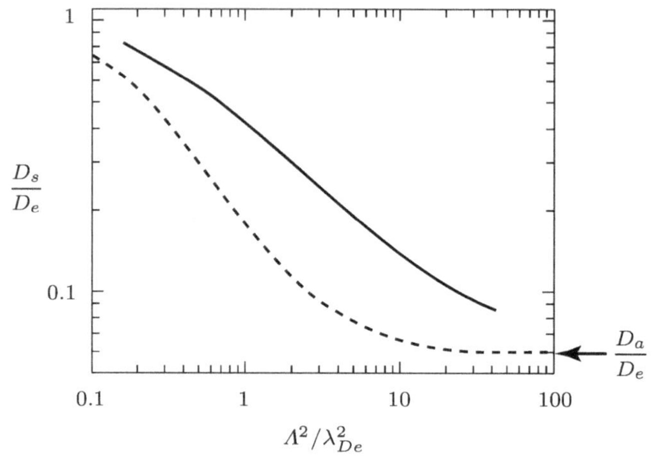

Fig. 3.5 Comparison of the analytic solution (dashed line) with that obtained by a numerical calculation (solid line) from the diffusion equations, for the case $T_e = T_i$ (after [1], American Physical Society, all rights reserved).

$$D_s = D_a \left[\frac{1 + (\Lambda/\lambda_{De})^2}{(D_a/D_e) + (\Lambda/\lambda_{De})^2} \right] , \tag{3.275}$$

where $(\Lambda/\lambda_{De})^2$ is a parameter characterising the transition from free diffusion ($D_s = D_e$) to ambipolar diffusion ($D_a/D_e \ll 1$). Figure 3.5 shows that the analytic solution (3.275) underestimates the value of D_s obtained by the numerical solution of the diffusion equations, which does not require the aforementioned approximations. In particular, the value of D_s obtained analytically only tends to that of the numerical calculation for large values of $(\Lambda/\lambda_{De})^2$.

Figure 3.6 shows that, for a planar discharge configuration, the plasma situated in the neighbourhood of the axis can be in the ambipolar regime, but tends to the free diffusion regime near the walls.

3.10.5 Necessary conditions for a discharge to be in the ambipolar regime

Provided the discharge is effectively governed by diffusion, to determine whether the regime is ambipolar or not we can use either one of the following two criteria:

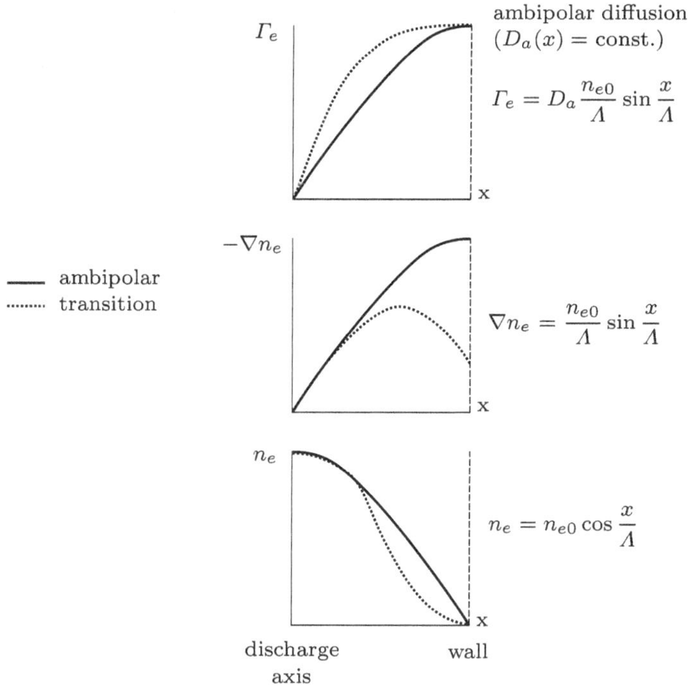

Fig. 3.6 Numerically calculated values of Γ_e and n_e as functions of spatial position x in planar geometry, compared to those obtained when assuming a constant ambipolar diffusion coefficient: the calculated diffusion coefficient changes from truly ambipolar on the axis ($x = 0$) to that of the transition regime as density decreases towards the wall (after Brown).

Density – diffusion length criterion

We can generally say that the field intensity $\boldsymbol{E}_D$ required for ambipolar diffusion is reached when $\Lambda^2\sigma_0$ is sufficiently large (see (3.274)), or more specifically when:

$$n_{e0}\Lambda^2 > 10^7\,\text{cm}^{-1}\,, \tag{3.276}$$

where n_{e0}, the electron density on the axis, is expressed in cm^{-3} and Λ in cm. For values of $n_{e0}\Lambda^2$ less than $10^5\,\text{cm}^{-1}$, the plasma is in the free diffusion regime, provided that the conditions are actually those of a diffusion regime (collision mean free path $< R$) and not that of free fall (the definition of free fall will be discussed later, in Sect. 3.12).

Remarks:

1. The average electron density (over the cross-section) of a classical fluorescent lamp tube, maintained by an alternating current, varies between

10^{10} and 10^{11} cm^{-3} in the course of a period of 60 Hz; its radius is 12 mm. This discharge is manifestly in the ambipolar diffusion regime, except near the walls, where it can be in the free diffusion regime, as we have already underlined (Fig. 3.6).

2. For values of $n_{e0}\Lambda^2 \gg 10^7$ cm^{-1}, the plasma density can be sufficiently large for volume recombination to be significant, or even dominant. It is then necessary to compare the losses of charged particles by volume recombination and by diffusion, in order to know which loss mechanism is the most important.

Debye length – diffusion length criterion

Recalling that $\lambda_{De}^2 = \epsilon_0 k_B T_e / n_e e^2$ and from the Einstein relation (3.212) $D_e/\mu_e = -k_B T_e/e$, we can write:

$$\lambda_{De}^2 = -\frac{\epsilon_0 D_e}{n_e e \mu_e} . \tag{3.277}$$

The departure from charge neutrality on the axis as a function of λ_{De}^2 can be obtained from (3.273), by substituting therein μ_e from (3.277), which gives:

$$\rho_0 \equiv (n_{i0} - n_{e0})e = \epsilon_0 \frac{(D_s - D_e)}{\Lambda^2 \mu_e} = -\epsilon_0 \frac{(D_s - D_e)}{\Lambda^2} \frac{\lambda_{De}^2 n_{e0} e}{\epsilon_0 D_e} ,$$

and finally:

$$\frac{n_{i0} - n_{e0}}{n_{e0}} = \frac{(D_e - D_s)}{D_e} \frac{\lambda_{De}^2}{\Lambda^2} . \tag{3.278}$$

We can then distinguish two cases depending on the ratio λ_{De}/Λ:

- $\lambda_{De} \gg \Lambda$: free diffusion
  A Debye length longer than Λ implies a departure from macroscopic neutrality over the distance Λ. This means that the intensity of the space-charge field is manifestly insufficient to couple the ions and electrons. In other words, there is free diffusion, ions and electrons moving independently of each other. Moreover, $\lambda_{De} \gg \Lambda$ implies, from (3.278), that the density n_{e0} should be small[132].
- $\lambda_{De} \ll \Lambda$: ambipolar diffusion
  This condition corresponds to high electron density ($1/n_{e0}$ is small in (3.278)) and the space charge field will have an intensity such that the motion of the ions and electrons are coupled. These conditions correspond precisely to the ambipolar diffusion regime ($D_s = D_a$ in (3.275)).

---

[132] This condition is similar to the first condition of remark 10 of Sect. 1.6 to the extent that $L \simeq \Lambda$.

Remarks:

1. A commonly used approximate value for the coefficient D_a. It is easy to see that:
$$\frac{D_i \mu_e}{D_e \mu_i} = \frac{k_B T_i}{m_i \nu_{in}} \frac{m_e \nu_{en}}{k_B T_e} \frac{-e}{m_e \nu_{en}} \frac{m_i \nu_{in}}{e} = -\frac{T_i}{T_e} \,. \tag{3.279}$$
Since, as a rule, $|\mu_e| \gg \mu_i$, the general expression (3.256) for D_a can be written:
$$D_a \approx D_i - D_e(\frac{\mu_i}{\mu_e})$$
and equivalently
$$D_a \approx D_i \left[1 - \frac{D_e}{D_i} \frac{\mu_i}{\mu_e}\right] \,.$$
Substituting (3.279), this becomes:
$$D_a \approx D_i \left[1 + \frac{T_e}{T_i}\right] \,. \tag{3.280}$$
Finally, since generally $T_e \gg T_i$ (non LTE electric discharges):
$$D_a \approx D_i \frac{T_e}{T_i} = \frac{k_B T_i}{m_i \nu_{in}} \frac{T_e}{T_i} = \left(\frac{k_B T_e}{e}\right) \mu_i \,. \tag{3.281}$$
D_a can thus be expressed as the product of the electron temperature (in eV) and the ion mobility.
2. The initial regime (or breakdown) of a discharge is necessarily governed by free diffusion, due to the fact that the densities at the beginning are small ($n_{e0} \Lambda^2 \to 0$, on condition that it is not in free fall regime, Sect. 3.12), but the resulting stationary regime can be governed by ambipolar diffusion. In practice, the breakdown of the discharge requires a much greater electric field $\boldsymbol{E}$ than that required to maintain the discharge, as is shown in Fig. 3.7.

3.11 Ambipolar diffusion in a static magnetic field

We would like to study the effect of an axial magnetic field on the ambipolar diffusion regime in a **long plasma column**[133] in the absence of an external electric field.

As in the case for free diffusion, a field $\boldsymbol{B}$ cannot produce an effect in its own direction. On the other hand, the influence in the perpendicular direction can be significant.

[133] If R/h is not much less than 1, see Chen, Sect. 5.5.1.

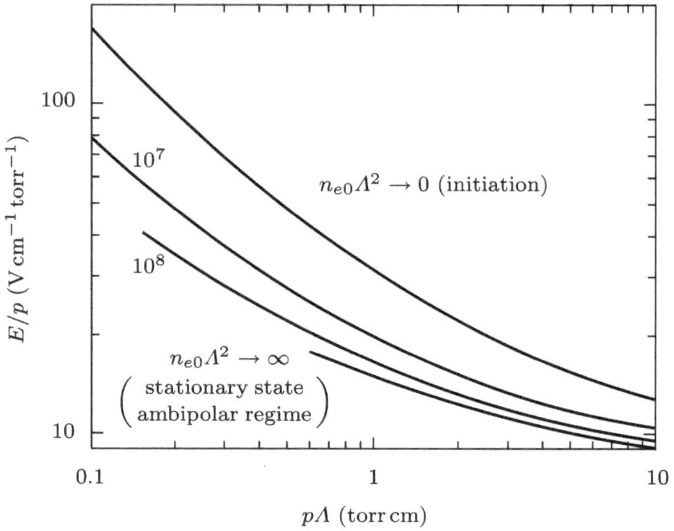

Fig. 3.7 Field intensity $\boldsymbol{E}$ (normalised to the pressure p) as a function of the parameter $p\Lambda$ in a microwave discharge in H_2 (after [32]). The values of the product $n_{e0}\Lambda^2$ are expressed in cm^{-1}.

Coefficient $D_{a\perp}$

In a long cylindrical column, diffusion is principally radial. The corresponding ion and electron fluxes in the presence of a magnetic field can be obtained by analogy with the $\boldsymbol{\Gamma}_e$ and $\boldsymbol{\Gamma}_i$ fluxes in the absence of a magnetic field, using (3.251) and (3.252):

$$\Gamma_{ir} = -D_{i\perp}\frac{\partial n_i}{\partial r} + \mu_{i\perp} n_i E_{Dr} , \tag{3.282}$$

and:

$$\Gamma_{er} = -D_{e\perp}\frac{\partial n_e}{\partial r} + \mu_{e\perp} n_e E_{Dr} , \tag{3.283}$$

where E_{Dr} is the radial component of the space-charge field (designated as the *ambipolar field*). The expressions for $D_{e\perp}$ and $D_{i\perp}$ follow the definition of $D_\perp$ in (3.200), and those of $\mu_{e\perp}$ and $\mu_{i\perp}$ those of $\mu_\perp$ in (3.186). Supposing that the diffusion is ambipolar ($\Gamma_{er} = \Gamma_{ir} = \Gamma_r$), then by analogy with (3.257), we can write a common expression for electrons and ions, in the form:

$$\boldsymbol{\Gamma}_r = -D_{a\perp}\boldsymbol{\nabla}_r n , \tag{3.284}$$

where $D_{a\perp}$ is given by:

$$D_{a\perp} = \frac{D_{e\perp}\mu_{i\perp} - D_{i\perp}\mu_{e\perp}}{\mu_{i\perp} - \mu_{e\perp}} . \tag{3.285}$$

Analysis of the relative values of $D_{e\perp}$ and $D_{i\perp}$

The calculation of the ratio of the two values leads to:

$$\frac{D_{e\perp}}{D_{i\perp}} = \frac{D_{e\parallel}\nu_{en}^2}{\omega_{ce}^2+\nu_{en}^2}\frac{\omega_{ci}^2+\nu_{in}^2}{D_{i\parallel}\nu_{in}^2} = \frac{T_e}{T_i}\frac{m_i}{m_e}\frac{\nu_{in}}{\nu_{cn}}\frac{\nu_{en}^2}{\nu_{in}^2}\left(\frac{\omega_{ci}^2+\nu_{in}^2}{\omega_{ce}^2+\nu_{en}^2}\right) . \tag{3.286}$$

Consider the particular case of a field B_0, which is sufficiently large such that $\omega_{ci} \gg \nu_{in}$. Since a fortiori $\omega_{ce} \gg \nu_{en}$ (because $\omega_{ce} \gg \omega_{ci}$ and $\nu_{en} \gtrsim \nu_{in}$), (3.286) can be simplified:

$$\frac{D_{e\perp}}{D_{i\perp}} = \frac{m_i}{m_e}\frac{\omega_{ci}^2}{\omega_{ce}^2}\frac{T_e}{T_i}\frac{\nu_{en}}{\nu_{in}} = \frac{m_e}{m_i}\frac{T_e}{T_i}\frac{\nu_{en}}{\nu_{in}} . \tag{3.287}$$

Although $T_e\nu_{en}$ is in general greater than $T_i\nu_{in}$[134], the mass ratio dominates to the extent that $D_{e\perp} \leq D_{i\perp}$. The ions diffuse faster, and the electrons retard the ions in the ambipolar process, in contrast to the situation without a magnetic field.

Remarks:

1. If we assume $\nu_{en} \approx \nu_{in}$, in the case where the field $\boldsymbol{B}$ is sufficiently large ($\nu_{in} \ll \omega_{ci}$), we find a useful approximate form for $D_{a\perp}$:

$$D_{a\perp} \approx D_{e\perp}[1 + T_i/T_e] , \tag{3.288}$$

   a result worthy of comparison with the diffusion coefficient without magnetic field (3.280), for which $D_a = D_i[1 + T_e/T_i]$: the role of electrons and ions in the diffusion process is thus reversed.
2. The congruence approximation for the derivation of $D_{a\perp}$, the ambipolar diffusion coefficient in the presence of a magnetic field $\boldsymbol{B}$, is in general not valid if the vessel walls are conductors [39]. In the direction perpendicular to $\boldsymbol{B}$, the electrons diffuse more slowly than the ions when $\boldsymbol{B}$ is sufficiently large ($D_{e\perp} \leq D_{i\perp}$, (3.287)), while the electrons diffuse much more rapidly than the ions along the lines of $\boldsymbol{B}$ ($D_e \gg D_i$). Thus the walls (at the ends) which cut the field lines mainly receive electrons, while the walls parallel to $\boldsymbol{B}$ will mainly collect the ions. There is therefore no ambipolar diffusion, because $\boldsymbol{\Gamma}_e \neq \boldsymbol{\Gamma}_i$. On the other hand, if the walls of the vessel are insulated, they do impose the ambipolar diffusion regime ($\boldsymbol{\Gamma}_e = \boldsymbol{\Gamma}_i$).

[134] In fact, $\nu_{en} \gtrsim \nu_{in}$ and $T_e/T_i \geq 1$.

3.12 Diffusion regime or free fall regime

Free fall: definition

When the dominant collision frequency between particles is sufficiently small, such that the corresponding mean free path is longer than the diffusion length, i.e, $\ell > \Lambda$, the plasma is in the *free fall* regime.

These conditions are opposite to those for diffusion because collisions mainly occur at the walls of the vessel, rather than in the volume: the phenomenon of directed velocity resulting in a density gradient cannot occur, because this requires a very large number of collisions between charged particles over their diffusion length.

Diffusion criterion

Consequently, there is diffusion if the time τ_D to move a charged particle to the walls is much greater than the time between two collisions in the volume (mostly electron-neutral collisions in a weakly ionised gas), i.e. $\tau_D \gg 1/\nu_{en}$. We wish to express this condition as a function of the mean free path ℓ of electrons, and the radius R, for the case of a long cylindrical vessel.

Recall that $\nu_D = D/\Lambda^2$, such that $\tau_D \equiv 1/\nu_D = \Lambda^2/D \approx R^2/D$. Furthermore, we know (Sect. 1.7.7) that $\ell \approx v_{th}/\nu_{en}$. We can then write:

$$\tau_D \nu_{en} \approx \frac{R^2}{D}\left(\frac{v_{th}}{\ell}\right) . \tag{3.289}$$

Since $D = k_B T_e/m_e \nu_{en} \approx v_{th}^2/\nu_{en} = v_{th}\,\ell$, we obtain from (3.289):

$$\tau_D \nu_{en} \approx \frac{R^2}{v_{th}}\frac{v_{th}}{\ell^2} = \frac{R^2}{\ell^2} . \tag{3.290}$$

The condition $\tau_D \nu_{en} \gg 1$ leads to $R/\ell \gg 1$ as a diffusion criterion.

In conclusion, before using any particular set of diffusion equations, it is imperative to verify that the characteristic dimensions for diffusion are much larger than the predominant mean free path. If this is the case, it is then necessary to determine if this is a free diffusion regime or an ambipolar diffusion regime, or a transition between the two, which can be determined, for instance, by examining the ratio between the Debye length and the diffusion length (Sect. 3.10.5).

Remarks:

1. In the free fall regime, the radial distribution of the electron density in a cylindrical column is parabolic: $n(r) = n(0)(1 - \bar{\alpha} r^2/R^2)$: the parameter $\bar{\alpha}$ depends on λ_{De}, as illustrated in Fig. 3.8.

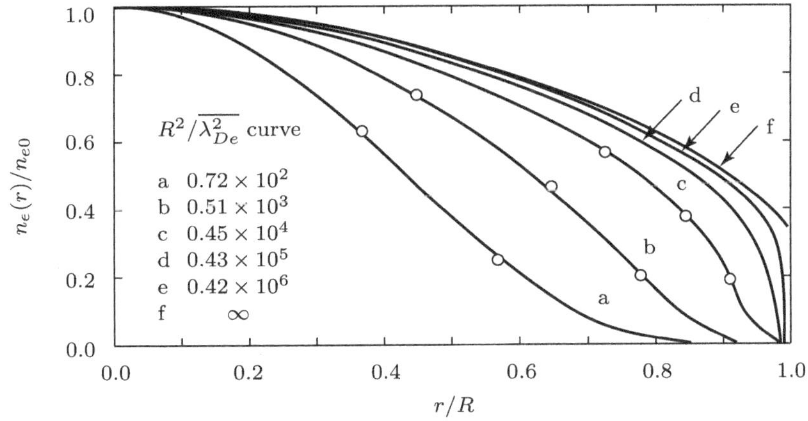

Fig. 3.8 Radial profile of electron density in the free fall regime: R is the internal radius of the discharge tube and n_{e0} the value of the electron density on the tube axis. $\overline{\lambda^2_{De}}$ represents the average value of λ^2_{De} over the radius of the plasma (adapted with permission from [28], all rights reserved, American Physical Society).

2. Summary of conditions for ambipolar diffusion ($\boldsymbol{B} = 0$)
   - $\ell \ll R$: we have diffusion and not free fall
   - $\lambda_{De} \ll \Lambda$: the diffusion is ambipolar, if volume recombination is small with respect to recombination at the walls (Sect. 1.8).

3.13 Electron temperature of a long plasma column governed by ambipolar diffusion: scaling law $T_e(pR)$

In this section, we will derive the relation for the electron temperature T_e as a function of the product cpR, where p is the gas pressure, R the plasma radius, and c, a specific constant for a given gas. This dependence as a function of pR is called a *scaling law*: such a law is one that regroups the product (or quotient) of two (or more) variables. The present scaling law, formulated by Von Engel and Steenbeck for the positive column (Sect. 4.2.1), has been largely confirmed experimentally, particularly in rare gases for both the positive column of a DC discharge and HF discharges.

The equation we are seeking is based on the equation for particle charge balance (3.236) in the ambipolar diffusion regime:

$$\langle \nu_i \rangle = D_a/\Lambda^2 \,, \tag{3.291}$$

where the mean ionisation frequency[135] $\langle\nu_i\rangle$ depends on the electron velocity distribution function, as indicated by the $\langle\rangle$ symbols. In the following, we will assume that this function is Maxwellian. In such a case, the diffusion coefficient D_a can be written explicitly as a function of T_e (see (3.281)); however, T_e only appears implicitly in the expression defining the mean frequency $\langle\nu_i\rangle$, so (3.291) cannot be used to express T_e explicitly.

From (3.291), we will derive an analytic expression, as a first approximation, for the dependence of T_e as a function of N, the neutral atom density, and Λ, the characteristic diffusion length. For a more practical application, we will obtain a function $T_e(p_0R)$, where p_0 is the reduced gas pressure (Sect. 1.7.5) and R, the internal radius of the discharge vessel, assumed to be long and cylindrical.

3.13.1 Assumptions of the model

To obtain an analytic solution, the following assumptions are required:

- The plasma is stationary (sufficient time has elapsed since the breakdown of the discharge).
- The discharge tube is a long cylinder: the charged particles escape by radial ambipolar diffusion to the walls, where they fully recombine ($n(R) = 0$).
- The intensity of the electric field sustaining the discharge is weak enough for the velocity distribution to be isotropic, and it does not vary radially.
- The electron velocity distribution function is Maxwellian.
- The plasma is far from LTE ($T_e \gg T_i$).
- Ionisation results from electron-neutral collisions (degree of ionisation $\leq 10^{-4}$).
- The ionisation of each atom occurs as a result of a single electron collision with an atom in the ground state, such that $\langle\nu_i\rangle = \mathcal{N}_0\langle\hat{\sigma}_{ti}w\rangle$.

Remark: The assumption that the charged particles are lost solely by ambipolar diffusion, together with the assumption of ionisation exclusively due to electron impact with atoms in the ground state, are referred to as Schottky *conditions* (see also Sect. 4.2.4).

3.13.2 Derivation of the relation $T_e(p_0R)$

We will calculate successively the LHS and the RHS of (3.291) for the balance of particles, which will finally lead to the $T_e(p_0R)$ scaling law.

---

[135] We have returned to our initial notation (Sect 1.7.8) for the mean collision frequencies, writing $\langle\nu_i\rangle$ explicitly rather than ν_i.

Expression for the ionisation frequency as a function of the mean electron energy

In general, the number of ionising collisions is small compared to the number of elastic collisions, because the energy threshold $\mathcal{E}_i$ for ionisation is high compared to the mean electron energy $3k_B/2T_e$: this implies that most electrons have energy smaller than the corresponding threshold energy and do not contribute to ionisation of the plasma. The calculation of the ionisation frequency $\langle \nu_i \rangle$ will be undertaken in stages.

1. The expression for the mean ionisation frequency
   For a separable and isotropic velocity distribution function (Sect. 3.3)[136], we obtain:

$$\langle \nu_i \rangle = \int_w \nu_i(w) f(w)\, 4\pi w^2 \, \mathrm{d}w \,, \tag{3.292}$$

   where:

$$\nu_i(w) \equiv \mathcal{N}_0 \hat{\sigma}_{ti}(w) w \,. \tag{1.152}$$

   Note that $\langle \nu_i \rangle$ is the average number of ionising collisions experienced by **one** electron.
2. The expression for $\langle \nu_i \rangle$ in terms of the electron energy expressed in eV

   - The velocity distribution function, assumed to be Maxwellian, can be written (Appendix I):

$$f(w) = \left(\frac{m_e}{2\pi k_B T_e} \right)^{3/2} \exp\left(-\frac{w^2}{v_{th}^2} \right) . \tag{3.293}$$

   To translate the electron kinetic energy into eV, U_{eV}, we set:

$$U_{eV} = \frac{m_e w^2}{2e} \,, \tag{3.294}$$

   from which:

$$w = \sqrt{\frac{2eU_{eV}}{m_e}} \,. \tag{3.295}$$

   Since:

$$\bar{U}_{eV} = \frac{3}{2} \frac{k_B T_e}{e} \,, \tag{3.296}$$

   where we have used $\bar{U}_{eV} \equiv \langle U_{eV} \rangle$ to simplify the notation, and since $\frac{1}{2} m_e v_{th}^2 = k_B T_e$, then:

$$v_{th} = \sqrt{\frac{4}{3} \frac{e}{m_e} \bar{U}_{eV}} \tag{3.297}$$

---

[136] We can use the modulus of the velocity because we have made the assumption that the electron energy is only weakly anisotropic, despite the presence of the field $\boldsymbol{E}$.

and after some calculation (Appendix XVII):

$$f\left(\sqrt{\frac{2eU_{eV}}{m_e}}\right) = \left(\frac{3}{4\pi}\frac{m_e}{e}\right)^{3/2} \frac{1}{\bar{U}_{eV}^{3/2}} \exp\left(-\frac{3}{2}\frac{U_{eV}}{\bar{U}_{eV}}\right) . \tag{3.298}$$

- Conversion of $\langle \nu_i(w) \rangle$ to $\langle \nu_i(U_{eV}) \rangle$
  Substituting (3.294) and (3.298) in (3.292), and noting that $P_i(w) = N\hat{\sigma}_{ti}(w)$, where $P_i(w)$ is the total macroscopic (Sect. 1.7.5) ionisation cross-section (Sect. 1.7.9), $\langle \nu_i \rangle$ can be expressed as a function of the energy U_{eV}, i.e.:

$$\langle \nu_i \rangle = \int_{\mathcal{E}_i} 3\sqrt{\frac{3e}{m_e \pi}} \frac{U_{eV}}{\bar{U}_{eV}^{3/2}} P_i(U_{eV}) \exp\left(-\frac{3}{2}\frac{U_{eV}}{\bar{U}_{eV}}\right) \mathrm{d}U_{eV} , \tag{3.299}$$

  where the lower limit of integration is the ionisation threshold $\mathcal{E}_i$.

3. Analytic approximation for $P_i(u)$
   Experimentally, $P_i(U_{eV})$ is observed to be linear in the domain $U_{eV} \leq 2\mathcal{E}_i$, and since $P_i(U_{eV}) = 0$ for $U_{eV} < \mathcal{E}_i$, to a very good approximation (Sect. 1.7.9), $P_i(U_{eV})$ can be written as:

$$P_i(U_{eV}) = a_i(U_{eV} - \mathcal{E}_i) \text{ for } U_{eV} \geq \mathcal{E}_i \tag{1.142}$$

   where a_i is the *ionisation coefficient*, a specific constant for each gas considered.
   Substituting (1.142) in (3.299), we obtain:

$$\langle \nu_i \rangle = 3\sqrt{\frac{3e}{m_e \pi}} \frac{1}{\bar{U}_{eV}^{3/2}} \int_{\mathcal{E}_i}^{\infty} a_i(U_{eV} - \mathcal{E}_i) U_{eV} \exp\left(-\frac{3}{2}\frac{U_{eV}}{\bar{U}_{eV}}\right) \mathrm{d}U_{eV} . \tag{3.300}$$

   Recalling that in Sect. 1.7.5 we introduced the reduced "pressure" p_0 such that $P_i = p_0 P_{i0}$, where P_{i0} is the total macroscopic ionisation cross-section at 0°C and 1 torr, we proceed in similar fashion by setting $a_i = p_0 a_{i0}$, a parameter which is expressed with respect to the reference conditions of the cross-sections; the units are $\mathrm{cm}^{-1}\,\mathrm{V}^{-1}$. We then obtain:

$$\langle \nu_i \rangle = 3\sqrt{\frac{3e}{m_e \pi}} \frac{p_0}{\bar{U}_{eV}^{3/2}} a_{i0} \int_{\mathcal{E}_i}^{\infty} (U_{eV} - \mathcal{E}_i) U_{eV} \exp\left(-\frac{3}{2}\frac{U_{eV}}{\bar{U}_{eV}}\right) \mathrm{d}U_{eV} . \tag{3.301}$$

4. Normalisation of the energy with respect to the mean energy $\bar{U}_{eV}$, which simplifies the integration of (3.301)
   Setting:

$$\mathcal{U} \equiv \frac{3}{2}\frac{U_{eV}}{\bar{U}_{eV}}, \qquad \mathcal{U}_i \equiv \frac{3}{2}\frac{\mathcal{E}_i}{\bar{U}_{eV}} , \tag{3.302}$$

then (3.301) can be replaced by (Appendix XVII):

$$\langle \nu_i \rangle = \left(\frac{4}{3}\right)^{3/2} \sqrt{\frac{e}{m_e \pi}}\, a_{i0}\, p_0\, \bar{U}_{eV}^{3/2} \int\limits_{\mathcal{U}_i}^{\infty} (\mathcal{U} - \mathcal{U}_i)\mathcal{U} \exp(-\mathcal{U})\, \mathrm{d}\mathcal{U} \,. \tag{3.303}$$

Integration by parts of (3.303) is easily achieved and, when returning to the initial quantities, it yields:

$$\langle \nu_i \rangle = 2 \left(\frac{4}{3}\right)^{3/2} \sqrt{\frac{e}{m_e \pi}}\, a_{i0} p_0 \bar{U}_{eV}^{3/2} \left[\frac{3}{4}\frac{\mathcal{E}_i}{\bar{U}_{eV}} + 1\right] \exp\left(-\frac{3}{2}\frac{\mathcal{E}_i}{\bar{U}_{eV}}\right) . \tag{3.304}$$

We can therefore conclude that $\langle \nu_i \rangle$ is highly sensitive to the value of the mean energy $\bar{U}_{eV}$, because it appears in the argument of the exponential term.

Charged particle balance in the discharge

We will now evaluate the RHS of equation (3.291), which describes the balance of charged particles. In a long cylindrical discharge, we know, from (3.226), that $\Lambda \approx R/2.405$. In addition, because $\mu_i \ll |\mu_e|$, and assuming that $T_e \gg T_i$ (valid for plasmas at reduced pressure, which is the case in the present model), we can write (from (3.281)):

$$D_a \approx \frac{k_B T_e}{e} \mu_i = \frac{2}{3} \bar{U}_{eV} \mu_i \,, \tag{3.305}$$

such that:

$$\frac{D_a}{\Lambda^2} = \frac{2}{3} \bar{U}_{eV} \mu_i \left(\frac{2.405}{R}\right)^2 \tag{3.306}$$

and from (3.291):

$$\langle \nu_i \rangle = \frac{2}{3} \bar{U}_{eV} \mu_i \left(\frac{2.405}{R}\right)^2 . \tag{3.307}$$

Equating expressions (3.304) and (3.307) for $\langle \nu_i \rangle$, we can obtain a completely analytic equation for the loss-creation balance of charged particles in the plasma.

An expression for $T_{eV}/\mathcal{E}_i$ as a function of $p_0 R$

The von Engel and Steenbeck approximation

The need to find an expression which is easier to evaluate (at the time, there were no computers) led von Engel and Steenbeck to make the approximation:

$$\frac{3}{4}\frac{\mathcal{E}_i}{\bar{U}_{eV}} \gg 1 \tag{3.308}$$

in (3.304) (it is a fact that $\bar{U}_{eV} \ll \mathcal{E}_i$, except when the pressure becomes so low that the plasma is in the free fall regime where, in any case, the present model is not valid). In the context of this approximation, the RHS of (3.304) and (3.307) may be equated to give:

$$\frac{2}{3}\bar{U}_{eV}\mu_i\frac{(2.405)^2}{R^2} = 2\left(\frac{4}{3}\right)^{3/2}\sqrt{\frac{e}{m_e\pi}}\,a_{i0}\,p_0\,\bar{U}_{eV}^{3/2}\left(\frac{3}{4}\frac{\mathcal{E}_i}{\bar{U}_{eV}}\right)\exp\left(-\frac{3}{2}\frac{\mathcal{E}_i}{\bar{U}_{eV}}\right), \tag{3.309}$$

and, returning to the variable $\mathcal{U}_i$ from (3.302), this becomes (Appendix XVII):

$$\mathcal{U}_i^{-1/2}\,\exp\mathcal{U}_i = \frac{2}{(2.405)^2}\sqrt{\frac{2e}{m_e\pi}}c_0^2\,p_0^2\,R^2\,, \tag{3.310}$$

where the coefficient c_0, specific for each gas, is defined by:

$$c_0^2 \equiv \frac{a_{i0}\sqrt{\mathcal{E}_i}}{\mu_i p_0}\,. \tag{3.311}$$

Note that the product $\mu_i p_0$, because of p_0, appears as a reduced mobility relative to 0°C and 1 torr; remember, however, that the reference values for mobility are usually given at 0°C and 760 torr (3.194).

Finally, we can extract the numerical value of $T_{eV}/\mathcal{E}_i$ from (3.310) as a function of $c_0 p_0 R$, where p_0 is the dimensionless reduced "pressure" with respect to 0°C and 1 torr, and $\mathcal{E}_i$ the ionization energy threshold. The units of c_0^2 are $(\text{kg/coulomb})^{1/2}\,\text{m}^{-2}$ or $\text{V}^{1/2}\,\text{s}\,\text{m}^{-3}$, while the units of $c_0 p_0 R$ are $(\text{V}^{1/2}\,\text{s}\,\text{m}^{-1})^{1/2}$.

Exact expression

Expanding (3.304) and (3.307) in terms of the reduced ionization energy $\mathcal{U}_i$ leads to an exact expression in a useful form:

$$\frac{\exp\mathcal{U}_i}{3\sqrt{\frac{2}{3}}\mathcal{U}_i^{1/2} + 4\sqrt{\frac{3}{2}}\mathcal{U}_i^{-1/2}} = \frac{2}{(2.405)^2}\sqrt{\frac{e}{3m_e\pi}}c_0^2\,p_0^2\,R^2\,. \tag{3.312}$$

The numerical evaluation of (3.312) with (3.302) defining $\mathcal{U}_i = e\mathcal{E}_i/k_BT_e$ enables us to plot $T_{eV}/\mathcal{E}_i$ as a function of $c_0 p_0 R$ (Fig. 3.9).

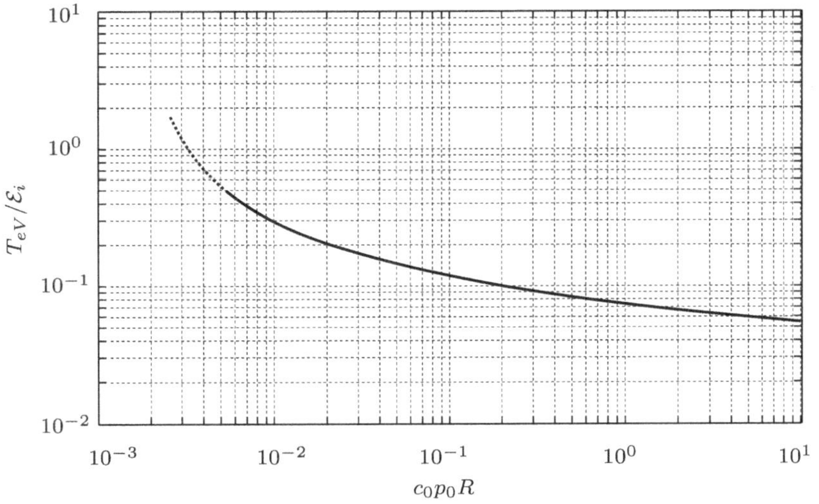

Fig. 3.9 Variation of electron temperature (normalised to the ionisation energy threshold of the gas considered) as a function of the operating conditions (type of gas (through the coefficient c_0) and its (reduced) pressure p_0, and radius R of the (long) cylindrical plasma column, assumed to be in the ambipolar regime). The dashes indicate the range where the application of (3.312) is no longer valid. R is in metres, and the value of c_0 is that of Tab. 3.1, with appropriate units.

Consequences

Figure 3.9 demonstrates that the electron temperature T_e of a discharge, in the ambipolar diffusion regime, depends only on the dimensions of the vessel (radius R for a long plasma column), the type of gas (ionisation energy threshold $\mathcal{E}_i$ and coefficient c_0) and the pressure (expressed as a reduced pressure p_0).

Value of the coefficient c_0

The value of a_{i0} appearing in the expression for c_0 can be obtained from the published ionisation cross-sections $P_{i0}(U_{eV})$ as the slope of the linear section near the threshold. The value of the ion mobility μ_i in (3.307) corresponds to the actual pressure and temperature of the gas considered; because it is multiplied by p_0 in (3.311), it takes the form of a reduced mobility at 0°C, 1 torr, although the reference values are given at 0°C and 1 atmosphere: it is necessary to make the appropriate conversion. Moreover, the published values of μ_{i0} depend on the ratio of E/p (another scaling law); usually, the value of μ_{i0} used in the calculation of c_0 is that extrapolated for $E/p \to 0$.

Table 3.1 gives the values of c_0 obtained for rare gases by different authors, the most recent (1980) being that of Zakrzewski, which are our recommended values. The units of c_0 allow us to use the graph of $T_{eV}/\mathcal{E}_i$ as a function of $c_0 p_0 R$ in Fig. 3.9 directly, where the internal radius of the discharge R is in metres; $\mathcal{E}_i$, the ionisation threshold energy, is given in Tab. 3.2.

Table 3.1 Values of c_0 obtained (chronologically from left to right) for different rare gases (to be used with Fig. 3.9 where R is expressed in m). The units of c_0^2 are in $\mathrm{V}^{1/2}\,\mathrm{C}^{1/2}\,\mathrm{m}^{-3}\,\mathrm{s}$.

	Von Engel	Brown[a]	Moisan	Zakrzewski
Helium	4	3.93	5.3	4.68
Neon	6	5.9	9.0	7.94
Argon	40	53	48	50.1
Krypton			68	68.2
Xenon			111	113

[a] Brown, Chap. 14, Sect. 2.2.

Table 3.2 Energy threshold $\mathcal{E}_i$ (eV) of the first ionisation state of rare gas atoms.

Helium	24.58
Neon	21.56
Argon	15.756
Krypton	13.996
Xenon	12.127

Example of the calculation of T_{eV}: for $R = 2\,\mathrm{cm}$ and $p_0 = 1$, we have $c_0 p_0 R = 1$ for argon. From Fig. 3.9, $T_{eV}/\mathcal{E}_i = 7.54 \times 10^{-2}$ (dimensionless). Since $\mathcal{E}_i = 15.76\,\mathrm{eV}$, we obtain $T_{eV} = 1.2\,\mathrm{eV}$.

Remark: The parameters of the discharge, fixed by the operator, are the type of gas and its pressure, the dimensions of the discharge vessel and, when applicable, the frequency of the HF field maintaining the discharge. These constitute the *operating conditions* of the plasma.

3.14 Formation and nature of sheaths at the plasma-wall interface: particle flux to the walls and the Bohm criterion

In a non-ionised gas, the flux of particles incident on a wall, per unit area of the surface, is equal to the random flux (Appendix I):

$$\Gamma = \frac{1}{4} n v \,, \tag{3.313}$$

where $v \equiv \langle w \rangle$ is the mean velocity of the Maxwell-Boltzmann distribution function. In an ionised gas, the situation is different in the vicinity of a wall (or a probe) because the surface can be brought to a given potential by the operator, but it can also become charged electrically due to the sole presence of the plasma particles: in addition, these charged particles can recombine on it (Sect. 1.8). A transition zone then forms between the plasma and the wall, called a *sheath*, which we will now study. In the following, we will consider a plasma far from LTE, such that $T_e \gg T_i \simeq 0$ (with a background of stationary ions) and we will assume that there are no collisions in the sheath that develops at the plasma-wall interface.

3.14.1 Positive wall-potential with respect to the plasma potential: electron sheath

This case is simple, compared to that of a wall at a negative potential. The variation of the potential is illustrated in Fig. 3.10.

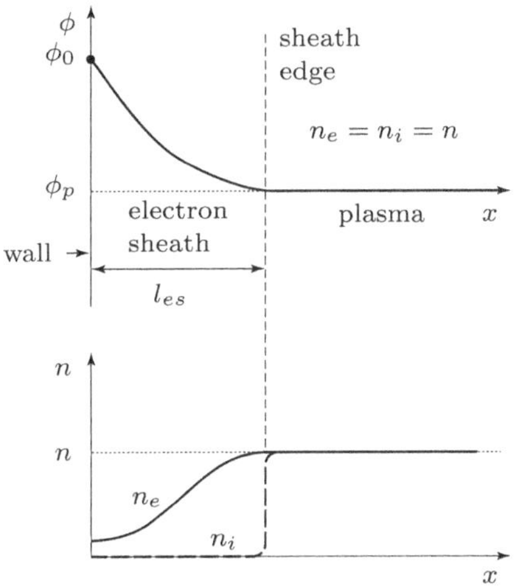

Fig. 3.10 Variation of the potential $\phi(x)$, the ion density $n_i(x)$ and the electron density $n_e(x)$ at the plasma-wall interface in the case of an electron sheath (x is the distance to the wall, l_{es} the thickness of the electron sheath, ϕ_p the plasma potential and ϕ_0 the potential applied on the wall).

We can distinguish two regions: on the right part of the figure, the plasma, characterised by its macroscopic neutrality ($n_e = n_i$), a zero space-charge electric field and a *plasma potential* ϕ_p, and on the left side, a pure electron sheath where ions, with low energy ($k_B T_i \simeq 0$), are completely reflected towards the plasma by the repulsive field which develops naturally at the plasma-wall interface. The boundary separating the macroscopically neutral plasma and the electron sheath, completely free of ions, is called the *sheath boundary*: departure from charge neutrality starts at this location. In the case of a planar surface, the electron flux at the wall is equal to the flux reaching the sheath edge (conservation of flux in the collisionless sheath), i.e from (3.313):

$$\Gamma_{es} = \frac{1}{4} n_e v_e \tag{3.314}$$

or, explicitly:

$$\Gamma_{es} = n_e \left(\frac{k_B T_e}{2\pi m_e} \right)^{\frac{1}{2}} . \tag{3.315}$$

An approximate value for the *sheath thickness* l_{es} can be deduced from the Child-Langmuir law[137], which stipulates that the current density that a planar diode can provide is limited by the space charge due to the electrons and varies as $\phi_0^{3/2}$, where ϕ_0 is the potential difference between the two plates. In the case of a plasma, the sheath thickness adjusts to the current density j carried by the plasma, and to the potential difference $\phi_0 - \phi_p$. For an electron sheath, the Child-Langmuir law leads to:

$$j_e = -\frac{4\epsilon_0}{9} \left(\frac{2e}{m_e} \right)^{\frac{1}{2}} \frac{(\phi_0 - \phi_p)^{\frac{3}{2}}}{l_{es}^2} , \tag{3.316}$$

i.e. an electron sheath thickness:

$$l_{es} = \frac{2\sqrt{2}\pi^{\frac{1}{4}}}{3} \lambda_{De} \left[\frac{e(\phi_0 - \phi_p)}{k_B T_e} \right]^{\frac{3}{4}} . \tag{3.317}$$

Remark: In (3.315), the electron flux to the wall is fixed by the plasma (T_e and n_e): it is independent of the potential applied to the wall (planar wall).

3.14.2 Negative wall-potential with respect to the plasma potential: ion sheath

The second case is much more complicated because, unlike the ions, electrons have a much higher mean energy ($k_B T_e \gg k_B T_i \simeq 0$). It follows, therefore, that if the wall has a potential which attracts ions from the plasma, this

---

[137] For a planar diode, the electron current density j_e that can be extracted from the emitting surface (for example a tungsten ribbon) is given by:

$$j_e = 2.34 \times 10^{-6} \phi_0^{3/2} / d^2 \; (\mathrm{A/m^2})$$

where d is the distance between the two plates and ϕ_0 the corresponding potential difference.

potential is only partially repulsive for the electrons. However, the higher the potential barrier to be crossed by the electrons, the less important the electron flux collected by the walls. In the case of a Maxwell-Boltzmann distribution, the effective electron current collected by the wall ($\phi_0 < \phi_p$) can be written:

$$\Gamma_{es} = \frac{1}{4} n_e v_e \exp \frac{e(\phi_0 - \phi_p)}{k_B T_e} \,. \tag{3.318}$$

It is clear, therefore, that in the case of an ion sheath, a considerable number of electrons, depending on their energy, penetrate the ion sheath that forms at the plasma-wall interface. This time, the boundary where the departure from neutrality between the plasma and the wall occurs is less well defined and extends over quite a large thickness, as is shown in Figure 3.11. To overcome this difficulty, the transition zone can be divided in two sections, the true *ion sheath*, where departure from charge neutrality occurs, and the *pre-sheath* which, as the name indicates, precedes the sheath, and begins at the point where the ions start to be accelerated by the space charge electric field. This purely artificial division enables us to define the *sheath boundary*, as the location between a region of quasi-neutrality (the pre-sheath) in which only a small number of electrons are reflected, and a non neutral region (the ion sheath), where the ions are in the majority.

The evolution of the potential $\phi(x)$ is governed by Poisson's equation:

$$\frac{\partial^2 \phi}{\partial x^2} = \frac{e}{\epsilon_0}(n_e - n_i) \,. \tag{3.319}$$

We denote by n_g, v_i, and ϕ_g the plasma density, the velocity of the ions and the potential at the sheath edge, respectively. The electron density in the sheath is given by the Boltzmann equation (I.14):

$$n_e(x) = n_g \exp \left[\frac{e(\phi(x) - \phi_g)}{k_B T_e} \right] \,. \tag{3.320}$$

The velocity of the ions $v_i(x)$ as a function of v_g, the velocity with which they enter the sheath, can be deduced from the conservation of total energy over the distance travelled in the sheath:

$$\frac{m_i}{2} \left(v_i^2(x) - v_g^2 \right) = e(\phi_g - \phi(x)) \,. \tag{3.321}$$

The conservation of flux in the sheath can be written:

$$n_i(x) v_i(x) = n_g v_g \,. \tag{3.322}$$

From (3.321), we obtain the ion density in the sheath:

$$n_i(x) = n_g \left[1 - \frac{2e(\phi(x) - \phi_g)}{m_i v_g^2} \right]^{-\frac{1}{2}} \,. \tag{3.323}$$

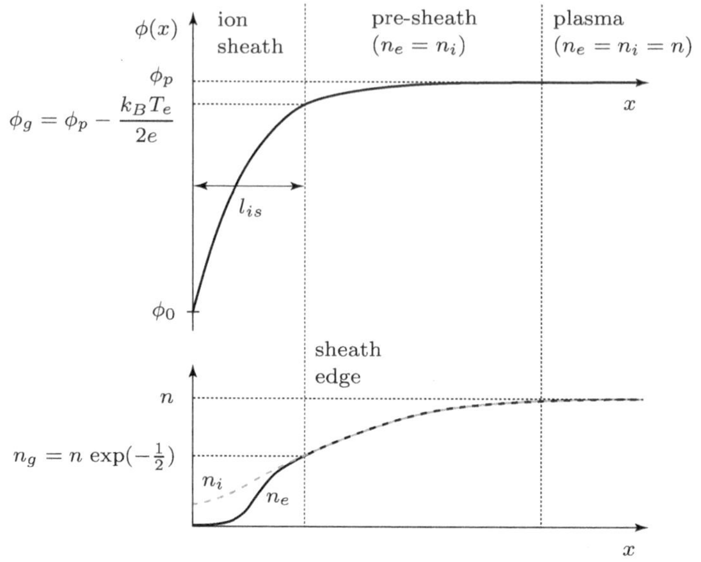

Fig. 3.11 Variation of the potential $\phi(x)$, the ion density $n_i(x)$ and the electron density $n_e(x)$ at the plasma-wall interface in the case of an ion sheath (x is the distance to the wall, l_{is} the thickness of the ion sheath, ϕ_p the plasma potential and ϕ_0 the potential applied on the wall while ϕ_g and n_g are the potential and plasma density at the sheath edge, respectively).

At each point of the ion sheath, we require:

$$n_i(x) > n_e(x) \ , \tag{3.324}$$

and this condition must be fulfilled, in particular, in the region near the sheath edge, i.e for small values of $\phi(x) - \phi_g$. A second order Taylor series expansion of (3.323) and (3.320) yields:

$$n_i(x) = n_g \left[1 + \frac{e(\phi(x) - \phi_g)}{m_i v_g^2} + \frac{3}{2} \frac{e^2(\phi(x) - \phi_g)^2}{(m_i v_g^2)^2} + \cdots \right] \ , \tag{3.325}$$

$$n_e(x) = n_g \left[1 + \frac{e(\phi(x) - \phi_g)}{k_B T_e} + \frac{e^2(\phi(x) - \phi_g)^2}{2(k_B T_e)^2} + \cdots \right] \ . \tag{3.326}$$

Since $\phi(x) - \phi_g$ is negative, condition (3.324) implies, to first order of the expansion:

$$v_g \geq v_B \equiv \left(\frac{k_B T_e}{m_i} \right)^{\frac{1}{2}} \ . \tag{3.327}$$

This criterion is known as the *Bohm criterion*. It means that the border between the macroscopically neutral zone (pre-sheath) and the zone where there is departure from neutrality (sheath) is situated at the point where the velocity of the ions, accelerated in the pre-sheath, is equal[138] to the *ion acoustic velocity* v_B, also called the *Bohm velocity*. Assuming a collisionless pre-sheath[139] and applying equation (3.321) between the plasma ($v_i(\phi_p) = 0$) and the sheath edge, the potential ϕ_g then becomes[140]:

$$e(\phi_p - \phi_g) = \frac{1}{2} m_i (v_g^2) = \frac{k_B T_e}{2} \,, \tag{3.328}$$

from which:

$$\phi_g = \phi_p - \frac{k_B T_e}{2e} \,. \tag{3.329}$$

From (3.320) and (3.328), we can extract the ion density at the sheath edge:

$$n_g = n \exp\left(-\frac{1}{2}\right) \tag{3.330}$$

and the value of the ion flux, $n_g v_g = \Gamma_{is}$, collected at the wall:

$$\Gamma_{is} = n \left(\frac{k_B T_e}{m_i}\right)^{\frac{1}{2}} \exp\left(-\frac{1}{2}\right) \,. \tag{3.331}$$

In applying the Child-Langmuir law (3.316) to the ion current, we can calculate the thickness of the ion sheath (assuming the space charge due to the electrons to be negligible), i.e.:

$$l_{gi} = \frac{2^{\frac{5}{4}}}{3\exp(-\frac{1}{4})} \lambda_{De} \left[\frac{e(\phi_p - \phi_0)}{k_B T_e}\right]^{\frac{3}{4}} \,. \tag{3.332}$$

Remark: In most texts, the Bohm criterion is deduced from a mathematical condition resulting from the integration of the Poisson equation (3.319), an operation that calls for simplified boundary conditions ($\partial\phi_g/\partial x = 0$). In the development presented here, the Bohm criterion is defined uniquely from the conditions at which departure from neutrality occurs.

---

[138] For $v_g = v_B$, it is easy to verify, from the second order expansion of (3.325) and (3.326), that the condition (3.324) is fulfilled.

[139] In reality, the pre-sheath is collisional because its thickness corresponds to a fraction of the ion mean free path in the presence of neutrals.

[140] Note that the potential at the sheath edge is sufficient to repel all electrons having an energy $\frac{1}{2} m_e w_x^2 < k_B T_e/2$.

3.14.3 Floating potential

The floating potential corresponds to the equality of ion and electron currents collected on a surface. This is the potential taken by an isolated surface (dielectric or conductor) in contact with the plasma. In fact, if these surfaces were to receive more charges of one sign or another, their potential would increase indefinitely. In the stable regime, this potential adjusts such that the surface collects an equal number of positive and negative charges. This potential ϕ_f, called the *floating potential*, is obtained by equating (3.318) and (3.331), i.e.:

$$\phi_f - \phi_p = -\frac{k_B T_e}{2e}\left(1 + \ln\frac{m_i}{2\pi m_e}\right) . \tag{3.333}$$

The floating potential adjusts to a negative value with respect to the plasma potential, such that a sufficient number of electrons are repelled, to equilibrate the ion and electron currents.

Remarks:

1. The directed energy acquired by the ions in the sheath is used in many surface treatment processes (etching, deposition, chemical modification). The ion bombardment energy can be increased by applying a *voltage* ϕ_0, referred to as the *bias voltage*, at the surface in contact with the plasma. If $\phi_0 = \phi_f$(ion sheath without an applied bias or natural sheath), $l_{is} \simeq \lambda_{De}$. On the other hand, if $\phi_p - \phi_0 \gg k_B T_e$, then $l_{is} \gg \lambda_{De}$.
2. From the point of view of a wave, the sheath can appear as a region of vacuum if the electron density is sufficiently weak, such that $\omega_{pe} \ll \omega$, for then $\epsilon_p \simeq 1$ (3.69).

Problems

3.1. There are two main methods for depicting a plasma: the kinetic model and the hydrodynamic model. Indicate the origin of these two models, their relationship and their respective domain of interest. Pose the problem of closure of the system of hydrodynamic equations and suggest how this can be resolved; give a concrete example of closure.

Answer

Kinetic model

This model treats, within a statistical framework, the individual motion of particles, the micro-fields induced by the motion of the charged particles and the collisions of particles.

The model is based on the Boltzmann equation (itself the result of integrating the Liouville equation), which describes the evolution of the single-point velocity distribution function f_1 for particles of a given species. This differential equation, which includes binary collisions between particles that are expressed through a two-point function, f_{12}, can be solved (assuming only weak correlations between particles) by replacing f_{12} by the product of the single-point functions $f_1$and f_2, for particles of species 1 and 2 respectively.

The Boltzmann differential equation is the basis for describing the physics of low-density plasmas (binary interactions).

Hydrodynamic model

The continuous fluid model makes use of the average values of molecular properties, induced fields and collisional interactions. These average values are obtained from the velocity distribution function, the evolution of which is described by the Boltzmann equation of the kinetic model. The hydrodynamic model provides a good approximation for describing most of the plasma properties, notably those concerned with the motion of charged particles, and the characteristics of the great majority of propagating waves.

The kinetic description is more exact and more complete (although there are only few phenomena which cannot be treated using the hydrodynamic model), but much more mathematically demanding and more complicated to interpret than the hydrodynamic model.

Structure of the hydrodynamic equations and need for their closure

The hydrodynamic equations are, in principle, infinite in number, as shown in Sect. 3.5. This series of equations, ordered according to the increasing moment of order m in $\boldsymbol{w}^m$, lead to a set of equations of increasing tensor order m: thus the continuity equation (moment of order zero in $\boldsymbol{w}^0$) is a zeroth order tensor, or scalar equation; the momentum transport equation (moment $\boldsymbol{w}^1$) is a first order tensor equation, while the kinetic pressure equation (moment $\boldsymbol{w}^2$) is a second order tensor equation (see Appendix VII for more on tensors). Each of these equations contains a variable, whose variation is described by the hydrodynamic equation of the next order moment, and thus this variable is a tensor of one order higher than that of the equation being considered. However, since this higher order tensor appears under a divergence operator in the equation under consideration, the tensor order of the equation itself is preserved. This is the case of the velocity $\boldsymbol{v}$ (a vector, i.e. a first-order tensor) that appears in the continuity equation, which is a zeroth order tensor equation: the quantity $\boldsymbol{v}$ actually appears in the term $\boldsymbol{\nabla} \cdot n\boldsymbol{v}$, which is of zeroth order. In terms of tensor formalism, this corresponds to a two-index contraction (generalized scalar product) between the components of the gradient operator (a vector) and the corresponding ones of the velocity vector, yielding a scalar (more details are given in Appendix VII).

Taking account of the underlying elements of the physical problem we are treating, we generally retain only the first two or three hydrodynamic equations. We must make sure, therefore, that the last equation considered does not include a variable depending on the next order moment equation. This process, referred to as closure of the system, is particularly well illustrated in the limit cases of the warm plasma and cold plasma approximations. Consider the hydrodynamic equation for the second order moment (of tensor order unity), which contains the term $\boldsymbol{\nabla} \cdot \underline{\boldsymbol{\Psi}}$, where $\underline{\boldsymbol{\Psi}}$ is the kinetic pressure tensor, a 2nd order tensor. This term, representing the contraction of $\underline{\boldsymbol{\Psi}}$ with the divergence operator, is of order 1. In the warm plasma approximation, $\boldsymbol{\nabla} \cdot \underline{\boldsymbol{\Psi}}$ is replaced by $\boldsymbol{\nabla} p_\alpha$, where p_α is the (scalar) kinetic pressure of particles of species α: $\boldsymbol{\nabla} p_\alpha$ is a first order tensor, and the link with the moment $\boldsymbol{w}^2$ describing the variation of $\underline{\boldsymbol{\Psi}}$ is cut. We can also neglect all thermal motion $T_\alpha = 0$ by setting either $\underline{\boldsymbol{\Psi}} = \underline{\boldsymbol{0}}$ directly, or $p_\alpha = 0$; this is the cold plasma approximation. In both these approximations, we only need to retain the first two hydrodynamic equations, the continuity equation and the momentum transport equation.

3.2. Consider the two following hydrodynamic equations:

$$m_e \dot{\boldsymbol{v}}_e = -e\boldsymbol{E} - m_e \nu (\boldsymbol{v}_e - \boldsymbol{v}_i) \,, \tag{3.334}$$

$$m_i \dot{\boldsymbol{v}}_i = e\boldsymbol{E} - m_e \nu (\boldsymbol{v}_i - \boldsymbol{v}_e) \,, \tag{3.335}$$

where m_e and m_i are the mass of the electrons and ions respectively, v_e and v_i their respective velocities and ν, the collision frequency for momentum transfer, assumed constant.

a) For the case of a cold plasma, show that, for a periodic electric field $\boldsymbol{E} = \boldsymbol{E}_0 \mathrm{e}^{-\mathrm{i}\omega t}$, the total current density, can be written in the form:

$$\boldsymbol{J} = \frac{ne^2 \boldsymbol{E}}{m_e \nu} \frac{1 + m_e/m_i}{(1 + m_e/m_i) - i(\omega/\nu)} \,, \tag{3.336}$$

where $n = n_e = n_i = n$ represents the density of the charged particles and of the plasma.

b) Calculate the relative permittivity of the medium and obtain approximations corresponding to the cases where $m_e/m_i \ll 1$, and then for $m_e/m_i \ll 1$ with the particular case $\nu/\omega \ll 1$; draw conclusions.
c) Can you justify the origin and the form of the collision term in (3.334) and (3.335)?

Answer

a) In a cold plasma subjected to a periodic electric field, we assume that the particle fluid velocity is purely periodic such that (2.28) $\boldsymbol{v}_e = \boldsymbol{v}_{0e}\, \mathrm{e}^{-\mathrm{i}\omega t}$ and $\boldsymbol{v}_i = \boldsymbol{v}_{0i}\, \mathrm{e}^{-\mathrm{i}\omega t}$. Taking this into account, and adding (3.334) and (3.335), we obtain:

$$-\mathrm{i}\omega(m_e\boldsymbol{v}_e + m_i\boldsymbol{v}_i) = 0 \,, \tag{3.337}$$

so that, for $\omega \neq 0$:

$$\boldsymbol{v}_e = -\frac{m_i}{m_e}\boldsymbol{v}_i \,. \tag{3.338}$$

Substituting (3.338) in (3.334), we can eliminate one of the two charged particle velocities, which, in this case, leads to:

$$m_i\mathrm{i}\omega\boldsymbol{v}_i = -e\boldsymbol{E} + \boldsymbol{v}_i(m_i\nu + m_e\nu) \,, \tag{3.339}$$

from which:

$$\boldsymbol{v}_i = -\frac{e\boldsymbol{E}}{m_i\mathrm{i}\omega - m_i\nu - m_e\nu} = \frac{e\boldsymbol{E}}{m_i\nu\left[1 + \dfrac{m_e}{m_i} - \dfrac{\mathrm{i}\omega}{\nu}\right]} \,. \tag{3.340}$$

Substituting (3.338) into (3.335) this time, we find:

$$\boldsymbol{v}_e = -\frac{e\boldsymbol{E}}{m_e\nu\left[1 + \dfrac{m_e}{m_i} - \dfrac{\mathrm{i}\omega}{\nu}\right]} \,. \tag{3.341}$$

The total current density can therefore be written:

$$\boldsymbol{J} \equiv -ne\boldsymbol{v}_e + ne\boldsymbol{v}_i = \frac{ne^2\boldsymbol{E}}{\nu}\left[1 + \frac{m_e}{m_i} - \frac{\mathrm{i}\omega}{\nu}\right]^{-1}\left(\frac{1}{m_e} + \frac{1}{m_i}\right) \tag{3.342}$$

and:

$$\boldsymbol{J} = \frac{ne^2\boldsymbol{E}}{m_e\nu}\frac{1 + m_e/m_i}{1 + m_e/m_i - \mathrm{i}\omega/\nu} \,. \tag{3.343}$$

b) From (3.343), by identification from $\boldsymbol{J} = \sigma\boldsymbol{E}$, we can extract the total conductivity:

$$\sigma = \frac{ne^2}{m_e \nu} \frac{1 + m_e/m_i}{1 + m_e/m_i - \mathrm{i}\omega/\nu} . \tag{3.344}$$

Knowing that the expression for the relative permittivity in vacuum of a fluid of charged particles in a field $E_0 \, \mathrm{e}^{-\mathrm{i}\omega t}$ is given by:

$$\epsilon_p = 1 + \frac{\sigma}{\mathrm{i}\omega\epsilon_0} , \tag{2.40}$$

we obtain from (3.344):

$$\epsilon_p = 1 - \frac{\omega_{pe}^2}{\omega\nu} \frac{1 + m_e/m_i}{\mathrm{i}(1 + m_e/m_i - \mathrm{i}\omega/\nu)} , \tag{3.345}$$

where we have set $ne^2/m_e\epsilon_0 = \omega_{pe}^2$. Moreover, since $\omega_{pe}^2/\omega_{pi}^2 = m_i/m_e$, this becomes:

$$\epsilon_p = 1 - \frac{\omega_{pe}^2}{\omega\nu} \frac{1 + \omega_{pi}^2/\omega_{pe}^2}{\omega/\nu + \mathrm{i}(1 + \omega_{pi}^2/\omega_{pe}^2)} , \tag{3.346}$$

which can be decomposed into its real and imaginary parts, i.e. $\epsilon_p = \epsilon_r + \mathrm{i}\epsilon_i$.

Interesting approximations:

1. For $m_e/m_i \ll 1$ ($\omega_{pi}^2/\omega_{pe}^2 \simeq 0$), from (3.346):

$$\epsilon_p = 1 - \frac{\omega_{pe}^2}{\omega\nu} \frac{1}{\omega/\nu + \mathrm{i}} = 1 - \frac{\omega_{pe}^2}{\omega(\omega + \mathrm{i}\nu)} , \tag{3.347}$$

   which is equivalent to (2.41) for the case where the field $\boldsymbol{E}$ varies as $\mathrm{e}^{-\mathrm{i}\omega t}$.
2. For $m_e/m_i \ll 1$ and $\nu/\omega \ll 1$, from (3.347):

$$\epsilon_p = 1 - \frac{\omega_{pe}^2}{\omega^2} , \tag{3.348}$$

   i.e. (2.42), which is purely real, in contrast to (3.347)

c) In (3.334) and (3.335), only the relative velocity of the fluids of electrons and ions, $\boldsymbol{v}_e - \boldsymbol{v}_i$, appears. This means that only electron-ion collisions are taken into account: there are no collisions with neutral atoms, which are in effect neglected.
Consider the collisional term (3.121) from the momentum transport equation:

$$\boldsymbol{\mathcal{P}}_\alpha = -\sum_{\beta\neq\alpha} \nu_{\alpha\beta} n_\alpha \mu_{\alpha\beta} (\boldsymbol{v}_\alpha - \boldsymbol{v}_\beta) = -\sum_{\beta\neq\alpha} \boldsymbol{\mathcal{P}}_{\alpha\beta} . \tag{3.349}$$

In the case where the particles α are electrons and the particles β are ions, the reduced mass $\mu_{\alpha\beta}$ is approximately m_α ($m_e/m_i \ll 1$). Setting $n_i = n_e = n$, then $\nu_{ei} = \nu_{ie} = \nu$ and we can then write (3.349), for electron-ion collisions:

$$\mathcal{P}_{ei} = -\nu n m_e (\boldsymbol{v}_e - \boldsymbol{v}_i) \; . \tag{3.350}$$

Since $\mathcal{P}_{ei} = -\mathcal{P}_{ie}$ (3.127):

$$\mathcal{P}_{ie} = -\nu n m_e (\boldsymbol{v}_i - \boldsymbol{v}_e) \; , \tag{3.351}$$

which are the collisional terms of (3.334) and (3.335) respectively. Equations (3.334) and (3.335) are, in fact, the momentum transport equations for electrons and ions, respectively, in which the convective term and the term containing $\boldsymbol{\nabla} p_\alpha$ have been neglected. The density n, which appears explicitly in these equations (for example (3.118)), is a common factor that can be cancelled out.

3.3. Consider an electron fluid, subject to a static magnetic field $\boldsymbol{B}$ and in which there is a pressure gradient $\boldsymbol{\nabla} p$.

a) Show that this results in a current density $\boldsymbol{J}$, perpendicular to the field $\boldsymbol{B}$, given by:

$$\boldsymbol{J} = \frac{\boldsymbol{B} \wedge \boldsymbol{\nabla} p}{B^2} \; . \tag{3.352}$$

b) Consider the case of a cylindrical plasma subjected to a magnetic field directed along the axis of the cylinder (the z axis), which is uniform in this direction and axially symmetric. For a pressure gradient directed radially towards the axis of the cylinder, show that:

$$B(r) - B(a) = \mu_0 \int_r^a \frac{1}{B(r')} \frac{\mathrm{d}p}{\mathrm{d}r'} \mathrm{d}r' \; , \tag{3.353}$$

where a is the plasma radius. To do this, make use of the Maxwell equation $\boldsymbol{\nabla} \wedge \boldsymbol{B} = \mu_o \boldsymbol{J} + \mu_0 \epsilon_0 \partial \boldsymbol{E} / \partial t$, where $\partial \boldsymbol{E} / \partial t = 0$ in the present case.

c) Show that (3.353) is equivalent to:

$$\frac{B^2(r)}{2\mu_0} + p(r) = C_1 \; , \tag{3.354}$$

where C_1 is a constant with respect to r.

Answer

a) The Lorentz model for plasma electrons (Sect. 3.7) leads to the momentum transport equations for the fluid of electrons, with velocity $\boldsymbol{v}_e$ and density n_e, of the form:

$$m_e \frac{\partial \boldsymbol{v}_e}{\partial t} = \boldsymbol{F} \equiv q_e[\boldsymbol{E} + \boldsymbol{v}_e \wedge \boldsymbol{B}] - \frac{\boldsymbol{\nabla} p_e}{n_e} - m_e \nu \boldsymbol{v}_e \ . \tag{3.170}$$

This equation shows that the term $-\boldsymbol{\nabla} p_e/n_e$ is a component of the total force $\boldsymbol{F}$ acting on the fluid, thus it has the nature of a force.
Furthermore, we have seen, in the context of the study of particle trajectories, that the expression for the drift velocity $\boldsymbol{w}_D$ due to the action of a force $\boldsymbol{F}_D$, acting on a charged particle subjected to a field $\boldsymbol{B}$ is (Appendix XII):

$$\boldsymbol{w}_D = \frac{\boldsymbol{F}_D \wedge \boldsymbol{B}}{qB^2} \ . \tag{XII.2}$$

Using this equation to describe the drift velocity $\boldsymbol{v}_D$ of the electron fluid subject to a force $\boldsymbol{F}_D = -\boldsymbol{\nabla} p_e/n_e$, we obtain:

$$\boldsymbol{v}_D = -\frac{\boldsymbol{\nabla} p_e \wedge \boldsymbol{B}}{n_e q_e B^2} = \frac{\boldsymbol{B} \wedge \boldsymbol{\nabla} p_e}{n_e q_e B^2} \ . \tag{3.355}$$

The drift current density, $\boldsymbol{J} \equiv nq\boldsymbol{v}_D$, is then given by the equation:

$$\boldsymbol{J} \equiv n_e q_e \boldsymbol{v}_D = \frac{\boldsymbol{B} \wedge \boldsymbol{\nabla} p_e}{B^2} \ , \tag{3.352}$$

where the drift velocity responsible for $\boldsymbol{J}$ is perpendicular to $\boldsymbol{B}$ and $\boldsymbol{\nabla} p_e$.

b) The non uniformity of the field $\boldsymbol{B}$ in the direction perpendicular to it is such that the B_z component depends on the different components perpendicular to the axis, as was shown in Sect. 2.2.3. This leads us to consider the following Maxwell equation:

$$\boldsymbol{\nabla} \wedge \boldsymbol{H} = \boldsymbol{J} + \epsilon_0 \frac{\partial \boldsymbol{E}}{\partial t} \ , \tag{3.356}$$

which, written in terms of $\boldsymbol{B}$ instead of $\boldsymbol{H}$, gives ($\mu_0 \epsilon_0 = 1/c^2$):

$$\boldsymbol{\nabla} \wedge \boldsymbol{B} = \mu_0 \boldsymbol{J} + \frac{1}{c^2} \frac{\partial \boldsymbol{E}}{\partial t} \ . \tag{3.357}$$

Since we are not in the framework of individual trajectories, the RHS of this equation is, in general, non zero. In the present case, since we are dealing with a constant field $\boldsymbol{B}$, (3.352) implies that $\boldsymbol{J}$ is constant and, so is $\boldsymbol{E}$ since $\boldsymbol{J} = \sigma \boldsymbol{E}$. Therefore, $\partial \boldsymbol{E}/\partial t = 0$ and only $\mu_0 \boldsymbol{J}$ remains on the RHS of (3.357).
Expanding the different components of $\boldsymbol{\nabla} \wedge \boldsymbol{B}$ in cylindrical coordinates (Appendix XX):

$$\frac{1}{r}\left(\frac{\partial B_z}{\partial \varphi} - \frac{\partial (rB_\varphi)}{\partial z}\right)\hat{\mathbf{e}}_r + \left(\frac{\partial B_r}{\partial z} - \frac{\partial B_z}{\partial r}\right)\hat{\mathbf{e}}_\varphi + \frac{1}{r}\left(\frac{\partial (rB_\varphi)}{\partial r} - \frac{\partial B_r}{\partial \varphi}\right)\hat{\mathbf{e}}_z \ , \tag{3.358}$$

since $\boldsymbol{B}$ is directed along $\hat{\mathbf{e}}_z$, the B_φ and B_r components are zero (Sect. 2.2.3). Furthermore, from axial symmetry, $\partial B_z/\partial \varphi = 0$, so that finally there remains only:

$$\nabla \wedge \boldsymbol{B} = \hat{\mathbf{e}}_\varphi \left(-\frac{\partial B_z}{\partial r} \right) . \tag{3.359}$$

Then, multiplying (3.352) by μ_0, we obtain:

$$\mu_0 \boldsymbol{J} = \mu_0 \frac{\boldsymbol{B} \wedge \nabla p}{B^2} . \tag{3.360}$$

From (3.357), with $\partial \boldsymbol{E}/\partial t = 0$, and (3.360):

$$\nabla \wedge \boldsymbol{B} = \frac{\mu_0 \boldsymbol{B} \wedge \nabla p}{B^2} \tag{3.361}$$

and, after the scalar multiplication of (3.359) by $\hat{\mathbf{e}}_\varphi$, we find:

$$-\frac{\partial B_z}{\partial r} = \left(\frac{\mu_0 \boldsymbol{B} \wedge \nabla p}{B^2} \right) \cdot \hat{\mathbf{e}}_\varphi \equiv \frac{\mu_0 B_z}{B_z^2} \frac{\mathrm{d}p}{\mathrm{d}r} , \tag{3.362}$$

since $\nabla p = -\hat{\mathbf{e}}_r \mathrm{d}p/\mathrm{d}r$ in this b) section.
Integrating (3.362) over r' from a to r indeed reproduces (3.353):

$$[B_z(r) - B_z(a)] = \int_a^r \frac{\mu_0}{B_z} \frac{\mathrm{d}p}{\mathrm{d}r'} \mathrm{d}r' . \tag{3.353}$$

c) Equation (3.354) can be validated by differentiating it, which leads to:

$$\frac{B_z}{\mu_0} \frac{\mathrm{d}B_z}{\mathrm{d}r} = -\frac{\mathrm{d}p}{\mathrm{d}r} , \tag{3.363}$$

which after rearrangement leads to (3.362) and to (3.353) by integration from a to r.

3.4. Assume a Maxwell-Boltzmann velocity distribution function for the electrons, modified in the following way:

$$f(\boldsymbol{w}) = n_e \left(\frac{m_e}{2\pi k_B T_\perp} \right) \left(\frac{m_e}{2\pi k_B T_\parallel} \right)^{1/2} \exp \left\{ -\frac{m_e}{2k_B} \left[\frac{w_x^2 + w_y^2}{T_\perp} + \frac{w_z^2}{T_\parallel} \right] \right\} , \tag{3.364}$$

where $\boldsymbol{w} = (w_x, w_y, w_z)$ is the velocity vector for individual electrons of density n_e and of mass m_e, and k_B is Boltzmann's constant.

a) Verify that, in this equation, n_e represents the electron density, which in the present case, is assumed independent of time and position.
b) Calculate the components of the kinetic pressure tensor:

$$\underline{\boldsymbol{\Psi}} \equiv n_e m_e \langle \boldsymbol{u}\boldsymbol{u} \rangle \tag{3.365}$$

in the case of a separable distribution function, where $\boldsymbol{u} = \boldsymbol{w} - \boldsymbol{v}$ with $\boldsymbol{v}(\boldsymbol{r})$ is the mean electron velocity, and $\boldsymbol{uu}$ represents the tensor product of the individual velocities of the electrons, which are strictly thermal (random) in nature.

c) Give a physical significance to (3.364) by suggesting how such a distribution might be reproduced in the laboratory.

Answer

a) Recall firstly the equations defining the hydrodynamic parameters, also called the macroscopic parameters $\langle \varUpsilon(\boldsymbol{r},t)\rangle$:

$$\langle \varUpsilon(\boldsymbol{r},t)\rangle \equiv \frac{1}{n_e(\boldsymbol{r},t)} \int_{\boldsymbol{w}} \varUpsilon(\boldsymbol{r},\boldsymbol{w},t) f(\boldsymbol{r},\boldsymbol{w},t)\, \mathrm{d}^3 w \; , \tag{3.39}$$

where the brackets $\langle\ \rangle$ represent the mean value of the microscopic property $\varUpsilon(\boldsymbol{r},\boldsymbol{w},t)$ taken over the distribution function $f(\boldsymbol{w},\boldsymbol{r},t)$. By setting $\varUpsilon = 1$, we therefore obtain the expression for the density:

$$n_e(\boldsymbol{r},t) = \int_{\boldsymbol{w}} f(\boldsymbol{r},\boldsymbol{w},t)\, \mathrm{d}^3 w \; . \tag{3.38}$$

Remember that the function $f(\boldsymbol{r},\boldsymbol{w},t)$ is separable if we can express it in the following form (Sect. 3.3):

$$f(\boldsymbol{r},\boldsymbol{w},t) = n(\boldsymbol{r},t) g(\boldsymbol{w}) \; . \tag{3.366}$$

The form of (3.364) indicates that the function f has been separated and we know, in addition, from our assumption in the present case, that n_e is independent of $\boldsymbol{r}$. Therefore, we need to verify that the function $g(\boldsymbol{w})$, defined by $f(\boldsymbol{w})/n_e$, integrated over all velocities, i.e.:

$$\int_{\boldsymbol{w}} \left(\frac{m_e}{2\pi k_B T_\perp}\right) \left(\frac{m_e}{2\pi k_B T_\parallel}\right)^{\frac{1}{2}} \exp\left\{ -\frac{m_e}{2k_B}\left[\frac{w_x^2 + w_y^2}{T_\perp} + \frac{w_z^2}{T_\parallel}\right]\right\} \mathrm{d}^3 w \tag{3.367}$$

is really equal to 1 (normalisation condition: (3.45)).
We can evaluate the term:

$$\int_{-\infty}^{\infty} \exp\left\{ -\left[\frac{m_e}{2k_B}\frac{w_x^2}{T_\perp}\right]\right\} \mathrm{d}w_x \tag{3.368}$$

by setting:

$$\frac{m_e w_x^2}{2k_B T_\perp} = x^2 \; , \tag{3.369}$$

such that:

$$\frac{2m_e w_x \mathrm{d}w_x}{2k_B T_\perp} = 2x\mathrm{d}x \tag{3.370}$$

and:

$$\begin{aligned}\mathrm{d}w_x &= \frac{x}{(m_e/2k_BT_\perp)^{\frac{1}{2}}}\frac{1}{(m_e/2k_BT_\perp)^{\frac{1}{2}}}\frac{\mathrm{d}x}{w_x} \\ &= \frac{x\mathrm{d}x}{(m_e/2k_BT_\perp)^{\frac{1}{2}}x} \, .\end{aligned} \tag{3.371}$$

The term (3.368) can then be rewritten:

$$\left(\frac{2k_BT_\perp}{m_e}\right)^{\frac{1}{2}} \int_{-\infty}^{\infty} \exp(-x^2)\,\mathrm{d}x = 2\left(\frac{2k_BT_\perp}{m_e}\right)^{\frac{1}{2}} \int_0^{\infty} \exp(-x^2)\,\mathrm{d}x \, . \tag{3.372}$$

Knowing that (Appendix XX):

$$\int_0^{\infty} \exp(-x^2)\,\mathrm{d}x = \frac{1}{2}\sqrt{\pi} \, , \tag{3.373}$$

then (3.372) becomes:

$$\left(\frac{2k_BT_\perp}{m_e}\right)^{\frac{1}{2}} \int_{-\infty}^{\infty} \exp(-x^2)\,\mathrm{d}x = 2\left(\frac{2k_BT_\perp}{m_e}\right)^{\frac{1}{2}} \frac{\sqrt{\pi}}{2} \, . \tag{3.374}$$

In a similar fashion, we find that:

$$\int_{-\infty}^{\infty} \exp\left\{-\frac{m_e}{2k_B}\frac{w_y^2}{T_\perp}\right\}\mathrm{d}w_y = \sqrt{\pi}\left(\frac{2k_BT_\perp}{m_e}\right)^{\frac{1}{2}} \, , \tag{3.375}$$

$$\int_{-\infty}^{\infty} \exp\left\{-\frac{m_e}{2k_B}\frac{w_z^2}{T_\parallel}\right\}\mathrm{d}w_z = \sqrt{\pi}\left(\frac{2k_BT_\parallel}{m_e}\right)^{\frac{1}{2}} \, , \tag{3.376}$$

such that the expression (3.367) is indeed unity.

b) In the presence of a directed electron velocity $\boldsymbol{v}$, it is necessary to consider a Maxwell-Boltzmann distribution centred about this velocity $\boldsymbol{v}$ (Appendix I). In this case, the distribution function can be written:

$$\begin{aligned}f(\boldsymbol{w}) &= n_e\left(\frac{m_e}{2\pi k_BT_\perp}\right)\left(\frac{m_e}{2\pi k_bT_\parallel}\right)^{\frac{1}{2}} \\ &\times \exp\left\{-\frac{m_e}{2k_B}\left[\frac{(w_x-v_x)^2+(w_y-v_y)^2}{T_\perp}+\frac{(w_z-v_z)^2}{T_\parallel}\right]\right\} \, .\end{aligned} \tag{3.377}$$

By introducing the change of variable $\boldsymbol{u} = \boldsymbol{w} - \boldsymbol{v}$, we can write for the component $i = x, y, z$:

$$u_i = w_i - v_i \ , \tag{3.378}$$

and since the fluid velocity is constant:

$$\mathrm{d}u_i = \mathrm{d}w_i \tag{3.379}$$

from which:

$$f(\boldsymbol{w}) = n_e \left(\frac{m_e}{2\pi k_B T_\perp}\right)\left(\frac{m_e}{2\pi k_b T_\parallel}\right)^{\frac{1}{2}} \exp\left\{-\frac{m_e}{2k_B}\left[\frac{u_x^2 + u_y^2}{T_\perp} + \frac{u_z^2}{T_\parallel}\right]\right\} . \tag{3.380}$$

Firstly consider the off diagonal components of $\underline{\boldsymbol{\Psi}}$, for example Ψ_{xy}. Following (3.365), this term can be written:

$$\Psi_{xy} = m_e \int u_x u_y f(\boldsymbol{r}, \boldsymbol{w}, t)\mathrm{d}^3 w \ , \tag{3.381}$$

then, taking account of the fact that the function $f(\boldsymbol{r}, \boldsymbol{w}, t)$ is separable (3.364):

$$\Psi_{xy} = n_e m_e \int u_x u_y f(\boldsymbol{w})\mathrm{d}^3 w \ . \tag{3.382}$$

After the change of variable $\boldsymbol{w} = \boldsymbol{u} + \boldsymbol{v}$:

$$\Psi_{xy} = n_e m_e \int u_x u_y f(\boldsymbol{u})\mathrm{d}^3 u \ , \tag{3.383}$$

$$\Psi_{xy} = n_e m_e \left(\frac{m_e}{2\pi k_B T_\perp}\right)\left(\frac{m_e}{2\pi k_B T_\parallel}\right)^{\frac{1}{2}} \tag{3.384}$$
$$\times \int_{-\infty}^{\infty}\int_{-\infty}^{\infty}\int_{-\infty}^{\infty} \exp\left\{-\frac{m_e}{2k_B}\left[\frac{u_x^2 + u_y^2}{T_\perp} + \frac{u_z^2}{T_\parallel}\right]\right\} u_x u_y \,\mathrm{d}u_x \mathrm{d}u_y \mathrm{d}u_z \ .$$

This integral is zero, because the integrand is odd in u_x and u_y, and that these velocity values extend from $-\infty$ to $+\infty$. Similarly, this is true of all the other off diagonal elements.

For the diagonal elements of the matrix representing the tensor $\underline{\boldsymbol{\Psi}}$, consider, for example:

$$\Psi_{xx} = n m_e \left(\frac{m_e}{2\pi k_B T_\perp}\right)\left(\frac{m_e}{2\pi k_B T_\parallel}\right)^{\frac{1}{2}} \tag{3.385}$$
$$\times \int_{-\infty}^{\infty}\int_{-\infty}^{\infty}\int_{-\infty}^{\infty} \exp\left\{-\frac{m_e}{2k_B}\left[\frac{u_x^2 + u_y^2}{T_\perp} + \frac{u_z^2}{T_\parallel}\right]\right\} u_x^2 \,\mathrm{d}u_x \mathrm{d}u_y \mathrm{d}u_z \ .$$

The calculation of the term:

$$\int_{-\infty}^{\infty} \exp\left[-\frac{m_e u_x^2}{2k_B T_\perp}\right] u_x^2 \, \mathrm{d}u_x \tag{3.386}$$

can be performed in an analogous fashion to that of (3.368) by setting $m_e u_x^2/2k_B T_\perp = x^2$, such that this term can be rewritten:

$$2\left(\frac{2k_B T_\perp}{m_e}\right)^{\frac{1}{2}} \int_0^{\infty} \left(\frac{2k_B T_\perp}{m_e}\right) \exp(-x^2)x^2 \, \mathrm{d}x = 2\left(\frac{2k_B T_\perp}{m_e}\right)^{\frac{3}{2}} \int_0^{\infty} \exp(-x^2)x^2 \, \mathrm{d}x \, . \tag{3.387}$$

From Appendix XX:

$$\int_0^{\infty} \exp(-x^2)x^2 \, \mathrm{d}x = \frac{\sqrt{\pi}}{4}$$

such that the expression (3.386) becomes:

$$\int_{-\infty}^{\infty} \exp\left[-\frac{m_e}{2k_B T_\perp} u_x^2\right] u_x^2 \, \mathrm{d}u_x = 2\left(\frac{2k_B T_\perp}{m_e}\right)^{\frac{3}{2}} \frac{\sqrt{\pi}}{4} \, , \tag{3.388}$$

and finally, from (3.385) and taking account of (3.375) and (3.376):

$$\Psi_{xx} = n_e m_e \left(\frac{m_e}{2\pi k_B T_\perp}\right)\left(\frac{m_e}{2\pi k_B T_\parallel}\right)^{\frac{1}{2}} 2\left(\frac{2k_B T_\perp}{m_e}\right)^{\frac{3}{2}} \frac{\sqrt{\pi}}{4} \times \underbrace{\left(\frac{2k_B T_\perp}{m_e}\right)^{\frac{1}{2}} \sqrt{\pi}}_{\text{after integration over } u_y} \times \underbrace{\left(\frac{2k_B T_\parallel}{m_e}\right)^{\frac{1}{2}} \sqrt{\pi}}_{\text{after integration over } u_z} \tag{3.389}$$

and:

$$\Psi_{xx} = \frac{n_e m_e}{2} \frac{2k_B T_\perp}{m_e} = n_e k_B T_\perp \, . \tag{3.390}$$

We then find for the other two diagonal terms:

$$\langle \Psi_{yy} \rangle = n_e k_B T_\perp \tag{3.391}$$

and:

$$\langle \Psi_{zz} \rangle = n_e k_B T_\parallel \tag{3.392}$$

such that we can write (Appendix VII):

$$\underline{\boldsymbol{\Psi}} = nk_BT_\perp(\hat{\mathbf{e}}_x\hat{\mathbf{e}}_x + \hat{\mathbf{e}}_y\hat{\mathbf{e}}_y) + nk_BT_\parallel(\hat{\mathbf{e}}_z\hat{\mathbf{e}}_z) \,. \tag{3.393}$$

c) We know that, for electrons satisfying a Maxwell-Boltzmann distribution, characterised by a temperature T_e, the scalar pressure along each axis of the coordinate system is given by:

$$p_x = p_y = p_z = p = n_e k_B T_e \,. \tag{3.394}$$

In the present case, the velocity distribution is anisotropic, since the mean energy along the z axis is described by a temperature $T_\parallel$ that is different from $T_\perp$, along x and y. Such anisotropy can be obtained with the aid of a magnetic field.
In a long cylindrical vessel ($L \gg R$), the radial diffusion is less important if we apply a homogeneous, axial magnetic field (along z) (Sect. 3.8). The radial diffusion then depends on the temperature $T_\perp$ rather than $T_\parallel$, the temperature that prevails in the absence of a magnetic field. Let us show that $T_\perp < T_\parallel$.
The diffusion coefficients, in the presence of a magnetic field, are given by (3.197) and (3.200) respectively:

$$D_\parallel = \frac{k_BT_\parallel}{m_e\nu} \,, \tag{3.395}$$

$$D_\perp = \frac{D_\parallel \nu^2}{\nu^2 + \omega_c^2} \,. \tag{3.396}$$

Then, in analogy with (3.395), we only need to set $D_\perp = k_BT_\perp/m_e\nu$, hence:

$$T_\perp = T_\parallel \frac{\nu^2}{\nu^2 + \omega_c^2} \,, \tag{3.397}$$

such that $T_\perp \lesssim T_\parallel$, provided that $\omega_{ce} \gtrsim \nu$.

3.5. Consider a plasma immersed in a uniform magnetic field directed along the z axis (Cartesian coordinates), $\boldsymbol{B} = \hat{\mathbf{e}}_z B$, and subject to a uniform HF electric field $\boldsymbol{E}_0\mathrm{e}^{\mathrm{i}\omega t}$, in an arbitrary direction. The plasma is cold and collisionless, and the field $\boldsymbol{B}$ is off cyclotron-resonance conditions.

a) Calculate the dielectric permittivity tensor $\underline{\boldsymbol{\epsilon}}_p$ (relative to vacuum) of the electron fluid, by introducing the electron plasma and cyclotron frequencies ω_{pe} and ω_{ce}.
b) Calculate the tensor $\underline{\boldsymbol{\epsilon}}_p$ describing both the electron and ion (singly ionised) fluids introducing explicitly the ion plasma and cyclotron frequencies ω_{pi} and ω_{ci}.

Answer

a) We know that the permittivity tensor $\underline{\boldsymbol{\epsilon}}_p$ is related to the conductivity tensor $\underline{\boldsymbol{\sigma}}$ by the 2nd order tensor relation:

$$\underline{\boldsymbol{\epsilon}}_p = \underline{\boldsymbol{I}} + \frac{\underline{\boldsymbol{\sigma}}}{\mathrm{i}\omega\epsilon_0}, \tag{3.398}$$

where $\underline{\boldsymbol{I}}$ is the unit tensor (unit matrix). The conductivity $\underline{\boldsymbol{\sigma}}$ can be obtained from the current density $\boldsymbol{J} = nq\boldsymbol{v}$, since $\boldsymbol{J} = \underline{\boldsymbol{\sigma}} \cdot \boldsymbol{E}$.
We can describe the motion of the electron fluid by assuming that its motion is that of an individual electron (see (2.6)–(2.8)), thereby merely using the velocity components of an individual electron, i.e.:

$$\mathrm{i}\omega v_x = -\frac{e}{m_e}E_x - \omega_{ce}v_y, \tag{3.399}$$

$$\mathrm{i}\omega v_y = -\frac{e}{m_e}E_y + \omega_{ce}v_x, \tag{3.400}$$

$$\mathrm{i}\omega v_z = -\frac{e}{m_e}E_z. \tag{3.401}$$

Eliminating v_y from (3.399) by substituting into it its value taken from (3.400), then successively:

$$\mathrm{i}\omega v_x = -\frac{e}{m_e}E_x - \frac{\omega_{ce}}{\mathrm{i}\omega}\left[-\frac{e}{m_e}E_y + \omega_{ce}v_x\right], \tag{3.402}$$

$$v_x = -\frac{e}{m_e\mathrm{i}\omega}E_x - \frac{\omega_{ce}}{\omega^2}\frac{e}{m_e}E_y + \frac{\omega_{ce}^2}{\omega^2}v_x, \tag{3.403}$$

$$v_x = \frac{-\dfrac{e}{m_e}\left[-\dfrac{\mathrm{i}\omega}{\omega^2}E_x + \dfrac{\omega_{ce}}{\omega^2}E_y\right]}{1 - \dfrac{\omega_{ce}^2}{\omega^2}} = \frac{\dfrac{e}{m_e}\left[\dfrac{\mathrm{i}\omega E_x - \omega_{ce}E_y}{\omega^2}\right]}{\dfrac{\omega^2 - \omega_{ce}^2}{\omega^2}}, \tag{3.404}$$

from which, finally:

$$v_x = \frac{e}{m_e}\left[\frac{\mathrm{i}\omega E_x - \omega_{ce}E_y}{\omega^2 - \omega_{ce}^2}\right]. \tag{3.405}$$

Similarly, we would find:

$$v_y = \frac{e}{m_e}\left[\frac{\mathrm{i}\omega E_y + \omega_{ce}E_x}{\omega^2 - \omega_{ce}^2}\right] \tag{3.406}$$

and:

$$v_z = \frac{\mathrm{i}}{\omega}\frac{e}{m_e}E_z. \tag{3.407}$$

The component J_x of the electron current (2.122) can now be expressed as:

$$J_x \equiv -n_e e v_x = -\frac{n_e e^2}{m_e}\left[\frac{\mathrm{i}\omega E_x - \omega_{ce}E_y}{\omega^2 - \omega_{ce}^2}\right] = \sigma_{xx}E_x + \sigma_{xy}E_y + \sigma_{xz}E_z\ . \tag{3.408}$$

The component σ_{xx} of the tensor $\underline{\boldsymbol{\sigma}}$ is the one which, in the expansion of $J_x \equiv -n_e e v_x$, comprises the E_x component of the electric field, hence, from (3.405) and (3.408):

$$\sigma_{xx} = -\frac{ne^2}{m_e}\left[\frac{\mathrm{i}\omega}{\omega^2 - \omega_{ce}^2}\right]\ , \tag{3.409}$$

then the corresponding component of the permittivity (3.398) is given by:

$$\epsilon_{pxx} = 1 + \frac{\sigma_{xx}}{\mathrm{i}\omega\epsilon_0} = 1 - \frac{n_e e^2}{m_e\epsilon_0}\frac{1}{\omega^2 - \omega_{ce}^2} = 1 - \frac{\omega_{pe}^2}{\omega^2 - \omega_{ce}^2}\ . \tag{3.410}$$

For the σ_{xy} component (the term in the expansion of J_x containing E_y), we obtain, from (3.405) and (3.408):

$$\sigma_{xy} = \frac{n_e e^2}{m_e}\left[\frac{\omega_{ce}}{\omega^2 - \omega_{ce}^2}\right] = \epsilon_0\left[\frac{\omega_{pe}^2\,\omega_{ce}}{\omega^2 - \omega_{ce}^2}\right] \tag{3.411}$$

and from (3.398), remembering that the off-diagonal elements of $\underline{\boldsymbol{I}}$ are zero:

$$\epsilon_{pxy} = -\mathrm{i}\left[\frac{\omega_{ce}}{\omega}\frac{\omega_{pe}^2}{\omega^2 - \omega_{ce}^2}\right]\ . \tag{3.412}$$

Similarly we would find that $\epsilon_{pyy} = \epsilon_{pxx}$ and $\epsilon_{pyx} = -\epsilon_{pxy}$. For ϵ_{pzz}, since from (3.407):

$$\sigma_{zz} = -\frac{\mathrm{i}ne^2}{\omega m_e}\ , \tag{3.413}$$

then:

$$\epsilon_{pzz} = 1 - \frac{\omega_{pe}^2}{\omega^2}\ . \tag{3.414}$$

The matrix representing $\underline{\boldsymbol{\epsilon}}_p$ has the form:

$$\underline{\boldsymbol{\epsilon}}_p = \begin{pmatrix} \epsilon_{pxx} & \epsilon_{pxy} & 0 \\ -\epsilon_{pxy} & \epsilon_{pxx} & 0 \\ 0 & 0 & \epsilon_{pzz} \end{pmatrix}\ . \tag{3.415}$$

This symmetry is quite general (Onsager *relations*) for a plasma subjected to a uniform field $\boldsymbol{B}$, directed axially (along z). By substituting the values of the components of $\underline{\boldsymbol{\epsilon}}_p$ in (3.415), we obtain:

$$\underline{\epsilon}_p = \begin{pmatrix} 1 - \dfrac{\omega_{pe}^2}{\omega^2 - \omega_{ce}^2} & -\mathrm{i}\left[\dfrac{\omega_{ce}}{\omega}\dfrac{\omega_{pe}^2}{\omega^2 - \omega_{ce}^2}\right] & 0 \\ +\mathrm{i}\left[\dfrac{\omega_{ce}}{\omega}\dfrac{\omega_{pe}^2}{\omega^2 - \omega_{ce}^2}\right] & 1 - \dfrac{\omega_{pe}^2}{\omega^2 - \omega_{ce}^2} & 0 \\ 0 & 0 & 1 - \dfrac{\omega_{pe}^2}{\omega^2} \end{pmatrix} , \tag{3.416}$$

where we have introduced the electron and cyclotron frequencies ω_{pe} and ω_{ce}, respectively.

b) To calculate the combined electrical conductivity from both electrons and ions, $\underline{\sigma}_{\text{total}}$, we simply need to remember that the ion and electron currents are additive. In the equations for the components of the current density, the charge appears in the form e^2 (see (3.408), for example) so that the two currents actually add. Then, with $n_e = n_i = n$:

$$\sigma_{xx} = -\frac{ne^2}{m_e}\left[\frac{\mathrm{i}\omega}{\omega^2 - \omega_{ce}^2}\right] - \frac{ne^2}{m_i}\left[\frac{\mathrm{i}\omega}{\omega^2 - \omega_{ci}^2}\right] , \tag{3.417}$$

where $\omega_{ci} = -eB/m_i$, so that finally:

$$\epsilon_{pxx} = 1 - \frac{\omega_{pe}^2}{\omega^2 - \omega_{ce}^2} - \frac{\omega_{pi}^2}{\omega^2 - \omega_{ci}^2} . \tag{3.418}$$

3.6. From the equation for conservation of particles, for a stationary plasma containing only one kind of ions, we have $\boldsymbol{\nabla} \cdot \boldsymbol{\Gamma}_e = \boldsymbol{\nabla} \cdot \boldsymbol{\Gamma}_i$, where $\boldsymbol{\Gamma}_e$ and $\boldsymbol{\Gamma}_i$ denote the flux of electrons and ions respectively. Consider the case of a discharge in the ambipolar regime with no applied magnetic field:

a) Show that:

$$\boldsymbol{\nabla} \wedge \boldsymbol{\Gamma}_e = \mu_e \boldsymbol{\nabla}\phi \wedge \boldsymbol{\nabla} n \tag{3.419}$$

and:

$$\boldsymbol{\nabla} \wedge \boldsymbol{\Gamma}_i = \mu_i \boldsymbol{\nabla}\phi \wedge \boldsymbol{\nabla} n , \tag{3.420}$$

where μ_e and μ_i denote the electron and ion mobilities, n_e and n_i the electron and ion densities ($n_e = n_i = n$) and ϕ is the potential induced by the space charge.

b) Assuming that

$$\boldsymbol{\nabla}\phi \wedge \boldsymbol{\nabla} n = 0 , \tag{3.421}$$

show that the flux difference $\boldsymbol{\Gamma}_e - \boldsymbol{\Gamma}_i$ is independent of position.

c) Consider the particular case in which the particle density satisfies a Boltzmann distribution:

$$n(\boldsymbol{r}) = n_0 \exp\left[\frac{q\phi(\boldsymbol{r})}{k_B T}\right] , \tag{3.422}$$

where the temperature is assumed to be independent of $\boldsymbol{r}$. Show that the assumption in (3.421) is then valid.

Answer

a) For ambipolar diffusion, we know that the electron particle flux can be written as (3.267):

$$\boldsymbol{\Gamma}_e = \mu_e n_e \boldsymbol{E}_D - D_e \boldsymbol{\nabla} n_e \,. \tag{3.423}$$

Taking the curl of this equation, we obtain:

$$\boldsymbol{\nabla} \wedge \boldsymbol{\Gamma}_e = \mu_e \boldsymbol{\nabla} \wedge (n_e \boldsymbol{E}_D) - D_e \boldsymbol{\nabla} \wedge \boldsymbol{\nabla} n_e \,, \tag{3.424}$$

where we have taken account of the assumption, implicit in (3.423), that D_e is independent of position. The term $D_e \boldsymbol{\nabla} \wedge \boldsymbol{\nabla} n$ is, in fact, zero, because the curl of a gradient is always zero (Appendix XX). The term $\boldsymbol{\nabla} \wedge (n_e \boldsymbol{E}_D)$, with $\boldsymbol{E}_D = -\boldsymbol{\nabla}\phi(r)$, can be expanded to:

$$\begin{aligned} \boldsymbol{\nabla} \wedge \boldsymbol{\Gamma}_e &= \mu_e (n_e \boldsymbol{\nabla} \wedge \boldsymbol{E}_D + \boldsymbol{\nabla} n_e \wedge \boldsymbol{E}_D) \\ &= -\mu_e (n_e \boldsymbol{\nabla} \wedge \boldsymbol{\nabla}\phi + \boldsymbol{\nabla} n_e \wedge \boldsymbol{\nabla}\phi) \,. \end{aligned} \tag{3.425}$$

Finally, since the term $\boldsymbol{\nabla} \wedge \boldsymbol{\nabla}\phi$ is also zero, this becomes:

$$\boldsymbol{\nabla} \wedge \boldsymbol{\Gamma}_e = -\mu_e \boldsymbol{\nabla} n_e \wedge \boldsymbol{\nabla}\phi \,, \tag{3.426}$$

and since $n_e = n_i = n$, we obtain:

$$\boldsymbol{\nabla} \wedge \boldsymbol{\Gamma}_e = \mu_e \boldsymbol{\nabla}\phi \wedge \boldsymbol{\nabla} n \,. \tag{3.419}$$

Similarly for the ions, we have:

$$\boldsymbol{\nabla} \wedge \boldsymbol{\Gamma}_i = \mu_i \boldsymbol{\nabla}\phi \wedge \boldsymbol{\nabla} n \,. \tag{3.420}$$

b) Returning to assumption (3.421), Eqs. (3.419) and (3.420) give:

$$\boldsymbol{\nabla} \wedge \boldsymbol{\Gamma}_e = \boldsymbol{\nabla} \wedge \boldsymbol{\Gamma}_i = 0 \tag{3.427}$$

from which:

$$\boldsymbol{\nabla} \wedge (\boldsymbol{\Gamma}_e - \boldsymbol{\Gamma}_i) = 0 \tag{3.428}$$

therefore, $\Gamma_e - \Gamma_i =$ constant (spatially).

c) From (3.422):

$$\boldsymbol{\nabla} n(\boldsymbol{r}) = \boldsymbol{\nabla}\phi \left(\frac{q}{k_B T}\right) n(\boldsymbol{r}) \,, \tag{3.429}$$

so that:

$$\boldsymbol{\nabla}\phi \wedge \boldsymbol{\nabla} n = \left[\left(\frac{q}{k_B T}\right) n(\boldsymbol{r})\right] \boldsymbol{\nabla}\phi \wedge \boldsymbol{\nabla}\phi = 0 \,, \tag{3.430}$$

from the definition of the vector product.

3.7. Consider a cold plasma, represented by an electron fluid, subjected to a periodic electric field $\boldsymbol{E} = \boldsymbol{E}_0 \mathrm{e}^{\mathrm{i}\omega t}$. The momentum equation for the fluid is given by:

$$m_e \frac{\mathrm{d}\boldsymbol{v}}{\mathrm{d}t} = -e\boldsymbol{E} - m_e \nu \boldsymbol{v} \,, \tag{3.431}$$

where m_e is the mass of the electron, e its charge in absolute units and $\boldsymbol{v}$ its mean velocity, and ν is the mean collision frequency for momentum transfer; the quantities $\boldsymbol{v}$ and $\boldsymbol{E}$ are expressed as complex quantities, but $\boldsymbol{E}_0$, the amplitude of the field, is real.

a) Show that the mean velocity of this fluid, and its displacement in the periodic field, are given respectively (within a constant) by:

$$\boldsymbol{v} = -\frac{e\boldsymbol{E}_0}{m_e} \frac{1}{\nu + \mathrm{i}\omega} \mathrm{e}^{\mathrm{i}\omega t} \,, \tag{3.432}$$

$$\boldsymbol{r} = \frac{e\boldsymbol{E}_0}{m_e} \frac{1}{\omega(\omega - \mathrm{i}\nu)} \mathrm{e}^{\mathrm{i}\omega t} \,. \tag{3.433}$$

b) Find the expression for the complex electrical conductivity of this fluid.
c) Show that the real part of (3.432) can be written in the form:

$$\boldsymbol{v}(t) = -\frac{e\boldsymbol{E}_0}{m} \frac{1}{(\nu^2 + \omega^2)^{\frac{1}{2}}} \cos(\omega t + \varphi) \,, \tag{3.434}$$

where:

$$\varphi = \arctan\left(-\frac{\omega}{\nu}\right) \,. \tag{3.435}$$

d) Calculate $\bar{\mathcal{E}}_{\mathrm{kin}}$, the mean kinetic energy over a period of the field $\boldsymbol{E}$.
e) Finally, show that θ_a, the average power absorbed per electron from the field $\boldsymbol{E}$, is given by:

$$\theta_a = \frac{1}{2} \frac{e^2 E_0^2}{m_e} \frac{\nu}{\nu^2 + \omega^2} = 2\nu \bar{\mathcal{E}}_{\mathrm{kin}} \,. \tag{3.436}$$

Answer

a) The cold plasma approximation (Sect. 3.6) assumes that the electrons only move under the influence of the field $\boldsymbol{E} = \boldsymbol{E}_0 \mathrm{e}^{\mathrm{i}\omega t}$, such that:

$$\boldsymbol{v} = \boldsymbol{v}_0 \mathrm{e}^{\mathrm{i}\omega t} \,. \tag{3.437}$$

The solution to (3.431) is trivial, since (3.437) substituted into (3.431) gives:

$$(m_e \boldsymbol{v}_0 \mathrm{i}\omega)\mathrm{e}^{\mathrm{i}\omega t} = -e\boldsymbol{E}_0 \mathrm{e}^{\mathrm{i}\omega t} - m_e \nu \boldsymbol{v} \,, \tag{3.438}$$

i.e.:

$$\boldsymbol{v} = -\frac{e\boldsymbol{E}_0}{m_e}\frac{1}{\nu + \mathrm{i}\omega}\mathrm{e}^{\mathrm{i}\omega t} \,. \tag{3.432}$$

Integrating (3.432), we have, to within a constant:

$$\boldsymbol{r} = -\frac{e\boldsymbol{E}_0}{\mathrm{i}\omega m_e}\frac{1}{\nu + \mathrm{i}\omega}\mathrm{e}^{\mathrm{i}\omega t} = \frac{e\boldsymbol{E}_0}{m_e\omega}\frac{1}{\omega - \mathrm{i}\nu}\mathrm{e}^{\mathrm{i}\omega t} \,. \tag{3.433}$$

b) By definition, the current density is $\boldsymbol{J} = nq\boldsymbol{v} = \sigma\boldsymbol{E}$ where $q = -e$ for electrons. From (3.432), we find directly:

$$\boldsymbol{J} = \frac{ne^2\boldsymbol{E}_0}{m_e}\frac{1}{\nu + \mathrm{i}\omega}\mathrm{e}^{\mathrm{i}\omega t} \,, \tag{3.439}$$

where the expression for the complex conductivity is:

$$\sigma = \frac{ne^2}{m_e}\frac{1}{\nu + \mathrm{i}\omega} \,. \tag{3.440}$$

c) Multiplying the numerator and denominator of (3.432) by $(\nu - \mathrm{i}\omega)$, we obtain:

$$\boldsymbol{v} = -\frac{e\boldsymbol{E}_0}{m_e}\frac{\nu - \mathrm{i}\omega}{\nu^2 + \omega^2}\,\mathrm{e}^{\mathrm{i}\omega t} \,. \tag{3.441}$$

Since a complex number z can be written in the form:

$$z = |z|\mathrm{e}^{\mathrm{i}\varphi} = |z|(\cos\varphi + \mathrm{i}\sin\varphi) \,, \tag{3.442}$$

this suggests that, in order to demonstrate the validity of (3.434), we should write the term $(\nu - \mathrm{i}\omega)/(\nu^2 + \omega^2)^{\frac{1}{2}}$ in the form:

$$\frac{1}{(\nu^2 + \omega^2)^{\frac{1}{2}}}\left[\frac{\nu}{(\nu^2 + \omega^2)^{\frac{1}{2}}} - \frac{\mathrm{i}\omega}{(\nu^2 + \omega^2)^{\frac{1}{2}}}\right] \,, \tag{3.443}$$

where, by identification from (3.442), $\cos\varphi = \nu/(\nu^2 + \omega^2)^{\frac{1}{2}}$ and $\sin\varphi = -\omega/(\nu^2 + \omega^2)^{\frac{1}{2}}$, so that $\boldsymbol{v}$ can be re-written:

$$\boldsymbol{v} = -\frac{e\boldsymbol{E}_0}{m}\frac{1}{(\nu^2 + \omega^2)^{\frac{1}{2}}}\,\mathrm{e}^{\mathrm{i}(\varphi + \omega t)} \,, \tag{3.444}$$

where:

$$\varphi = \arctan\left(-\frac{\omega}{\nu}\right) \,. \tag{3.445}$$

The expression for the real part of $\boldsymbol{v}$ is then:

$$\Re(\boldsymbol{v}) = -\frac{e\boldsymbol{E}_0}{m_e(\nu^2 + \omega^2)^{\frac{1}{2}}}\cos(\varphi + \omega t) \,. \tag{3.446}$$

Remark: Eqs. (3.445) and (3.446) show that, in the absence of collisions ($\nu = 0$), $\boldsymbol{v}$ and $\boldsymbol{E}$ are $\pi/2$ out of phase. This de-phasing tends to zero when ν is large with respect to ω, i.e. $\omega/\nu \to 0$.

d) The mean kinetic energy $\bar{\mathcal{E}}_{\text{kin}}$, in the case where the velocity is varying periodically in time (3.437), can be calculated, using complex algebra, in the form:

$$\bar{\mathcal{E}}_{\text{kin}} \equiv \frac{1}{2} m_e \overline{v^2} = \frac{m_e}{4} \Re(vv^*) = \frac{1}{4} m_e vv^* \,. \tag{3.447}$$

From (3.432), this then becomes:

$$\bar{\mathcal{E}}_{\text{kin}} = \frac{m_e}{4} \left[\frac{e^2 E_0^2}{m_e^2} \frac{1}{\nu^2 + \omega^2} \right] = \frac{1}{4} \frac{e^2 E_0^2}{m_e} \left[\frac{1}{\nu^2 + \omega^2} \right] \,. \tag{3.448}$$

e) Since the instant power is $\boldsymbol{F} \cdot \boldsymbol{v}$, the average power transferred from the field $\boldsymbol{E}$ to an electron is (2.33):

$$\theta_a = \frac{1}{2} \Re(-e\boldsymbol{E} \cdot \boldsymbol{v}^*) = \frac{1}{2} \Re \left[\frac{e^2 E_0^2}{m_e} \frac{1}{\nu - \mathrm{i}\omega} \right] = \frac{1}{2} \frac{e^2 E_0^2}{m_e} \frac{\nu}{\nu^2 + \omega^2} \,, \tag{3.449}$$

and comparing (3.449) with (3.448), we thus have:

$$\theta_a = 2\nu \bar{\mathcal{E}}_{\text{cin}} \,. \tag{3.436}$$

Remark: Equation (3.436) is only valid for $\nu =$ constant with respect to v.

3.8. Consider the conductivity tensor $\underline{\boldsymbol{\sigma}}$ of the plasma electrons. Show, for the case where $\boldsymbol{E}$ is a periodic electric field, that this field transfers an energy per unit volume to the electrons, averaged over a period of oscillation, given by:

$$\bar{W} = \left[\frac{1}{2} \Re(\underline{\boldsymbol{\sigma}}) \cdot \boldsymbol{E}_0 \right] \cdot \boldsymbol{E}_0 \,, \tag{3.450}$$

where $\Re(\underline{\boldsymbol{\sigma}})$ denotes the real part of $\underline{\boldsymbol{\sigma}}$ and $\boldsymbol{E}_0$ is the amplitude of the field.

Answer

The work per unit time and per electron in the field $\boldsymbol{E}$ is $\boldsymbol{F} \cdot \boldsymbol{v}$, where $\boldsymbol{F} = -e\boldsymbol{E}$. The average value of the work over a period of the alternating field $\boldsymbol{E} = \boldsymbol{E}_0 \mathrm{e}^{\mathrm{i}\omega t}$ is (2.33):

$$\theta_a = -\frac{\Re}{2} [e\boldsymbol{E} \cdot \boldsymbol{v}^*] \,. \tag{3.451}$$

The average energy transferred to the electrons per unit volume is then:

$$\bar{W} = -\frac{\Re}{2} [e\boldsymbol{E} \cdot \boldsymbol{v}^*] n_e \,, \tag{3.452}$$

where n_e is the electron density.

Remembering that, by definition:

$$\boldsymbol{J} = -n_e e \boldsymbol{v} \ , \tag{3.453}$$

we can express (3.452) in the form:

$$\bar{W} = \frac{1}{2}\Re[\boldsymbol{E}\cdot\boldsymbol{J}^*] = \frac{1}{4}[\boldsymbol{E}\cdot\boldsymbol{J}^* + \boldsymbol{E}^*\cdot\boldsymbol{J}] \tag{3.454}$$

since the sum of two conjugated complex quantities is real. Then:

$$\begin{aligned}\bar{W} &= \frac{1}{4}[\boldsymbol{E}\cdot\underline{\boldsymbol{\sigma}}^*\cdot\boldsymbol{E}^* + \boldsymbol{E}^*\cdot\underline{\boldsymbol{\sigma}}\cdot\boldsymbol{E}] = \frac{1}{4}[\boldsymbol{E}\cdot\underline{\boldsymbol{\sigma}}^*\cdot\boldsymbol{E}^* + \boldsymbol{E}\cdot\underline{\boldsymbol{\sigma}}\cdot\boldsymbol{E}^*] \\ &= \frac{1}{4}[\boldsymbol{E}\cdot(\underline{\boldsymbol{\sigma}}^* + \underline{\boldsymbol{\sigma}})\cdot\boldsymbol{E}^*] = \frac{1}{4}\left[\boldsymbol{E}\cdot 2\Re(\underline{\boldsymbol{\sigma}})\cdot\boldsymbol{E}^*\right] .\end{aligned} \tag{3.455}$$

Since $\boldsymbol{E} = \boldsymbol{E}_0\mathrm{e}^{\mathrm{i}\omega t}$, $\boldsymbol{E}^* = \boldsymbol{E}_0\mathrm{e}^{-\mathrm{i}\omega t}$, then:

$$\bar{W} = \left[\frac{1}{2}\Re(\underline{\boldsymbol{\sigma}})\cdot\boldsymbol{E}_0\right]\cdot\boldsymbol{E}_0 \ . \tag{3.450}$$

Note in (3.450) that there are two scalar products in a row: starting from a 2nd order tensor, this leads to W, a scalar. This operation is known as a double contraction (see Appendix VII).

3.9. Consider a long, stationary, cylindrical plasma column in the perfect ambipolar diffusion regime, containing positive (singly ionised) and negative ions, with local densities $n_i(\boldsymbol{r})$ and $n_-(\boldsymbol{r})$ respectively, and electrons with density $n_e(\boldsymbol{r})$.

a) Show that the diffusion fluxes of the three charged species are linked by the relation:

$$n_e\boldsymbol{v}_e + n_-\boldsymbol{v}_- = n_i\boldsymbol{v}_i \ . \tag{3.456}$$

b) Show the physical significance of the assumption of proportionality which is, in the present case:

$$\frac{\boldsymbol{\nabla} n_i}{n_i} = \frac{\boldsymbol{\nabla} n_-}{n_-} = \frac{\boldsymbol{\nabla} n_e}{n_e} \ . \tag{3.457}$$

Is there a link between this relationship and particle charge neutrality? Draw the radial profiles for the case of a long plasma column with zero species densities at the wall. This assumes that there is no ion sheath at the wall.

c) Show that the positive ion ambipolar diffusion coefficient is given by:

$$D_{ai} = \frac{n_e\left[D_i\mu_e - D_e\mu_i\right] + n_-\left[D_i\mu_- - D_-\mu_i\right]}{\mu_e n_e + \mu_- n_- - \mu_i n_i} \ , \tag{3.458}$$

where D_k is the free diffusion coefficient and μ_k is the mobility of the various particles ($k = e, i, -$).

Answer

a) In the stationary state, the continuity equation for each type of particle is written ((3.243)–(3.234)):

$$\nabla \cdot (n_e \boldsymbol{v}_e) = S_e \; , \tag{3.459}$$

$$\nabla \cdot (n_i \boldsymbol{v}_i) = S_i \; , \tag{3.460}$$

$$\nabla \cdot (n_- \boldsymbol{v}_-) = S_- \; , \tag{3.461}$$

where the S_k are source terms representing the net number of charged particles of species k ($k = e, i, -$) created per second, per unit volume. The ions and electrons are generally created by electron-neutral collisions with atoms (molecules) and negative ions are created by attachment of an electron to a neutral particle (Sect. 1.7.1). Whatever the mechanisms for creation (and for volume losses, if they occur), the source terms are related such that there is an equal number of negative and positive charges created, i.e. that $S_e + S_- = S_i$, so from (3.459) and (3.461):

$$n_e \boldsymbol{v}_e + n_- \boldsymbol{v}_- = n_i \boldsymbol{v}_i \; . \tag{3.456}$$

b) The assumption of proportionality (Sect. 3.10) in the present case implies:

$$n_i = C_1 n_e \; , \tag{3.462}$$

$$n_i = C_2 n_- \tag{3.463}$$

at each point r, the constants C_j being independent of r. It then follows that:

$$\frac{\nabla n_i}{n_i} = \frac{\nabla n_e}{n_e} = \frac{\nabla n_-}{n_-} \; . \tag{3.457}$$

The radial density profile is the same for the three charged species: in fact, each of the terms $\nabla n_k / n_k$ in (3.457) is equal to the same constant $\mathcal{C}$, and we thus have three equations in the form $\nabla n_k = \mathcal{C} n_k$, with the same eigen value $\mathcal{C}$ (Sect. 3.9).

In a long cylindrical column with zero densities at the wall, the density profiles can be approximated by the following curves:

In addition, macroscopic charge neutrality requires:

$$n_i = n_e + n_- \; , \tag{3.464}$$

and from (3.462), (3.463) and (3.464) we have:

$$n_i = \frac{n_i}{C_1} + \frac{n_i}{C_2} \tag{3.465}$$

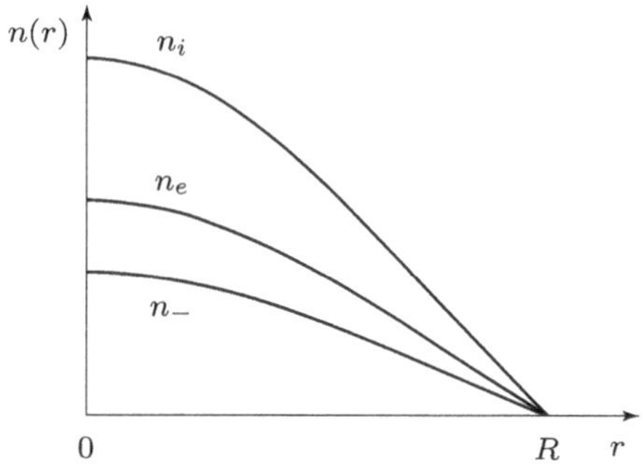

Fig. 3.12 Approximate profile of the density of charged species in the presence of positive and negative ions, assuming $n_k(R) = 0$.

from which:

$$1 = \frac{C_2 + C_1}{C_2 C_1} , \tag{3.466}$$

which shows that the assumption of proportionality is not sufficient to calculate the distribution of negative charges into electrons and negative ions. At each point r, the absolute values of the three densities, which are indeterminate in the present problem, must satisfy macroscopic neutrality.

Remark: Remember that the previous results have been obtained for the case of perfect ambipolar diffusion and the assumption that the density of each species at the wall is zero (no ion sheath). In the case where an ion sheath is established, only the most energetic electrons can escape from the plasma to the walls. In contrast, under these conditions ($k_B T_- = k_B T_i \ll k_B T_e$), the negative ions are confined in the plasma.

c) From (3.456), we know that the positive ion flux must be equal to the total electron and negative ion flux, from which, if the common value is designated $\boldsymbol{\Gamma}$ (3.252):

$$\boldsymbol{\Gamma} \equiv n_i \boldsymbol{v}_i = -D_i \boldsymbol{\nabla} n_i + \mu_i n_i \boldsymbol{E}_D , \tag{3.467}$$

$$\boldsymbol{\Gamma} = -D_e \boldsymbol{\nabla} n_e + \mu_e n_e \boldsymbol{E}_D - D_- \boldsymbol{\nabla} n_- + \mu_- n_- \boldsymbol{E}_D , \tag{3.468}$$

where $\boldsymbol{E}_D$ is the space-charge ambipolar field.
After multiplying (3.467) by $\mu_e n_e + \mu_- n_-$ and (3.468) by $-\mu_i n_i$ and then adding these two expressions, we obtain:

$$\begin{aligned} &\boldsymbol{\Gamma}(\mu_e n_e + \mu_- n_- - \mu_i n_i) = \\ &- D_i \mu_e n_e \boldsymbol{\nabla} n_i - D_i \mu_- n_- \boldsymbol{\nabla} n_i + D_e \mu_i n_i \boldsymbol{\nabla} n_e + D_- \mu_i n_i \boldsymbol{\nabla} n_- . \end{aligned} \tag{3.469}$$

Then, making use of (3.457), we obtain from (3.469):

$$\begin{aligned} &\boldsymbol{\Gamma}(\mu_e n_e + \mu_- n_- - \mu_i n_i) = \\ &\qquad - \{ n_e \left[D_i \mu_e - D_e \mu_i \right] + n_- \left[D_i \mu_- - D_- \mu_i \right] \} \boldsymbol{\nabla} n_i \end{aligned} \tag{3.470}$$

from which, by setting:

$$\boldsymbol{\Gamma} = -D_{ap}\boldsymbol{\nabla} n_i \ , \tag{3.471}$$

we find the expression for D_{ap}:

$$D_{ap} = \frac{n_e[D_i\mu_e - D_e\mu_i] + n_-[D_i\mu_- - D_-\mu_i]}{\mu_e n_e + \mu_- n_- - \mu_i n_i} \ . \tag{3.458}$$

3.10. Consider a plasma composed of electrons, neutral atoms and ions (with a single positive charge). The electrons and ions are in the perfect ambipolar diffusion regime and move, therefore, under the influence of the space charge field as two fluids, each described by the corresponding Langevin equation.

a) In the framework of this description, which assumes that the collision frequencies are independent of the particle velocities, the collisional term for particles of type α and density n_α, is given by (3.134):

$$\frac{\mathcal{P}_\alpha}{n_\alpha} = \sum_{\beta\neq\alpha} \frac{\mathcal{P}_{\alpha\beta}}{n_\alpha} = -\sum_{\beta\neq\alpha} \mu_{\alpha\beta}\nu_{\alpha\beta}(\boldsymbol{v}_\alpha - \boldsymbol{v}_\beta) \ , \tag{3.472}$$

where $\mu_{\alpha\beta}$ is the reduced mass of the particles of species α and β, $\nu_{\alpha\beta}$ is the collision frequency, α, $\beta = e, i, n$, for electrons, ions and neutral atoms). Show that, to a first approximation, this term can be reduced to:

$$\frac{\mathcal{P}_e}{n_e} = -\mu_{en}\nu_{en}\boldsymbol{v}_e \ , \tag{3.473}$$

$$\frac{\mathcal{P}_i}{n_i} = -\mu_{in}\nu_{in}\boldsymbol{v}_i \ , \tag{3.474}$$

for the electron and ion fluids respectively.

b) In the stationary regime, in the absence of a magnetic field, but with a (macroscopic) electric field $\boldsymbol{E}$, show that the existence of spatial gradients in the density n_α and temperature T_α of species α introduces a directed velocity of the particles, that can be written in the form:

$$\boldsymbol{v}_\alpha = \mu_\alpha \boldsymbol{E} - \frac{D_\alpha}{n_\alpha}\boldsymbol{\nabla} n_\alpha - \frac{D_\alpha^T}{T_\alpha}\boldsymbol{\nabla} T_\alpha \ , \tag{3.475}$$

where the particle mobility is $\mu_\alpha = q_\alpha/\mu_{\alpha n}\nu_{\alpha n}$, their diffusion coefficient $D_\alpha = k_B T_\alpha/\mu_{\alpha n}\nu_{\alpha n}$ and their thermal diffusion coefficient$D_\alpha^T = k_B T_\alpha/\mu_{\alpha n}\nu_{\alpha n}$ (the equality $D_\alpha = D_\alpha^T$ assumes that $\nu_{\alpha\beta}$ is independent of the particle velocities), and k_B is Boltzmann's constant.

c) For the case of perfect ambipolar diffusion, calculate the value of the space charge field $\boldsymbol{E}_D$ as a function of the coefficients in (3.475).

Answer

a) For the fluid of electrons, the collision term (3.472) can be expanded to give:

$$\frac{\mathcal{P}_e}{n_e} = -\mu_{en}\nu_{en}(\boldsymbol{v}_e - \boldsymbol{v}_n) - \mu_{ei}\nu_{ei}(\boldsymbol{v}_e - \boldsymbol{v}_i) \,. \tag{3.476}$$

In the perfect ambipolar diffusion regime, ions and electrons diffuse with exactly the same velocity, such that the second term in the RHS of (3.476) is zero. Moreover, the neutral atom fluid is not influenced by the space charge field, and has no directed velocity ($\boldsymbol{v}_n = 0$). It follows that (3.476) actually reduces to:

$$\frac{\mathcal{P}_e}{n_e} = -\mu_{en}\nu_{en}\boldsymbol{v}_e \,. \tag{3.473}$$

For the fluid of ions, the expression for the collision term is:

$$\frac{\mathcal{P}_i}{n_i} = -\mu_{in}\nu_{in}(\boldsymbol{v}_i - \boldsymbol{v}_n) - \mu_{ie}\nu_{ie}(\boldsymbol{v}_i - \boldsymbol{v}_e) \,. \tag{3.477}$$

For reasons already mentioned, $\boldsymbol{v}_e - \boldsymbol{v}_i = 0$, and $\boldsymbol{v}_n = 0$, such that the collisional term for the ion fluid leads to:

$$\frac{\mathcal{P}_i}{n_i} = -\mu_{in}\nu_{in}\boldsymbol{v}_i \,. \tag{3.474}$$

We can then write the collision terms for the electron and ion fluids as a common term for the two fluids:

$$\frac{\mathcal{P}_{\alpha n}}{n_\alpha} = -\mu_{\alpha n}\nu_{\alpha n}\boldsymbol{v}_\alpha \,, \tag{3.478}$$

where $\alpha = e$ or i.

b) The stationary Langevin equation for particles of species α (3.134), taking account of (3.478) and neglecting the convective term (!), can be written:

$$q_\alpha \boldsymbol{E} = \frac{1}{n_\alpha}\boldsymbol{\nabla} p_\alpha + \mu_{\alpha n}\nu_{\alpha n}\boldsymbol{v}_\alpha \,, \tag{3.479}$$

where $p_\alpha = n_\alpha k_B T_\alpha$. Substituting p_α explicitly in (3.479), with n_α and T_α both depending on position, and after rearrangement of the terms, we obtain:

$$\boldsymbol{v}_\alpha = \frac{1}{\mu_{\alpha n}\nu_{\alpha n}}\left[q_\alpha \boldsymbol{E} - k_B \boldsymbol{\nabla} T_\alpha - \frac{k_B T_\alpha}{n_\alpha}\boldsymbol{\nabla} n_\alpha\right] \,. \tag{3.480}$$

In Sect. 3.8, we treated the case of mobility and diffusion of charged particles when $\boldsymbol{\nabla} T_\alpha = 0$. We then found two particular solutions. All that is needed now is to find the particular contribution to be added to this solution (3.210) by the term $\boldsymbol{\nabla} T_\alpha$, which, from (3.480) (with $\boldsymbol{E} = 0$ and

$\nabla n_\alpha = 0$, can be written:

$$\boldsymbol{v}_{\alpha T} = -\frac{k_B T_\alpha}{\mu_{\alpha n}\nu_{\alpha n}}\frac{\nabla T_\alpha}{T_\alpha} = -\frac{D_\alpha^T}{T_\alpha}\nabla T_\alpha \, . \tag{3.481}$$

c) In sum, we then have:

$$\boldsymbol{v}_\alpha = \mu_\alpha \boldsymbol{E} - \frac{D_\alpha}{n_\alpha}\nabla n_\alpha - \frac{D_\alpha^T}{T_\alpha}\nabla T_\alpha \, . \tag{3.475}$$

To calculate the value of the space charge electric field ($\boldsymbol{E} = \boldsymbol{E}_D$), we follow the same procedure as in Sect. 3.8. From (3.475), the respective electron and ion fluxes are given by:

$$\boldsymbol{\Gamma}_e = -D_e \nabla n_e - \frac{D_e^T}{T_e} n_e \nabla T_e + \mu_e n_e \boldsymbol{E}_D \, , \tag{3.482}$$

$$\boldsymbol{\Gamma}_i = -D_i \nabla n_i - \frac{D_i^T}{T_i} n_i \nabla T_i + \mu_i n_i \boldsymbol{E}_D \, . \tag{3.483}$$

The congruence approximation enables us to set $\boldsymbol{\Gamma}_e = \boldsymbol{\Gamma}_i = \boldsymbol{\Gamma}$, and the assumption of proportionality, in the perfect ambipolar regime, leads to $n_e = n_i = n$. Subtracting (3.483) from (3.482) and dividing by n gives:

$$-(D_e - D_i)\frac{\nabla n}{n} - D_e^T\frac{\nabla T_e}{T_e} + D_i^T\frac{\nabla T_i}{T_i} + (\mu_e - \mu_i)\boldsymbol{E}_D = 0 \tag{3.484}$$

such that:

$$\boldsymbol{E}_D = \frac{D_e - D_i}{\mu_e - \mu_i}\frac{\nabla n}{n} + \frac{D_e^T}{\mu_e - \mu_i}\frac{\nabla T_e}{T_e} - \frac{D_i^T}{\mu_e - \mu_i}\frac{\nabla T_i}{T_i} \, . \tag{3.485}$$

3.11. To describe a long, cylindrical plasma column in the ambipolar diffusion regime, we have used the following hydrodynamic equations:

$$\nabla \cdot (n\boldsymbol{v}_r) = \nu_i n \, , \tag{3.486}$$

$$\boldsymbol{v}_r = -\mu_e \boldsymbol{E}_D - \frac{1}{n}\nabla(D_e n) \, , \tag{3.487}$$

$$(\boldsymbol{v}_r \cdot \nabla)\boldsymbol{v}_r + \nu_i \boldsymbol{v}_r = \frac{e}{m_i}\boldsymbol{E}_D - \frac{k_B T_i}{m_i}\frac{\nabla n}{n} - \nu_{in}\boldsymbol{v}_r \, , \tag{3.488}$$

where the indices e, i and n represent the electrons, ions and neutral atoms respectively; $n = n_e = n_i$, $\boldsymbol{v}_r$ is the radially directed velocity, ν_i is the ionisation frequency, μ_e is the absolute value of the electron mobility and D_e is the electron free diffusion coefficient; ν_{in} is the ion-neutral collision frequency for momentum transfer, m_i is the mass of the ions and T_i is their temperature. k_B is Boltzmann's constant and $\boldsymbol{E}_D$ is the space charge electric field associated with the ambipolar diffusion.

a) Indicate what each of these equation represents and which species of particle they refer to.
b) Obtain equation (3.488) by judiciously using the hydrodynamic transport equations.

Answer

a) From the fact that the plasma is cylindrical and assumed to be infinitely long, the system is described in cylindrical coordinates (reduced to $\hat{\mathbf{e}}_r$) and the drift and diffusion phenomena in the axial direction are neglected.
Equation (3.486) is the stationary continuity equation ((3.95) and (3.96)), describing both electrons and ions: indeed, under ambipolar diffusion, there is macroscopic charge neutrality ($n = n_e = n_i$), such that the radial diffusion velocity $\boldsymbol{v}_r$ is the same for the electron and ion fluids, conforming with perfect ambipolar diffusion. The term $\nu_i n$ represents the ionisation frequency per unit volume, for electron-neutral collisions with atoms in the ground state (Sect 1.8).
The first term of the RHS of (3.487) describes the radial drift of the electron fluid in the ambipolar field $\boldsymbol{E}_D$. The second term comprises of the radial diffusion of electrons, due to the gradient in $n(r)$ and the radial transport of energy, because the electron temperature $T_e(r)$ appears in the expression for the coefficient D_e under the gradient operator. Equation (3.487) allows us to calculate $\boldsymbol{v}_r$. Note that here, μ_e is positive, in contrast to our usual convention (compare with (3.210) for the sign before μ_e in (3.487)).
Equation (3.488) is the stationary momentum transport equation for singly charged ions, where the convective term ($(\boldsymbol{v}_r \cdot \boldsymbol{\nabla})\boldsymbol{v}_r$ (3.134)) has been retained. We will explain the reason for the presence of the volume ionisation term in b) below. There are no external fields ($\boldsymbol{E}$ or $\boldsymbol{B}$) to act on the particles, but only the ambipolar field $\boldsymbol{E}_D$.
b) To recover (3.488), return to Sect. 3.5, where the momentum transport equation for particles of species α (3.134) can be written:

$$m_\alpha \frac{\mathrm{d}}{\mathrm{d}t}\boldsymbol{v}_\alpha = q_\alpha[\boldsymbol{E} + \boldsymbol{v}_\alpha \wedge \boldsymbol{B}] - \frac{1}{n_\alpha}\boldsymbol{\nabla} p_\alpha - \sum_{\beta \neq \alpha} \mu_{\alpha\beta}\nu_{\alpha\beta}[\boldsymbol{v}_\alpha - \boldsymbol{v}_\beta] \,. \quad (3.489)$$

In the absence of external fields $\boldsymbol{E}$ and $\boldsymbol{B}$, but including the ambipolar electric field $\boldsymbol{E}_D$, assuming that the ion mass m_i is equal to that of the neutrals, we obtain:

$$m_i \left[\frac{\partial}{\partial t} + \boldsymbol{v}_r \cdot \boldsymbol{\nabla}\right] \boldsymbol{v}_r =$$

$$e\boldsymbol{E}_D - \frac{1}{n}\boldsymbol{\nabla}(nk_BT_i) - m_e\nu_{ie}(\boldsymbol{v}_r - \boldsymbol{v}_r) - \frac{m_i}{2}\nu_{in}(\boldsymbol{v}_r - \boldsymbol{v}_n) \,, \quad (3.490)$$

where, if there is no time dependence (as can be seen from (3.486)), we can set:

$$\frac{\partial \boldsymbol{v}_r}{\partial t} = 0 \, . \tag{3.491}$$

Assuming that T_i is independent of position and that $\boldsymbol{v}_n = 0$ (the fluid of neutrals is not influenced by the field $\boldsymbol{E}_D$, therefore it is stationary), we obtain from (3.490), after dividing by m_i:

$$(\boldsymbol{v}_r \cdot \boldsymbol{\nabla})\, \boldsymbol{v}_r = \frac{e}{m_i}\boldsymbol{E}_D - \frac{k_B T_i}{m_i}\frac{\boldsymbol{\nabla} n}{n} - \frac{\nu_{in}}{2}\boldsymbol{v}_r \, , \tag{3.492}$$

where the term $\nu_i \boldsymbol{v}_r$ in (3.488) is missing.
The absence of this term comes from the fact that, in the derivation of (3.489) in Sect. 3.5, the continuity equation was written in the form:

$$\frac{\partial n}{\partial t} + \boldsymbol{\nabla} \cdot (n\boldsymbol{v}) = 0 \, , \tag{3.493}$$

i.e. without any RHS. In contrast, (3.486), which is the stationary continuity equation, includes a term for volume ionisation, and this is the only term in the RHS because there is no volume recombination.
To obtain (3.488), we must thus return to the 1st order moment equation in $\boldsymbol{w}$ (3.109) which, assuming that $\boldsymbol{\nabla} \cdot \underline{\boldsymbol{\Psi}}$ reduces to $\boldsymbol{\nabla} p$, can be written:

$$m_\alpha \frac{\partial}{\partial t}(n_\alpha \boldsymbol{v}_\alpha) + m_\alpha \boldsymbol{v}_\alpha (\boldsymbol{\nabla} \cdot n_\alpha \boldsymbol{v}_\alpha) + n_\alpha m_\alpha (\boldsymbol{v}_\alpha \cdot \boldsymbol{\nabla})\boldsymbol{v}_\alpha + \boldsymbol{\nabla} p_\alpha - n\boldsymbol{F}_\alpha = - \sum_{\beta \neq \alpha} \nu_{\alpha\beta} n_\alpha \mu_{\alpha\beta} (\boldsymbol{v}_\alpha - \boldsymbol{v}_\beta) \, . \tag{3.494}$$

Applying this equation to ions, assuming $m_i = m_n$, we have $\mu_{in} = m_i/2$, and making the same assumptions as for (3.492), we obtain, after dividing by m_i:

$$\boldsymbol{v}_r(\boldsymbol{\nabla} \cdot n\boldsymbol{v}_r) + n(\boldsymbol{v}_r \cdot \boldsymbol{\nabla})\boldsymbol{v}_r + \frac{k_B T_i}{m_i}\boldsymbol{\nabla} n - n\frac{e}{m_i}\boldsymbol{E}_D = -\frac{\nu_{in}}{2}\boldsymbol{v}_r n \, . \tag{3.495}$$

Then, this time taking account of the continuity equation (3.486), we obtain from (3.495), after dividing by n:

$$\nu_i \boldsymbol{v}_r + \boldsymbol{v}_r \cdot \boldsymbol{\nabla} \boldsymbol{v}_r = \frac{e}{m_i}\boldsymbol{E}_D - \frac{k_B T_i}{m_i}\frac{\boldsymbol{\nabla} n}{n} - \frac{\nu_{in}}{2}\boldsymbol{v}_r \, , \tag{3.488'}$$

which is the same as (3.488), except for the factor $1/2$, associated with ν_{in}.

Remark: The difference in the factor associated with ν_{in} comes from the fact that the value of this frequency in (3.488) is not defined by (3.124), but rather by:

$$\nu^*_{\alpha\beta} = n_\beta w_{\alpha\beta} \frac{m_\beta}{m_\alpha + m_\beta} \int_0^\pi 2\pi\hat{\sigma}(\theta)(1 - \cos\theta)\sin\theta \, \mathrm{d}\theta \ , \tag{3.496}$$

the *effective collision frequency*, which allows us to write the collisional term $\boldsymbol{\mathcal{P}}_{\alpha\beta}$ in the form:

$$\boldsymbol{\mathcal{P}}_{\alpha\beta} = m_\alpha \int_{\boldsymbol{w}_\alpha} \int_{\boldsymbol{w}_\beta} (\boldsymbol{w}_\beta - \boldsymbol{w}_\alpha) \frac{\nu^*_{\alpha\beta}}{n_\beta} f_\alpha(\boldsymbol{w}_\alpha) f_\beta(\boldsymbol{w}_\beta) \, \mathrm{d}\boldsymbol{w}_\alpha \mathrm{d}\boldsymbol{w}_\beta \ , \tag{3.123}$$

or alternatively (making the same assumptions as for (3.124)):

$$\boldsymbol{\mathcal{P}}_{\alpha\beta} = -m_\alpha n_\alpha \nu^*_{\alpha\beta} (\boldsymbol{v}_\alpha - \boldsymbol{v}_\beta) \ . \tag{3.497}$$

Within this formalism, the collision term in (3.488) may effectively be written:

$$\boldsymbol{\mathcal{P}}_{in} = -m_i n_i \nu^*_{in} \boldsymbol{v}_i \ . \tag{3.498}$$

where the mass of the neutrals does not appear explicitly.

3.12. Using the spherical harmonics expansion in velocity space for the unseparated distribution function $f(\boldsymbol{r}, \boldsymbol{w})$, show that the electron diffusion coefficient D_e and the mobility μ_e, in a continuous current discharge, are given by:

$$D_e = \frac{1}{n_e(\boldsymbol{r})} \int_0^\infty \frac{w^2}{3\nu(w)} f_0(\boldsymbol{r}, w) \, 4\pi w^2 \mathrm{d}w \ , \tag{3.499}$$

$$\mu_e = \frac{1}{n_e(\boldsymbol{r})} \int_0^\infty \frac{ew}{3m_e \nu(w)} \frac{\partial f_0(\boldsymbol{r}, w)}{\partial w} \, 4\pi w^2 \mathrm{d}w \ , \tag{3.500}$$

where $n_e(\boldsymbol{r})$ is the electron density, $\boldsymbol{w}$ the velocity of individual electrons, $\nu(w)$ the microscopic electron-neutral collision frequency for momentum transfer, $f_0(\boldsymbol{r}, w)$ the zero order (isotropic) term of the distribution function in spherical harmonics, e and m_e, the charge and mass of the electron. Describe the approximations used in developing this calculation. Limit the expansion to second order.

Answer

1. The general expression for D_e, the electron diffusion coefficient, is (3.208):

$$D_e = \frac{1}{3} \left\langle \frac{w^2}{\nu(w)} \right\rangle \ , \tag{3.501}$$

where the brackets $\langle\,\rangle$ represent the average over the distribution function in velocity space. Remembering that the mean value of a molecular variable represented by the function $\Upsilon(\boldsymbol{r},\boldsymbol{w},t)$ is defined, in the hydrodynamic description, by:

$$\langle \Upsilon(\boldsymbol{r},t)\rangle = \frac{1}{n(\boldsymbol{r},t)} \int\limits_{\boldsymbol{w}} \Upsilon(\boldsymbol{r},\boldsymbol{w},t) f(\boldsymbol{r},\boldsymbol{w},t)\,\mathrm{d}\boldsymbol{w} \tag{3.39}$$

in which we use the complete function $f(\boldsymbol{r},\boldsymbol{w},t)$ (rather than the separated function $f(\boldsymbol{w},t)$). In the present case, D_e which is an average parameter, is given by:

$$D_e = \frac{1}{3n_e(\boldsymbol{r})} \int\limits_{\boldsymbol{w}} \frac{w^2}{\nu(w)} f(\boldsymbol{r},\boldsymbol{w})\,\mathrm{d}\boldsymbol{w}\ , \tag{3.502}$$

where $w^2 \equiv \boldsymbol{w}\cdot\boldsymbol{w}$ is a scalar.
We use an expansion in spherical harmonics, in a spherical coordinate system in velocity space, with the z axis along the direction of anisotropy resulting either from the diffusion gradient of the particles or from the field $\boldsymbol{E}$ inducing the particle drift. We set $\mathrm{d}\boldsymbol{w} = 2\pi w^2 \sin\theta\,\mathrm{d}\theta \mathrm{d}w$, which implies that we have already integrated over the angle φ (assuming of isotropy about the z axis). Limiting the expansion to second order, the expression for $f(\boldsymbol{r},\boldsymbol{w})$ is:

$$f(\boldsymbol{r},\boldsymbol{w}) = f_0(\boldsymbol{r},w) + f_1(\boldsymbol{r},w)\cos\theta + f_2(\boldsymbol{r},w)\frac{3\cos^2\theta - 1}{2}\ . \tag{3.503}$$

Substituting (3.503) in (3.502) gives:

$$D_e = \frac{1}{3n_e(\boldsymbol{r})} \int\limits_0^\infty \frac{w^2}{\nu(w)} \Bigg[f_0(\boldsymbol{r},w) + f_1(\boldsymbol{r},w)\cos\theta + f_2(\boldsymbol{r},w)\frac{3\cos^2\theta - 1}{2} \Bigg] 2\pi w^2 \sin\theta\,\mathrm{d}\theta \mathrm{d}w\ . \tag{3.504}$$

Of the different terms in the expansion of $f(\boldsymbol{r},\boldsymbol{w})$, only that containing the function f_0 contributes to the integral for D_e, because the 1st order term in θ vanishes since:

$$-\int\limits_0^\pi \cos\theta\,\mathrm{d}(\cos\theta) = -\frac{1}{2}\sin^2\theta\Big|_0^\pi = 0\ , \tag{3.505}$$

and similarly for the 2nd order terms, because:

$$-\int\limits_0^\pi \frac{3\cos^2\theta - 1}{2}\,\mathrm{d}(\cos\theta) = \frac{1}{2}\left[\cos^3\theta - \cos\theta\right]_0^\pi = 0\ , \tag{3.506}$$

from which, finally:

$$D_e = \frac{1}{n(\boldsymbol{r})} \int_0^\infty \frac{w^2}{3\nu(w)} f_0(\boldsymbol{r}, w)\, 4\pi w^2 \,\mathrm{d}w \,. \tag{3.499}$$

2. The mobility is related to the electrical conductivity σ by (3.190):

$$\sigma = nq\mu \,, \tag{3.507}$$

where $q = -e$ for electrons.
We have already calculated σ for electrons in a HF electric field, but for the case of a separable distribution function $f(\boldsymbol{r}, \boldsymbol{w})$, i.e. one which can be written as the product of a function in $\boldsymbol{r}$ and a function in $\boldsymbol{w}$ (3.43):

$$f(\boldsymbol{r}, \boldsymbol{w}) = n(\boldsymbol{r}) f(\boldsymbol{w}) \,. \tag{3.508}$$

We then obtained (Sect 3.4)[141]:

$$\sigma_e = -\frac{4\pi n_e e^2}{3m_e} \int_0^\infty \frac{1}{\nu(w) + \mathrm{i}\omega} \frac{\partial f_0(w)}{\partial w} w^3 \,\mathrm{d}w \,. \tag{3.63}$$

To obtain the expression for σ_e that depends on the complete distribution function $f(\boldsymbol{r}, \boldsymbol{w})$, we use the inverse of (3.508) to replace $f_0(w)$ by $f(\boldsymbol{r}, w)$. Then, setting $\omega = 0$ for the case of a continuous current discharge, we find:

$$\sigma_e = -\frac{4\pi e^2}{3m_e} \int_0^\infty \frac{w^3}{\nu(w)} \frac{\partial f_0(\boldsymbol{r}, w)}{\partial w} \,\mathrm{d}w \,. \tag{3.509}$$

To obtain μ_e from σ_e, from (3.507), we only need to divide (3.509) by $-n_e(\boldsymbol{r})e$, i.e.:

$$\mu_e = \frac{4\pi e}{3m_e n_e(\boldsymbol{r})} \int_0^\infty \frac{w^3}{\nu(w)} \frac{\partial f_0(\boldsymbol{r}, w)}{\partial w} \,\mathrm{d}w \,, \tag{3.510}$$

such that, after rearrangement:

$$\mu_e = \frac{1}{n_e(\boldsymbol{r})} \int_0^\infty \frac{we}{3m_e \nu(w)} \frac{\partial f_0(\boldsymbol{r}, w)}{\partial w} 4\pi w^2 \,\mathrm{d}w \,. \tag{3.500}$$

Remark: The electrical conductivity, for either electrons or ions, is always positive. The fact that (3.509) is preceded by a minus sign comes from the

[141] $f_0(\boldsymbol{w})$ is the zeroth order (isotropic) term in the spherical harmonics expansion of the separated function $f(\boldsymbol{w})$.

fact that the mobility μ_e defined by (3.510) is negative, in accord with our usual convention.

3.13. Consider a long column of helium plasma, with radius $R = 1\,\mathrm{cm}$ and pressure $p = 0.4\,\mathrm{torr}$. The electron temperature T_e is $1\,\mathrm{eV}$, while the temperature of the gas, T_g, and that of the ions, T_i, are both about $300\,\mathrm{K}$. The plasma is weakly ionised and, for these values of T_e and p, the mean electron-neutral collision frequency ν_{en} is $10^9\,\mathrm{s}^{-1}$ and the reduced ion mobility μ_{i0}, i.e. at 760 torr and $T_g = 273\,\mathrm{K}$, is $10.4\,\mathrm{cm}^2\,\mathrm{V}^{-1}\,\mathrm{s}^{-1}$. The plasma density on axis is 10^{16} particles m^{-3}. At $t = 0$, the source driving the discharge is cut off.

a) Derive the temporal development of the plasma for $t > 0$, assuming that T_e is constant, and estimate the characteristic time τ_D for the decay of the plasma density.
b) Extend your calculations to the case where a stationary axial magnetic field of 0.1 T is applied.
c) Continue with the same data and operating conditions as in a), except for a pressure of 0.1 mtorr, assuming that the collision frequency ν is simply proportional to pressure.

Answer

a) In Sect. 3.9, we examined the mechanism for the decay of the post-discharge. If this occurs in the diffusion regime, the decay of the density is exponential, and it is eventually controlled by the fundamental diffusion mode, which for a long plasma column, corresponds to the characteristic diffusion length $\Lambda = R/2.405$. The characteristic exponential decay time τ is related to Λ and the diffusion coefficient D (3.224). Therefore, we only need to determine whether the coefficient D is that of free or ambipolar diffusion. Note that when the mean free path is small compared to Λ, the volume recombination regime is equally possible but, since the plasma density is relatively small, we will see that this possibility has to be eliminated in the present case. On the other hand, if the mean free path of electrons, ℓ, is larger than Λ, the decay will be that of a plasma in free fall, a situation not considered in Sect. 3.9 and that needs to be verified.
The plasma at 0.4 torr is not in the free fall regime. Recalling that $\ell = v_{th}/\nu$ (1.134) and, numerically for the present case, $v_{th}/\nu = 5.9 \times 10^5\,\mathrm{m\,s}^{-1}$ (for 1 eV)$/10^9\,\mathrm{s}^{-1}$, we find $\ell = 6 \times 10^{-4}\,\mathrm{m}$, which is much smaller than R ($10^{-2}\,\mathrm{m}$), and therefore the free fall regime is excluded.
The plasma is actually in the ambipolar diffusion regime by verifying that the criterion $n_{e0}\Lambda^2 \geq 10^7\,\mathrm{cm}^{-1}$ is satisfied. From the input parameters of the problem, $n_{e0} = 10^{10}\,\mathrm{cm}^{-3}$, $\Lambda = R/2.405$ such that $n_{e0}\Lambda^2 = 1.7 \times 10^9$. The diffusion remains in the ambipolar regime, even after the density has decreased to 1/e (37%) of the initial density.

We will now calculate the characteristic decay time of the plasma, τ_D, which appears in the expression for the decay:

$$n(\boldsymbol{r},t) = n(\boldsymbol{r},0)\exp-(\nu_{D1}t) \tag{3.222}$$

with $\tau_D \equiv \nu_{D1}^{-1}$, where ν_{D1} is the diffusion-loss frequency in the fundamental mode:

$$\nu_{D1} = \frac{D_a}{\Lambda^2}\,. \tag{3.511}$$

We must now calculate D_a: since $T_e \gg T_i$, we will use the simplified expression:

$$D_a \simeq \frac{k_B T_e}{e}\mu_i\,, \tag{3.281}$$

and, numerically:

$$\mu_i = \frac{760\,(\text{torr})}{p\,(\text{torr})}\frac{T_g\ (\text{K})}{273\ (\text{K})}\mu_{i0} = 2.2\,\text{m}^2\,\text{V}^{-1}\text{s}^{-1}\,, \tag{3.512}$$

$$\begin{aligned} D_a &= \frac{1.38\times10^{-23}(\text{m}^2\,\text{kg}\,\text{s}^{-2}\,\text{K}^{-1})\times 11600\ (\text{K})\times 2.2\ (\text{C kg}^{-1}\,\text{s})}{1.6\times10^{-19}\ (\text{C})} \\ &= 2.2\,\text{m}^2\,\text{s}^{-1}\,, \end{aligned} \tag{3.513}$$

such that (3.511) becomes:

$$\tau_D = \frac{\Lambda^2}{D_a} = \left(\frac{10^{-2}}{2.405}\right)^2(\text{m}^2)\,\frac{1}{2.2(\text{m}^2\,\text{s}^{-1})} = 7.9\,\mu\text{s}\,. \tag{3.514}$$

b) We are still in the ambipolar diffusion regime. This time, we need to calculate $D_{a\perp}$:

$$D_{a\perp} = \frac{D_{e\perp}\mu_{i\perp} - D_{i\perp}\mu_{e\perp}}{\mu_{i\perp} - \mu_{e\perp}}\,. \tag{3.285}$$

We have $\omega_{ce} = 2\pi f_{ce}$, where $f_{ce}(\text{Hz}) = 2.8\times10^6 B_0\,(\text{gauss}) = 2.8\,\text{GHz}$, $f_{ci} = \omega_{ce}m_e/m_i = 3.8\times10^5\,\text{Hz} = 0.38\,\text{MHz}$, from which $\omega_{ce} = 1.76\times 10^{10}\,\text{s}^{-1}$ and $\omega_{ci} = 2.4\times10^6\,\text{s}^{-1}$, so, taking (3.186) and (3.200) into account:

$$\begin{aligned} D_{e\perp} &\simeq \frac{k_B T_e}{m_e\nu}\left[\frac{\nu^2}{\omega_{ce}^2}\right] \\ &= \frac{1.38\times10^{-23}\times 11600\times10^9}{9.11\times10^{-31}\times(1.76\times10^{10})^2} = 0.58\,\text{m}^2\,\text{s}^{-1}\,, \end{aligned} \tag{3.515}$$

$$\begin{aligned} D_{i\perp} &\simeq \frac{k_B T_i}{e}\mu_i \\ &= \frac{1.38\times10^{-23}\times 300\times 2.2}{1.6\times10^{-19}} = 5.7\times10^{-2}\,\text{m}^2\,\text{s}^{-1}\,, \end{aligned} \tag{3.516}$$

$$\mu_{e\perp} \simeq -\frac{e}{m_e \nu} \frac{\nu^2}{\omega_{ce}^2} \tag{3.517}$$

$$= \frac{-1.6 \times 10^{-19} \times 10^9}{9.11 \times 10^{-31} \times (1.76 \times 10^{10})^2} \simeq -0.57\,\mathrm{m^2\,V^{-1}\,s^{-1}}\,,$$

$$\mu_{i\perp} = \mu_i = 2.2\,\mathrm{m^2\,V^{-1}\,s^{-1}}\,, \tag{3.518}$$

from which:

$$D_{a\perp} = \frac{0.58 \times 2.2 + 5.7 \times 10^{-2} \times 0.57}{2.2 + 0.57} \tag{3.519}$$

$$\simeq \frac{1.28 + 0.03}{2.77} \simeq 0.47\,\mathrm{m^2\,s^{-1}}\,.$$

We can conclude that the inclusion of the magnetic field results in a value of $D_{a\perp}$, smaller than D_a by a factor 4. This result is consistent with the fact that $\omega_{ce} > \nu$.

c) The particles are in the free fall regime at 0.1 mtorr. To show this, we must recalculate the mean free path ℓ for electrons by determining the value for ν at 0.1 mtorr. Since:

$$\nu = N\langle\hat{\sigma}(w)w\rangle, \tag{1.132}$$

where N is proportional to the pressure, the collision frequency will be reduced by the same factor as the pressure, i.e. a factor of 4000. We then obtain $\nu(0.1\,\mathrm{mtorr}) = 2.5 \times 10^5\,\mathrm{s^{-1}}$, and $\ell = 6 \times 10^{-4}\,\mathrm{m} \times 4000 = 2.4\,\mathrm{m}$, i.e. indeed $\ell \gg \Lambda$.

3.14. Consider an argon plasma with an electron density $10^{18}\,\mathrm{m^{-3}}$, an electron temperature 10 eV, a temperature of the ions and neutrals of 300 K and a 10^{-4} torr gas pressure. The cylindrical plasma column (along the z axis) is contained in a machine limited at both ends by magnetic mirrors.

The plasma (already created by other means) is to be heated by applying electron cyclotron resonance (ECR) in the uniform region of the magnetic field. To do this, an electromagnetic field with frequency $\omega/2\pi$ is applied, such that $\omega/2\pi = 10 f_{pe}$, where f_{pe} is the electron plasma frequency. Assume that the individual trajectory description can be applied to the electron fluid.

a) Calculate the magnetic field intensity required to enable plasma heating by ECR in the region where the field is uniform.
b) Calculate the initial Larmor radius of electrons in the same region, before the application of the HF field.
c) Calculate the mirror ratio $\mathcal{R}$ for which the loss of particles is less than 20%, assuming that the velocity distribution is initially isotropic.
d) Draw an approximate schematic of $B(z)$, taking c) into account.
e) A non-linear effect (parametric instability), due to the action of the HF electric field, can occur if x_E, the amplitude of oscillation of the electrons in

the HF field (also called the *excursion parameter*), exceeds one tenth of the electron Debye length. Calculate the threshold of the electric field intensity for this non linearity to occur (neglect collisions in this calculation).

f) Make an approximate calculation of the initial electron-neutral collision frequency in the plasma. Does this type of collision hinder the cyclotron heating?

Answer

a) From (1.27), $f_{pe}(\text{Hz}) \simeq 9000\sqrt{n_e(\text{cm}^{-3})}$, so that, in the present case, $f_{pe} = 9000 \times 10^6\,\text{Hz} = 9\,\text{GHz}$, from which $f_{ce} = 90\,\text{GHz}$. From (2.69), $f_{ce}(\text{Hz}) = 2.8 \times 10^6 B(\text{gauss})$, so that $B = 9 \times 10^{10}/2.8 \times 10^6 = 32.1\,\text{kG}$ (3.2 tesla).

b) Recall that $r_B = v_{\perp 0}/\omega_c$ (see (2.66)). In the present case $T_{eV} = k_B T_e/e = 10\,\text{eV}$, and since the velocity $v_\perp$ of the electron fluid is the same along both axes perpendicular to $\boldsymbol{B}$, we have:

$$\frac{1}{2} m_e v_{\perp 0}^2 = k_B T_e \,. \tag{3.520}$$

We obtain:

$$v_\perp = \sqrt{\frac{2 \times 10 \times e}{m_e}} = 1.87 \times 10^6\,\text{m}\,\text{s}^{-1} \,. \tag{3.521}$$

Finally:

$$r_B = \frac{v_{\perp 0}}{\omega_{ce}} = \frac{1.87 \times 10^6}{9 \times 10^{10} \times 2\pi} = 3.3 \times 10^{-3}\,\text{mm} \,. \tag{3.522}$$

c) The reflection coefficient of a magnetic mirror is given by (2.196):

$$C_r = 1 - \mathcal{R}^{-1} \,. \tag{3.523}$$

Since we require $C_r \geq 0.8$:

$$\mathcal{R} = \frac{1}{0.2} = 5 \,. \tag{3.524}$$

Knowing that:

$$\mathcal{R} \equiv \frac{B_{\max}}{B_0} \,, \tag{2.188}$$

we need to have $B_{max} = 5 \times 32(\text{kG}) = 160\,\text{kG}$ (16 T); superconducting coils will be required.

d)

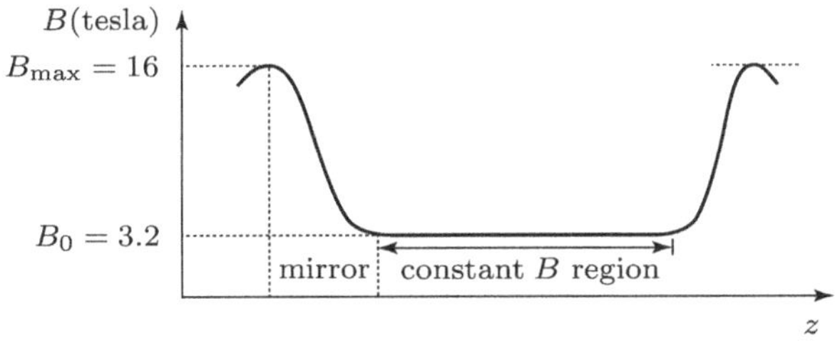

Fig. 3.13 Approximate distribution of the magnetic field intensity for confinement in a linear machine (Sect. 2.2.3).

e) The excursion parameter, in the absence of collisions ($\omega \gg \nu$), is given by (2.31):

$$x_E = \frac{|e|E_0}{m_e\omega^2} \tag{3.525}$$

and the Debye length, from (1.55), is:

$$\begin{aligned}\lambda_{De}(\text{cm}) &= 740\left[\frac{T(\text{eV})}{n_e(\text{cm}^{-3})}\right]^{\frac{1}{2}} = 740\left[\frac{10}{10^{12}}\right]^{\frac{1}{2}} \\ &= 2.3\times 10^{-3}\,\text{cm} = 2.3\times 10^{-5}\,\text{m}\,.\end{aligned} \tag{3.526}$$

Since the threshold for the appearance of instabilities is assumed to be:

$$x_E \geq 0.1\lambda_{De}\,, \tag{3.527}$$

the threshold intensity of the HF electric field (in V/m) is found to be:

$$E_0 \geq \frac{m_e\omega^2(2.3\times 10^{-5})\times 0.1}{e} = 4.2\,\text{MV/m}\,. \tag{3.528}$$

f) The (microscopic) electron-neutral collision frequency for momentum transfer is given by (1.132):

$$\nu_x \simeq N_0\hat{\sigma}_{tx}(10\,\text{eV})w = p_0P_{x0}w\,. \tag{3.529}$$

We will use this approximation, instead of calculating the mean value $\langle\nu\rangle$. From Fig. 1.14, at 10 eV, $P_{x0} = 53\,\text{cm}^{-1}$. The value of the reduced pressure p_0 is:

$$p_0 = \frac{p(\text{torr})273}{273+T_C} = \frac{10^{-4}\,273}{300} = 9.1\times 10^{-5}\ (\text{no units})\,. \tag{3.530}$$

We will take v_{th}, which defines the plasma energy as $k_B T_e$, as the value for v (Sect. 1.4.3). In the present case, $v_{th} = v_\perp$ (3.521) and:

$$\nu = 9.1 \times 10^{-5} \times 53 \times 1.87 \times 10^8 = 0.64 \times 10^6 \,\mathrm{s}^{-1} \,. \tag{3.531}$$

Given that $\omega = 90 \times 10^9 \times 2\pi$ (Hz), we obtain:

$$\frac{\nu}{\omega} = 1.1 \times 10^{-6} \,. \tag{3.532}$$

A large number of periods of the HF field will occur before there is a collision, so they will not significantly hinder the ECR heating.

3.15. Consider a continuous-current electric discharge in helium at a pressure of 1 torr. The plasma parameters of the (long) positive column, with a discharge current density of 200 mA/cm^2, are the following:

- temperatures: $T_{eV} = 2\,\mathrm{eV}$, $T_i = T_n = 300\,\mathrm{K}$,
- densities (uniform): $n_n = 3.2 \times 10^{16}\,\mathrm{cm}^{-3}$, $n_e = n_i = 10^{10}\,\mathrm{cm}^{-3}$,

where the indices n, e and i denote the neutral atoms, electrons and ions respectively. The internal radius R of the discharge tube is 12 cm.

The total microscopic cross-sections for momentum transfer by collisions between electrons and neutrals and between ions and neutrals, under the present conditions, have average values respectively given by:

$$\langle \hat{\sigma}_{en}(w) \rangle (T_{eV} = 2\,\mathrm{eV}, T_n = 300K) = 5 \times 10^{-16}\,\mathrm{cm}^2 \,,$$
$$\langle \hat{\sigma}_{in}(w) \rangle (T_i = T_n = 300\,\mathrm{K}) = 1 \times 10^{-14}\,\mathrm{cm}^2 \,.$$

a) Calculate the electric field intensity along the positive column.
b) Calculate the value of the diffusion coefficient.

Explain your reasoning and clearly indicate the underlying assumptions. It is unnecessary to develop the equations shown in the book, but you must justify their use.
Remark: Two significant figures are adequate for the present calculations.

Answer

a) The electric field $\boldsymbol{E}$ in a discharge is related to the total current density $\boldsymbol{J}$ through the electrical conductivity $\underline{\boldsymbol{\sigma}}$, which, in the absence of a magnetic field imposed on the discharge, is a scalar: we then have the well-known expression (2.38):

$$\boldsymbol{J} = \sigma \boldsymbol{E} \,. \tag{3.533}$$

Since $T_e \gg T_n = T_i$ and the degree of ionisation is small and given by:

$$\frac{n_e}{(n_n + n_i)} \simeq 2 \times 10^{-7} , \tag{3.534}$$

we can apply the "Lorentz plasma" model (Sect. 3.7), i.e. a single fluid, that of electrons, can be used to describe all the plasma properties. For these conditions, we only need to calculate the (real part of the) electron conductivity (2.39):

$$\sigma = \frac{n_e e^2}{m_e \nu} \tag{3.535}$$

to determine $|E|$.
In order to calculate (3.521), it is necessary to evaluate ν. We can do this by setting:

$$\nu \simeq n_n \langle \hat{\sigma}_{en} \rangle \langle w \rangle , \tag{3.536}$$

where $\langle w \rangle$ is the mean electron velocity. Equation (3.536) is then an approximation for the exact relation (1.140):

$$\nu = n_n \langle \hat{\sigma}_{en}(w) w \rangle , \tag{3.537}$$

where the brackets $\langle \ \rangle$ denote the average over the velocity distribution function. Since we are using temperatures to characterise the energy of the particles, this implies a Maxwell-Boltzmann velocity distribution.
The value of $\langle w \rangle$ is then:

$$\langle w \rangle = \sqrt{\frac{8}{\pi} \frac{k_B T_e}{m_e}} , \tag{I.9}$$

where k_B is Boltzmann's constant, and m_e is the electron mass; $\langle w \rangle = 1.13 v_{th}$ where v_{th} is the most probable value of the velocity for a Maxwell-Boltzmann velocity distribution (Appendix I):
We know from (1.127) that, numerically:

$$v_{th}(2\,\mathrm{eV}) = \sqrt{2} \times 5.93 \times 10^5\,\mathrm{m\,s^{-1}} , \tag{3.538}$$

from which:

$$\langle w \rangle \equiv 1.13 v_{th} = 9.5 \times 10^5\,\mathrm{m\,s^{-1}} . \tag{3.539}$$

Then from (3.536), expressing everything in cm:

$$\nu \simeq 3.2 \times 10^{16} (\mathrm{cm^{-3}}) \times 5 \times 10^{-16} (\mathrm{cm^2}) \times 9.5 \times 10^7 (\mathrm{cm\,s^{-1}}) = 1.5 \times 10^9\,\mathrm{s^{-1}} . \tag{3.540}$$

Finally, from (3.533) and (3.535), we obtain (this time in MKS units):

$$E = \frac{J}{\sigma} = \frac{200 \times 10^{-3} \times 10^4\,(\mathrm{A/m^2}) \times 9.1 \times 10^{-31}\,(\mathrm{kg}) \times 1.5 \times 10^9\,(\mathrm{s^{-1}})}{10^{16}\,(\mathrm{m^{-3}}) \times (1.6 \times 10^{-19})^2\,(\mathrm{C^2})}$$
$$= 11\,\mathrm{kV\,m^{-1}} . \tag{3.541}$$

Remarks:

1. Units: the volt $\mathrm{V} = \mathrm{m^2 kg\, s^{-3} A^{-1}}$ and the coulomb $\mathrm{C} = \mathrm{A\ s}$, hence from (3.541):
$$\frac{\mathrm{A\ kg\ m^3}}{\mathrm{m^2\, s\, C^2}} = \frac{(\mathrm{m^2 kg\ s^{-3} A^{-1}})\mathrm{A^2 s^2}}{\mathrm{m\ A^2\ s^2}} = \mathrm{V\, m^{-1}}\ ,$$
which is the usual unit for an electric field.
2. We can verify that the ion conductivity is effectively very much smaller than the electron conductivity ($\sigma_i \ll \sigma_e$) because, even if ν_{in} is much smaller than ν in (3.535) (compare (3.540) and (3.548)), $m_e \ll m_{\mathrm{He}}$.

b) *Diffusion regime*

We need to verify that the charged particle losses are governed by diffusion, i.e. that the plasma is sufficiently collisional that it is not in the free fall regime.
The mean free path ℓ is given by:

$$\ell = \frac{\langle w \rangle}{\nu} \tag{1.141}$$

because the electron-neutral collisions dominate (compare (3.540) and (3.548)) and from (3.538) and (3.540):

$$\ell(\mathrm{cm}) \simeq \frac{9.5 \times 10^7\ \mathrm{cm/s}}{1.5 \times 10^9\ \mathrm{s^{-1}}} = 0.06\, \mathrm{cm}\ , \tag{3.542}$$

which demonstrates that $\ell \ll R$: thus the plasma is indeed in the diffusion regime.

Ambipolar diffusion

To determine whether the plasma is in the free or ambipolar regime, we need to examine the product $n_e(0)\Lambda^2$ (3.276): if it is greater than $10^7\ \mathrm{cm^{-1}}$, the diffusion is ambipolar. We find:

$$n_e(0)\Lambda^2 = 10^{10}\ \mathrm{cm^{-3}} \left(\frac{12\, \mathrm{cm}}{2.405}\right)^2 \gg 10^7\ \mathrm{cm^{-1}}\ , \tag{3.543}$$

confirming that the diffusion is ambipolar.
Remark: We could use another equivalent criterion, $\lambda_{De} \ll \Lambda$, to verify that the diffusion is ambipolar. Evaluating (1.55) numerically, we obtain:

$$\lambda_{De} = 740 \left[\frac{T_e(\text{eV})}{n_e(\text{cm}^{-3})} \right]^{\frac{1}{2}} \simeq 0.01\,\text{cm}\ , \tag{3.544}$$

such that $\lambda_{De} \ll \Lambda$, and the diffusion is ambipolar.

Calculation of the value of the coefficient D_a

Since $T_e \gg T_i$, we can use the approximate expression (3.281):

$$D_a \simeq \frac{k_B T_e}{m_i \nu_{in}}\ . \tag{3.545}$$

To evaluate ν_{in}, we can set, as for ν in (3.536):

$$\nu_{in} \simeq n_n \langle \hat{\sigma}_{in} \rangle \langle w_{in} \rangle\ . \tag{3.546}$$

From (1.12) modified to describe the ions, we have:

$$\langle w_i \rangle = \sqrt{\frac{8 \times 1.38 \times 10^{-23} \times 300}{4\pi\ \times 1.7 \times 10^{-27}}} = 1.2 \times 10^3\,\text{m}\,\text{s}^{-1} \tag{3.547}$$

and, expressing everything in cm:

$$\nu_{in} = 3.2 \times 10^{16} \times 1 \times 10^{-14} \times 1.2 \times 10^5 = 3.8 \times 10^7\,\text{s}^{-1}\ , \tag{3.548}$$

so that, from (3.545):

$$D_a \simeq \frac{1.38 \times 10^{-23} \times 2 \times 11600}{4 \times 1.7 \times 10^{-27} \times 3.8 \times 10^7} = 1.24 \left[\frac{\text{J}\,\text{K}^{-1}\text{K}}{\text{kg s}^{-1}} \right]\ , \tag{3.549}$$

and knowing that J is expressed in m^2 kg s^{-1}, we obtain:

$$D_a \simeq 1.2\,\text{m}^2\,\text{s}^{-1}\ , \tag{3.550}$$

which are the usual units for a diffusion coefficient.

3.16. Consider a long, cylindrical plasma column of helium with a diameter of 20 mm and gas pressure 0.9 torr. The temperature of the ions and neutral atoms, obtained from Doppler broadening of emission spectral lines, is 500 K. The electron density, measured on the axis, is 10^{17} electrons m^{-3}.

a) Assuming that the electron velocity distribution function is Maxwellian and the diffusion is ambipolar, estimate the electron temperature T_e.
b) To what extent is the assumption of ambipolar diffusion, implicit in the calculation of part a), justified?
c) Calculate the approximate values for the electron free diffusion coefficient (D_e) and that of ambipolar diffusion (D_a) (don't forget their units), then compare them and discuss.

d) At time $t = 0$, the electric field maintaining the discharge is suppressed. Describe the evolution of the plasma for time $t > 0$. Calculate the characteristic decay time for the electron density on the axis (assuming that T_e does not decay significantly during this period of time).

Data:

1. The approximate mean electron-neutral collision frequency for momentum transfer in helium, at the reduced pressure p_0, is:

$$\frac{\nu}{p_0} = 2.4 \times 10^9\,\mathrm{s}^{-1}\,. \tag{3.551}$$

2. The reduced ion mobility, i.e. referring to the standard conditions of 760 torr and 273 K, namely for 2.69×10^{25} atoms m^{-3}, of He^+ in He is:

$$\mu_{i0} = 10.4 \times 10^{-4}\,\mathrm{m}^2\,\mathrm{V}^{-1}\,\mathrm{s}^{-1}\,. \tag{3.552}$$

Answer

a) To calculate T_e, we will make use of the results developed for this purpose in Sect. 3.13, for a long cylindrical plasma column, assumed to be in the ambipolar diffusion regime: we will verify in b) that this condition is satisfied.
Let us first determine p_0, the "reduced pressure" associated with the pressure and temperature of the gas T_g through the relation (1.122):

$$p_0 = \frac{273}{T_g(\mathrm{K})}\,p(\mathrm{torr}) \tag{3.553}$$

from which, in the present case:

$$p_0 = \frac{273}{500}\,0.9 = 0.49\,. \tag{3.554}$$

For helium, the constant c_0 of the model (Tab. 3.1) is 4.68, and hence the product $c_0 p_0 R$ gives:

$$c_0 p_0 R = 4.68 \times 0.49 \times 10^{-2} = 2.3 \times 10^{-2}\,. \tag{3.555}$$

This value of $c_0 p_0 R$, from Fig. 3.9, corresponds to:

$$T_{eV}/\mathcal{E}_i \simeq 0.2 \text{ (no units)} \tag{3.556}$$

and, since $\mathcal{E}_i = 24.9\,\mathrm{eV}$ for helium (Tab. 3.2), $T_{eV} = 4.9\,\mathrm{eV}$, or alternatively $T_e\ (\mathrm{K}) = T_{eV}e/k_B = 56500\,\mathrm{K}$.

b) *Diffusion regime*
We will first verify that the plasma is in this regime and not in free fall, i.e. that the mean free path, ℓ, is smaller than the radius of the discharge R. We know that $\ell \simeq v_{th}/\nu$ (Sect. 1.7.8) where $v_{th} = (2k_B T_e/m_e)^{\frac{1}{2}}$ (1.9). The collision frequency ν is given by (3.551). We can then calculate ℓ:

$$\ell \simeq \sqrt{\frac{2k_B T_e}{m_e}}\frac{1}{\nu} = \sqrt{\frac{2 \times 1.38 \times 10^{-23} \times 4.9 \times 11600}{9.1 \times 10^{-31}}}\frac{1}{2.4 \times 10^9 \times 0.49}$$
$$= 1.1 \times 10^{-3}\,\text{m}\,, \tag{3.557}$$

such that $\ell \ll R$: the plasma is indeed in the diffusion regime.

Ambipolar diffusion
To find whether the diffusion is ambipolar rather than free diffusion, we will use one of the two criteria given in Sect. 3.10. We will use the requirement that $n(0)\Lambda^2$ should be greater than 10^7 (cm^{-1}) for the diffusion to be ambipolar. Since the electron density on the axis is $n_{e0} = 10^{11}\,\text{cm}^{-3}$ and $\Lambda = R/2.405 \simeq 0.42\,\text{cm}$:

$$n_{e0}\Lambda^2 = 1.7 \times 10^{10}\,\text{cm}^{-1} > 10^7\,\text{cm}^{-1}\,, \tag{3.558}$$

and the criterion is verified (except possibly very close to the wall, where n is much smaller than on the axis).

Remark: It can be shown that volume recombination in helium is no longer negligible if the pressure is greater than 5 torr and $n_{e0} > 10^{12}\,\text{cm}^{-3}$.

c) *Calculation of the coefficient D_e*
The expression for D_e is given by (3.197):

$$D_e = \frac{k_B T_e}{m_e \nu} = \frac{1.38 \times 10^{-23} \times 56500}{9.1 \times 10^{-31} \times 2.4 \times 10^9 \times 0.49}$$
$$= 728.6 \simeq 730\,\text{m}^2\,\text{s}^{-1}\,. \tag{3.559}$$

Calculation of the coefficient D_a

We know from a) that $T_e \gg T_i$, the ion temperature, because $T_e = 56500\,\text{K}$ and $T_i = 500\,\text{K}$. For these conditions, a simple, approximate expression for D_a is:

$$D_a \simeq \frac{k_B T_e}{e}\mu_i\,. \tag{3.281}$$

The ion mobility, μ_i, for an atom density N, is obtained from the reduced mobility μ_{i0}, i.e. for the reference conditions (760 torr, 0°C) according to:

$$\mu_i = \mu_{i0}\frac{N_L}{N}\,, \tag{3.185}$$

where N_L, the Lochsmidt number, is equal to 2.69×10^{25} at m^{-3}, for the reference conditions. In addition, the perfect gas law:

$$N = \frac{p}{k_B T} , \tag{3.560}$$

allows us to calculate N for the operating conditions (p, T_g), from $N_L = p_A/k_B\, 273$ ($p_A = 760\,\text{torr}$, $T = 273\,\text{K}$). Finally, from (3.194) and (3.560), we obtain:

$$\mu_i = \mu_{i0} \frac{p_A}{k_B\, 273} \frac{k_B T}{p} , \tag{3.561}$$

which is, numerically:

$$\mu_i = \frac{10.4 \times 10^{-4} \times 760 \times 500}{0.9 \times 273} = 1.6\,\text{m}^2\,\text{s}^{-1}\,\text{V}^{-1} \tag{3.562}$$

from which:

$$D_a = \frac{k_B T_e}{e} \mu_i = \frac{1.38 \times 10^{-23} \times 56500}{1.6 \times 10^{-19}} \times 1.6 = 7.8\,\text{m}^2\,\text{s}^{-1} , \tag{3.563}$$

and thus $D_a \ll D_e$, as expected: the free diffusion of electrons is faster than when the ions and electrons diffuse together.

d) This situation corresponds to the case of a time-dependent post discharge in the diffusion regime (Sect. 3.9). Then, the charged particle density at a given point decays exponentially with time, according to the relation:

$$n(\boldsymbol{r}, t) = n(\boldsymbol{r}, t = 0) \exp(-\nu_D t) , \tag{3.222}$$

where $\tau_D = \nu_D^{-1}$ is the characteristic decay time of the plasma density due to diffusion (ambipolar in the present case). This regime persists roughly until the density has decayed to 1/e of its initial value (the required time for our present analysis), since the initial density is sufficiently high. We must first calculate ν_{Da}. We know, from (3.236), that:

$$\nu_i = \frac{D_a}{\Lambda^2} , \tag{3.564}$$

and since:

$$\nu_{Da} = \nu_i , \tag{3.565}$$

then:

$$\nu_{Da} = \frac{D_a}{\Lambda^2} = \frac{7.8}{(10^{-2}/2.405)^2} = 451 \times 10^3\,\text{s}^{-1} , \tag{3.566}$$

from which $\nu_{Da}^{-1} = \tau_D \simeq 2.2\,\mu\text{s}$. We can conclude that over a period of 10 to 20 times τ_D, i.e. 20 to 40 µs, the charged particle density has become negligible.

3.17. In general, the characteristic diffusion length Λ of a plasma is related to plasma dimensions by coefficients that depend on the plasma geometry, (see Sect. 3.9.1), but also on the boundary conditions chosen. Therefore, we would like to calculate the characteristic diffusion length in a plasma for different geometrical configurations (planar and cylindrical) in the case where the assumption $n(r = R) = 0$ is no longer valid, i.e. when the losses to the walls of the experiment are through a flux across an ion sheath. Assume that the plasma is in the ambipolar diffusion regime and that the neutrality condition ($n_i = n_e$) is applicable up to the edge of the collisionless ion sheath, whose thickness can be neglected when compared to the plasma dimensions.

a) Calculate the characteristic diffusion length $\Lambda = L/a$ (where a is a dimensionless coefficient) of a planar configuration that is infinite along y and z, and with a width along x of $L = 2\,\text{cm}$, for the two following cases:

   1. Argon plasma, $p = 0.5\,\text{torr}$, $T_{eV} = 1.7\,\text{eV}$, $T_g = 300\,\text{K}$, mobility of Ar^+ ions in Ar: $\mu_{i0} = 1.52\,\text{cm}^2\,\text{V}^{-1}\,\text{s}^{-1}$ at 760 torr and 273 K.
   2. Helium plasma, $p = 0.5\,\text{torr}$, $T_{eV} = 5.8\,\text{eV}$, $T_g = 700\,\text{K}$, mobility of He^+ ions in He: $\mu_{i0} = 10.4\,\text{cm}^2\,\text{V}^{-1}\,\text{s}^{-1}$ at 760 torr and 273 K.

   To calculate a, use the graph $a \tan a = f(a)$.

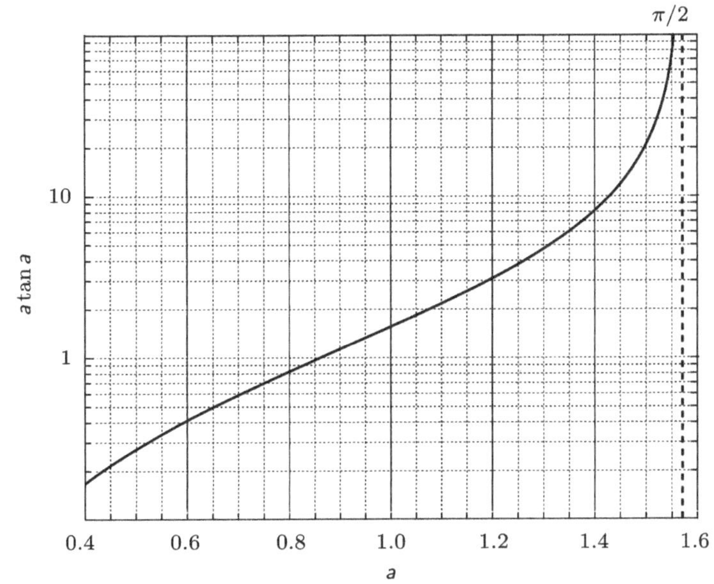

Fig. 3.14 Variation of the function $f(a) = a \tan a$ with a.

b) Calculate the characteristic diffusion length $\Lambda = L/b$ (where b is a dimensionless coefficient) of an infinite cylindrical plasma column of radius $R = 1\,\text{cm}$ for the two sets of plasma conditions defined above.

To calculate b, use the graph $bJ_1(b)/J_0(b) = f(b)$. The derivative of the zeroth order Bessel function of the first kind is:

$$J_0'(z) = -J_1(z) , \tag{3.567}$$

where J_1 is the first order Bessel function of the first kind.

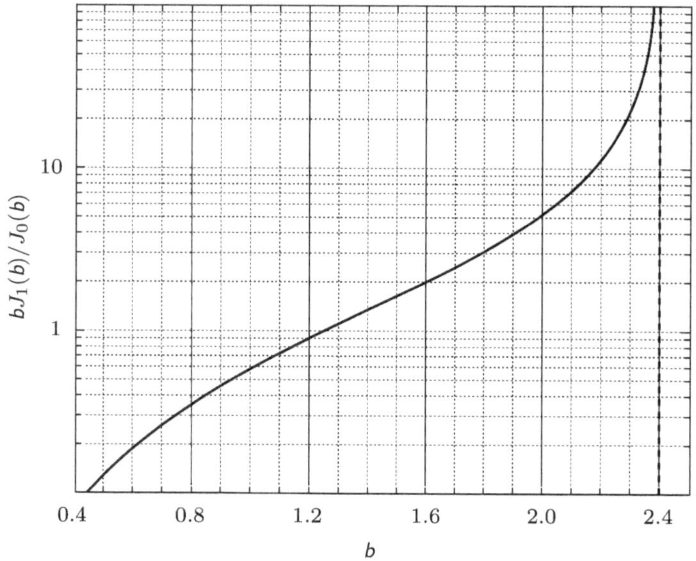

Fig. 3.15 Variation of the function $f(b) = bJ_1(b)/J_0(b)$ with b.

Answer

In the ambipolar diffusion regime, the plasma is described by the equations of continuity (3.216):

$$\frac{\partial n}{\partial t} + \boldsymbol{\nabla} \cdot n\boldsymbol{v} = \nu_i n \tag{3.568}$$

and of the diffusion flux (3.257):

$$\boldsymbol{\Gamma} = n\boldsymbol{v} = -D_a \boldsymbol{\nabla} n , \tag{3.569}$$

such that in the stationary state:

$$\nabla^2 n = -\left(\frac{\nu_i}{D_a}\right) n \tag{3.570}$$

or, alternatively:

$$\nabla^2 n = -\frac{1}{\Lambda^2} n . \tag{3.571}$$

a) In the case of an infinite plane-plasma configuration along y and z, the solution of (3.570) in Cartesian coordinates:

$$\frac{\partial^2 n}{\partial x^2} = -\frac{1}{\Lambda^2} n \tag{3.572}$$

is given by:

$$n(x) = n(0) \cos\left(\frac{x}{\Lambda}\right) = n(0) \cos\left(\frac{ax}{L}\right) . \tag{3.573}$$

The boundary conditions at the wall suppose, for $x = \pm L/2$, that the electron and ion fluxes are equal and that the ion velocity at the sheath edge is:

$$\boldsymbol{v} = \boldsymbol{v}_B , \tag{3.574}$$

where $\boldsymbol{v}_B$ is the Bohm velocity of scalar value (3.328):

$$v_B = \sqrt{\frac{k_B T_e}{m_i}} . \tag{3.575}$$

Taking account of (3.569), the conservation of ion flux at the sheath edge can then be written:

$$- D_a \left(\frac{\partial n}{\partial x}\right)_{\pm \frac{L}{2}} = n\left(x = \pm\frac{L}{2}\right) v_x . \tag{3.576}$$

For $x = +L/2$,

$$v_x = + v_B \tag{3.577}$$

so that, from (3.573), (3.576) can be written:

$$\left(\frac{\partial n}{\partial x}\right)_{+\frac{L}{2}} = -\frac{a}{L} n(0) \sin\left(\frac{a}{2}\right) = -\frac{v_B}{D_a} n(0) \cos\left(\frac{a}{2}\right) \tag{3.578}$$

from which:

$$\frac{L v_B}{2 D_a} = \frac{a}{2} \tan\frac{a}{2} , \tag{3.579}$$

where an approximate value for D_a is given by (3.281):

$$D_a \simeq \frac{k_B T_e}{e} \mu_i , \tag{3.580}$$

where μ_i depends on the pressure.

Remark: In contrast to the usual boundary conditions $n(x = \pm L/2) = 0$, where the characteristic length is independent of the geometric configuration and dimensions of the plasma, the diffusion length also depends, in the present case, on the operating conditions of the plasma, namely the type of gas and the pressure.

Numerically, we obtain:

1. For argon:

$$v_B = \left(\frac{1.7\ (\mathrm{V}) \times 1.6 \times 10^{-19}\ (\mathrm{C})}{9.1 \times 10^{-31}\ (\mathrm{kg}) \times 1836 \times 40}\right)^{\frac{1}{2}} = 2.03 \times 10^3\,\mathrm{m\,s^{-1}}\ ,$$

$$\begin{aligned}\mu_i &= \mu_{i0}\frac{760\,T_g\ (\mathrm{K})}{p\ (\mathrm{torr})\ 273} = \frac{1.52 \times 760 \times 300}{0.5 \times 273} \\ &= 2.54 \times 10^3\,\mathrm{cm^2\,V^{-1}\,s^{-1}} = 25.4 \times 10^{-2}\,\mathrm{m^2\,V^{-1}\,s^{-1}}\ , \\ D_a &= \frac{k_B T_e}{e}\mu_i = 1.7\ (\mathrm{V}) \times 25.4 \times 10^{-2}\ (\mathrm{m^2\,V^{-1}\,s^{-1}}) \\ &= 43.2 \times 10^{-2}\,\mathrm{m^2\,s^{-1}}\ ,\end{aligned}$$

$$\frac{a}{2}\tan\frac{a}{2} = \frac{Lv_B}{2D_a} = \frac{2 \times 10^{-2} \times 2.03 \times 10^3}{2 \times 43.2 \times 10^{-2}} = 47\ .$$

We can find a from the graph of $a \tan a$, i.e.:

$$\frac{a}{2} \simeq 1.54\ ,$$

from which: $\Lambda \simeq L/3.08$.
The numerical value thus obtained is close to the value of $\Lambda = L/\pi$ resulting from the boundary conditions $n(x = \pm L/2) = 0$.

2. For helium:

$$v_B = \left(\frac{5.8\ (\mathrm{V}) \times 1.6 \times 10^{-19}\ (\mathrm{C})}{9.1 \times 10^{-31}\ (\mathrm{kg}) \times 1836 \times 4}\right)^{\frac{1}{2}} = 11.8 \times 10^3\,\mathrm{m\,s^{-1}}\ ,$$

$$\begin{aligned}\mu_i &= \mu_{i0}\frac{760\,T_g\ (\mathrm{K})}{p\ (\mathrm{torr})\ 273} = \frac{10.4 \times 760 \times 300}{0.5 \times 273} \\ &= 17.4 \times 10^3\ \mathrm{cm^2\,V^{-1}\,s^{-1}} = 1.74\,\mathrm{m^2\,V^{-1}\,s^{-1}}\ ,\end{aligned}$$

$$D_a = \frac{k_B T_e}{e}\mu_i = 5.8\ (\mathrm{V}) \times 1.74\ (\mathrm{m^2\,V^{-1}\,s^{-1}}) = 10.1\,\mathrm{m^2\,s^{-1}}\ ,$$

$$\frac{a}{2}\tan\frac{a}{2} = \frac{Lv_B}{2D_a} = \frac{2 \times 10^{-2} \times 11.8 \times 10^3}{2 \times 10.1} = 11.7\ .$$

We can find a from the $a \tan a$ graph, i.e.:

$$\frac{a}{2} \simeq 1.45\ ,$$

from which: $\Lambda \simeq L/2.9$, which is a slightly different value compared to $\Lambda = L/\pi$.

b) In the case of an infinitely long cylindrical configuration, the solution to (3.570) in cylindrical coordinates:

$$\frac{1}{r}\frac{\partial}{\partial r}nr = -\frac{1}{\Lambda^2} , \tag{3.581}$$

is given by:

$$n(r) = n(0)J_0\left(\frac{r}{\Lambda}\right) = n(0)J_0\left(\frac{br}{R}\right) , \tag{3.582}$$

where J_0 is the zeroth order Bessel function of the first kind. Since:

$$\frac{\partial J_0(br)}{\partial r} = -bJ_1(br) , \tag{3.583}$$

where J_1 is the first order Bessel function of the first kind, the boundary conditions at $r = R$ can this time be written, taking account of (3.582):

$$\left(\frac{\partial n}{\partial r}\right)_R = -\frac{b}{R}n(0)J_1(b) = -\frac{v_B}{D_a}n(0)J_0(b) , \tag{3.584}$$

from which:

$$\frac{Rv_B}{D_a} = b\frac{J_1(b)}{J_0(b)} . \tag{3.585}$$

Numerically, we obtain:

1. For argon:

$$b\frac{J_1(b)}{J_0(b)} = \frac{Rv_B}{D_a} = 47 .$$

   We can find b from the graph $bJ_1(b)/J_0(b)$, i.e.:

$$b \simeq 2.36 ,$$

   from which:

$$\Lambda \simeq \frac{R}{2.36} .$$

   The value obtained is close to the value $\Lambda = R/2.405$ resulting from the boundary condition $n(R) = 0$.

2. For helium:

$$b\frac{J_1(b)}{J_0(b)} = \frac{Rv_B}{D_a} = 11.7 .$$

   We can find b from the graph $bJ_1(b)/J_0(b)$, i.e.:

$$b \simeq 2.21 ,$$

   from which:

$$\Lambda = \frac{R}{2.21} ,$$

   which is a slightly different value compared to $\Lambda = R/2.405$.

Chapter 4
Introduction to the Physics of HF Discharges

4.1 Preamble

This chapter discusses plasmas produced by a high frequency (HF) periodic electric field, at both radio ($\cong$ 1–300 MHz) and microwave (0.3–300 GHz) frequencies. These discharges, originally used mainly in research laboratories, have only comparatively recently found a very important place in industrial applications (e.g., micro-electronics, destruction of greenhouse gases Sect. 1.2). Thus, an understanding of the mechanisms that sustain them and a knowledge of the physico-chemical phenomena that are specific to these discharges (for example, the influence of the HF frequency on the plasma properties), can lead to improved designs for HF devices and more efficient processes.

For a number of reasons, the most commonly used HF plasmas operate at low-pressure ($<$ 10–20 torr):

- the implementation of a plasma source is, in this case, much simpler than at atmospheric pressure, due to much lower gas temperatures;
- the modelling of HF plasmas has now attained maturity, which is not yet the case for plasmas at atmospheric pressure [22,26].

Our discussion will distinguish between low-pressure plasmas and high (essentially atmospheric) pressure plasmas, the latter being characterised, as we shall see, by specific phenomena such as the contraction and filamentation of the discharge.

In order to better understand the physics of HF discharges, where the electric field varies periodically in time, we will also consider *direct current discharges* (DC), in which the electric field intensity is constant in time. This entire group of discharges is designated by the term *electric discharges*.

Compared to DC discharges, HF discharges present a number of advantages. This is particularly true if the discharge vessel is made from a dielectric material, because its transparency to electromagnetic (EM) waves allows the

M. Moisan, J. Pelletier, *Physics of Collisional Plasmas*, DOI 10.1007/978-94-007-4558-2_4,

electric field to penetrate from the exterior and to ionise the gas it contains. In such cases, in contrast to DC discharges, there are no electrodes in contact with the gas: electrodes constitute a source of gas contamination and lead to deposits on the walls of the discharge tube and, more generally, they limit the life of the discharge tube. The device used to impose the EM field on the discharge vessel is called an *HF field applicator*. Another advantage of HF discharges, with respect to DC discharges, is the possibility of modifying the plasma parameters by tuning the frequency of the EM field: varying the field frequency, in certain cases, modifies the electron energy distribution function (EEDF), which can be used to optimise the kinetics of a process. With regard to the cost of apparatus, DC discharges are, in general, less expensive, although now the reduced cost of magnetron microwave generators operating at 2450 MHz[142] has made them competitive. In addition, the advance of HF generators based on power transistors encourages us to believe that, in the near future, more compact designs, with improved security (no high voltage in the circuit) and improved reliability, will become available.

In the preceding chapters, we have developed the basic rudiments of plasma physics, with a view to their application to HF discharges. In the following sections, we will make abundant use of these fundamentals, to describe and model HF plasmas. This chapter contains three sections that consider successively:

1. the power transfer from the electric field $\boldsymbol{E}$ to the discharge. To this end, we will use the power θ_a absorbed per electron as the characteristic parameter, for both collisional and non collisional (electron cyclotron resonance, Sect. 4.2) energy transfer.
2. the influence of the frequency of the $\boldsymbol{E}$ field on the properties of the plasma, with some examples of the application of this effect (Sect. 4.3). This study will be principally applied to low-pressure plasmas.
3. the phenomena of contraction and filamentation encountered in high-pressure plasmas (Sect. 4.4). To take account of the contraction effect, we need to examine the ionisation-recombination kinetics of molecular ions.

[142] A certain number of frequencies from the EM spectrum are reserved for industrial, medical and scientific applications (ISM frequencies). To this effect, the earth is subdivided in three regions: region 1 (Europe, Africa, the Middle-East, the former USSR and Mongolia), region 2 (The Americas, Greenland), region 3 (Asia, outside the former USSR, Oceania). In all regions, the frequencies 13.56, 27.12 and 40.68 MHz, as well as 2.45 and 5.8 GHz, are authorised [26]. The frequency 433 MHz is only authorised in region 1, and 915 MHz only in region 2. In addition to these ISM frequencies, the 2.65 MHz frequency is a standard for RF lighting devices.

4.2 Power transfer from the electric field to the discharge

Traditionally, the transfer of power in a DC discharge is characterised by the ratio E/p where E is the field intensity and p the gas pressure. In the following, we will use the parameter θ_a (or θ_a/p)[143], the average power absorbed per electron (2.39), instead. On one hand, this enables us to unify the DC and HF descriptions of the plasma and, on the other hand, it emphasises a fundamental aspect, which is generally ignored, namely that the intensity of the electric field sustaining the electric discharge is not set by the operator, but depends on the loss mechanisms for the charged particles (the parameter θ_l, defined by (4.1)).

4.2.1 Direct current discharges

Figure 4.1 is a schematic of a discharge referred to as a *cold cathode discharge* (without a thermo-electric emitting filament). The constant voltage U, applied to the leads of the two electrodes, creates an electric field of intensity E, which acts on the electrons initially present in the gas, either from cosmic rays or natural radiation, or from external excitation, for example, produced by a spark from a Tesla coil, or a piezo-electric gas lighter directed against the dielectric walls of the vessel. These initial electrons are accelerated by a force $\boldsymbol{F} = -e\boldsymbol{E}$, where e is the absolute value of the electronic charge, and they continue to gain energy until they experience a collision with another electron, or with an atom (molecule). At the instant of impact, the "incident" electron either gains or loses energy (Sect. 1.7.2). The collision can always be elastic (conservation of energy), but can only lead to an excitation or ionisation if the energy of the electron is equal to or greater than the threshold energy V_j for excitation of the atom into the state j, or the threshold ionisation energy V_i of the atom (Sect. 1.7.9); following an inelastic collision, the internal energy of the atom is increased, depending on the case, by an energy V_j or V_i released by the electron. The energy transferred to the ions from the electric field is negligible compared to that of the electrons, due to the mass ratio of these two types of particles (Sect. 2.2).

After a transition stage, in which the density of the charged particles increases, the stationary state is reached. As shown in Fig. 4.2, we can observe different luminous and dark zones along the discharge tube; the intensity E now varies axially, in contrast to the situation before ignition (the horizontal dashed line in the figure). Macroscopic electron-ion neutrality is found only in the zone referred to as the *positive column*, which thus corresponds to the definition of a plasma.

[143] In discharges, the ratio E/p is a fundamental parameter, entering into the scaling laws governing the plasma. Its relation to θ_a/p will be elaborated in Sect. 4.3.3.

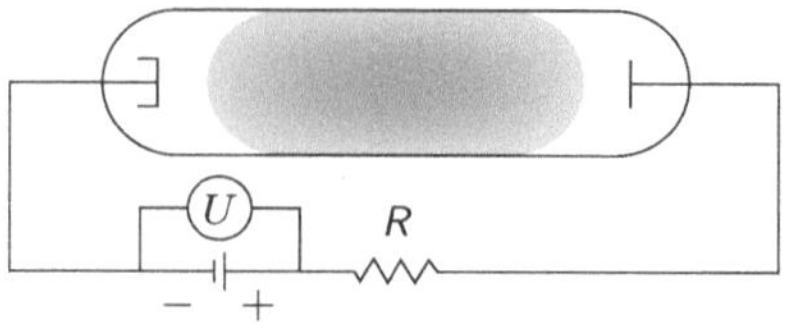

Fig. 4.1 Schematic of an electric discharge maintained at constant current, usually referred to as a direct current (DC) discharge. The resistance R (ballast) ensures the stability of the discharge.

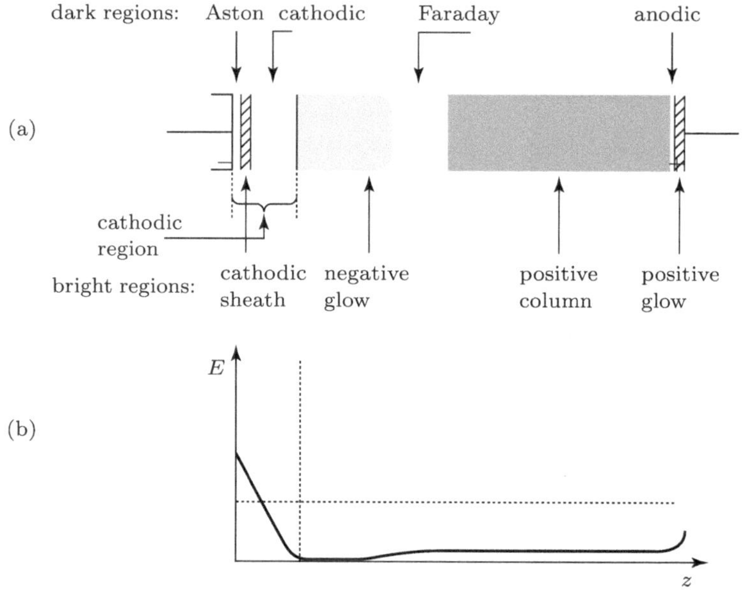

Fig. 4.2 a Representation of the different dark and luminous zones in a direct current discharge, together with **b** the qualitative variation of the electric field intensity E in the stationary state, along the length of the discharge. The horizontal dashed line indicates the electric field intensity E before ignition.

To characterise the transfer of power from the field $\boldsymbol{E}$ to the positive column plasma by means of the electrons, we will establish the balance between the power taken (on average) by an electron from the electric field, referred to as the absorbed power θ_a (Sect. 2.2.1), and the power that the electron (on average) transfers to the heavy particles as a result of collisions, referred to as the power lost θ_l.

The average *power* θ_l *lost per electron*, and transferred to the plasma following the various types of collision of electrons with heavy particles [144] can be written [26]:

[144] Equation (4.1) can be obtained from the homogeneous Boltzmann equation by considering the isotropic part $F_0(U)$ of the EEDF. This equation is then multiplied by U and integrated over all values of U. The integral of $F(U)$ is described in Appendix XVII (see note at the bottom of the page).

$$\theta_l(\langle U_{eV}\rangle) = \frac{2m_e}{M}\langle\nu(U_{eV})U_{eV}\rangle + \sum_j\langle\nu_j(U_{eV})\rangle V_j + \langle\nu_i(U_{eV})\rangle V_i\ , \qquad (4.1)$$

where m_e/M is the mass ratio of the electron to that of the atom (molecule), $\nu(U_{eV})$ represents the microscopic collision frequency for an electron of energy U_{eV}, which results in a transfer of momentum (Sect. 1.7.7), ν_j and ν_i are the collision frequencies (also microscopic) for the excitation to the level j (threshold energy V_j) or ionisation (threshold energy V_i) respectively; finally, the symbol $\langle\,\rangle$ represents the average taken over the EEDF. In the case where the EEDF is Maxwellian, the average values in (4.1) are completely determined by the electron temperature T_e and the gas pressure.

In general, $\theta_l(\langle U_{eV}\rangle)$ is an increasing function of $\langle U_{eV}\rangle$, as is shown in Fig. 4.3, for the case of argon and a Maxwellian EEDF ($\frac{3}{2}\langle U_{eV}\rangle = T_{eV}$). We observe that, if $T_{eV} \geq 1\,\mathrm{eV}$, the value of θ_l is essentially determined by inelastic collisions for excitation and ionisation, while if $T_{eV} < 1\,\mathrm{eV}$, the value of θ_l is due to elastic collisions. Furthermore, we know that the value of T_{eV} is less than 1 eV in an argon plasma at atmospheric pressure, whilst it is equal or greater than 1 eV at low-pressure[145]. Clearly, collisions at atmospheric pressure are much more numerous than at low-pressure, and it is necessary to take multi-step ionisation processes into account in order to calculate θ_l (which is not the case in Fig. 4.3), as we will see in Sect. 4.2.4.

The charged particles, which are thus created in the volume, tend to disappear from the discharge by two main mechanisms (Sect. 1.8):

- by diffusion towards the walls of the vessel, on which ions and electrons readily recombine to form neutral atoms;
- by electron-ion recombination in the volume of the plasma.

In the following treatment, for simplification, we will assume that only diffusion is responsible for the loss of charged particles (Sect. 3.10–3.12), which is generally the case for low-pressure rare-gas discharges (0.5–10 torr).

The average *power* θ_a *absorbed per electron*, taken from the field $\boldsymbol{E}$, is related to the work effected by the electron in the field. In the absence of collisions, the energy of the electron will increase during its entire path from the cathode to the anode, its velocity evolving with time according to (Sect. 2.2.1):

$$w = \frac{eE}{m_e}t\ . \qquad (4.2)$$

Of course, if there are no collisions, then there is no transfer of energy to the plasma, hence no discharge. In the presence of collisions, the motion of the electrons is hindered by electron-neutral collisions: this results in an average velocity for progression in the field $\boldsymbol{E}$, called the electric drift (Sect. 3.8.2):

$$\boldsymbol{v}_d = \mu_e\boldsymbol{E} \qquad (4.3)$$

145 The value of T_e above which elastic collisions dominate depends on the discharge gas and the threshold energy V_j for excitation of the first excited level.

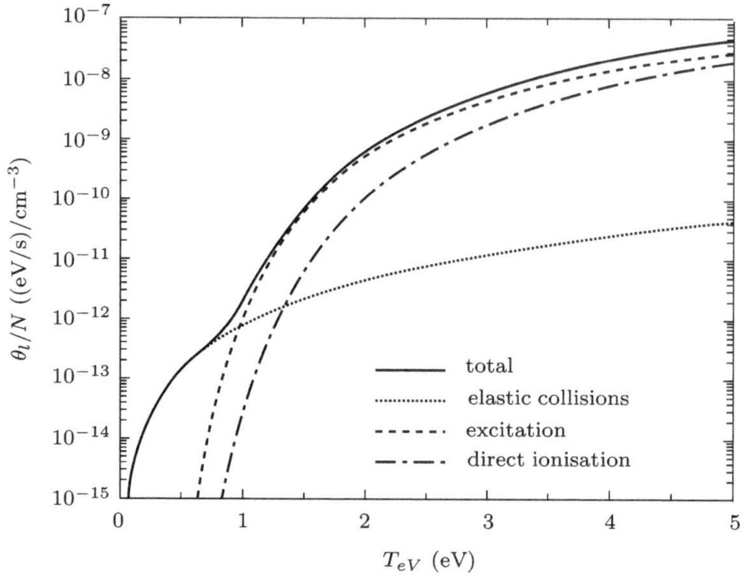

Fig. 4.3 Value of the power per electron θ_l, normalised to the density N of neutral atoms, as a function of T_{eV}, calculated for a Maxwellian distribution in the case of an argon discharge. We assume direct excitation from the ground state (no stepwise excitation or ionisation).

where, as we have seen, $\mu_e = -e/m_e\nu$ is the electron mobility and ν the electron-neutral mean collision frequency for momentum transfer. The power P_a taken from the field $\boldsymbol{E}$ by the electrons, per unit volume, in the case of a plasma with a uniform electron density n_e, is given by (Sect. 2.2.1):

$$P_a \equiv n_e\theta_a = \boldsymbol{J} \cdot \boldsymbol{E} \,, \tag{4.4}$$

which constitutes a generalisation of Ohm's Law. We know that the expression for $\boldsymbol{J}$, the current density vector of the electron fluid, is given by:

$$\boldsymbol{J} = -n_e e \boldsymbol{v}_d \,. \tag{4.5}$$

Thus, combining (4.3), (4.4) and (4.5), we obtain:

$$\theta_a(E) = \frac{e^2}{m_e\nu} E^2 \,. \tag{4.6}$$

In the stationary state, the power absorbed θ_a adjusts to compensate for the power lost θ_l and:

$$\theta_a(E) = \theta_l(\langle U_{eV} \rangle) \,. \tag{4.7}$$

In fact, if the power θ_a was less than θ_l, the discharge would extinguish. If, on the contrary, θ_a was greater than θ_l, the plasma density would increase, contradicting the assumption of a stationary state. Equation (4.7) thus constitutes the equation of the *electron power balance* in the plasma. Their common value will be designated from now on by θ. It is important to note that (4.7) implies that the electric field intensity E in the plasma[146] (connected to θ_a by (4.6)) adapts to exactly compensate for the power lost θ_l.

In the particular case where the charged particles disappear from the plasma by diffusion to the walls where they recombine, these losses increase with the average energy of the electrons, hence, in the case of a Maxwellian EEDF, with the electron temperature: the diffusion coefficients (given by (3.284) or (3.308)) increase with T_e. The value of $\theta_l(\langle U_{eV} \rangle)$, which increases with the average electron energy (Fig. 4.3), will therefore increase when the charged particle losses increase.

In the ambipolar diffusion regime, the electron temperature of the discharge is only a function of the vessel dimensions, the type of gas and the pressure (Sect. 3.13), i.e. it entirely depends on the operating conditions (4.7). As a result, θ_l, relative to θ_a, is the dominant quantity in the electron power balance, θ_a adjusting to the value of θ_l. Therefore, once the stationary state is reached the field intensity $\boldsymbol{E}$ present in the positive column is not related to the potential difference applied between the two electrodes, an observation confirmed by experiment. The "surplus" from this potential difference is found in the cathode and anode falls and other zones bordering the positive column (Fig. 4.2).

4.2.2 HF discharges

To set the scene, we will consider an HF discharge that is simple to implement. An HF power generator feeds an EM field applicator consisting of a conducting coil wound around a discharge tube[147], as is shown in Fig. 4.4.

In the case of an HF discharge, the question of energy transfer must be posed differently than in DC discharges, because the electric field is periodic. At the beginning of the cycle, the electron is accelerated in one direction under the influence of the field for the first half of the cycle, then in the opposite

---

[146] This field is called the *maintenance field* (meaning that of the discharge) in contrast to the *applied field* set by the operator (for example, that applied between the two electrodes in a DC discharge just before ignition).

[147] The material of the discharge tube is chosen such that it absorbs the least possible of the HF power. From this point of view, fused silica (incorrectly called quartz) is particularly advantageous if the gas temperature is not too high ($\lesssim 900\,^{\circ}$C) and in the absence of reactions with by-products from the discharge, such as fluorine. Ceramics, such as Al_2O_3 and AlN, are more resistant to temperature but absorb more of the HF power than silica.

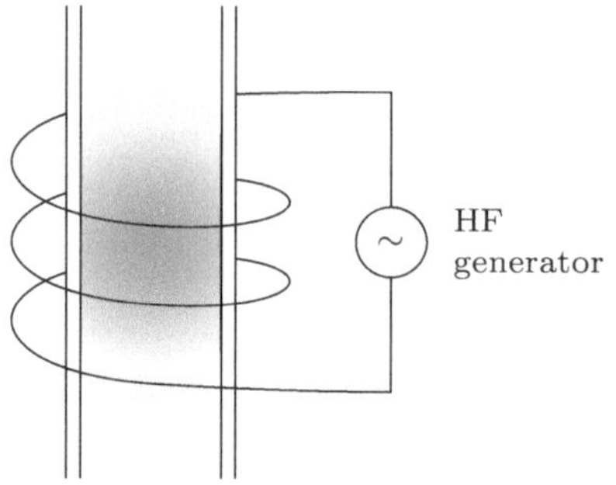

Fig. 4.4 Schematic of an HF discharge, referred to as an *inductively coupled plasma (ICP)*. The excitation frequencies are generally much lower than 100 MHz, the most commonly used being the ISM frequencies of 13.56 MHz and 27.12 MHz[142].

direction[148]: on average over a cycle, the work effected by an electron in an HF field is zero (Sect. 2.2.1). Only collisions can interrupt the periodic motion and allow the electron to take energy from the HF electric field, as we have seen in considering a cold plasma (Sect. 2.2.1). We then obtain (2.37):

$$\theta_a = \frac{e^2}{m_e} \frac{\nu}{\nu^2 + \omega^2} \overline{E^2} , \tag{4.8}$$

where $\overline{E^2}$, the average quadratic value of the electric field intensity, is equal to $E_0^2/2$, E_0 being the maximum amplitude of the field in the course of a period of oscillation, and ω the angular frequency of the field. Equation (4.8) reduces to (4.6) for $\omega = 0$.

The loss of charged particles is influenced by the same mechanisms as in the positive column. On the other hand, in contrast to a DC discharge, the electric field intensity E in an HF plasma is not radially constant, but decreases from the tube walls to the axis: this phenomenon is analogous to the attenuation of an EM wave entering a conductive material (*skin effect*). Under these conditions, the integration of θ_a (4.8) across a transverse section of the discharge should lead to an average value of θ_a, such that $\theta_a = \theta_l$.

Since it is easier to measure θ_a than E in an HF discharge[149], the parameter θ appears as the natural reference parameter for comparing electric discharges. This consideration reinforces the preeminence, mentioned earlier, of θ over E (Sect. 4.2.1): the electric field intensity E adjusts to satisfy the

---

[148] We make the implicit assumption that the amplitude of oscillation (or excursion) of the electron in an HF field (2.34) is smaller than the discharge vessel dimensions (for example, the radius R in a long cylindrical column). This is generally the case for fields with frequencies above 1 MHz.

[149] The measurement of the electric field in an HF discharge is generally very imprecise because of the perturbations caused by the measurement antenna. On the other hand, the value of θ can be simply deduced from the power absorbed per unit volume, knowing the value of the average electron density in the plasma (4.4).

value of θ_l. This dominance of θ_l over θ_a and hence, on E, will be confirmed in the later treatment of HF magnetised plasmas.

Remark: The *characteristic penetration depth* δ_c *of an HF field* in a conductive medium is defined as the distance over which the field intensity of a plane EM wave reduces to 1/e of its initial value (e is used, exceptionally here, to denote the base of the natural logarithm). This value is given by (see Appendix XVIII):

$$\delta_c = \frac{(c/\omega)}{\Im\left(-\epsilon_p^{1/2}\right)}, \tag{4.9}$$

where c is the speed of light in vacuum, $\Im(A)$ denotes the imaginary part of a complex quantity A and ϵ_p the permittivity of the plasma relative to vacuum (2.41). In the case of a non-collisional plasma ($\omega \gg \nu$ and $\omega_{pe} > \omega$), the skin depth δ_c takes the value c/ω_{pe}. On the other hand, for a collisional plasma ($\nu \gg \omega$ and $\omega_{pe} \gg \omega$), it can be shown (Appendix XVIII) that $\delta_c = (c/\omega_{pe})(2\nu/\omega)^{\frac{1}{2}}$.

4.2.3 HF discharges in the presence of a static magnetic field

For certain kinds of plasma applications, it can be advantageous to operate at the lowest possible gas pressure, in order to minimise the collision frequency in the discharge. This is the case for *anisotropic etching*, where the goal is to use ion bombardment to "excavate" perfectly vertical trenches in a given material (Fig. 1.4). The acceleration of the ions, obtained by polarisation of the substrate holder or the material itself, leads to an ion flux directed perpendicular to the surface being treated: the less the trajectory is modified by collisions, the more perfectly anisotropic (vertical) the etching. However, how can we create an HF plasma with a sufficiently low collision frequency when, as we have noted in the previous section, the existence of collisions is essential for sustaining the HF discharge? In this case, it is necessary to subject the plasma to a static magnetic field $\boldsymbol{B}_0$, as we will now demonstrate.

Two principal phenomena characterise the action of a static field $\boldsymbol{B}_0$ on an HF discharge: reduction of the diffusion losses of the charged particles to the walls and, when applicable, resonant transfer of the energy of the HF field to the electrons at $\omega_{ce} = \omega$.

Reduction of charged particle diffusion losses to the walls

In the case of a field $\boldsymbol{B}_0$ directed axially in a cylindrical vessel, the charged particles are trapped in either a purely cyclotron motion, or a helicoidal motion, about the field lines of $\boldsymbol{B}_0$, depending on whether the axial velocity is zero or non zero (Sect. 2.2.2). The radial diffusion of charged particles, hence their loss to the walls, is found to be the more reduced the smaller is the cyclotron gyration radius imposed upon them is small (much smaller than the vessel radius), i.e., as the magnetic field intensity is increased sufficiently. For the magnetic confinement of the electrons to be efficient, it is necessary that there be several cyclotron gyrations between two collisions, which imposes $\nu \ll \omega_{ce}$ (Sect. 3.11)[150]. Note that the reduction of charged particle losses means that T_e and, hence, θ_l[151] decrease such that, for the same value of absorbed power P_a (cf. to (4.4)), a larger value of electron density is obtained.

Furthermore, for constant B_0, θ_l increases as the pressure decreases, because as diffusion losses increase, the mean electron energy increases (Fig. 3.9) and, as a result, the value of θ_l also increases. There is a maximum value of θ_l after which the power θ_a becomes less than θ_l (see further, in Fig. 4.5, the non-resonant regime). This maximum value of θ_l corresponds to a minimum pressure, below which it is impossible to maintain the discharge.

In the following, in order to examine the influence of electron cyclotron resonance (ECR) on the maintenance of the discharge, we will consider two cases in succession: a discharge in an infinite medium in which a planar EM wave propagates in the same direction as the magnetic field $\boldsymbol{B}_0$, then a discharge in a bounded medium, in this case a plasma column maintained by an EM surface wave, which also propagates in the direction of $\boldsymbol{B}_0$.

The case of a planar wave in an infinite medium

Comparison of the regime of power absorption by collisions with that of electron cyclotron resonance (ECR): variation of the value of θ_a as a function of pressure, for an assumed constant field E_0

The method of calculating θ_a in HF plasmas in the ECR regime is analogous to that used in the purely collisional regime. It is sufficient to introduce the term $\boldsymbol{B}_0$ in the hydrodynamic equation for momentum transfer (Sect. 3.7), which can be written, in the cold plasma approximation ($T_e = 0$):

$$m_e \frac{\partial \boldsymbol{v}}{\partial t} = q[\boldsymbol{E} + \boldsymbol{v} \wedge \boldsymbol{B}_0] - m_e \nu \boldsymbol{v} , \tag{4.10}$$

where the electron velocity, in this context, is purely periodic, i.e.:

$$v_x = v_{0x} \exp(\mathrm{i}\omega t) , \tag{4.11}$$

$$v_y = v_{0y} \exp(\mathrm{i}\omega t) . \tag{4.12}$$

[150] To put this in context, in argon at 1 torr, for $T_{eV} = 2\,\mathrm{eV}$, $\nu \simeq 2 \times 10^9\,\mathrm{s}^{-1}$. Also, recall that $\omega_{ce}/2\pi(\mathrm{Hz}) = 2.8 \times 10^{10} B_0$ (tesla).

[151] Remember that Fig. 4.3 shows that θ_l grows monotonically with T_e.

The static magnetic field $\boldsymbol{B}_0$ and the planar wave are directed along z while the HF electric field of the wave $\boldsymbol{E} = \boldsymbol{E}_0 \mathrm{e}^{\mathrm{i}\omega t}$ is taken along y. Following the coordinates x and y in the Cartesian frame, we then have:

$$\mathrm{i}\omega v_{0x} = \frac{qB}{m_e} v_{0y} - \nu v_{0x} \,, \tag{4.13}$$

$$\mathrm{i}\omega v_{0y} = \frac{q}{m_e} E_0 - \frac{qB}{m_e} v_{0x} - \nu v_{0y} \,. \tag{4.14}$$

Eliminating v_{0x} between (4.13) and (4.14), we obtain:

$$v_{0y} = -\frac{eE_0}{m_e} \frac{(\nu + \mathrm{i}\omega)[\omega_{ce}^2 - \omega^2 + \nu^2 - 2\mathrm{i}\omega\nu]}{(\omega_{ce}^2 - \omega^2 + \nu^2)^2 + 4\omega^2\nu^2} \,, \tag{4.15}$$

which gives, after factorising the real and imaginary terms:

$$v_{0y} = -\frac{eE_0}{m_e} \frac{\nu\left(\nu^2 + \omega^2 + \omega_{ce}^2\right) - \mathrm{i}\omega\left(\nu^2 + \omega^2 - \omega_{ce}^2\right)}{\left[(\omega - \omega_{ce})^2 + \nu^2\right]\left[(\omega + \omega_{ce})^2 + \nu^2\right]} \,. \tag{4.16}$$

The average power (over a period) per unit volume, P_a, absorbed by the electrons, is given, in general (Eqs. (2.37) and (2.40)), by:

$$P_a = \frac{1}{2} \Re(\boldsymbol{J} \cdot \boldsymbol{E}^*) \,, \tag{4.17}$$

where $\Re$ represents the real part of the product in parenthesis and the asterisk indicates the complex conjugate of a given quantity. This expression can also be written in terms of the conductivity tensor $\underline{\boldsymbol{\sigma}}$ (Eqs. (2.124) to (2.126)), i.e.:

$$P_a = \frac{1}{2}[(\underline{\boldsymbol{\sigma}} \cdot \boldsymbol{E}) \cdot \boldsymbol{E}^*] \,. \tag{4.18}$$

Since E_y is the only non-zero component of the electric field, and the component σ_{zy} of $\underline{\boldsymbol{\sigma}}$ is zero (see (3.189) and (3.191)), the expression developed for the current density (2.126) reduces to:

$$\boldsymbol{J} = \sigma_{xy} E_y \hat{\mathbf{e}}_x + \sigma_{yy} E_y^* \hat{\mathbf{e}}_y \,. \tag{4.19}$$

However, only the component of $\boldsymbol{J}$ along y contributes to the contracted (scalar) product with $\boldsymbol{E} = \hat{\mathbf{e}}_y E_y$ in (4.17). Resorting to complex algebra, we can represent the absorbed power per unit volume, averaged over a period of the HF field, in the form:

$$P_a = \frac{1}{2} \Re(J_y E_y^*) \,. \tag{4.20}$$

Since $J_y = -n_e e v_{0y} \mathrm{e}^{-\mathrm{i}\omega t}$ and $E_y = E_0 \mathrm{e}^{-\mathrm{i}\omega t}$, the average power absorbed per electron can be expressed as:

$$\theta_a \equiv \frac{P_a}{n_e} = -\frac{e}{2}\Re\left[v_y E_y^*\right]$$
$$= \frac{e^2E_0^2}{2\nu m_e}\left[\frac{1}{2}\frac{\nu^2}{\left(\omega-\omega_{ce}\right)^2+\nu^2}+\frac{1}{2}\frac{\nu^2}{\left(\omega+\omega_{ce}\right)^2+\nu^2}\right] . \tag{4.21}$$

From (4.20) and (4.21), we can also write that:

$$\theta_a \equiv \frac{P_a}{n_e} = \frac{1}{2n_e}\Re\left(\sigma_{yy}E_0^2\right) , \tag{4.22}$$

or equivalently:

$$\theta_a = \frac{E_0^2}{2n_e}\Re(\sigma_{yy}) . \tag{4.23}$$

The result expressed in (4.21) can be interpreted in a relatively simple fashion if we allow, on the one hand, that the oscillating field $\boldsymbol{E}$ decomposes into a field $\boldsymbol{E}_r$ of a right circular wave and a field $\boldsymbol{E}_l$ of a left circular wave, with equal and constant amplitude, rotating about the magnetic field lines, and, on the other hand, that the electrons, under the influence of the magnetic field, revolve around the same magnetic field lines with a rotating motion to the right (with respect to the direction of the magnetic field). The right circular component of the electric field then turns in the same direction as the electrons in their cyclotron motion. The result is that an electron, in its own frame of reference, "sees" the electric field $\boldsymbol{E}_r$ oscillating at a reduced frequency $\omega_r = \omega - \omega_{ce}$, whereas it "sees" the electric field $\boldsymbol{E}_l$ oscillating at an increased frequency $\omega_l = \omega + \omega_{ce}$. Equation (4.21) includes both contributions: the first term of the RHS corresponds to the right circular wave and the second to that of the left circular wave. It should be noted that the presence of collisions prevents the occurrence of a singularity at $\omega = \omega_{ce}$.

In the case $\omega_{ce} = 0$, we recover the expression given by (4.8). On the other hand, if we adjust the magnetic field intensity such that $\omega = \omega_{ce}$, we then obtain the condition of electron cyclotron resonance, and the electrons, in their own frame of reference, "see" a constant electric field intensity which accelerates them continuously. The energy thus acquired by the electrons is transferred by elastic and inelastic collisions to the heavy particles in the discharge (4.1).

The mechanism of electron heating by ECR is fundamentally different from the mechanism of collisional transfer (4.8). In the case of collisional transfer, the energy imparted at the instant of collision is that acquired during only a fraction of the HF period, because the work done by the electron in the electric field over one or more complete periods is zero. In other words, for collisional transfer, the longer the time between two collisions (i.e. the smaller the collision frequency ν), the less efficient the transfer of energy to the discharge: for $\nu \ll \omega$, the term $\nu/(\nu^2+\omega^2)$, appearing in the definition of θ_a (4.8) reduces to ν/ω^2, which decreases as ν decreases. In contrast, for resonant transfer by ECR, the energy taken from the electric field increases continuously between two collisions, such that the mechanism becomes more effective as the time between collisions increases, which requires $\nu \ll \omega$. Resonant transfer, in fact, reduces to collisional transfer for a static electric field ($\omega = 0$). Indeed, at ECR ($\omega = \omega_{ce}$), the expression for θ_a (4.21), when $\omega \gg \nu > 0$, reduces to the form obtained in the case of a static electric field (4.6), within a factor 1/2 [152].

The interest in ECR and the disappearance of this effect when the plasma becomes more and more collisional are illustrated in Fig. 4.5, which shows the variation, for a constant amplitude field E_0, of the power absorbed per electron, successively, from $\omega_{ce} = 0$, to the resonance condition $\omega_{ce} = \omega$ and above at $\omega_{ce} = 1.5\omega$, as a function of ν/ω (we can assume, to put things in context, that ν, and hence the ratio ν/ω, is proportional to pressure). We also observe that, with respect to the case $\omega_{ce} = 0$, confining the plasma even with a strong field B_0 ($\omega_{ce} = 1.5\omega$) does not result in a significant increase in the power which can be taken by an electron from the HF field. We also see that, for ECR at low-pressure, the power θ_a absorbed per electron is many orders of magnitude greater than that obtained by simple collisional transfer, which is the reason for the interest in ECR for plasma applications at low and very low-pressures (typically below 20 mtorr) because only ECR leads to a sufficiently large value of θ_a to attain $\theta_a = \theta_l$. Such a scheme is implemented, for example, in multi-charged ion sources, where the pressures are even less than 10^{-5} torr, permitting the heating of the electrons to average energies of a few keV. Figure 4.6 illustrates the variation of θ_a as a function of ω_{ce}/ω, still for a constant amplitude field E_0, for different values of the parameter ν/ω (to put this in context, in argon at 1 torr and for a Maxwellian electron distribution function with temperature $T = 2\,\text{eV}$, we have $\nu = 2\times10^9\,\text{s}^{-1}$). Here again, we can verify that resonance is strongly damped for higher collision frequencies, or when moving away from the resonance condition.

Variation of the electric field intensity E_0 around the resonance condition $\omega_{ce} = \omega$

All the preceding calculations and reasoning have been done under the assumption of a constant amplitude electric field. In fact, as was indicated at the beginning of this chapter (Sect. 4.2.1), the electric field intensity constitutes an adjustable parameter, allowing the power balance per electron in the plasma, $\theta_a = \theta_l$ to be satisfied. It is thus important to understand how this adjustable parameter varies in passing from ECR to non-resonance.

[152] The factor 2 results from the decomposition (in the magnetic field) of the planar wave into a right and a left circular wave, only the right circular wave contributing (significantly) to the transfer of power to electrons. In the opposite approximation to that of ECR, namely that $\nu \gg \omega$ (collisional power transfer), each circular wave contributes equally to the power transfer and the result is exactly (4.6).

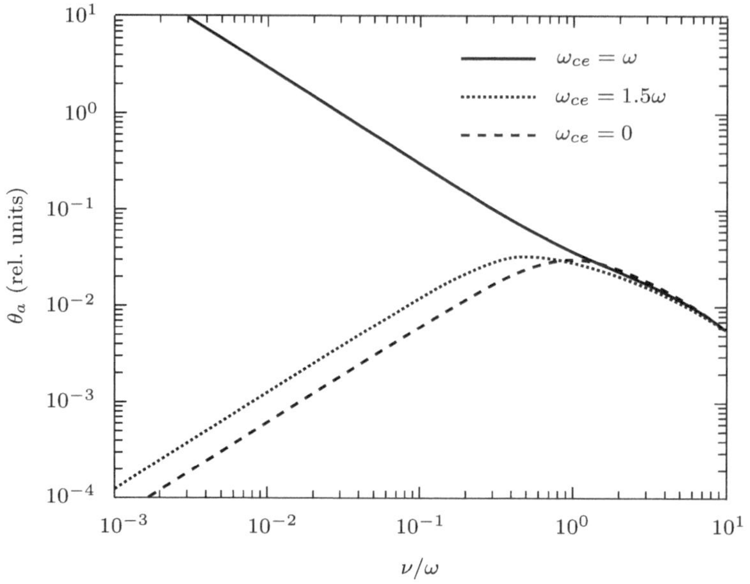

Fig. 4.5 Variation of θ_a as a function of ν/ω, assuming a constant amplitude field E_0, calculated from (4.21) for three values of the angular cyclotron frequency: $\omega_{ce} = 0$, $\omega_{ce} = 1.5\omega$ and $\omega_{ce} = \omega$. Recall that the stationary state requires that $\theta_a = \theta_l$ (for a given ν/ω), which supposes that θ_a can reach values such that $\theta_a \geq \theta_l$. In the opposite case, the discharge cannot exist.

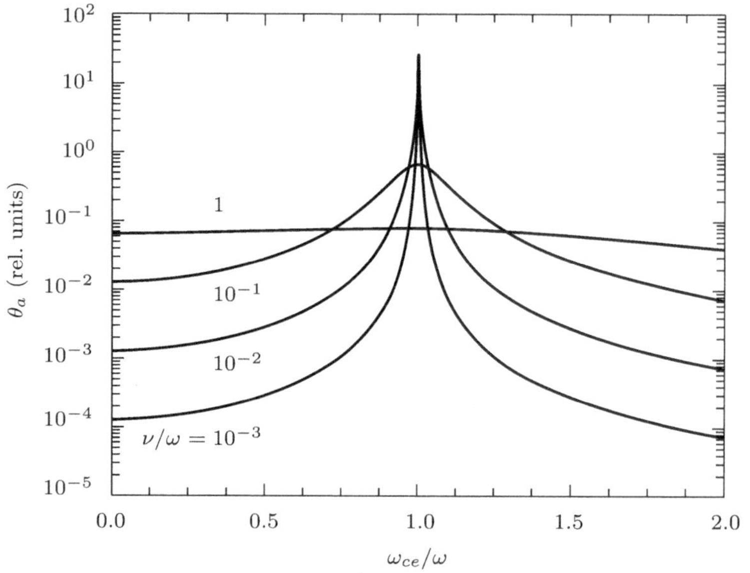

Fig. 4.6 Variation of θ_a (cf. Eq. (4.21)) as a function of ω_{ce}/ω, for constant amplitude field E_0, at four values of the ratio ν/ω.

In extending the reasoning developed above to a stationary discharge in the diffusion regime (Sect. 4.3.3), to which is added a field $\boldsymbol{B}_0$, we claim, for the same reasons, that the parameter θ_l cannot depend on the $\boldsymbol{E}$ field intensity, nor a fortiori on the characteristics of the wave generating the field, but solely on the mechanism for the loss of charged particles. In other words, the value of θ_l does not pass through an extremum at $\omega_{ce}/\omega = 1$, which is confirmed by experiment: θ_a, which is equal to θ_l, decreases progressively and slowly when the magnetic field increases, as is shown in Fig. 4.14 below. On the other hand, we demonstrate in the next paragraph that the electric field intensity $\boldsymbol{E}$ in the discharge, as a function of ω_{ce}/ω, both in the ignition stage and stationary state, shows a minimum at $\omega_{ce} = \omega$, precisely because the efficiency of the resonant transfer of energy, in contrast to collisional transfer, is a maximum in these conditions.

Considering first (4.21), we see that the expression within brackets for θ_a passes through a maximum at $\omega = \omega_{ce}$, the corresponding variation of the electron density n_e, as observed experimentally, being small. The terms within the brackets are, within a constant factor, equal to σ_{yy} in (4.22)–(4.23). As already mentioned, θ_a decreases experimentally very slowly and monotonically with increasing $\boldsymbol{B}_0$, including at ECR it must be concluded that E_0^2 goes through a minimum at ECR. This is contrary to the accepted idea that the intensity $\boldsymbol{E}$ passes through a maximum, which naturally comes to mind when dealing with a "resonance"[153].

In addition, Fig. 4.5 suggests that, at low-pressure, since the value of ν/ω is smaller at 2.45 GHz than at 100 MHz, it might not be possible to ignite the plasma away from resonance for 2.45 GHz, while it is possible at 100 MHz, due to collisional absorption: this is confirmed by experiment. The preceding discussion shows that, when the power lost per electron is very large and ν/ω is small, only ECR enables us to reach a sufficiently high value of θ_a to establish the equilibrium $\theta_a = \theta_l$.

The case of a plasma column maintained by an EM surface wave

Conductivity of a plasma in a bounded medium

The preceding development considered a wave propagating in an infinite, uniform medium. The same type of calculation can be performed in a finite medium for a guided wave, in this case an EM surface wave said to be "generalised" [19]. Fig. 4.7 shows that the *effective conductivity*[154] passes through a

[153] Some authors, such as W. P. Allis [2], claimed that at ECR, the value of the electric field intensity in the plasma should be amplified, although in fact, it passes through a minimum.

[154] Due to the presence of $\boldsymbol{B}_0$, the plasma is an anisotropic medium, and the conductivity is therefore a second-order tensor and no longer a scalar (2.124) to (2.126). The concept of effective electrical conductivity [19] allows this difficulty to be overcome, for a given wave mode, by a scalar representation.

maximum at ECR, such that the electric field intensity E then passes through a minimum.

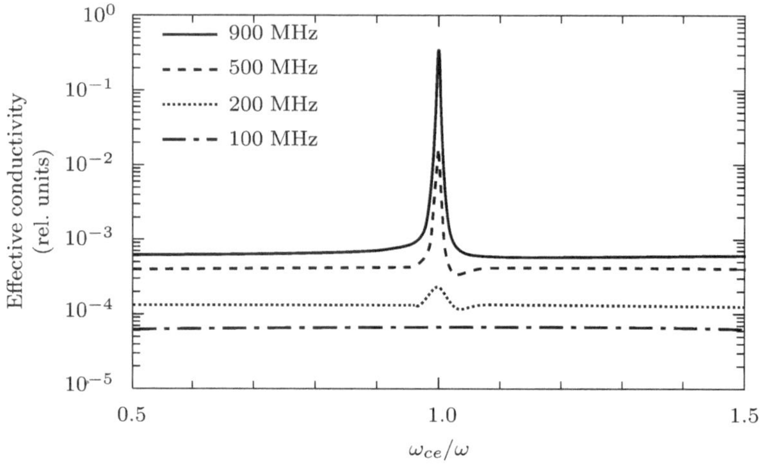

Fig. 4.7 Effective electrical conductivity (taking account of the wave polarisation) due to electrons, calculated for the fundamental mode HE_0, for different values of the frequency of the surface wave field [19].

4.2.4 Variation of the value of θ as a function of $\bar{n}_e$ for different plasma conditions

In the preceding discussion of the power θ we have adopted, for reasons of simplicity, the approximation that ionisation of the atom is due to a single electron impact on it in the ground state (direct ionisation), although we know that stepwise ionisation (using, for example, the metastable states as relays (Sect. 1.8)) cannot be neglected, if the plasma density is sufficiently high. In addition, we have assumed that the loss of particles is effected solely by diffusion (ambipolar), an assumption that is not necessarily valid when the density of charged species is sufficiently large for the recombination of charged species to take place in the plasma volume (Sect. 1.8) (rather than by diffusion to the walls). These two initial assumptions impose the Schottky condition.

Equation (4.24) represents the balance of losses and gains of charged particles in a more general form than the Schottky condition, which only retains the two first terms (the notation is that of Sect. 1.8) [16]:

$$\nabla \cdot (D_a \nabla n_e) + \nu_{id} n_e + \frac{\rho_{ie} n_e^2}{1 + \eta n_e} - \alpha_{ra} n_e^3 = 0 \tag{4.24}$$

The first term represents the loss of charged particles by ambipolar diffusion and the second, that of their creation through direct ionisation of an atom from its ground state. The third term takes account of two-step ionisation (numerator), but also of the possibility of saturation of the corresponding relay states (denominator), while the fourth term, in the absence of molecular and negative ions, represents the volume recombination of atomic ions (a three-body recombination, Sect. 1.8.1). With regard to the third term, we can conclude that when the value of n_e is sufficiently high, the number of available relay states per second decreases (characterised by the term η in the denominator) such that, finally, the frequency of two-step ionisation reaches a maximum, constant value, independent of n_e $(1 \ll \eta n_e)$, since in this case:

$$\nu_{ie} \equiv \frac{\rho_{ie} n_e}{1 + \eta n_e} \simeq \frac{\rho_{ie}}{\eta} \, . \tag{1.159}$$

In (4.24), we have represented the volume recombination by a three-body recombination. However, it should be noted that, in order for three-body recombination to manifest itself (dependence on n_e^3), very high electron densities are required and, from this fact, is less probable than the dissociative recombination of molecular ions (Sect. 1.8.1) formed in the discharge (dependence on n_e^2)[155]. Thus, although the density of molecular ions[156], for example in a rare gas discharge, might be less than that of atomic ions, their recombination frequency is many orders of magnitude larger than that for atomic ions, such that for $n_e \leq 10^{14}\,\mathrm{cm}^{-3}$, it is more correct to write the charged particle balance equation in the form:

$$\boldsymbol{\nabla} \cdot (D_a \boldsymbol{\nabla} n_e) + \nu_{id} n_e + \frac{\rho_{ie} n_e^2}{1 + \eta n_e} - \alpha_{rm} n_e^2 = 0 \, . \tag{4.25}$$

By taking account of the losses both by ambipolar diffusion and by dissociative recombination, the calculation of the power θ as a function of $\bar{n}_e$, the average density across a radial section of the discharge tube, leads to Fig. 4.8 [16,17,36]. For low electron density (region I), the Schottky condition applies, and the value of θ is that determined from the horizontal dotted line in Fig. 4.8: the loss of charged particles and their creation are, in effect, both linear in $\bar{n}_e$, such that their corresponding frequencies are independent of $\bar{n}_e$.

For values of $\bar{n}_e$ that are slightly larger (region II), the charged particles continue to be lost by ambipolar diffusion, but multi-step ionisation supplements the direct ionisation, thus reducing the power taken from the field to produce an electron-ion pair, hence the decrease in θ. For even larger values of $\bar{n}_e$ (region III), the loss regime is still ambipolar diffusion, but multi-step

155 In general, a three-body recombination is always less probable than a two body recombination, hence the importance of the dissociative recombination of molecular ions according to the reaction (1.152).

156 The kinetics of molecular ions is only considered in the high-pressure section of this chapter (Sect. 4.4).

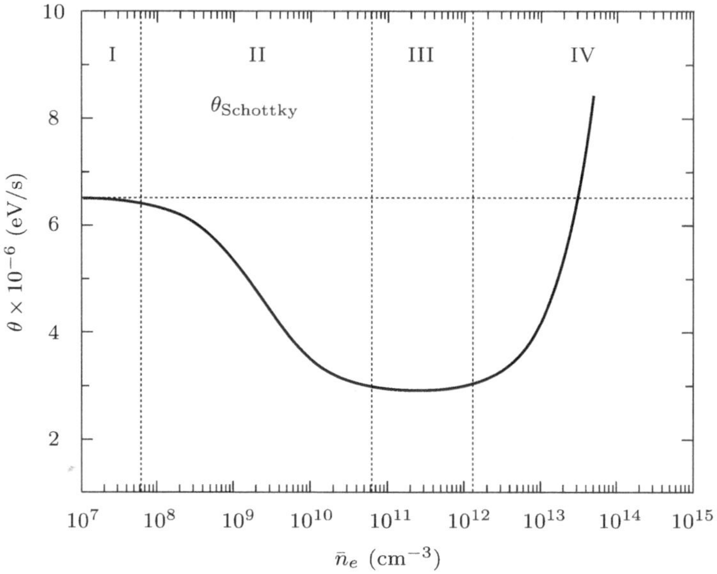

Fig. 4.8 Variation of θ as a function of $\bar{n}_e$, the average electron density across the radial section of the discharge tube, according to (4.25), calculated for $p = 0.5$ torr and $R = 1.4$ cm, in argon [16,17,36].

ionisation no longer allows the electron density to grow faster than $\bar{n}_e$, because the saturation regime (1.159) has been reached, so that the value of θ remains constant as a function of $\bar{n}_e$.

Finally, for even larger electron densities (region IV), the loss of charged particles results from both ambipolar diffusion and volume recombination, the latter progressively taking over from the former as $\bar{n}_e$ increases. Regarding the creation of these particles, it occurs in the saturated regime of multi-step ionisation, and the frequency ν_{ie} is independent of $\bar{n}_e$. In total, the frequency of charged particle loss increases with $\bar{n}_e$, and θ increases, as can be demonstrated, as the square root of $\bar{n}_e$ [17].

4.3 Influence of the frequency of the HF field on some plasma properties and on particular processes

One of the specific advantages of HF plasmas compared to DC discharges is the ability to vary the plasma properties considerably (notably the EEDF) by adjusting the frequency of the applied EM field. We will make use of a theoretical approach to describe this frequency effect, confirmed by some experimental examples.

4.3.1 Posing of the problem

Generally speaking, plasmas at reduced pressure are not in thermal equilibrium. In practical terms, the average electron energy is much greater than that of the ions and neutrals (Sect. 1.4.3). From this, we conclude that the electron collisions, and not those due to the heavy particles, are primarily responsible for the ionisation of atoms and molecules, so that the form of the *electron energy distribution function* $F_0(U)$ (EEDF)[157] determines[158] the relative repartition of the densities N_j of the different excited states of the atoms and molecules in the discharge. To see this, we will consider a simple but common case, where the excited or ionised species are produced by a single electron impact on an atom (molecule) in the ground state. The density of species thus formed, per second, in the state j (excited or ionised) is then given by:

$$\frac{\mathrm{d}\mathcal{N}_j}{\mathrm{d}t} \equiv \dot{\mathcal{N}}_j = \langle \nu_j \rangle n_e \equiv k_{0j}\mathcal{N}_0 n_e \,, \tag{4.26}$$

where $\mathcal{N}_0$ is the density of the neutral atoms in the ground state. The expression for the *excitation coefficient* k_{0j} is:

$$k_{0j} = \left(\frac{2}{m_e}\right)^{1/2} \int\limits_{V_j}^{\infty} \hat{\sigma}_j(U) F_0(U) U^{\frac{1}{2}} \,\mathrm{d}U \,, \tag{4.27}$$

where $\hat{\sigma}_j$ is the total microscopic excitation cross-section (Sect. 1.7.4), which is a function of the electron energy U, above the energy threshold V_j. Equation (4.27) shows that, depending on the shape of $F_0(U)$, the value of the coefficient k_{0j} changes, which leads to the different coefficients varying relative to each other. For certain applications (see Sect. 4.3.7 for examples), this permits an increase in the population of a specific excited or ionised level, thus optimising a process.

The reader should note that we are particularly interested in HF discharges where the angular frequency $\omega = 2\pi f$ is sufficiently large that the ions cannot respond to the periodic variation of the HF field, so only the electrons can take energy from the field (see Sect. 4.2.1). The ions effectively remain stationary if $\omega \gg \omega_{pi}$, where ω_{pi} is the ion plasma frequency (Sect. 1.5); for example, in

[157] See Appendix XVII for the distinction between $F_0(U)$ and $f_0(U)$. The normalisation condition for $F_0(U)$ is:

$$\int\limits_0^{\infty} F_0(U)\sqrt{U}\,\mathrm{d}U = 1 \,.$$

The zero-subscript implies an isotropic distribution (which, in the present case, is the first term of the development of the distribution function in spherical harmonics (Sect. 3.1)).

[158] Electron collisions completely determine the population of an excited state only if the population and depopulation of these levels by (electric dipole) radiative transitions are negligible.

low-pressure argon discharges ($n_e \simeq 10^{10}\,\mathrm{cm}^{-3}$), this occurs when the applied frequency f is greater than a few MHz.

In the following, the effect of the frequency of the field on the EEDF will lead us to distinguish between two typical cases, one in which the EEDF is stationary and the other where, on the contrary, it oscillates, totally or partially, as a function of the period of the HF field. For the EEDF to vary with the period of the field, it is necessary for the *frequency of transfer of energy from an electron* (of energy U) *to the heavy particles*[159]:

$$\nu_u(U) \equiv \frac{2m_e}{M}\nu(U) + \sum_j \nu_j(U) \tag{4.28}$$

to be such that $\nu_u(U) \gg \omega$. In this case, the total number of elastic[160] and inelastic collisions is so large during an HF period that the transfer of energy from the electric field to the heavy particles via the electrons takes place at each instant of the period of the HF field: the EEDF is thus subject to the **instantaneous** value of the amplitude of the HF field and therefore varies as a function of time over the period duration $\mathcal{T} = \in\pi/\omega$ (this effect typically manifests itself for frequencies below 100 MHz). On the contrary, for $\nu_u(U) \ll \omega$, the EEDF is stationary because the collisions, one or none during a period, occur at different times during the period, from one period to the next.

In the case of discharges in atomic gases, the value of $\nu_u(U)$ increases abruptly above the energy of the first excitation level, while for molecular gases, $\nu_u(U)$ reaches a high value already at low energies as a result of the transfer of energy through ro-vibrational states, as shown in Fig. 4.9.

4.3.2 The EEDF in the non-stationary regime

The transition from the non-stationary regime to the stationary regime is achieved, at constant pressure, by increasing ω. The EEDF becomes progressively more stationary, starting with the lowest energy electrons, because the condition $\nu_u(U) \ll \omega$ is first satisfied for low values of U, as can be seen in Fig. 4.9. When the angular frequency ω is sufficiently large, such that $\nu_u(U) < \omega$ for all values of U, the EEDF becomes stationary. To put this in context, in order to determine the value of ω for which

159 The total power transferred from an electron of energy U to heavy particles can be obtained by multiplying the first term on the RHS of (4.28) by U and the second term by V_j (excitation and ionisation). The corresponding mean total power value $\langle \nu_u(U) U \rangle$ transferred to heavy particles is then, according to (4.1), equal to θ_l.

160 In the expression for $\nu_u(U)$, the number of elastic collisions is weighted by the factor $2m_e/M$, the maximum fraction of kinetic energy which an electron can transfer to a heavy particle as the result of an elastic collision (1.100).

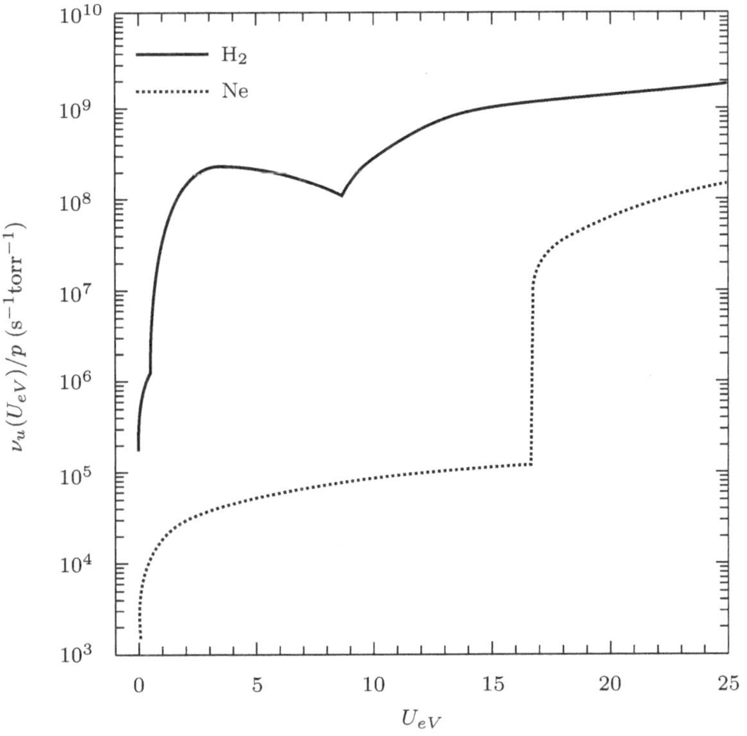

Fig. 4.9 Frequency $\nu_u(U)$ for the transfer of energy from electrons to heavy particles in the case of neon and of molecular hydrogen (following [42]). The gas pressure p is expressed at 0°C.

the EEDF becomes stationary, we can take $\max[\nu_u(U)] = \omega$ as a practical criterion for the lower bound of the stationary regime, such that from Fig. 4.9 for neon[161, 162] $\omega/p = 10^8\,\mathrm{s^{-1}\,torr^{-1}}$. According to this criterion, at $p = 0.2$ torr, the EEDF in neon would be stationary for $f > 3$ MHz, while for H_2 ($\omega/p = 1.5 \times 10^9\,\mathrm{s^{-1}\,torr^{-1}}$) this requires $f > 48$ MHz [12]: the stationary state is attained for a much lower value of ω in the case of an atomic gas.

Fig. 4.10 corresponds to an intermediate case of the ratio $\nu_u(U)/\omega$, such that the *bulk of the EEDF*, comprising the low energy electrons, is stationary while the *tail of the EEDF*, comprising the electrons with energies above the first excitation threshold ($V_j = 16.6$ eV for neon) is not; the tail of the EEDF varies significantly as a function of time within the period of the HF field. It can, however, be demonstrated [34] that the density of the electrons is, in practice, stationary because, as is suggested by Fig. 4.10, there are relatively

[161] Detailed calculations show that, for $\omega/p = \pi \times 10^5\,\mathrm{s^{-1}\,torr^{-1}}$, the EEDF is already stationary (see [41]).

[162] The pressure p is expressed relative to 0°C.

few (rapid) electrons affected by the periodic motion of the tail when the angular frequency ω is not too low: for argon, n_e is already stationary when $f > 100\,\mathrm{kHz}$!

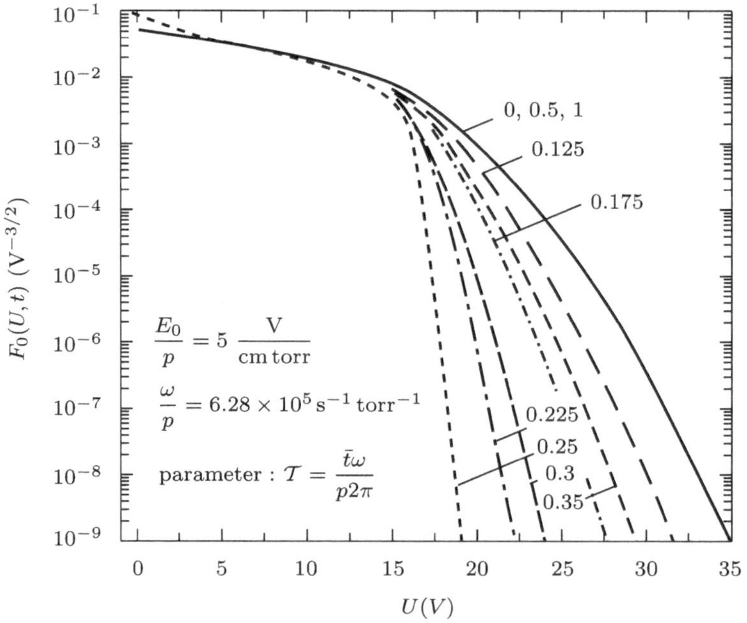

Fig. 4.10 Isotropic part of the energy distribution function of electrons calculated in the case of neon for $\omega/p = 2\pi 10^5\,\mathrm{s}^{-1}\,\mathrm{torr}^{-1}$ at different fractions of the period duration $\mathcal{T}$ of the HF field; $\bar{t} = pt$ is the reduced time and p the gas pressure expressed at 0°C (from [41]).

4.3.3 EEDF in the stationary regime

In order to understand the essential features of this problem, we will assume the EEDF is stationary, and also homogeneous. This EEDF can be obtained from the stationary, homogeneous Boltzmann equation, which can be written in the form [2]:

$$-\frac{2}{3}\frac{\mathrm{d}}{\mathrm{d}U}\left[U^{3/2}\nu(U)U_c(U)\frac{\mathrm{d}F_0}{\mathrm{d}U}\right] = S_0(F_0)\,, \tag{4.29}$$

where the collision operator $S_0(F_0)$ represents the influence of the collisions between electrons and other particles in the plasma; the quantity:

$$\nu(U)U_c(U) \equiv \frac{e^2}{m_e}\frac{\nu(U)}{\nu^2(U)+\omega^2}\overline{E^2}\,, \tag{4.30}$$

which has the units of energy per second (see (4.8)), represents the *power transferred* (on average over a period) *from the HF electric field to an electron of energy* U[163]. This transfer is a function both of U and ω: it is a maximum for the energy U of an electron such that $\nu(U) = \omega$ (from the derivative of (4.30) with respect to U). Thus, when the frequency maintaining the discharge is varied, the maximum power transfer occurs at an energy U which varies with the value of ω, so that it influences the LHS of (4.29), and modifies the form of the EEDF, unless it is Maxwellian and remains so for all values of $\nu(U)/\omega$. For a fixed value of ω (non-Maxwellian case), the EEDF depends on $\nu(U)$, which depends, in turn, on the collision cross-section for momentum transfer, a property specific to each gas. Some of these cross sections vary strongly as functions of the energy of electrons: this is the case, for example, for argon, and much less so for neon (Fig. 1.14). The case of helium is particularly interesting because the product $wP_m(w) = \nu(w)$, i.e. $\nu(U)$, is almost constant[164]. Therefore, for any frequency value ω, the power transferred from the HF electric field to an electron (4.30) is independent of its energy U and, consequently, the power density transferred to the electrons is independent of the EEDF: the effect of frequency on the EEDF of helium plasmas is expected to be negligible.

In the following sections, we will illustrate the influence of ω on the EEDF, for the case of an argon discharge. We will see that we can distinguish three limit cases, giving rise to three completely different EEDFs.

Remark: Since both the collision operator $S(F_0)$ and $\nu(U)$ are proportional to N [11], both sides of (4.29) can be divided by N, and if $\nu \gg \omega$, the LHS of (4.29) depends on the ratio E^2/N^2, thereby defining a *microscopic scaling law*. Varying E and N such that E/N is constant will lead to the same solution of the Boltzmann equation. As ν is also proportional to p (1.130), E/N and E/p thus constitute two variants of this scaling law.

Peforming an integration of (4.29) over the EEDF leads to the expression for θ (4.8), so that θ/p also constitutes a scaling law, this time in the macroscopic domain. Note that θ/p is proportional to $(E/\nu)^2$, or equivalently $(E/N)^2$, in the case where $\nu \gg \omega$ (see, for example, Fig. 4.13).

---

163 Unlike (4.8), where the value of ν is an average value taken over the EEDF, $\nu(U)$ is here a microscopic frequency, as indicated by its explicit dependence on U.

164 This allows us to write in addition, for the case of helium, that $\nu/p_0 \simeq 2.4 \times 10^9\ \mathrm{s}^{-1}$.

4.3.4 Three limit cases of the influence of ω on a stationary EEDF

As was shown in equations (4.29) and (4.30), the influence of ω on the EEDF occurs more exactly through the ratio $\nu(U)/\omega$. For this effect to be manifested, however, it is necessary that the number of electron-electron collisions be sufficiently small, such that the EEDF is non Maxwellian. Excluding therefore the case of a Maxwellian EEDF, there are two possible limit cases: $\nu/\omega = 0$, referred to as the *microwave* (MW) regime, and $\nu/\omega \to \infty$, referred to as the *direct current* (DC) regime, which corresponds effectively to the DC discharge condition ($\omega = 0$), but also to an HF discharge at sufficiently low frequency that $\nu/\omega \to \infty$ (recall that if the angular frequency ω is too low, the EEDF is not stationary)[165].

Figure 4.11 shows a comparison, for the same value of the product pR, of the EEDFs calculated in the two limits (DC and MW), together with a Maxwellian EEDF (M). We can see that the three EEDFs are completely different from each other. In particular, we can distinguish the electrons in the bulk of the distribution from those in the tail. Figure 4.11 shows that the section of the tail between V_1, the energy threshold for the first excited state, and V_i, the energy threshold for ionisation, i.e. the region bounded by the vertical lines in Fig. 4.11, is more populated in a discharge in the DC regime than when the EEDF is Maxwellian (M) or when it is in the MW regime. This signifies that the excitation coefficient k_{0j} (4.25) and hence N_j, the density of atoms in excited state j, at **constant electron density**[166], are larger in the DC regime; in consequence, there are relatively fewer low-energy electrons in a DC discharge. This translates into a higher average electron energy in the DC limit, which is supported by calculations which, in the present case, yield $\langle U_{eV} \rangle = 6.8$, 2.35 and 3.15 eV, for the DC and MW cases, and for the Maxwellian EEDF, respectively [24].

Figure 4.12 shows the behaviour of the EEDF in the DC and MW cases of Fig. 4.11 if account is taken of electron-electron collisions: as one might expect, the difference between the DC and MW regimes is reduced when the electron density increases (because the electron-electron collisions increase). The difference vanishes for sufficiently high values of n_e. Once again, for the frequency to have a significant effect on an HF discharge, the electron density

---

[165] In the following, since this concerns limit cases $\nu/\omega \to 0$ or $\nu/\omega \to \infty$, we can simply consider the average value of ν rather than the microscopic value $\nu(U)$.

[166] The EEDFs in Fig. 4.11 are normalised (the same area under each curve), the normalisation condition (see note 188 at the bottom of the page of Appendix XVII) being:

$$\int_0^\infty F(U) U^{\frac{1}{2}} \, \mathrm{d}U = 1 \,.$$

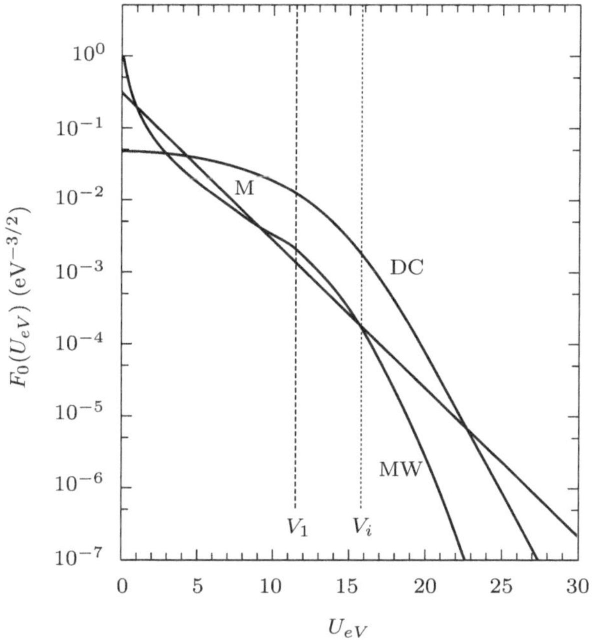

Fig. 4.11 Electron energy distribution function for an argon plasma in a long cylindrical tube of radius R, resulting from calculations using a self-consistent model assuming an ambipolar diffusion regime and direct ionisation from the ground state only ($pR = 0.15$ torr-cm) [24]. The curve M is for a Maxwellian EEDF (sufficiently numerous electron-electron collisions), while the curves DC and MW correspond to the limits $\nu/\omega \gg 1$ (direct current regime) and $\nu/\omega \ll 1$ (microwave regime) described in the text.

should not be too large. From a practical point of view, these calculations show that this is the case in most gases for[167] $n_e/N < 10^{-4}$.

Remark: In the case of a Maxwellian distribution, the average energy (in this case related to T_e) depends on the configuration and the dimensions of the discharge vessel, and the type and pressure of gas (Sect. 3.13). However, in the more general case, from Fig. 4.11, it is necessary to add the angular frequency ω of the field[168] to the set of *operating conditions*. More generally, we can say that the form and the mean energy of the EEDF are fixed by these operating conditions, but also by the absorbed power density in the discharge: although the power density is not included in the operating conditions, raising

---

[167] Thus, at $n_e/N = 10^{-3}$, in an argon discharge, the electron-electron collisions are such that the EEDF is almost Maxwellian, although the number of these collisions is often less than that of the electron-neutral collisions. This is because the electron-neutral collisions are less efficient in transferring energy, limited to a factor of m_e/M on each collision, while electron-electron collisions can transfer all of the energy of one electron to another (Sect. 1.7.2).

[168] This explains why the values of $\langle U \rangle$, for a given value of pR, can be different.

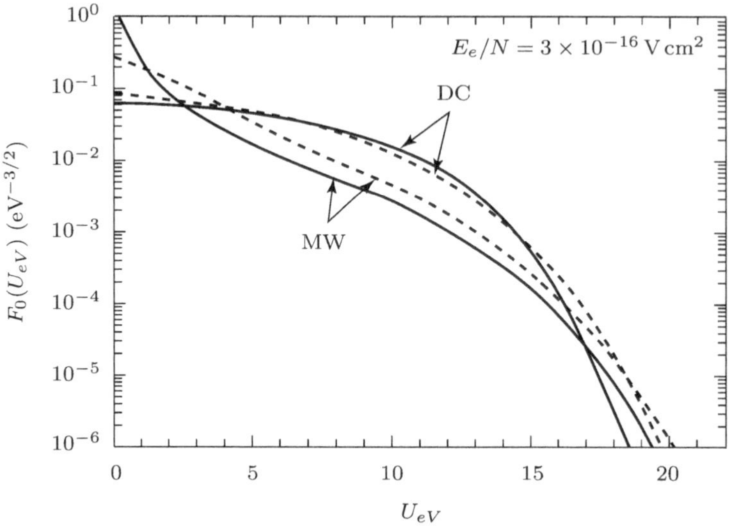

Fig. 4.12 Electron energy distribution functions calculated in argon, in the DC ($\nu/\omega = \infty$) and MW ($\nu/\omega \simeq 0.1$) limit cases for $n_e/N = 0$ (full curves) and $n_e/N = 10^{-4}$ (dashed curves) with direct ionisation from the ground state [33].

it increases the plasma density (see (4.4)), hence the electron-electron collision frequency [43].

4.3.5 Influence of ω on the power θ

We have seen that the values of $\langle U \rangle$ and θ in a low-pressure discharge depend only on the operating conditions and the absorbed HF power density, but we have not yet discussed the role of the power density. Increasing the power density, and hence the electron density n_e, not only increases the number of electron-electron collisions, but also the importance of multi-step ionisation with respect to direct ionisation from the ground state of the atom: this increase in n_e thus corresponds to a reduction in the average energy of the electrons, and hence of θ (Sect. 4.2.4). We will not consider this effect in the following sections, but only the effect of ω on the EEDF.

Figure 4.13 shows the theoretical dependence of θ (in fact θ_l) on pR for decreasing values of the ν/ω ratio, going from $\nu/\omega = \infty$ (DC case, curve A) to $\nu/\omega = 0$ (MW case, curve H) in the absence of electron-electron collisions, in argon: we can observe that, for a given pressure, the value of θ decreases when ω increases [24]. The curve M corresponds to the case when the electron-electron collisions are sufficiently numerous for the EEDF to be Maxwellian.

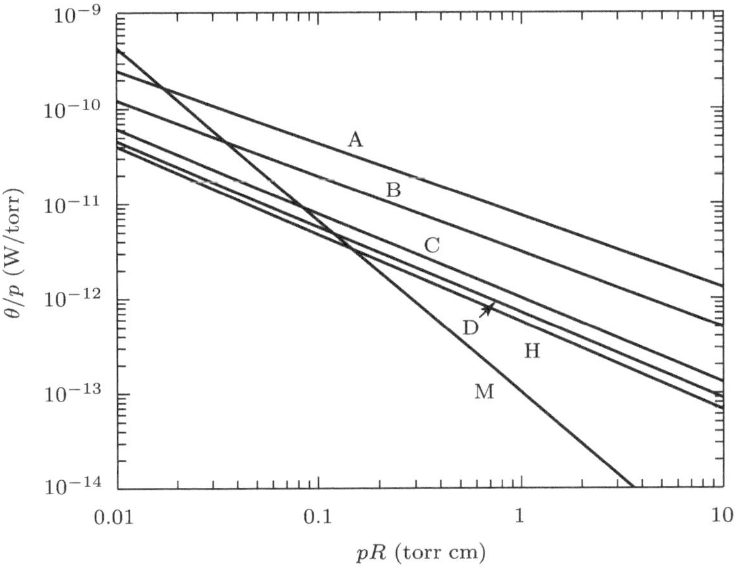

Fig. 4.13 Calculated values of θ/p as a function of pR for decreasing values of ν/ω, going from $\nu/\omega = \infty$ (DC case, curve A) to $\nu/\omega = 0$ (MW case, curve H) and for a Maxwellian EEDF (M) in the ambipolar diffusion regime and with direct ionisation from the ground state only [24].

4.3.6 Density of species produced per second for a constant absorbed power density: energy efficency of the discharge

In Sect. 4.3.4, we have seen that the excitation coefficient k_{0j} is the highest in the case of the DC discharge regime, and hence the density N_j of atoms (molecules) in the state j, produced per second, is also the largest, for discharges with a given electron density. Of the three limit cases examined, which is the one that, for a **given power density** P_a, leads to the largest value of N_j? To answer this question on the energy efficiency of the discharge, recall that $\dot{\mathcal{N}}_j = (k_{0j}\mathcal{N}_0)\,n_e$, where n_e is given by $P_a = n_e\theta$ ((4.26) and (4.4) respectively) and consider the group of curves in Fig. 4.13. We can see that the smallest value of θ is attained for a Maxwellian EEDF, provided $pR \geq 0.1$ torr-cm. Thus, for a given absorbed power density P_a in the plasma, a greater number of electron-ion pairs (assuming the atoms and molecules are ionised only once) is obtained in the MW case than in the DC case, and even a little more with a Maxwellian EEDF, when the product pR is sufficiently large. This signifies, among other things, that one should not use a DC discharge to operate an ion source in which one expects the best possible energy efficiency (highest ion density at a given P_a), at least when the electron density in this discharge is not sufficiently large for the EEDF to be Maxwellian.

The calculation of θ/p in Fig. 4.13 has been made for argon as the main (carrier) gas in the discharge. If we add another gas, assuming that this is at the trace level, such that the properties of the gas remain that of the carrier gas [24], we can use a perturbative approach to calculate $k_{0j'}$, the direct excitation coefficient for the state j' of the trace gas with energy threshold $V_{j'}$, with V_i the ionisation threshold energy of the main gas. We can draw the following conclusions:

- the DC regime ($\nu/\omega = \infty$) gives the lowest value of $\dot{\mathcal{N}}_{j'}$ for a given power P_a,
- the excitation to the state j' is generally more effective when the EEDF is Maxwellian, with some exceptions in favour of the microwave case for certain values of energy $V_{j'} < V_i$ (for more details, see [24]).

The influence of ω on the EEDF depends, as we have seen, on the gas considered. The example of argon treated above can thus be considered as a particular case, but nevertheless it shows the general characteristics required for understanding the influence of frequency (Sect. 4.3.3).

4.3.7 Experimental and modelling results

In the next section, we will report experimental measurements of the effect of a magnetic field $\boldsymbol{B}_0$ on the value of θ_a (recalling that $\theta_a = \theta_l$), which has strongly inspired our modelling of the ECR (Sect. 4.2.1). In a second part, we will firstly examine the influence of the frequency of the HF field on the intensity of the UV light emitted from a hydrogen discharge, and then on the coating and etching rates of polymers. We will endeavour to explain these experimental results within the theoretical framework that we have developed.

Influence of a stationary magnetic field on the value of θ

We consider an HF discharge in a cylindrical vessel subject to a magnetic field $\boldsymbol{B}_0$, directed axially[169]. Figure 4.14 shows that, for a given pressure, the value of θ_a decreases when ω_{ce}/ω increases; this effect is however reduced when the pressure increases, due to the corresponding increase in ν (see (4.8) and (4.21)), and disappears, in the present case, for $pR > 1$ torr-cm. In addition, we can see that the values of θ_a/p as a function of pR in the case where $\omega_{ce}/\omega = 1$ (the ECR condition) are hardly distinguishable from

[169] In the present case, this discharge is maintained by means of an electromagnetic surface wave (see [25] or Appendix XIX), but the results obtained, for these values of plasma density, are independent of the method used to create the discharge [23].

those for $\omega_{ce}/\omega \neq 1$: the evolution of θ_a/p as a function of ω_{ce} is monotonic. This result is in agreement with the model that we have developed above (Sect. 4.2.3): the value of θ_a is fixed by the loss of charged particles and is independent of the method of introducing the HF power into the discharge, which can be through resonant or collisional absorption [20].

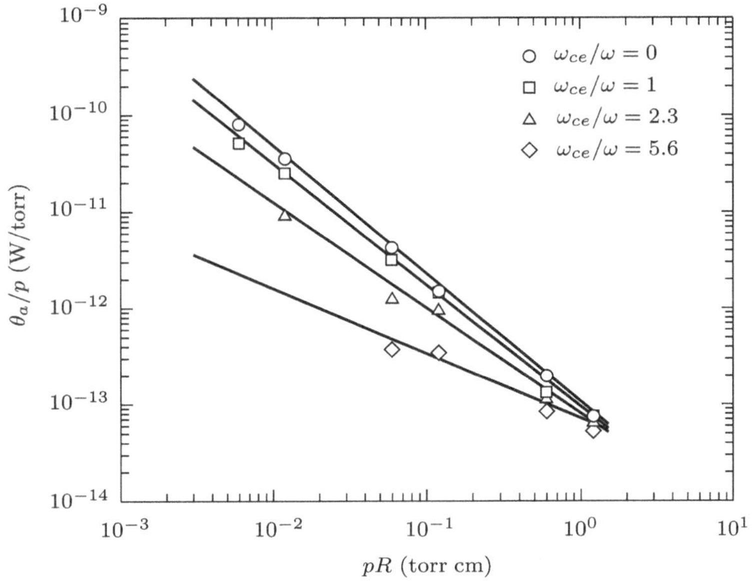

Fig. 4.14 Measured values of θ_a/p, as a function of pR, for different values of the magnetic field $\boldsymbol{B}_0$ (argon discharge sustained by the HF field from an EM surface wave with azimuthal symmetry ($m = 0$) [20], $\omega/2\pi = 600\,\text{MHz}$, $R = 13\,\text{mm}$).

Problems encountered when trying to show the influence of the frequency of the field $\boldsymbol{E}$ on the properties of the HF plasma

The existence of an optimum frequency for the efficiency of a given process has been observed experimentally, but in order to explain this in terms of the basic mechanisms that we have described, it is necessary to know some plasma parameters that are often difficult to measure (limits to plasma accessibility by the diagnostics, perturbations in the HF field due to the measuring probes). Furthermore, to conduct an experiment in which only the frequency is varied and the other operating parameters remain constant is difficult [20]. For example, it is usually impossible to generate HF plasmas with the same configuration of electric field $\boldsymbol{E}$ at radio and then at microwave frequencies. Only plasmas produced by electromagnetic surface waves allow such a parametric study (Appendix XIX). A further difficulty in interpreting the results occurs when the EEDF oscillates (non-stationary EEDF) combined with the variation in shape of the EEDF due to the effect of ν/ω (Sects. 4.3.2 and 4.3.3). The following experimental results show that it is possible to optimise a process by tuning ω.

Effect of the transition from a non-stationary EEDF to a stationary EEDF on the intensity of UV emission from a plasma

Figure 4.15 shows the influence of the field frequency on the intensity of emission of the Lyman α (Ly_α) line and the H_β, H_γ and H_δ lines of the Balmer series in a pure H_2 discharge [12]. The aim of this exercise is to maximise the UV radiation intensity, for a given absorbed HF power, of the Ly_α transition ($\text{N} = 1 \rightarrow \text{N} = 2$, N being the principal quantum number of the hydrogen atom), for irradiation of polymers. Examination of the 3 Balmer transitions as functions of the field frequency indicates that they behave similarly to the Ly_α line. The calculations in [12] show that the transition from the low intensity regime to the high intensity regime takes place for $f \geq 80\,\text{MHz}$, in other words, when the EEDF reaches a stationary state. This increase in intensity could be related to the increase in the number of electrons in the tail of the distribution, which are the only electrons capable of dissociating H_2 molecules and exciting hydrogen atoms.

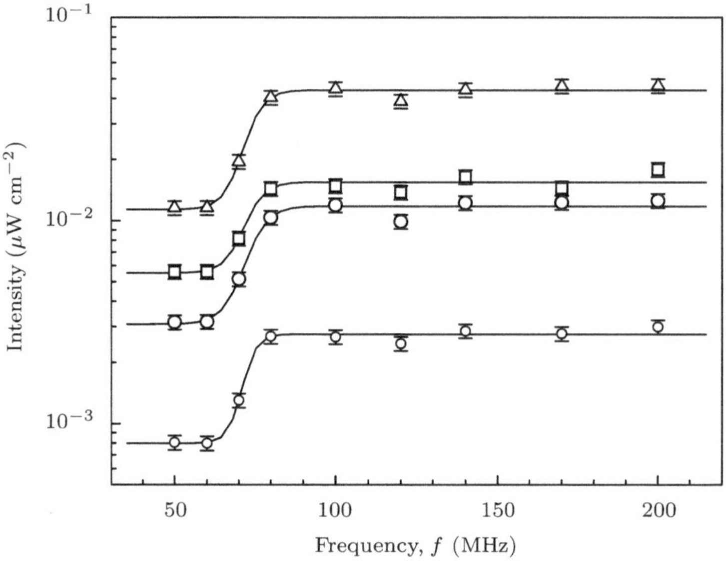

Fig. 4.15 Frequency dependence of the intensities of atomic hydrogen lines from the hydrogen plasma ($P_t = 50\,\text{W}$, $p = 0.5\,\text{torr}$, $R = 13\,\text{mm}$): □ Ly_α (121.5 nm), ◊ H_δ (410.2 nm), ○ H_γ (434.1 nm), △ H_β (486.1 nm) [12].

Effect of the transition from a non stationary EEDF to a stationary EEDF on polymer deposition

Figure 4.16 shows the variation of the deposition rate of thin films on polymers, normalised to P_t, the total HF power absorbed[170], as a function of the frequency of the applied HF field. These coatings have been obtained from C_4H_8 (isobutylene) or C_4F_8 (perfluorocyclobutane) using argon as the carrier gas, the supply of the monomer being (under standard temperature and pressure conditions) 3 standard cubic centimeter/sec (sccm) and that of argon 10 sccm, with a total pressure of 0.2 torr [8]. The transition to the upper plateau for $f \geq 100$ MHz corresponds to the transition, for a stationary EEDF, from the DC case to that of a Maxwellian [24]. However, taking into account the presence of molecules in the discharge (which cause the stationary EEDF to be reached at a higher frequency than for atomic gases), one might then think that this involves the transition from a non-stationary EEDF to a stationary EEDF (DC case). Since we do not have such calculations on the EEDF at our disposal, it appears difficult to answer this question.

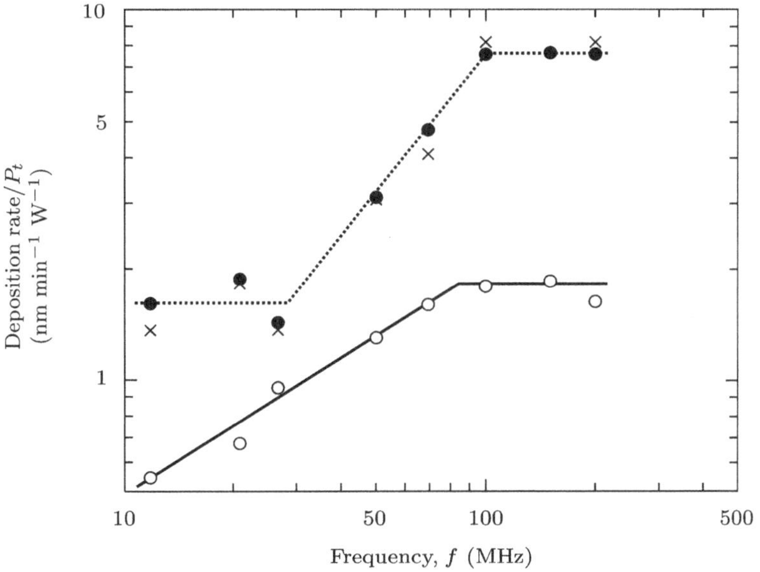

Fig. 4.16 Growth rate of polymer films, normalised to the total absorbed power P_t, as a function of the frequency of the applied HF field, $R = 30.5$ mm (○ C_4F_8, ● and × C_4H_8, for two values of P_t) [8].

[170] Since we are unable to maintain a constant power density in the present case, we normalise the deposition rate to P_t, the total power absorbed in the discharge.

An increased deposition rate is an obvious industrial objective. In the present case, the resulting optimisation allows not only a reduction in the power density used, but of the gas temperature in the discharge, which in the present case leads to a better quality of deposition.

Effect of frequency on etching rate

Figure 4.17 describes the etching rate of a polymer (polyamide), normalised to the total absorbed power P_t, as a function of the frequency f of the HF field. The etching takes place without intentional biasing of the substrate (that is, at the floating potential Sect. 3.14), in a plasma of O_2-CF_4 at 0.2 torr and with a total gas flow of 0.1 sccm [35]. In contrast to the preceding case (Figs. 4.15 and 4.16), the etching rate does not switch from one plateau to another, but goes through a maximum. This suggests that competitive phenomena take place simultaneously, making the interpretation more complicated. These effects are related to the characteristics of the EEDF. Thus, when f increases:

1. The EEDF becomes stationary, and a maximum number of electrons is obtained in the tail of the aforesaid EEDF, leading to a corresponding increase in the dissociation of the O_2-CF_4 molecules.
2. The EEDF (ν/ω effect) tends towards the MW case, giving rise to a reduction in θ (Fig. 4.13) and a correlated increase in the plasma density, leading finally to a Maxwellian EEDF.
3. If θ decreases, $\langle U \rangle$ (T_e in the case of a Maxwellian EEDF) also decreases (see Fig. 4.3), then the potential difference $V_p - V_f$ of the substrate sheath decreases (Sect. 3.14).

Consequently, when f increases, the energy of the ion bombardment on the surface to be etched decreases (3), although the flux of ions (2) and reactive species (1 and 2) increases, hence the possibility of finding a maximum in the etching rate.

4.3.8 Summary of the properties of low-pressure HF plasmas

The average power lost per electron through collisions with heavy particles is seen to be an essential parameter in the description of low-pressure ($\leq$ 15 torr in argon) electric discharges (we could equally show that this is also true for high-pressure discharges, including atmospheric). The operating parameters of these discharges can be chosen such that the frequency of the HF field modifies the shape of the EEDF, or produces either a stationary or non-stationary EEDF, allowing the optimisation of the kinetics of a given process. As a particular case of interest, we have seen that, under Schottky conditions, for a

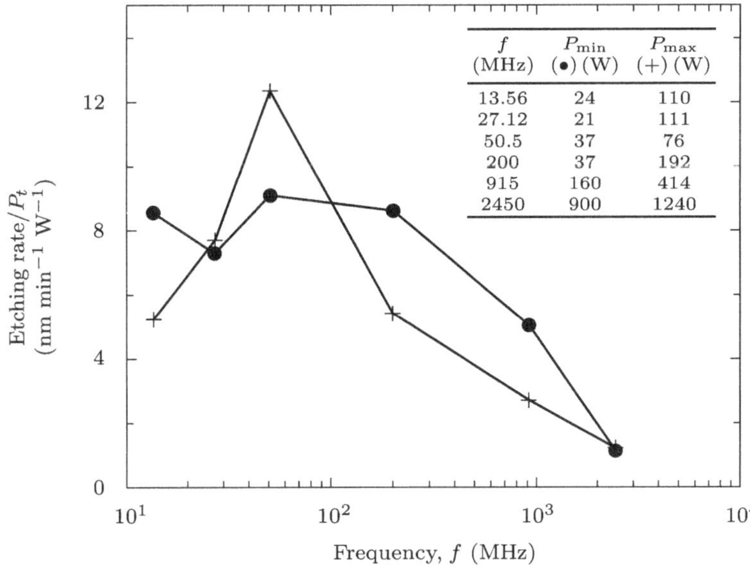

f (MHz)	P_{min} (•) (W)	P_{max} (+) (W)
13.56	24	110
27.12	21	111
50.5	37	76
200	37	192
915	160	414
2450	900	1240

Fig. 4.17 Etching rate of a polyamide, normalised to the total absorbed power P_t, for two series of values of P_t [35]. Discharge in O_2 containing 6% CF_4 (total pressure 200 mtorr, total gas flow of 0.1 sccm, $R = 26$ mm).

given power density absorbed by the plasma, microwave discharges produce a greater number of electron-ion pairs than DC discharges (if the electron density is not sufficiently high for the EEDF to be Maxwellian). Finally, the parameter θ is also instrumental in explaining how the electron cyclotron resonance favours the maintenance of HF discharges at gas pressures that are notably smaller than in the absence of the magnetic field.

4.4 High-pressure HF sustained plasmas

High-pressure plasmas are distinguished from low-pressure plasmas, not by their method of production[171], but by the mechanisms by which they lose charged particles. These mechanisms involve two specific factors: the increase in the gas temperature T_g due to the increase in elastic electron-neutral collisions and the formation of molecular ions, which may dominate over the atomic ions in determining the creation and loss of charged particles. This results in phenomena such as contraction and filamentation of the plasma.

---

[171] In both cases, ionisation results from electron collisions on the atom. At low pressure, ionisation is mainly achieved through a single collision on the atom in the ground state while at high pressure, it results from stepwise processes.

High-pressure plasmas are produced by DC discharges, by inductively-coupled discharges (ICP) in the radio-frequency domain, and by MW discharges (for instance, by surface waves). Compared to low-pressure plasmas, the modelling of high-pressure plasmas is more complex, taking account of the more varied thermal and kinetic phenomena occurring in these discharges and, for these reasons, is not so well established as at low pressures. In the following, we will firstly illustrate experimentally the phenomena of contraction and filamentation, which are characteristic of high-pressure HF plasmas. Then, we will establish the various hypotheses that are part of our model on contraction, in which inhomogeneous gas heating and molecular ion kinetics play key roles. Finally, we will show both experimentally and theoretically how and why a contracted discharge expands when traces of a rare gas with a lower ionisation potential is added to the carrier gas.

4.4.1 Experimental observation of contraction and filamentation at atmospheric pressure

If the gas pressure exceeds a few tens of torr ($\simeq 1\,\text{kPa}$), the radial cross-section of the plasma in a cylindrical discharge may contract as the pressure increases, or the discharge may become filamentary. These phenomena, referred to as *contraction* and *filamentation*, affect all electric discharges in most rare (noble) gases, namely neon, argon, krypton and xenon or certain molecular gases, particularly those which are electronegative. In the case of rare gases such as neon and argon, these phenomena begin at pressures as low as a few torr.

Figure 4.18 shows the light emitted by discharges produced by an EM surface wave in different gases, at atmospheric pressure, in a tube of 6 mm inner diameter. In such discharges, the electron density decreases (almost linearly) from the gap of the surface wave launcher (EM field applicator) toward the end of the column (Appendix XIX). We observe that the diameter of the discharges in helium and N_2 does not vary, to a first approximation, as a function of the distance from the launching gap. In contrast, the diameter of the plasma column is clearly reduced in the case of krypton and argon, but much less in neon. The contraction phenomenon is characterised by the fact that the plasma column shrinks radially, producing a dense, bright filament, oriented, generally speaking, along the direction of the electric field in the discharge [30]. In surface-wave discharges, the main E-field component is directed along the tube axis. Then, provided the tube is mounted vertically, the filament is directed along the tube axis and centred on it, as can be seen in Fig. 4.18[172]. Comparing the observed degree of contraction with the value

[172] When the discharge tube is oriented horizontally, natural convection pushes the filament above the axis, towards the top.

of the thermal conductivity κ of the gas at the temperature of the gas in the discharge (Tab. 4.1), we see that the weaker κ, the more pronounced is the contraction phenomenon.

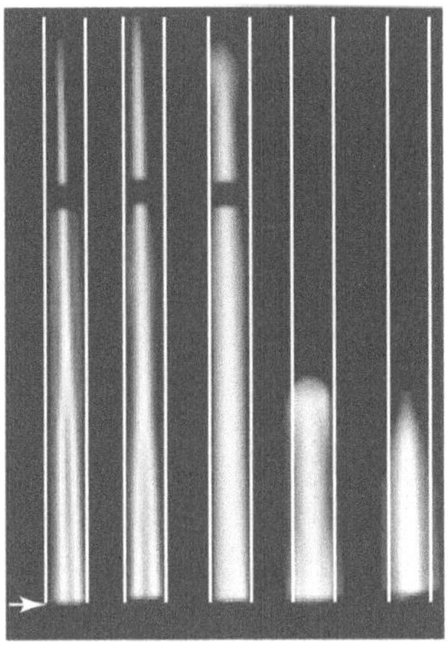

Fig. 4.18 Photograph of the upper part (with respect to the plane of the microwave field applicator, in this case a surfaguide) of a vertically oriented surface wave discharge in different gases, for a tube of 6 mm inner diameter (the tube walls are represented by the white lines) and a frequency of 2450 MHz. The black band, which cuts each of the three longest discharges, is due to the structure of the Faraday cage shielding the MW radiation coming from the discharge [15].

Table 4.1 Thermal conductivity κ (in $10^{-2}\,\mathrm{Wm^{-1}K^{-1}}$) of gases calculated at the ambient temperature and at the discharge temperature, which is that measured on the axis of the discharge.

Temperature (K)	He	N_2[173]	Ne	Ar	Kr	Xe
300	20.67	2.62	3.4	1.72	1	0.55
T_g (K)	(2700) 82	(5200) 88.7	(2000) 11.5	(2100) 6.45	(1900) 2.5	(1700) 2.27

[173] The thermal conductivity of the discharge in N_2 increases considerably when the temperature of the gas is sufficiently high to dissociate N_2 into atomic nitrogen, N atoms being much lighter (hence more mobile) than the N_2 molecule.

Filamentation is the breaking up of a single filament into two or many filaments. In Fig. 4.18, the plasma columns in krypton and argon, close to the wave launcher where the HF power flow is the highest, are separated into two filaments: this effect is observed to disappear when the intensity of the electric field becomes radially more homogeneous, which occurs when n_e decreases and ν/ω increases (towards the end of the column: upper part of the photograph)[174].

Variation of the apparent radius of the discharge as a function of the tube radius

We define the apparent radius of the filament, which we will regard as the plasma radius, as the radial position r_p at which the luminous intensity has reduced to half of the maximum filament intensity (on the axis): the *degree of contraction* can then be defined as the ratio R/r_p , where R is the internal radius of the discharge tube. Figure 4.19 shows the way r_p increases when R increases in an argon discharge. When contraction occurs, i.e. above a particular value of R, which is 3 mm in the present case, the filament radius ceases to increase. On the contrary, in the N_2 discharge, not only R/r_p remains constant when R increases, but the radial profile of the emitted light intensity remains the same, as can be seen in Fig. 4.20: this discharge is clearly not contracted (at least for the small values of R considered). The contraction of a discharge is accompanied by important changes in the radial profile of the plasma parameters, such as electron density and gas temperature as shown in Fig. 4.26 below [15].

Figure 4.21 shows that when the tube diameter is not too large and the field frequency not too high, there is contraction while, otherwise, the discharge is filamentary. The figure also indicates that both the contraction and filamentation effects can be progressively reduced by the addition to the carrier gas of traces of a rare gas with a lower ionisation potential than that of the carrier gas. Contraction and filamentation abruptly vanish when a specific percentage of the trace gas is reached [6], as shown for contraction in Fig. 4.22. At this optimum percentage value, the light emitted from the plasma is that of the added gas, not of the carrier gas. For larger percentages of the added gas, contraction/filamentation starts to reappear.

Remarks:

1. Electric (meaning DC and HF) discharges maintained at reduced pressure ($p <$1–10 torr) are relatively homogeneous and entirely fill the vessel containing them. They are generally designated as *luminous discharges* or *diffuse discharges* (to distinguish them from contracted discharges). In the

[174] The phenomenon of filamentation is related to the weak penetration of the HF field in the plasma. Its mechanism is not discussed further herein.

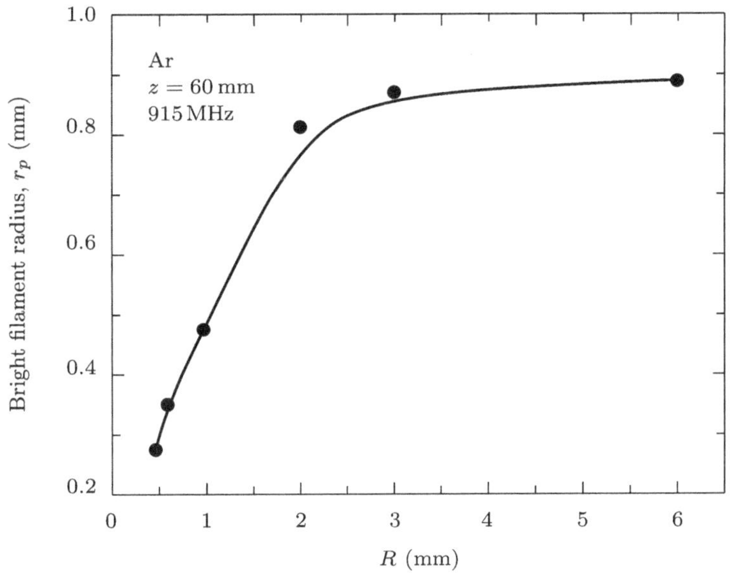

Fig. 4.19 Variation of the apparent radius of the plasma filament (see text) in a surface-wave argon discharge as a function of the internal radius R of the discharge tube, at the same distance z from the end of the column, at atmospheric pressure and 915 MHz (after [15]).

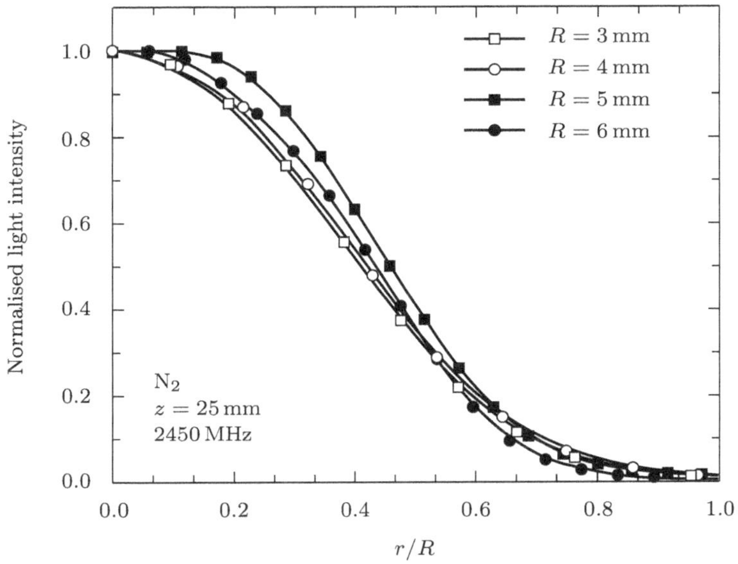

Fig. 4.20 Radial profile of the luminous intensity emitted as a function of the normalised radial position, in N_2 discharge for tubes of different radii (after [15]), showing that this discharge is not contracted: the radial profile appears independent of R.

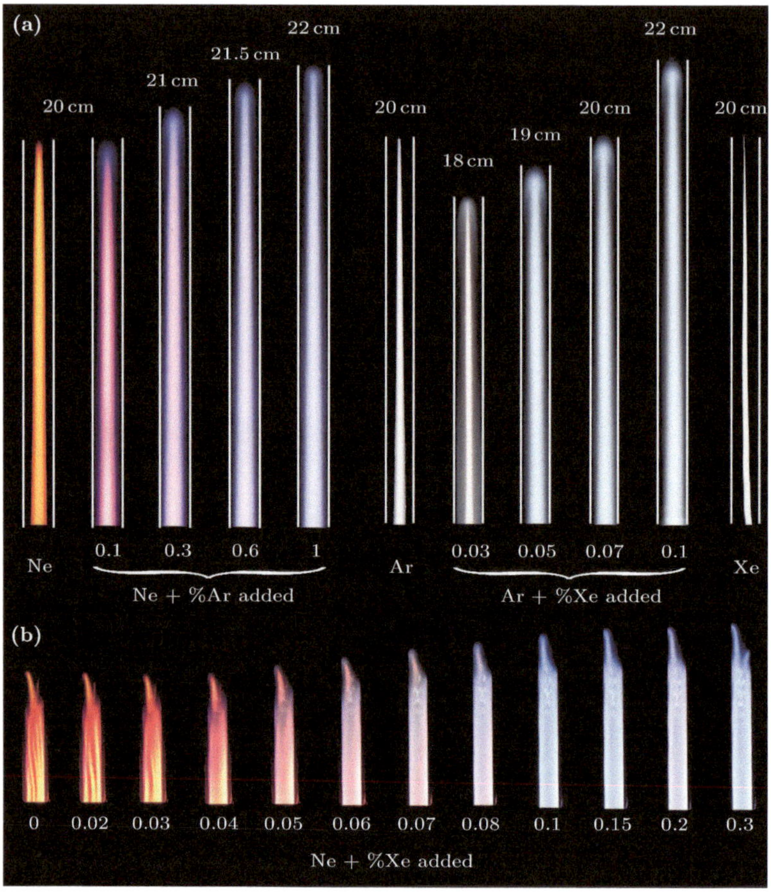

Fig. 4.21 Photographs of the upper section (with respect to the plane of the microwave field applicator, which is perpendicular to the tube axis) of a vertically oriented surface wave discharge in different noble gases at atmospheric pressure under a 0.5 slm gas flow: **a** for a tube of 12 mm inner diameter and a field frequency of 915 MHz, showing contraction in pure Ne and pure Ar. As traces of Ar are added to Ne and traces of Xe to Ar, there is a progressive expansion of the discharge; **b** for a tube of 20 mm inner diameter and a field frequency of 2450 MHz, showing filamentation. Progressively adding traces of Kr to Ne reduces filamentation, which ultimately vanishes [7].

case of a cylindrical discharge, the plasma, determined by its luminous section, occupies the total radial cross-section of the tube. This is related to the fact that the loss of charged particles (electrons and ions) takes place through diffusion to the walls of the tube where they recombine. Under these conditions, the radial distribution of electrons is determined only by the pressure and the radius of the discharge tube (Sect. 3.13). In contrast to the diffuse case, the electrons in a contracted discharge are confined to the filament region and the loss of charged particles occurs principally by volume recombination, as we will see (Sect. 4.4.2).

2. The transition from the diffuse regime to the contracted regime, which is particularly easy to observe in a DC discharge [15], leads to a large increase in the electron density and gas temperature, while the electron temperature decreases. However, the transition from the diffuse state to the contracted state, which is much more dense, does not necessarily imply a transition of the discharge towards a state of thermal equilibrium (Sect. 1.4.3): T_e remains generally higher than T_g. These contracted discharges are, in fact, in an intermediate state, between the state far from thermal equilibrium in a diffuse discharge and that of thermal equilibrium in a thermal arc or an ICP discharge at high power density. Thus, the properties of contracted discharges are neither those of cold luminous discharges nor those of thermal arcs.

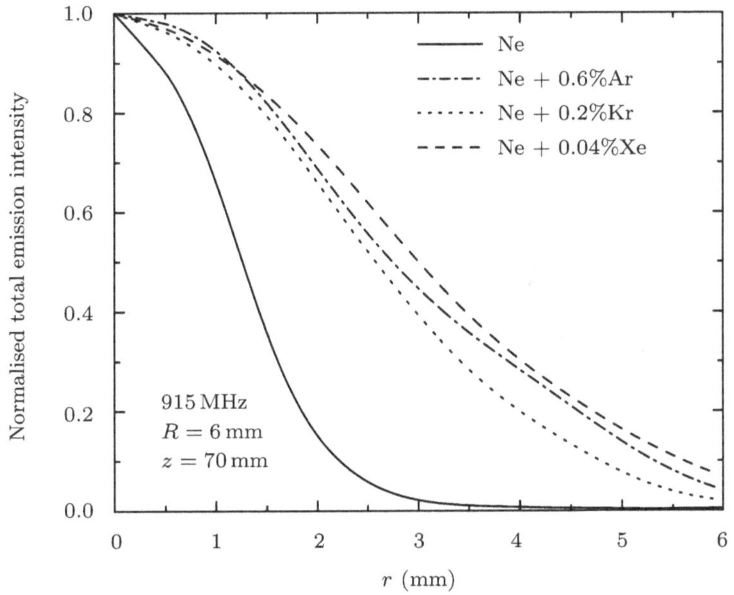

Fig. 4.22 Experimental radial profiles of the emitted-light total intensity in a discharge of pure Ne, and when adding to it traces of either Ar, Kr or Xe ($R = 6$ mm, $f = 915$ MHz at a fixed axial position $z = 70$ mm from the column end). The displayed percentages of the added rare gas correspond to the maximum radial expansion of the plasma. The recorded intensities are normalised at unity at the axis [6].

4.4.2 Modelling contraction at atmospheric pressure

The inhomogeneous heating of the discharge in the radial direction (resulting from the finite thermal conductivity of the gas) and the control of the creation and loss of charged particles by the molecular ions formed in high-pressure discharges are the basis of the contraction effect [6].

Inhomogeneous gas heating

Figure 4.23 shows the radial variation of the gas temperature T_g, obtained from optical emission spectroscopy,[175] in neon, helium and nitrogen discharges. We can observe that the gradient in the gas temperature is relatively steep in neon (contracted discharge), while it is relatively weak in N_2 discharges (not contracted). Nevertheless, although the helium discharge is not contracted (see Fig. 4.18), there is a large gradient in T_g. To understand the influence of the gradient in T_g on the contraction, we must additionally consider the influence of the molecular ion kinetics on the charged particle balance. Thus, at this point, the results in Fig. 4.23 suggest that the inhomogeneous heating of the gas is a necessary, but not sufficient condition for the contraction to occur.

Kinetics of molecular ions at atmospheric pressure

Charged-particle loss and creation through molecular ions

The loss of charged particles occurs by dissociative recombination of the molecular ions X_2^+ with electrons:

$$X_2^+ + \mathrm{e} \rightarrow X^m + X \,, \tag{4.31}$$

where X^m and X represent metastable[176] and ground state atoms of rare gases, respectively.

The creation of charged particles occurs through stepwise ionisation of the metastable state X^m:

$$X^m + \mathrm{e} \rightarrow X^+ + \mathrm{e} + \mathrm{e} \,. \tag{4.32}$$

---

[175] From a ro-vibrational band of the OH molecule (from water vapour introduced as a trace into the discharge), one can, with the help of a Boltzmann diagram (Appendix III), determine the rotational temperature T_{rot} from the recorded spectral intensity. In a discharge at atmospheric pressure with a sufficiently high electron density, the ro-vibrational energy of the thermometric molecule (OH) is in equilibrium with the translational energy of the carrier gas, from which $T_{\mathrm{rot}} = T_g$.

[176] In the case of rare gas atoms, with T_g in the range 300–2000 K, the dissociative recombination yields one atom in the np^6 atomic ground state and the second atom in the $np^5(n+1)s$ orbital configuration: two out of four of these energy levels are metastable states. For example in argon, less than ≈ 30% is in the Ar(4p) compared to the Ar(4s) configuration [29].

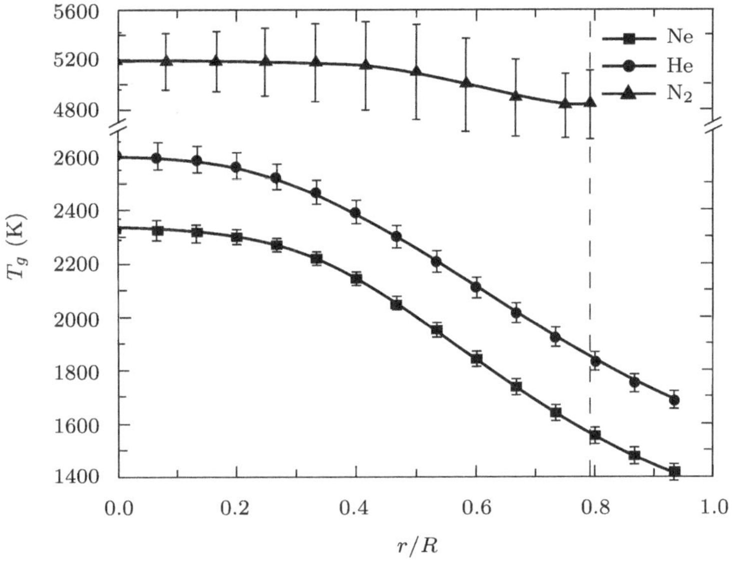

Fig. 4.23 Radial distribution of the gas temperature T_g, observed in neon, helium and nitrogen discharges at 2450 MHz in a tube of inner radius $R = 3$ mm (after [15]).

The metastable relay-state for ionisation can either follow directly from the dissociative recombination (4.31) of the molecular ions, or result from an electron collision on the ground state atom X:

$$\mathrm{e} + \mathrm{X} \rightarrow \mathrm{e} + \mathrm{X}^m \,. \tag{4.33}$$

However, in contracted discharges, this process contributes less to the formation of X^m than dissociative recombination, as shown in Fig. 4.27a below for He, Ne and Ar [37].

Creation and loss of molecular ions (without affecting the number of charged particles)

The atomic ions X^+ created by stepwise ionisation (4.32) are converted into molecular ions according to (atomic-ion association):

$$\mathrm{X}^+ + \mathrm{X} + \mathrm{X} \rightarrow \mathrm{X}_2^+ + \mathrm{X} \,, \tag{4.34}$$

where the third body serves to absorb part of the excess kinetic energy produced by the formation of the molecular ion X_2^+.

The loss of molecular ions can also occur by spontaneous thermal dissociation:

$$\mathrm{X}_2^+ \rightarrow \mathrm{X}^+ + \mathrm{X} \,, \tag{4.35}$$

by collisions with electrons (electron impact dissociation), yielding atomic ions:

$$X_2^+ + e \rightarrow X^+ + X + e \tag{4.36}$$

and by collision with atoms (atomic impact dissociation):

$$X_2^+ + X \rightarrow X^+ + X + X\,. \tag{4.37}$$

This process is also a thermal dissociation process, as its rate increases with the gas temperature T_g.

In summary, the molecular ions clearly control the creation and loss of charged particles in such discharges. The predominance of any one process for the loss of molecular ions depends mainly on the gas temperature T_g (see Fig. 4.24).

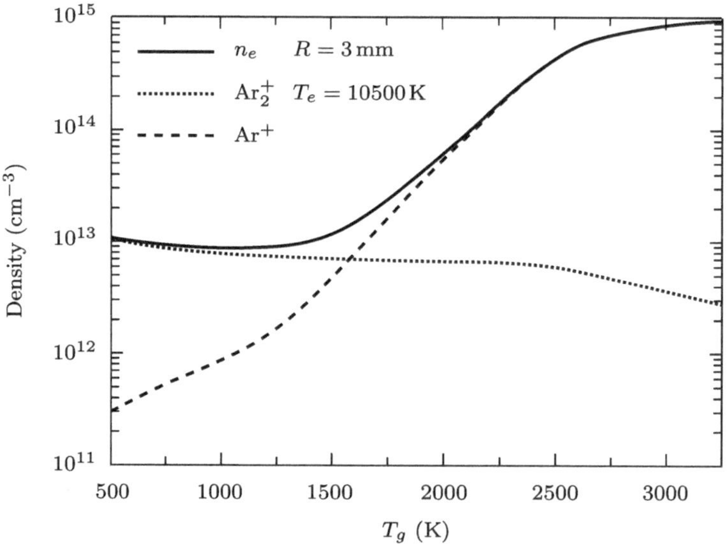

Fig. 4.24 Electron density n_e (full curve), atomic-ion density Ar^+ (dashed curve), and molecular-ion density Ar_2^+ (dotted curve) as functions of the gas temperature T_g, at a fixed value of the electron temperature T_e, at atmospheric pressure [4].

Discharge contraction: nonuniform gas heating and molecular ions

We have already mentioned that inhomogeneous gas heating plays an essential role in the discharge radial contraction. To show this, the electron density n_e as well as the atomic and molecular ion densities are calculated as functions of the gas temperature T_g in an argon discharge. This requires solving a balance equation for each type of charged particles as a function of T_g for a fixed value of T_e (in contracted discharges, the radial variation of T_e is much less important than the radial variation of T_g, and T_e can thus be considered, to a first approximation, constant radially). The reactions included in the balance equations are those given above, to which are added less important reactions under our conditions: ionisation of the atom in the ground state by electrons ($X + e \rightarrow X^+ + e + e$), three-body atomic-ion recombination ($X^+ + e + e \rightarrow X + e$) and diffusion of all charged particles and metastable-state atoms. The kinetics of the radiative states of the $4p$ orbital configuration has been neglected, because at pressures higher than 1 kPa, their density is lower than that of the metastable and resonant ($4s$) states.

Figure 4.24 shows the dependence of the electron and atomic- and molecular-ion densities on T_g, for an argon plasma sustained at atmospheric pressure, with $T_e = 10500\,\mathrm{K}$ [31]. Even though T_e is kept constant, as T_g varies between 1500 and 3000 K, the electron density n_e increases by two orders of magnitude, while the density of atomic ions increases by more than three orders of magnitude. On the other hand, the molecular ion density decreases by almost an order of magnitude over the 500–3000 K gas temperature interval. The increase of atomic-ion density with T_g is due to the strong (exponential) increase of molecular-ion dissociation ((4.35) and (4.37)). In such a case, since the density of molecular ions decreases, the charged-particle loss through dissociative recombination decreases (4.31) and, as a result, n_e increases. The increase of n_e with increasing T_g is further enhanced by the decrease of the density of molecular ions through electron impact (4.36) and the increasing contribution of step-wise ionisation (4.32). Due to the gas temperature gradient from the discharge axis to the wall (for example, Ne in Fig. 4.23), there is a significant decrease of electron density (Fig. 4.25) that leads to the radial contraction of the discharge.

4.4.3 Validation of the basic assumptions of contraction at atmospheric pressure, using a self-consistent model

Using a self consistent model, described in [4], applied to different gases (He, Ne, Ar), radial profiles of the parameters (T_g, T_e, n_e) in the plasma have been obtained. The modelling results show the existence of an inhomogeneous gas heating in the three gases studied (Fig. 4.26a). However, only discharges in argon and neon show a contraction (Fig. 4.26b), in agreement with experimental observations.

The model also enables us to compare: a) the frequency at which charged particles are created (direct ionisation ν_{id}, multi-step ionisation ν_{ie1}, where the relay states are populated from the ground state (4.33), and multi-step ionisation ν_{ie2}, where the metastable relay states are populated by disso-

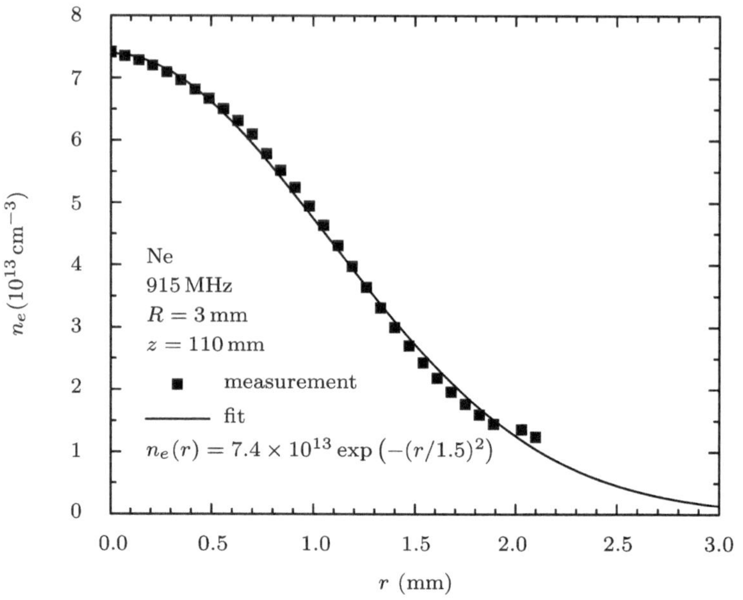

Fig. 4.25 Radial distribution of electron density (obtained through Stark broadening of the H_β line) in a 915 MHz neon surface-wave discharge sustained at atmospheric pressure, at axial position z = 110 mm from the plasma end, and in a 3 mm inner radius tube [5].

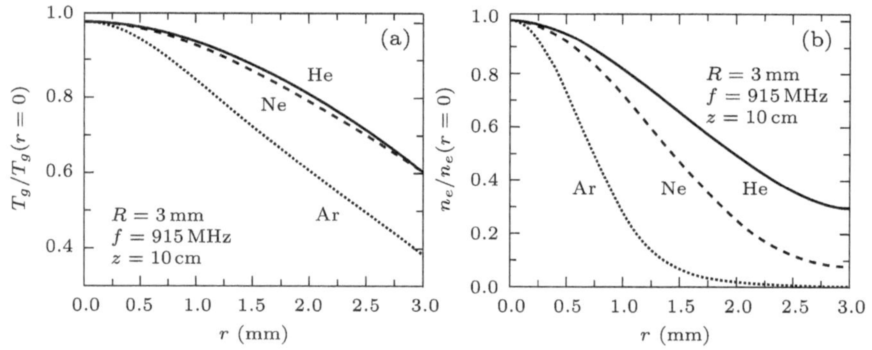

Fig. 4.26 Radial profiles **a** of the gas temperature and **b** the electron density in surface-wave discharges in helium, neon and argon at atmospheric pressure.

ciative recombination (4.32)): b) the frequency at which charged particles are lost (ambipolar diffusion ν_D (Sect. 3.12), dissociative recombination ν_{rm} (4.32) and three-body recombination ν_{ra} (1.151) of the charged particles). These results are shown in Fig. 4.27. Analysis of these various mechanisms of charged-particle loss and creation indicates that contraction only manifests itself in gases for which the kinetics of loss and gain of charged particles is completely controlled by molecular ions. Figure 4.27 shows that in argon and neon, the frequency for multi-step ionisation ν_{ie2} (process (4.32) involving molecular ions at the origin) and the frequency for dissociative recombination ν_{rm} (4.31) predominate for creation and loss repectively. In helium, the kinetics of molecular ions is not as important as in Ar and Ne: neither the creation nor the loss of charged particles are controlled by molecular ions in helium. Consequently, there is no non-linear dependence of n_e on T_g in the He discharge, and the presence of a gradient in T_g does not induce its contraction (Fig. 4.26b).

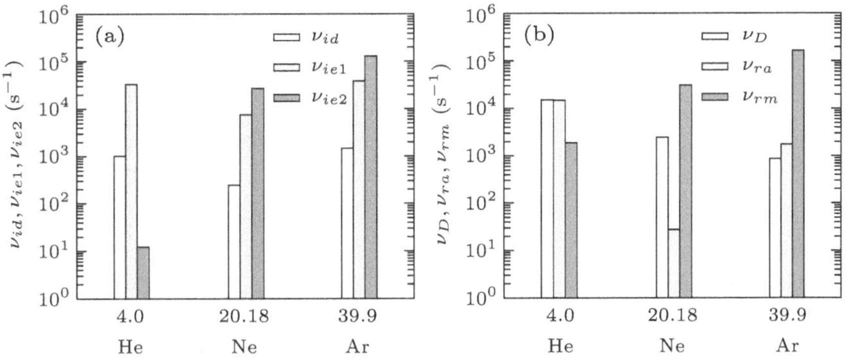

Fig. 4.27 Calculated frequencies **a** for the creation of charged particles (direct ionisation ν_{id} from the ground state, multi-step ionisation ν_{ie1}, where the relay states are populated from the ground state (4.33) and multi step ionisation ν_{ie2}, where the relay-states are populated by dissociative recombination (4.31)); **b** for the loss of charged particles (ambipolar diffusion ν_D (Sect. 3.12), dissociative recombination ν_{rm} (4.31) and three-body recombination ν_{ra} (1.151)).

Agreement between experiment and theory

Figure 4.28 compares experimental and theoretical results for the case of a neon discharge at atmospheric pressure. We can see, in the first place, the excellent fitting of the experimental n_e profile to $\exp(-(r/r_p)^2)$. The theoretical profile is obtained from numerical calculations, also for neon, using a self-consistent plasma model similar to that developed for the argon discharge [4]. The theoretical curve shown is for $T_g(r = R) = 1200\,\text{K}$, at the same axial position z with respect to the end of the column, as in the experimental case.

In summary, we have seen that contraction manifests itself by a n_e profile that exponentially decreases towards the walls. This rapid decrease in n_e results from a strong radial reduction in T_g: it takes place because the kinetics of the loss-creation of charged particles is dominated by the molecular ions. On the other hand, when T_g is sufficiently high for the atomic ions to govern these kinetics, then n_e hardly varies as a function of T_g (Fig. 4.24), and contraction is not possible. This seems to be the case in discharges with very

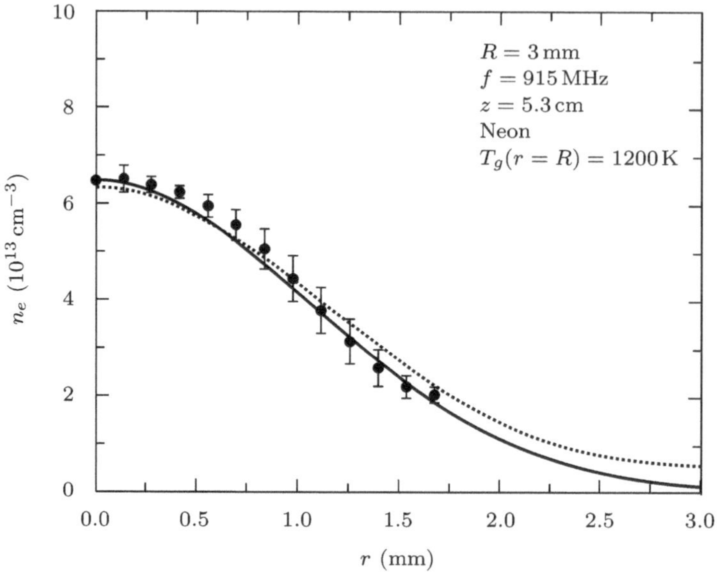

Fig. 4.28 Radial distribution of the electron density obtained (•) experimentally, from Stark broadening of the H_β line, smoothing by $\exp(-(r/r_p)^2)$ (—) and by calculations (...)[4].

high electron densities ($n_e > 10^{15}$ cm^{-3}) such as, for example, ICP discharges (Fig. 4.4) at atmospheric pressure. The calculations show, in addition, that the slope of rapid variation in n_e as a function of T_g in argon (Fig. 4.24) is less pronounced in the neon discharge, but very abrupt in the case of xenon: experimentally, the argon discharge is effectively more contracted than that of neon but less contracted than that of xenon. The good agreement between experiment and theory validates the model proposed for the contraction of atmospheric pressure discharges.

4.4.4 Kinetics of expanded discharges at atmospheric pressure as a result of adding traces of rare gases with a lower ionisation potential

The kinetics of the expanded discharge is characterised by the fact that the molecular ions of both the carrier gas and the added gas cease controlling the charged-particle loss and creation mechanisms (in contrast to the contracted discharge case, as we have just seen): their concentration is too low as we will show further on. Instead, the charged-particle creation of the discharges is governed, as expansion progresses, by Penning ionisation[177] of the ground-state atoms of the added gas colliding with metastable-state atoms of the carrier gas. When the discharge has fully expanded, the creation of charged particles is ensured by step-wise ionisation of the added gas atoms, through the relay of metastable-state atoms of this gas (4.33). Losses are controlled by ambipolar diffusion of the trace-added gas atomic ions and no longer by volume (dissociative) recombination. Changing from a volume recombination regime to one of diffusion causes the plasma to expand radially [6].

Molecular ion formation

Under expanded discharge conditions, the carrier and added gas molecular ions are not controlling the discharge kinetics because their density is too low. Recall that molecular ions are formed through a 3-body reaction (4.34). In the case of the carrier gas, this reaction is no longer efficient, because the density of its atomic ions, essential for creating molecular ions, has decreased significantly (due to quenching of the X^m metastable-state atoms); as for the added gas, the atomic-ion conversion into molecular ions is also inefficient because there are too little additional neutral atoms.

Ion density calculations for a Ne/Ar mixture

The preceding explanations are supported by a global (0-D) calculation of the atomic and molecular ion densities, reported below for Ne as the carrier gas and Ar as the added gas, as an example. The kinetics considered for the charged particles is that developed in [4], with further inclusion of Ar Penning ionisation from the neon metastable atoms:

$$\mathrm{Ne}^m + \mathrm{Ar} \rightarrow \mathrm{Ne} + \mathrm{Ar}^+ + \mathrm{e}\ , \tag{4.38}$$

where Ne^m indicates a metastable-state Ne atom [13], and charge transfer from Ne_2^+ to Ar^+ is included. The ion densities are obtained by solving a set of balance equations for Ne^+, Ne_2^+ and Ar^+, together with the charge neutrality relation. The input parameters of the model are the electron density n_e and gas temperature T_g, taken from experiments (through the H_β line Stark broadening and a N_2^+ ro-vibrational Boltzmann plot respectively) at the discharge axis, and electron temperature T_e calculated using the two-temperature Saha equation. Figure 4.29 presents the calculated ion densities as functions of the percentage of Ar added to Ne, when varied from 0 to 1%: experiments show that, over this interval, n_e and T_g can be considered

[177] Penning ionisation can be obtained when the energy level E_m of the metastable state atoms (e.g. $E_m = 16.60\,\mathrm{eV}$ for Ne) of the carrier gas is higher than the energy threshold E_i for ionisation of the atoms of the added gas (e.g. $E_i = 15.756\,\mathrm{eV}$ for Ar).

constant at the discharge axis ($n_e = 5 \times 10^{13}\,\mathrm{cm}^{-3}$ and $T_g = 2300\,\mathrm{K}$ at $\omega/2\pi = f = 915\,\mathrm{MHz}$, $R = 6\,\mathrm{mm}$, $z = 110\,\mathrm{mm}$). As Ar is added to Ne, Ar^+ rapidly becomes the dominant ion: the concentration of Ne^+ and Ne_2^+ decreases by more than three orders of magnitude, while that of Ar_2^+ increases, but nonetheless remains very low compared to that of Ar^+.

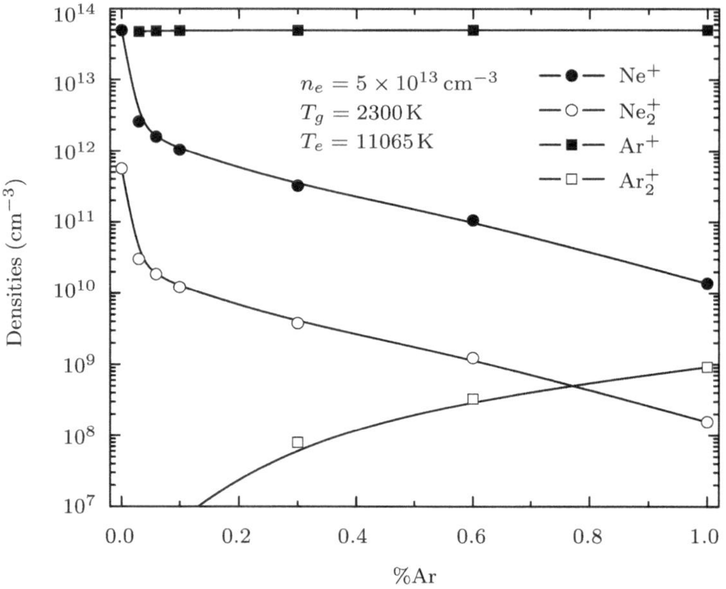

Fig. 4.29 Ne^+, Ne_2^+, Ar^+ and Ar_2^+ densities calculated for a Ne/Ar mixture discharge at atmospheric pressure. The values of n_e and T_g are those obtained experimentally at the discharge axis of the Ne/Ar mixture (over the 0–1%Ar range) with $f = 915\,\mathrm{MHz}$, $R = 6\,\mathrm{mm}$ and $z = 110\,\mathrm{mm}$ [6].

In a pure Ne or Ar contracted discharge, the density of molecular ions can be up to 100 times lower than that of atomic ions and still control the discharge kinetics, because of the high coefficient rate of the dissociative recombination. In the Ne/Ar expanded discharge ($0.3 < \%\mathrm{Ar} < 1$), the Ne_2^+ and Ar_2^+ molecular ion concentrations are much lower than in a pure rare gas, at least four orders of magnitude lower than that of Ar^+, i.e. much too low for molecular ions to control the discharge charged-particle kinetics.

In the case of maximum expansion (Ne + 1%Ar), the reaction rate of Ar^+ ambipolar diffusion is more than two orders of magnitude higher than the reaction rates of dissociative and three-body recombinations, indicating that the loss of charged particles is controlled by (ambipolar) diffusion. As a result, the charged particles can diffuse to the discharge-tube wall, ensuring the radial expansion of the plasma.

The modelling of the atmospheric pressure expansion is a further proof of the crucial role played by molecular ions in high-pressure plasmas, this time by showing what happens when their formation is hindered.

4.4.5 Summary of the properties of high-pressure HF plasmas

The fact that molecular ions control the creation and loss of charged particles is certainly an essential characteristic of high-pressure plasmas. In atomic rare gases, this feature remains valid even though the density of molecular ions can be as low as two orders of magnitude below that of the atomic ions. The dissociative recombination of rare-gas molecular ions provides a metastable-state atom, that ensures an efficient step-wise (re)ionisation process, another distinctive feature of high-pressure plasmas compared to low-pressure plasmas, where ionisation results from a single collision on the ground-state atom. In further contrast to low-pressure discharges, the electron-neutral collisions in high-pressure discharges are so numerous that they heat the gas to the point that, in discharge columns, a radial gradient of the gas temperature T_g appears: the lower the thermal conductivity κ of the discharge gas, the steeper is the radial gradient of T_g. This inhomogeneous heating of the discharge gas, combined with the dissociative recombination of molecular ions, have been found to be the two features responsible for discharge contraction: these are two necessary conditions for the occurrence of discharge contraction (and filamentation). As a matter of fact, expansion of a contracted discharge is obtained by breaking up the molecular-ion cycle of charged particle creation and loss. The model presented for the expansion phenomenon provides strong support to the explanations provided for the discharge contraction, emphasising the key role played by molecular ions in high-pressure discharges.

Appendix I
Some Properties of the Maxwell-Boltzman (M-B) Velocity Distribution

This distribution is related to the stationary state of a system with temperature T, provided the interactions between particles are sufficiently numerous. If the thermodynamic system is not in complete equilibrium, the minimum requirement is that collisions between particles of the same type be numerous enough for the Maxwell-Boltzmann distribution (in short a Maxwellian distribution) to be established.

M-B distribution in the absence of external fields

In one dimension, using the electrons as an example, the distribution function is given by the expression (Fig. I.1):

$$f(w) = \left(\frac{m_e}{2\pi k_B T}\right)^{1/2} \exp\left[-\frac{m_e w^2}{2k_B T}\right] , \tag{I.1}$$

where m_e is the electron mass, T their temperature, k_B the Boltzmann constant, and w, the microscopic (individual) electron velocity of thermal origin.

In three dimensions, for the case where the particles have a collective motion with velocity $\boldsymbol{v}$, the microscopic velocity distribution depends on the orientation of $\boldsymbol{w}$ with respect to $\boldsymbol{v}$:

$$f(\boldsymbol{w}) = \left(\frac{m_e}{2\pi k_B T}\right)^{3/2} \exp\left[-\frac{m_e}{2k_B T}(\boldsymbol{w}-\boldsymbol{v})^2\right] , \tag{I.2}$$

where $(\boldsymbol{w}-\boldsymbol{v})^2 \equiv w^2 + v^2 - 2\boldsymbol{w}\cdot\boldsymbol{v}$.

Unless otherwise stated, the normalisation condition used here is:

$$\int_{-\infty}^{\infty} f(\boldsymbol{w})\,\mathrm{d}^3 w = 1 \tag{I.3}$$

with $\mathrm{d}^3\, w = \mathrm{d}w_x\,\mathrm{d}w_y\,\mathrm{d}w_z$ in Cartesian coordinates.

M. Moisan, J. Pelletier, *Physics of Collisional Plasmas*,
DOI 10.1007/978-94-007-4558-2,

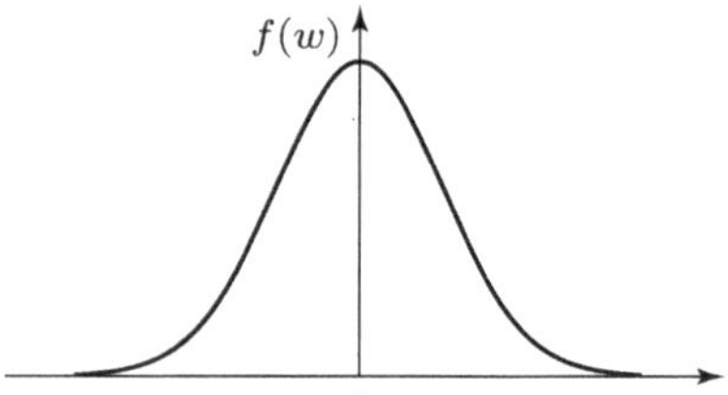

Fig. I.1 Maxwell-Boltzmann distribution in one dimension.

We could also have chosen the normalisation in terms of the electron density:

$$\int_{-\infty}^{\infty} f(\boldsymbol{w})\, \mathrm{d}^3 w = n_e \,. \tag{I.4}$$

In this case, n_e must be included as a factor in (I.2): this is discussed further in Sect. 3.3.

If the group velocity $\boldsymbol{v} = 0$, the distribution is isotropic:

$$f(w) = \left(\frac{m_e}{2\pi k_B T}\right)^{3/2} \exp\left[-\frac{m_e w^2}{2k_B T}\right] , \tag{I.5}$$

i.e. it is independent of the direction of the velocity $\boldsymbol{w}$. In this case, the isotropy leads to $\mathrm{d}^3 w = 4\pi w^2\, \mathrm{d}w$ in spherical coordinates, and the distribution of particles, travelling with a scalar (positive) velocity in the interval w, $w + \mathrm{d}w$, is then given by:

$$g(w) \equiv 4\pi w^2 f(w) \tag{I.6}$$

from which:

$$g(w) = \sqrt{\frac{2}{\pi}} \left(\frac{m_e}{k_B T}\right)^{3/2} w^2 \exp\left[-\frac{m_e w^2}{2k_B T}\right] , \tag{I.7}$$

which is illustrated in Fig. I.2. The normalisation condition under these conditions is:

$$\int_0^{\infty} g(w)\, \mathrm{d}w = \int_0^{\infty} 4\pi w^2 f(w)\, \mathrm{d}w = 1. \tag{I.8}$$

Characteristic velocities of the M-B distribution with zero group velocity ($\langle \boldsymbol{w} \rangle \equiv \boldsymbol{v} = \boldsymbol{0}$)

- the *most probable speed*

$$v_{th} = \left(\frac{2k_B T}{m_e}\right)^{1/2} , \tag{1.8}$$

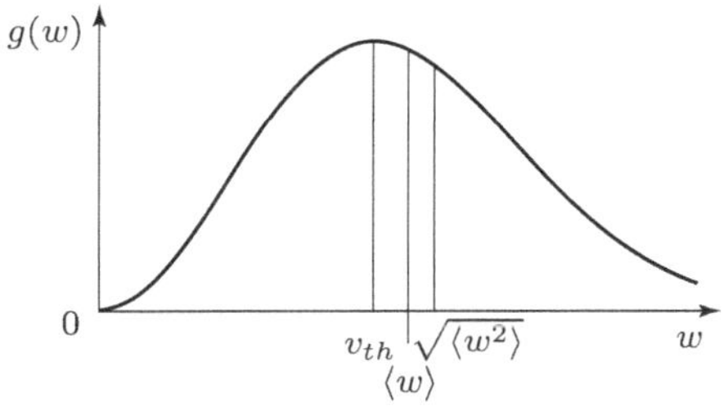

Fig. I.2 Isotropic Maxwell-Boltzmann distribution, represented in scalar form, showing characteristic velocities.

- the *average speed*

$$\langle w\rangle = \left(\frac{8k_BT}{\pi m_e}\right)^{1/2} = 1.128\ v_{th}\ , \tag{I.9}$$

- the *mean square speed* (related to the average energy):

$$\sqrt{\langle w^2\rangle} = \left(\frac{3k_BT}{m_e}\right)^{1/2} = 1.225\ v_{th}\ , \tag{I.10}$$

- the *average kinetic energy*

$$\frac{1}{2}m\langle w^2\rangle = \frac{3}{2}k_BT\ , \tag{I.11}$$

- the *random flux*, defined as the flux of particles traversing a surface in a single direction (in the positive z direction, for example, see exercise 1.2):

$$\langle nw_z\rangle = \frac{n\langle w\rangle}{4}\ . \tag{I.12}$$

M-B distribution in a conservative force field

If a conservative force field $\boldsymbol{F}$ acts on the particles, a prerequisite for a Maxwell-Boltzmann distribution is that this force obeys the relation:

$$\boldsymbol{F} = -\boldsymbol{\nabla}\Phi(\boldsymbol{r}) \tag{I.13}$$

where $\Phi(\boldsymbol{r})$ is the potential energy. The distribution function $f(\boldsymbol{r},\ \boldsymbol{w})$ can then be written in the form:

$$f(\boldsymbol{r},\boldsymbol{w}) = \hat{n}(\boldsymbol{r})\exp\left[-\frac{\Phi(\boldsymbol{r})}{k_BT}\right]f(\boldsymbol{w}) \tag{I.14}$$

where $\hat{n}(\boldsymbol{r})$ is the density of particles in the absence of applied or space-charge field $\boldsymbol{E}$.

Writing:

$$n(\boldsymbol{r}) = \hat{n}(\boldsymbol{r}) \exp\left[-\frac{\Phi(\boldsymbol{r})}{k_B T}\right] , \tag{I.15}$$

we obtain:

$$f(\boldsymbol{r}, \boldsymbol{w}) = n(\boldsymbol{r}) f(\boldsymbol{w}) , \tag{I.16}$$

which shows that the function $f(\boldsymbol{r}, \boldsymbol{w})$ is separable (Sect. 3.3). Including the normalisation (I.3) for the function $f(\boldsymbol{w})$, this leads to the normalisation condition for $f(\boldsymbol{r}, \boldsymbol{w})$:

$$\int_w f(\boldsymbol{r}, \boldsymbol{w}) \, \mathrm{d}^3 w = n(\boldsymbol{r}) \int_w f(\boldsymbol{w}) \, \mathrm{d}^3 w = n(\boldsymbol{r}) . \tag{I.17}$$

Note that we use the notation f for the velocity distribution function, whether it is separated or not: if the argument of f does not contain the position vector, we can conclude that it has been separated.

Druyvesteyn electron distribution

This distribution is often used in plasma physics, notably because it can be expressed in analytic form. Used conjointly with the M-B distribution function, it allows us to determine the conditions for which certain hydrodynamic parameters depend on the form of the electron energy distribution function (EEDF).

The Druyvesteyn distribution can be considered as an adequate description of the EEDF when the electrons satisfy the four following assumptions [10]:

1. Elastic electron-heavy particle collisions predominate: inelastic collisions (excitation and ionisation) are thus negligible;
2. Electron-electron collisions are negligible;
3. The total microscopic cross-sections for electron-neutral collisions are independent of electron energy for all types of collision;
4. The average electron energy is higher than that of heavy particles ($T_e > T_g$).

For an isotropic distribution, the Druyvesteyn EEDF may be written:

$$f_D(w) = \frac{1.04\pi^{\frac{1}{2}}}{2} \left(\frac{m_e}{2\pi k_B T_e}\right)^{\frac{3}{2}} \exp\left[-0.55 \left(\frac{m_e w^2}{2 k_B T_e}\right)^2\right] . \tag{I.18}$$

Figure I.3 compares the Druyvesteyn distribution with that of Maxwell-Boltzmann's for the same average electron energy and the same electron density: the Druyvesteyn distribution contains much fewer high energy electrons than the M-B distribution.

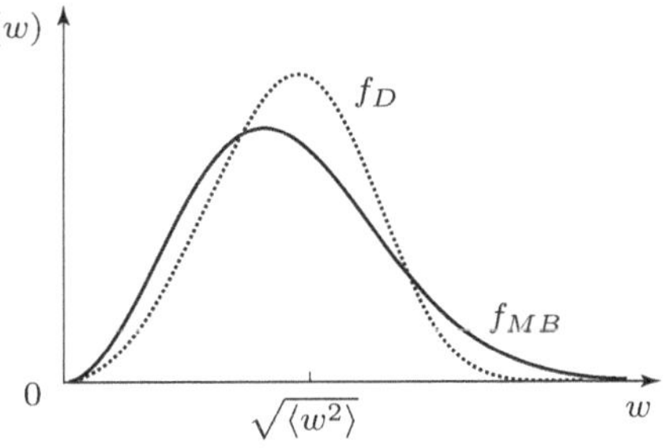

Fig. I.3 Comparison of isotropic Druyvesteyn's and Maxwell-Boltzmann's EEDFs with the same average energy.

M-B electron energy distribution

In Appendix XVII, we show that the Maxwell-Boltzmann distribution function in electron energy U_{eV} can be written:

$$F(U_{eV}) = \frac{2}{\pi^{\frac{1}{2}}(k_B T_e)^{\frac{3}{2}}} \exp\left(-\frac{eU_{eV}}{k_B T_e}\right) , \tag{I.19}$$

where $\bar{U}_{eV} = \frac{3}{2}\frac{k_B T_e}{2e}$ is the average energy.

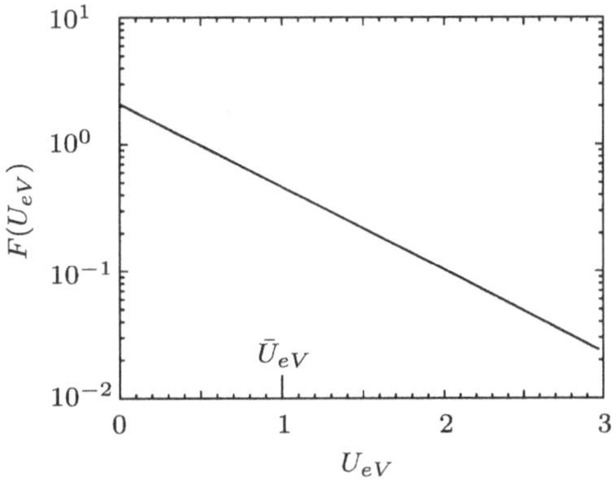

Fig. I.4 Distribution function for Maxwellian electrons of energy U_{eV} with an average energy $\bar{U}_{eV} = 1\,\text{eV}$.

Appendix II
The Complete Saha Equation

The complete Saha equation can be written in the form:

$$\frac{n_{it}n_e}{n_{0t}} = \frac{(2\pi m_e k_B T)^{3/2}}{h^3} 2 \frac{B'(T)}{B(T)} \exp\left[-\frac{e\phi_i}{k_B T}\right] \tag{II.1}$$

where ϕ_i is the first ionisation potential of the neutral atoms, n_{0t} and n_{it} are the *total* density of neutral atoms and the *total* density of ions respectively (total density includes the ground state and all the respective excited states of the neutral atoms and of the ions), n_e is the electron density ($n_e = n_i$ because the ions are only singly ionised); $B(T)$ and $B'(T)$ are the *partition functions* given by:

$$B(T) = \sum_{k=0}^{\infty} g_k \exp\left[-\frac{e\phi_k}{k_B T}\right] \tag{II.2}$$

where the sum over k includes all of the excited states of the atom ($k = 0$ represents the ground state) and:

$$B'(T) = \sum_{j=0}^{\infty} g_j \exp\left[-\frac{e\phi_j}{k_B T}\right] \tag{II.3}$$

where the sum over j includes the excited states of the singly-ionised (positive) ion ($j = 0$ is the ion ground state), g_k and g_j represent the degeneracies of the neutral atom and ion levels, respectively, and ϕ_k and ϕ_j are the potentials (value at threshold, see Sect. 1.7.9) corresponding to the excitation levels (measured with respect to the ground state of the neutral atom or that of the ion, whence $\phi_k(k = 0) = 0$ and $\phi_j(j = 0) = 0$).

M. Moisan, J. Pelletier, *Physics of Collisional Plasmas*,
DOI 10.1007/978-94-007-4558-2,

Significance of the partition function

The density n_m of the excited level m with respect to the density of its ground state (subscript zero) is given by (Boltzmann's law):

$$\frac{n_m}{n_0} = \frac{g_m}{g_0} \exp\left[-\frac{(\mathcal{E}_m - \mathcal{E}_0)}{k_B T}\right] . \tag{1.10}$$

One would like to express n_m as a function of the total density of the neutral atoms n_{0t} (or the ions n_{it}). From the Boltzmann law, the cumulative density n_{0t} becomes:

$$\begin{aligned} n_{0t} &\equiv \sum_{k=0}^{\infty} n_k \\ &= \frac{n_0}{g_0}\left(g_0 + g_1 \exp\left[-\frac{(\mathcal{E}_1 - \mathcal{E}_0)}{k_B T}\right] + \cdots + g_m \exp\left[-\frac{(\mathcal{E}_m - \mathcal{E}_0)}{k_B T}\right] + \cdots\right) . \end{aligned} \tag{II.4}$$

Substituting the definition of the partition function (II.2), the ratio n_m/n_{0t} can be written:

$$\frac{n_m}{n_{0t}} = \frac{g_m}{B(T)} \exp\left[-\frac{(\mathcal{E}_m - \mathcal{E}_0)}{k_B T}\right] . \tag{II.5}$$

Comparing (II.1), the exact form of the Saha law, with its simplified form (1.12), we note that this approximation assumes that $B(T) \simeq g_0$. It actually takes no account of the **excited** neutral atoms in the assessment of the total neutral atom density. This is possible in the case where T is sufficiently low; in this case, the density of the ground state is very large compared to the cumulative density of all the excited atoms.

Appendix III
Partial Local Thermodynamic Equilibrium

When introducing the concept of a two-temperature plasma (Sect. 1.4.3), we have shown that the population of the different energy levels of the neutral atoms are not, in this case, governed by the Boltzmann law (1.10): the neighbouring levels to the ground state have radiative lifetimes sufficiently short compared to the time between electron-neutral collisions, such that they depopulate radiatively rather than by electron-neutral collisions, and therefore fail to satisfy the electron kinetics (Saha's law and Boltzmann's law); on the contrary, the higher levels, those situated beneath the first level of the ionised atom, are in collisional equilibrium with the electrons, and satisfy the Boltzmann law because they experience a much greater number of inelastic collisions than the lower levels[178]; setting $T_{\mathrm{exc}} = T_e$, enables us to determine their population concentrations. On the other hand, the gas temperature (principally that of the atoms in the ground state, because they are more numerous), denoted by T_g, is such that $T_g \ll T_e$.

This situation is illustrated in Fig. III.1, where we have reproduced the diagram of the energy levels of the neutral argon atom; to simplify the discussion, these levels have been regrouped according to the orbital electronic configuration to which they belong. In this notation, for the ground state level we have $1s^2\,2s^2\,3s^2\,3p^6$, although in Fig. III.1 we have retained only the last term, for simplicity. The first excited configuration, denoted $4s$, contains 4 levels (see the inset) that will be treated as a block.

To verify the applicability of the Boltzmann law, we can draw the logarithm of the population concentration of the neutral atom levels as a function of their energy, with reference to the ground state (1.10), which is called a Boltzmann curve. If the law is satisfied, a straight line is obtained, proportional to T_{exc}^{-1}. Fig. III.2, obtained by optical emission spectroscopy in an argon plasma sustained by a microwave discharge at atmospheric pressure,

[178] The electron energy distribution (Sect. 1.4.2 and Appendix I) contains more low energy electrons than high energy electrons, while the collisional excitation of the first levels of the argon atom from the ground state, for example, requires very high energy electrons (above 11.55 eV).

M. Moisan, J. Pelletier, *Physics of Collisional Plasmas*,
DOI 10.1007/978-94-007-4558-2,

shows that the orbital electronic configurations $5p$ and above are in Boltzmann equilibrium among themselves, while the concentrations of the $4p$ configuration fall below this line: the Boltzmann equilibrium is thus only partial with $T_g \ll T_e$ (Sect. 1.4.3).

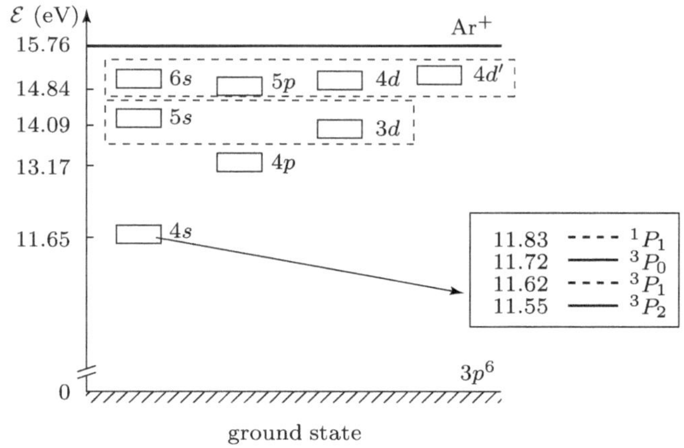

Fig. III.1 Energy diagram of the neutral argon atom, up to the first level of the ionised atom. The energy states have been regrouped according to the orbital electronic configuration to which they belong.

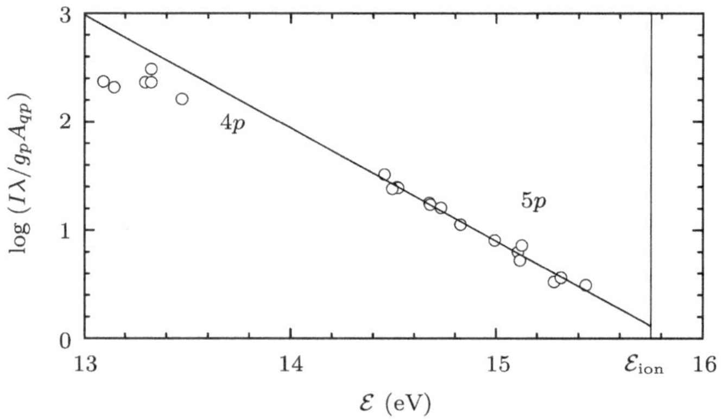

Fig. III.2 Diagram, referred to as a *Boltzmann diagram*, observed in a microwave-discharge in argon (surface wave plasma). The population concentration of the levels is proportional to the intensity I of the radiation emitted at the energy of the levels, expressed in eV, and referenced to the ground state energy; the coefficient A_{qp} represents the frequency of the spontaneous electric-dipolar radiation transition, from the state q to the state p [3].

Appendix IV
Representation of Binary Collisions in the Centre of Mass and Laboratory Frames

In the laboratory frame

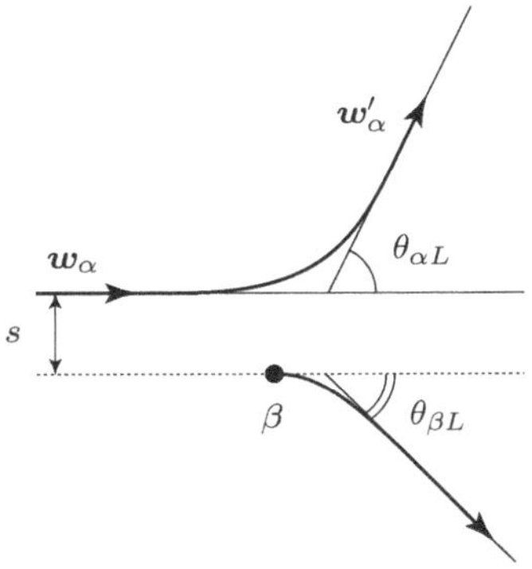

Fig. IV.1 Schematic of a binary collision in the laboratory frame, with impact parameter s: the particle β is assumed to be initially at rest.

In the center of mass frame

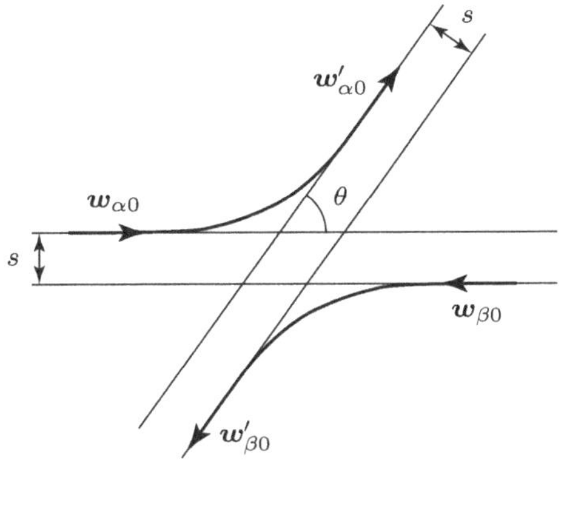

Fig. IV.2 Schematic of a binary collision in the centre of mass frame, showing the scattering angle θ. The distance s between the two pairs of asymptotes is the *impact parameter*.

M. Moisan, J. Pelletier, *Physics of Collisional Plasmas*,
DOI 10.1007/978-94-007-4558-2,

The velocities before and after the collision are parallel and anti-parallel respectively. The representation of the interaction in the centre of mass (CM) frame greatly simplifies the description of the collisional interaction, since a single angle θ completely describes the scattering process.

Relationship between the two frames

From Fig. IV.3, we find:

$$\tan\theta_{\alpha L} = \frac{w'_{\alpha 0}\sin\theta}{w'_{\alpha 0}\cos\theta + w_0} \,. \tag{IV.1}$$

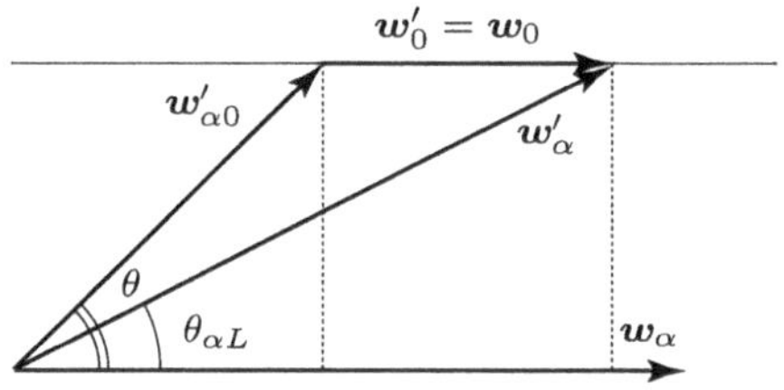

Fig. IV.3 Description, in the most general case, of the velocity of the particle α before and after collision, in the laboratory frame and that of the centre of mass, showing the relationship between the two frames.

The case where particle β is initially at rest:

- If $m_\alpha \ll m_\beta$, the velocity of the CM in the laboratory frame is the same as the velocity of the particle β (1.69). Since it is assumed at rest, then from (1.67) $\boldsymbol{w}_0 \simeq 0$. In this case (IV.1):

$$\theta_{\alpha L} = \theta \,. \tag{IV.2}$$

- If $m_\alpha = m_\beta$ and since $\boldsymbol{w}_\beta = 0$ in the present case, then from (1.72) $\boldsymbol{w}_{\alpha\beta} = \boldsymbol{w}_\alpha$. The velocity of the CM in the laboratory frame is given by (1.69) and can be written:

$$\boldsymbol{w}_0 = \frac{1}{2}\boldsymbol{w}_\alpha = \frac{1}{2}\boldsymbol{w}_{\alpha\beta} \,. \tag{IV.3}$$

Similarly, according to (1.73), the velocity $\boldsymbol{w}_{\alpha 0}$ of the particle α in the CM frame is given by:

$$\boldsymbol{w}_{\alpha 0} = \frac{1}{2}\boldsymbol{w}_{\alpha\beta} \,. \tag{IV.4}$$

Since $w'_{\alpha 0} = w_{\alpha 0}$ in the case of an elastic collision (1.98), (IV.3) and (IV.4) give $w'_{\alpha 0} = w_0$ and finally, from the trigonometric relation (IV.1):

$$\theta_{\alpha L} = \frac{\theta}{2} \,. \tag{IV.5}$$

Appendix V
Limiting the Range of the Coulomb Collisional Interactions: the Coulomb Logarithm

The ultimate objective of this Appendix is to calculate the collision frequencies of the Coulomb interactions, knowing that weak interactions (small scattering angles θ) prevent the corresponding integrals taken over θ from converging (Sect. 1.7.4). These weak interactions have, in fact, no physical importance, when their radius of influence is greater than the Debye length, λ_D: there is electrostatic screening. Accounting for this allows us to reduce the range of integration in θ and thus ensure the convergence of the integrals by introducing the concept of the Coulomb logarithm. To reach this goal, we will first determine the value of θ during a binary elastic collision due to an unspecified central force F.

General study of the trajectories of two particles (binary interactions) subjected to a central force field

Here, we only consider binary, elastic, electromagnetic[179] interactions. This is the case for Van Der Waals interactions between neutrals[180] (potential varies as r^{-6}), between neutral and charged particles[181] (potential varies as r^{-4}) and the Coulomb interactions between charged particles[182] (potential varies as r^{-1}). These electromagnetic interactions induce a conservative central field

[179] The use of the term "electromagnetic interaction" is justified in the two following notes at the foot of this page. On the other hand, the interactions are quantum in nature if the particles approach within a minimum distance that is of the same order of magnitude as the particle dimensions.

[180] Interaction between the instantaneous electric dipole of one of the particles, and the dipole that it induces in the second particle.

[181] Interaction between the charge of one particle, and the electric dipole that it induces in the neutral particle.

[182] We assume that the velocities of the charged particles are sufficiently small, such that the radiation due to the particle deceleration, when deflected by another charged particle (braking radiotion: bremsstrahlung) can be neglected.

M. Moisan, J. Pelletier, *Physics of Collisional Plasmas*,
DOI 10.1007/978-94-007-4558-2,

of force $\boldsymbol{F}$ (see Appendix I), collinear with $\boldsymbol{r}$, such that $\boldsymbol{F} = -\boldsymbol{\nabla}\Phi(r)$ (1.16), where $\Phi(r)$ is the potential energy of the interaction between the particles separated by a distance r.

The geometry of the elastic interaction between two particles α and β is described in Fig. V.1a for a repulsive interaction and in Fig. V.1b for the case of an attractive interaction. In the centre of mass frame (where the centre of gravity G of α and β moves at a constant velocity in the laboratory frame, Sect. 1.7.2), the trajectories of α and β approaching from infinity are two similar curves with respect to G (hyperbola in the particular case of a Coulomb interaction[183]), each having two asymptotes (trajectories a long time before and a long time after the collision). The distance s between the two pairs of asymptotes is the impact parameter (see Fig. V.1). This is also the distance of closest approach in the abscence of interaction.

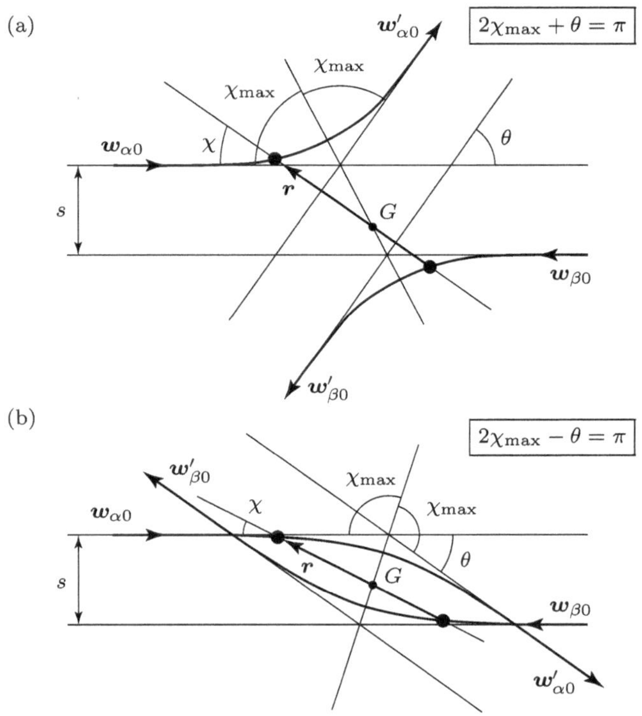

Fig. V.1 Geometric representation of a binary interaction in the barycentric frame (centre of mass): **a** repulsive interaction; **b** attractive interaction. The polar coordinates r and χ describe the position of the particle α with respect to particle β. The value $\chi_{\max}$ corresponds to the minimum value of the distance r.

---

[183] In the case of $1/r^2$ interaction forces (gravitation and Coulomb forces), the trajectories are either elliptic or hyperbolic.

The study of the motion of the particles is conducted in the centre of mass frame in polar coordinates, with r the distance between the particles α and β (one of the particles taken to be at the origin) and χ the angle between the vector $\boldsymbol{r}$ and the relative velocity $\boldsymbol{w}_{\alpha\beta} = \boldsymbol{w}_{\alpha 0} - \boldsymbol{w}_{\beta 0}$ of the particles α and β before the interaction: the velocities $\boldsymbol{w}_{\alpha 0}$ and $\boldsymbol{w}_{\beta 0}$ of the particles α and β before collision are collinear with $\boldsymbol{w}_{\alpha\beta}$ (1.71) and the velocities $\boldsymbol{w}'_{\alpha 0}$ and $\boldsymbol{w}'_{\beta 0}$ of the particles α and β after collision are collinear with $\boldsymbol{w}'_{\alpha\beta}$. In this system of coordinates, the components of the relative velocity $\boldsymbol{w}$ during the interaction (before the interaction $\boldsymbol{w} = \boldsymbol{w}_{\alpha\beta}$, after the interaction $\boldsymbol{w} = \boldsymbol{w}'_{\alpha\beta}$) are expressed respectively by:

$$w_r = \frac{\mathrm{d}r}{\mathrm{d}t} \,, \quad w_\chi = r\frac{\mathrm{d}\chi}{\mathrm{d}t} \,, \quad w_z = 0 \,, \tag{V.1}$$

where the z axis is perpendicular to the plane (r,χ) containing the trajectories.

In the centre of mass frame, the total kinetic energy related to the sole relative motion of the particles α and β is simply expressed by:

$$\mathcal{E}_c = \frac{\mu_{\alpha\beta} w^2}{2} \,, \tag{1.79}$$

or as a function of the different components of the relative velocity:

$$\mathcal{E}_c = \frac{\mu_{\alpha\beta}}{2}\left[\left(\frac{\mathrm{d}r}{\mathrm{d}t}\right)^2 + r^2\left(\frac{\mathrm{d}\chi}{\mathrm{d}t}\right)^2\right] . \tag{V.2}$$

We can easily verify that the kinetic moment of the relative motion, defined by:

$$\boldsymbol{L} = \boldsymbol{r} \wedge \mu_{\alpha\beta}\boldsymbol{w} \tag{V.3}$$

is an invariant of the motion. In fact:

$$\frac{\partial \boldsymbol{L}}{\partial t} = \underbrace{\boldsymbol{w} \wedge \mu_{\alpha\beta}\boldsymbol{w}}_{0} + \boldsymbol{r} \wedge \mu_{\alpha\beta}\frac{\mathrm{d}\boldsymbol{w}}{\mathrm{d}t} \,, \tag{V.4}$$

or again:

$$\frac{\partial \boldsymbol{L}}{\mathrm{d}t} = \boldsymbol{r} \wedge \boldsymbol{F} \equiv 0 \,, \tag{V.5}$$

since $\boldsymbol{r}$ and $\boldsymbol{F}$ are collinear. The calculation of the vector product of the relative kinetic moment $\boldsymbol{L}$ shows that the sole non-zero component is L_z:

$$L_z = \mu_{\alpha\beta} r^2 \frac{\mathrm{d}\chi}{\mathrm{d}t} \tag{V.6}$$

which, from (V.5), is therefore constant during the motion (first invariant of the motion). The value of L_z:

$$L_z = \left(\boldsymbol{r} \wedge \mu_{\alpha\beta}\boldsymbol{w}\right)_z \tag{V.7}$$

is then easily obtained from the initial conditions at infinity in (V.3), such that:

$$L_z = s\mu_{\alpha\beta}w_{\alpha\beta} = s\mu_{\alpha\beta}(w_{\alpha 0} - w_{\beta 0}) \;, \tag{V.8}$$

where $w_{\alpha\beta}$ is the modulus of the relative velocity $\boldsymbol{w}$ at infinity and, following (V.7), where s, the impact parameter, is the projection of $\boldsymbol{r}$ perpendicularly to $\boldsymbol{w}_{\alpha\beta}$ in the plane containing the trajectories.

A second invariant of the motion is the total energy $\mathcal{E}$ (kinetic energy plus potential energy) which is conserved during the course of the interaction, i.e.:

$$\mathcal{E} = \mathcal{E}_c(r) + \Phi(r) = \text{constant} \;. \tag{V.9}$$

The value of $\mathcal{E}$ is given by the initial conditions before the interaction, when the potential energy is zero, thus:

$$\mathcal{E} = \frac{\mu_{\alpha\beta}}{2} w_{\alpha\beta}^2 \;. \tag{V.10}$$

The equation for the trajectory $\chi = \chi(r)$, which we will now calculate, can be simply deduced from the two invariants. From (V.2) and (V.6), we can write:

$$\mathcal{E} = \frac{\mu_{\alpha\beta}}{2}\left(\frac{\mathrm{d}r}{\mathrm{d}t}\right)^2 + \frac{L_z^2}{2\mu_{\alpha\beta}r^2} + \Phi(r) \;, \tag{V.11}$$

from which:

$$\frac{\mathrm{d}r}{\mathrm{d}t} = \pm\sqrt{\frac{2}{\mu_{\alpha\beta}}\Big[\mathcal{E} - \Phi(r)\Big] - \left(\frac{L_z}{\mu_{\alpha\beta}r}\right)^2} \;. \tag{V.12}$$

Since:

$$\frac{\mathrm{d}r}{\mathrm{d}t} = \frac{\mathrm{d}r}{\mathrm{d}\chi}\frac{\mathrm{d}\chi}{\mathrm{d}t} \;, \tag{V.13}$$

the differential equation for the trajectory can be directly deduced from (V.6) and (V.12), such that:

$$\frac{\mathrm{d}\chi}{\mathrm{d}r} = \pm\frac{\dfrac{L_z}{\mu_{\alpha\beta}r^2}}{\sqrt{\dfrac{2}{\mu_{\alpha\beta}}\Big[\mathcal{E} - \Phi(r)\Big] - \left(\dfrac{L_z}{\mu_{\alpha\beta}r}\right)^2}} \tag{V.14}$$

or, by replacing $\mathcal{E}$ and L_z by their values ((V.8) and (V.10)):

$$\frac{\mathrm{d}\chi}{\mathrm{d}r} = \pm\frac{s}{r^2\sqrt{1 - \dfrac{s^2}{r^2} - \dfrac{2\Phi(r)}{\mu_{\alpha\beta}w_{\alpha\beta}^2}}} \;. \tag{V.15}$$

The equation of the trajectory can be obtained by a simple integration over r, provided that the form of the potential energy $\Phi(r)$ of the interaction is known.

The domain of possible values for r is defined by the quantity under the square root, which must remain positive. In particular, the minimum distance $r_{\min}$ between the particles during their interaction is obtained when $dr/d\chi = 0$, i.e., when the quantity under the square root in (V.15) is zero:

$$\Phi(r_{\min}) = \frac{\mu_{\alpha\beta} w_{\alpha\beta}^2}{2}\left(1 - \frac{s^2}{r_{\min}^2}\right) . \tag{V.16}$$

The minimum value $r_{\min}$ corresponds to the maximum angle χ, $\chi_{\max}$. In fact, during the motion, while r decreases from infinity to the minimum value $r_{\min}$, the angle χ increases from 0 to $\chi_{\max}$. The angle $\chi_{\max}$ is half the angle between the asymptotes before and after the collision. In the case of a repulsive interaction (Fig. V.1a), the angle $\chi_{\max}$ is linked to θ by the relation:

$$\theta = \pi - 2\chi_{\max} , \tag{V.17}$$

where the scattering angle θ, together with the impact parameter s, is one of the important characteristics of a binary collision. The angle $\chi_{\max}$ is obtained by integration of (V.15) along r from infinity to $r_{\min}$:

$$\chi_{\max} = \int_{\infty}^{r_{\min}} \frac{s \, dr}{r^2 \sqrt{1 - \left(\frac{s}{r}\right)^2 - \frac{2\Phi(r)}{\mu_{\alpha\beta} w_{\alpha\beta}^2}}} . \tag{V.18}$$

We have now established the general relations describing the trajectories (repulsive and attractive) of the interaction in the case of any central force. To apply these results to specific cases, we need to know the expression for the central force, or $\Phi(r)$, which will allow us to calculate $\chi_{\max}(s, w_{\alpha\beta})$, then finally the scattering angle θ.

Remark: All the preceding calculations have been performed in the centre of mass frame. However, the trajectory in the laboratory frame is almost the same as that calculated in the centre of mass frame if $m_\alpha \ll m_\beta$. In this case, the centre of mass is practically indistinguishable from the case in which the particle β is assumed to be stationary and the scattering angle θ remains unchanged from one frame to the other. On the other hand, if the masses m_α and m_β are similar, the scattering angle approaches $\theta/2$ in the laboratory frame (see Appendix IV).

We will now calculate the angle θ explicitly for the case of a Coulomb interaction.

Scattering angle θ for the specific case of a Coulomb interaction

The electrostatic interaction potential created by the particle β of charge Z_β is:

$$\phi(r) = \frac{1}{4\pi\epsilon_0}\frac{eZ_\beta}{r} \tag{V.19}$$

and the potential energy for the interaction of particle α of charge Z_α with the particle β has the value:

$$\Phi(r) = eZ_\alpha\phi(r) , \tag{V.20}$$

hence:

$$\Phi(r) = \frac{Z_\alpha Z_\beta\, e^2}{4\pi\epsilon_0 r} . \tag{V.21}$$

We can then define the *critical impact parameter* s_0 (the significance of which will become apparent later) such that:

$$\frac{s_0}{r} = \frac{\Phi(r)}{2\mathcal{E}} \equiv \frac{\Phi(r)}{\mu_{\alpha\beta}w_{\alpha\beta}^2} . \tag{V.22}$$

For $Z_\alpha = Z_\beta = 1$, the repulsive case resulting from two positive charges, we have:

$$s_0 = \frac{e^2}{4\pi\epsilon_0\mu_{\alpha\beta}w_{\alpha\beta}^2} . \tag{V.23}$$

Substituting (V.22), (V.18) can be written explicitly:

$$\chi_{\max} = \int_\infty^{r_{\min}} \frac{s\,\mathrm{d}r}{r^2\sqrt{1+\left(\frac{s_0}{s}\right)^2 - \left(\frac{s}{r}+\frac{s_0}{s}\right)^2}} . \tag{V.24}$$

By a change of variable:

$$\xi = \frac{\frac{s}{r}+\frac{s_0}{s}}{\sqrt{1+\left(\frac{s_0}{s}\right)^2}} , \tag{V.25}$$

Eq. (V.24) takes the form:

$$\chi_{\max} = \int_{\xi_\infty}^{1} \frac{-\mathrm{d}\xi}{\sqrt{1-\xi^2}} , \tag{V.26}$$

where ξ_∞ is the value of ξ when r tends to infinity, such that:

$$\xi_\infty = \frac{s_0/s}{\sqrt{1+(s_0/s)^2}} . \tag{V.27}$$

It should be noted that the upper limit of integration $r_{\min}$ in (V.24) corresponds to $\mathrm{d}r/\mathrm{d}t = 0$, i.e. when the quantity under the square root in (V.12) ia zero, corresponding to $\xi = 1$ in (V.26). The integration of (V.26) thus leads to:

$$\chi_{\max} = \text{arc cos} \frac{s_0/s}{\sqrt{1+(s_0/s)^2}} , \tag{V.28}$$

hence:

$$\cos \chi_{\max} = \frac{s_0/s}{\sqrt{1+(s_0/s)^2}} . \tag{V.29}$$

Substituting (V.17), we obtain:

$$\sin(\theta/2) = \cos \chi_{\max} . \tag{V.30}$$

Finally, knowing that:

$$\sin^2(\theta/2) = \frac{1}{1+\cot^2(\theta/2)} = \frac{(s_0/s)^2}{1+(s_0/s)^2} , \tag{V.31}$$

we arrive at the formula:

$$\cot(\theta/2) = s/s_0 , \tag{V.32}$$

which gives the expression for the scattering angle θ for a Coulomb collision. This deflection value is a function of the impact parameter s, of s_0 (V.23), of the relative velocity $w_{\alpha\beta}$ of the particles α and β before their interaction. Note that $\theta = \pi/2$ if $s = s_0$, while $\theta = \pi$ if $s = 0$, and the collision is head-on. It follows that if $s < s_0$, the deflection, i.e. the interaction, is important ($\theta > \pi/2$), and weak if $s > s_0$ ($\theta < \pi/2$). We now begin to see the importance of the parameter s_0 for Coulomb interactions.

Total microscopic cross-section for a Coulomb interaction

The differential relation between the total microscopic collision cross-section $\hat{\sigma}_{tc}$ and the microscopic differential scattering cross-section $\hat{\sigma}(\theta)$ can be deduced from (1.110), i.e.:

$$\mathrm{d}\hat{\sigma}_{tc} = 2\pi\hat{\sigma}(\theta) \sin\theta \, \mathrm{d}\theta , \tag{V.33}$$

where $\mathrm{d}\hat{\sigma}_{tc}$ is the *microscopic cross-section element*. From Fig. V.2, this can be expressed as a function of the impact parameter, to give:

$$\mathrm{d}\hat{\sigma}_{tc} = 2\pi s \, \mathrm{d}s . \tag{V.34}$$

In order to make use of $\hat{\sigma}(\theta)$, we can express the microscopic cross-section element $2\pi s \mathrm{d}s$ as a function of the solid angle element $2\pi \sin\theta \mathrm{d}\theta$ from (V.32).

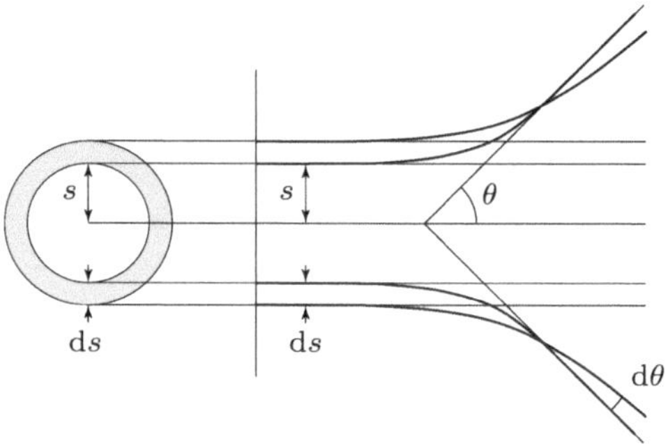

Fig. V.2 Schematic description of the geometric relation between the impact parameter s and the scattering angle θ for a Coulomb binary interaction, where the shaded surface represents the microscopic cross-section element $\mathrm{d}\hat{\sigma}_{tc} = 2\pi s\,\mathrm{d}s$.

After squaring of (V.32), and differentiating, we find:

$$\frac{2s\,\mathrm{d}s}{s_0^2} = -\frac{\left[\cos(\theta/2)\sin^3(\theta/2) + \cos^3(\theta/2)\sin(\theta/2)\right]\,\mathrm{d}\theta}{\sin^4(\theta/2)}\,, \tag{V.35}$$

hence:

$$\frac{2s\,\mathrm{d}s}{s_0^2} = -\frac{1}{2}\,\frac{\sin\theta\,\mathrm{d}\theta}{\sin^4(\theta/2)}\,. \tag{V.36}$$

By comparing (V.36) with (V.33) and (V.34), we deduce that:

$$\hat{\sigma}(\theta) = \left|-\frac{s_0^2}{4\sin^4(\theta/2)}\right|\,. \tag{V.37}$$

The negative sign in (V.37) simply implies that the scattering angle θ decreases as the impact parameter s increases, (V.32), but we shall use the absolute value of the RHS as an expression for the microscopic differential cross-section. Substituting (V.23) into (V.27), for Coulomb collisions, leads to the expression:

$$\hat{\sigma}(\theta,\, w_\infty) = \frac{\left(e^2/8\pi\epsilon_0\mu_{\alpha\beta}w_{\alpha\beta}^2\right)^2}{\sin^4(\theta/2)}\,, \tag{V.38}$$

which allows us, a priori, to calculate the effective total microscopic collision cross-section:

$$\hat{\sigma}_{tc} = \frac{\pi s_0^2}{2}\int_0^\pi \frac{\sin\theta\,\mathrm{d}\theta}{\sin^4(\theta/2)} \tag{V.39}$$

and the momentum transfer cross-section (1.111):

$$\hat{\sigma}_{tm} = \frac{\pi s_0^2}{2}\int_0^\pi \frac{(1-\cos\theta)\sin\theta\,\mathrm{d}\theta}{\sin^4(\theta/2)}\,. \tag{V.40}$$

Unfortunately, it is easy to verify that the integral in (V.39) diverges at $\theta = 0$. This is due to the fact that Coulomb forces are very long range, and in consequence, the distant particles, whose scattering angles approach $\theta = 0$ (un-deviated particles) all contribute to this integral, hence its divergence. The same is true for the integral in (V.40) which, as will be shown in the following calculation, also diverges for $\theta = 0$.

Total microscopic momentum cross-section for Coulomb's interactions

Concept of the Coulomb logarithm

The total cross-section for transfer of momentum (V.40) can also be written as a function of $\theta/2$, in the form:

$$\hat{\sigma}_{tm} = 4\pi s_0^2 \int_0^{\pi} \frac{\mathrm{d}\left[\sin(\theta/2)\right]}{\sin(\theta/2)} \,, \tag{V.41}$$

such that, after integration:

$$\hat{\sigma}_{tm} = 4\pi s_0^2 \ln\left[\sin(\theta/2)\right]\Big|_0^{\pi} \,. \tag{V.42}$$

We can verify that, as in the case of $\hat{\sigma}_{tc}$, the cross-section $\hat{\sigma}_{tm}$ diverges since the innumerable long-range collisions for which $\theta \simeq 0$, are taken into account. However, in a plasma, the range of the electric field created by a charged particle is reduced by the screening effect of the neighbouring charged particles and is, in fact, limited by the Debye sphere (Sect. 1.6). Consequently, any two charged particles in a plasma, separated by a distance $r > \lambda_D$, neither "sees" the field of the other particle, and therefore should not be included in the Coulomb interaction. The integration must therefore be limited to particles having an impact parameter s that is smaller than the Debye length ($s < \lambda_D$), i.e. to scattering angles θ greater than the minimum value $\theta_{\min}$ ($\theta < \theta_{\min}$) defined by:

$$\cot\frac{\theta_{\min}}{2} = \frac{\lambda_D}{s_0} \,, \tag{V.43}$$

that is, since $\theta_{\min}$ is small:

$$\frac{\theta_{\min}}{2} \simeq \frac{s_0}{\lambda_D} \,. \tag{V.44}$$

Applying this new limit of integration to (V.42), and setting:

$$\Lambda_c \equiv \frac{\lambda_D}{s_0} \,, \tag{V.45}$$

we obtain the well known expression for the total microscopic momentum transfer cross-section:

$$\hat{\sigma}_{tm} = 4\pi s_0^2 \ln \Lambda_c \ , \tag{V.46}$$

where the parameter $\ln \Lambda_c$ is referred to as the *Coulomb logarithm.*

Remark: Using the same integration limits as for $\hat{\sigma}_{tm}$, the total microscopic collision cross-section can be written:

$$\hat{\sigma}_{tc} = \pi \lambda_D^2 \ . \tag{V.47}$$

This result reflects the fact that **all** Coulomb interactions should be taken into account, provided they are within the Debye sphere ($s \leq \lambda_D$).

Coulomb logarithms for particles satisfying a Maxwellian distribution

The expression for the total microscopic momentum transfer cross-section in (V.46) requires the values of λ_D and s_0, which can be obtained by assuming that the population of charged particles α and β satisfy Maxwellian distributions, with temperatures T_α and T_β.

The Debye length λ_D is, in principle, the global Debye length defined by (1.41). In fact, following Delcroix (1959), the duration of a collision is too short for the screening action of the ions to have an effect, and it is preferable to use $\lambda_D = \lambda_{De}$, the electronic Debye length, in the expression for Λ_c, i.e.:

$$\lambda_{De} = \left(\frac{\epsilon_0 k_B T_e}{n e^2} \right)^{\frac{1}{2}} \ . \tag{1.41}$$

This expression is also consistent with the assumption that the ions constitute a neutralizing background for the electrons (see remark 8, Sect. 1.6).

For the critical impact parameter s_0, we need to calculate $\langle s_0 \rangle$, its mean averaged value $\mu_{\alpha\beta} w_{\alpha\beta}^2$, i.e.:

$$\langle \mu_{\alpha\beta} w_{\alpha\beta}^2 \rangle = \mu_{\alpha\beta} \langle (\boldsymbol{w}_{\alpha 0} - \boldsymbol{w}_{\beta 0})^2 \rangle \ , \tag{V.48}$$

knowing that:

$$\boldsymbol{w}_{\alpha 0} - \boldsymbol{w}_{\beta 0} = \boldsymbol{w}_{\alpha\beta} \ . \tag{1.69}$$

It follows that:

$$\langle \mu_{\alpha\beta} w_{\alpha\beta}^2 \rangle = \mu_{\alpha\beta} \langle w_{\alpha 0}^2 + w_{\beta 0}^2 - 2 \boldsymbol{w}_{\alpha 0} \cdot \boldsymbol{w}_{\beta 0} \rangle \ , \tag{V.49}$$

where the average of $\boldsymbol{w}_{\alpha 0} \cdot \boldsymbol{w}_{\beta 0}$ is zero since all the initial relative directions of the particles in the laboratory frame are equally probable. For Maxwellian distributions, we find, following (I.11):

$$\langle \mu_{\alpha\beta} w_{\alpha\beta}^2 \rangle = \frac{m_\alpha m_\beta}{m_\alpha + m_\beta} \left(\frac{3 k_B T_\alpha}{m_\alpha} + \frac{3 k_B T_\beta}{m_\beta} \right) \ . \tag{V.50}$$

If $T_e = T_i = T$, $\langle s_0 \rangle$ takes a unique value, independent of the nature of the collisions, electron-electron, ion-ion or ion-electron:

$$\langle s_0 \rangle = \frac{e^2}{12\pi\epsilon_0 k_B T} \, . \tag{V.51}$$

On the other hand, if $T_e \neq T_i$, we can distinguish three mean critical impact parameters:

- for electron-electron collisions:

$$s_{0_{ee}} = \frac{e^2}{12\pi\epsilon_0 k_B T_e} \, , \tag{V.52}$$

- for ion-ion collisions:

$$s_{0_{ii}} = \frac{e^2}{12\pi\epsilon_0 k_B T_i} \, , \tag{V.53}$$

- and, for ion-electron collisions:

$$s_{0_{ei}} = s_{0_{ie}} = \frac{e^2 \left(\dfrac{1}{m_e} + \dfrac{1}{m_i} \right)}{12\pi\epsilon_0 \left(\dfrac{k_B T_e}{m_e} + \dfrac{k_B T_i}{m_i} \right)} \, . \tag{V.54}$$

There are three corresponding Coulomb logarithms, $\ln \Lambda_{c_{ee}}$, $\ln \Lambda_{c_{ii}}$ and $\ln \Lambda_{c_{ei}}$. These can be written as functions of $\Lambda_{c_{ee}}$, from the expressions for $s_{0_{\alpha\beta}}$ ($m_e \ll m_i$):

$$\ln \Lambda_{c_{ii}} = \ln \Lambda_{c_{ee}} + \ln \frac{T_i}{T_e} \, , \tag{V.55}$$

and since:

$$s_{0_{ei}} \simeq s_{0_{ee}} \, , \tag{V.56}$$

then:

$$\ln \Lambda_{c_{ei}} \simeq \ln \Lambda_{c_{ee}} \, . \tag{V.57}$$

Coulomb collision frequencies and mean free paths

Collision frequencies for particles satisfying a Maxwellian distribution

Rigorously speaking, the average collision frequency of the species α with the (target) species β is defined by (1.140):

$$\langle \nu_{\alpha\beta}(w_{\alpha\beta}) \rangle = n_\beta \langle \hat{\sigma}_{\alpha\beta}(w_{\alpha\beta}) w_{\alpha\beta} \rangle \tag{V.58}$$

where $\hat{\sigma}_{\alpha\beta}$ is the total microscopic momentum transfer cross-section. However, in the present case of Coulomb collisions, the cross sections calculated above are already the result of an average taken over the relative velocities, and thus cannot be used to determine the exact expression for (V.58). To do this, it would be necessary to integrate (V.58) over the ensemble of velocities of the populations α and β (in the same way as in exercise 1.9), which leads to a very complex calculation. On the other hand, the collision frequency can be obtained, within an order of magnitude, from the approximate expression:

$$\langle \hat{\sigma}_{\alpha\beta} w_{\alpha\beta} \rangle \simeq \langle \hat{\sigma}_{\alpha\beta} \rangle \langle w_{\alpha\beta} \rangle \; , \tag{V.59}$$

where $\langle w_{\alpha\beta} \rangle$ is the mean velocity.

In fact, it is preferable to define what is called the *individual average collision frequency*, which corresponds to the *most probable relative velocity* $v_{\alpha\beta}$ (distinct from the mean relative velocity). The expression for $v_{\alpha\beta}$ is given by (exercise 1.9):

$$v_{\alpha\beta} = \sqrt{2k_B \left(\frac{T_\alpha}{m_\alpha} + \frac{T_\beta}{m_\beta} \right)} \; . \tag{V.60}$$

The collision frequency is then written:

$$\nu_{\alpha\beta}(v_{\alpha\beta}) = n_\beta \hat{\sigma}_{\alpha\beta}(v_{\alpha\beta}) \, v_{\alpha\beta} \; , \tag{V.61}$$

where $\hat{\sigma}_{\alpha\beta}$ is calculated from (V.46), substituting $v_{\alpha\beta}$ for $w_{\alpha\beta}$ in s_0 (V.23):

$$s_{0_{\alpha\beta}} = \frac{e^2}{4\pi\epsilon_0 \mu_{\alpha\beta} v_{\alpha\beta}^2} \; . \tag{V.62}$$

Assuming $T_e \gg T_i$, the collision frequencies can then be written:

$$\nu_{ee} = 4\pi n s_{0_{ee}}^2 \sqrt{\frac{4k_B T_e}{m_e}} \, \ln \left(\frac{\lambda_{De}}{s_{0_{ee}}} \right) \; , \tag{V.63}$$

$$\nu_{ei} = \nu_{ie} \simeq 4\pi n s_{0_{ee}}^2 \sqrt{\frac{2k_B T_e}{m_e}} \, \ln \left(\frac{\lambda_{De}}{s_{0_{ee}}} \right) \; , \tag{V.64}$$

$$\nu_{ii} = 4\pi n s_{0_{ii}}^2 \sqrt{\frac{4k_B T_i}{m_i}} \, \ln \left(\frac{\lambda_{De}}{s_{0_{ii}}} \right) \; . \tag{V.65}$$

It is possible to extract a number of simple relations from (V.63), (V.64) and (V.65). Thus, a first obvious relation:

$$\nu_{ee} \simeq \sqrt{2} \, \nu_{ei} \tag{V.66}$$

shows that the electron-electron and electron-ion collision frequencies are of the same order of magnitude. The second relation, obtained from (V.63):

$$\nu_{ee} = \omega_{pe} \frac{\ln \Lambda_{c_{ee}}}{\Lambda_{c_{ee}}} \tag{V.67}$$

allows us to relate the electron-electron collision frequency to the plasma electron angular frequency ω_{pe}.

Mean free paths for particles satisfying a Maxwellian distribution

Generally speaking, the mean free path of a particle α colliding with particles β can be defined by (1.39):

$$\ell_{\alpha\beta} = \frac{1}{n_\beta} \left\langle \frac{w_\alpha}{\hat{\sigma}_{\alpha\beta}(w_{\alpha\beta}) w_{\alpha\beta}} \right\rangle . \tag{V.68}$$

We define the *average mean free path* for Coulomb collisions, for *the most probable velocities*, in the same way as previously used for the collision frequency, i.e:

$$\ell_{\alpha\beta}(v_\alpha) = \frac{v_\alpha}{n_\beta \hat{\sigma}_{\alpha\beta} v_{\alpha\beta}} . \tag{V.69}$$

The different mean free paths can then be written ($T_e \gg T_i$):

$$\ell_{ee} = \frac{\sqrt{\dfrac{2k_B T_e}{m_e}}}{\nu_{ee}} , \tag{V.70}$$

$$\ell_{ei} = \frac{\sqrt{\dfrac{2k_B T_e}{m_e}}}{\nu_{ei}} , \tag{V.71}$$

$$\ell_{ie} = \frac{\sqrt{\dfrac{2k_B T_i}{m_i}}}{\nu_{ei}} , \tag{V.72}$$

$$\ell_{ii} = \frac{\sqrt{\dfrac{2k_B T_i}{m_i}}}{\nu_{ii}} . \tag{V.73}$$

The mean free path for electrons colliding with all charged particles (electrons and ions) can be written:

$$\ell_e = \frac{\sqrt{\dfrac{2k_B T_e}{m_e}}}{\nu_{ee} + \nu_{ei}} . \tag{V.74}$$

Remark: It is important to note that the Coulomb collision frequencies and the corresponding average mean free paths are independent of the density of the gas.

Appendix VI
Stepwise Ionisation

Two-step, and more generally, multi-step ionisation constitute mechanisms for creating charged particles, which become important whenever the gas pressure exceeds a few torr (a few hundred pascal). Such stepwise processes increase with increasing pressure and electron density to such an extent that they can supersede direct ionisation.

Stepwise ionisation starts with the excitation by an electron collision with the atom in its ground state:

$$\underline{\underline{\mathrm{e}}} + \mathrm{A} \rightarrow \mathrm{A}(j) + \underline{\mathrm{e}} \, . \tag{VI.1}$$

A second electron collision with this atom, which is now excited in the state j, can ionise the atom:

$$\underline{\underline{\mathrm{e}}} + \mathrm{A}(j) \rightarrow \mathrm{A}^{+} + \underline{\mathrm{e}} + \mathrm{e} \, . \tag{VI.2}$$

The excited atom thus serves as an intermediate stage, allowing ionisation with electrons of lower energy than that required for direct ionisation.

Population balance of intermediate (relay) state(s)

The ionisation frequency from the excited state is, by definition:

$$\nu_{ie} = \mathcal{N}_j \langle \hat{\sigma}_{ji}(w) w \rangle \, , \tag{VI.3}$$

where $\mathcal{N}_j$ is the density of atoms in the excited state (the targets for the electrons), $\hat{\sigma}_{ji}$ is the total microscopic cross-section for ionisation from the excited state j, and the square brackets refer to an integration conducted over the velocity distribution function of the particles. To calculate the stepwise ionisation frequency, one needs the density of the atoms excited in state j, which can be determined from the balance of the processes of creation and

M. Moisan, J. Pelletier, *Physics of Collisional Plasmas*,
DOI 10.1007/978-94-007-4558-2,

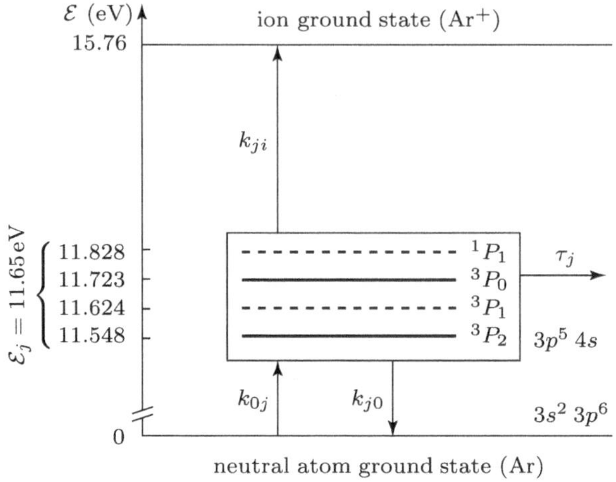

Fig. VI.1 Three-level energy diagram of the argon atom to characterise, in the present case, the two-step ionisation process: these levels are the ground state of the neutral atom, the intermediate state j (the 4 energy states $3p^5$ $4s$ orbital configuration are effectively considered as a single level) and the first level of the ionised atom. The $^3P_0$ and $^3P_2$ levels are metastable states, i.e. they have a comparatively lower probability of de-exciting through a (dipolar electric) radiative transition than the $^1P_1$ and $^3P_1$ levels, which are termed resonance radiation states (also referred to as quasi-metastable states).

loss of atoms in the intermediate state j. This can be established with the help of the three-level energy diagram shown in Fig. VI.1.

In the stationary state, the balance of creation and loss obviously requires that:

$$\frac{\mathrm{d}\mathcal{N}_j}{\mathrm{d}t} = 0\ , \tag{VI.4}$$

where $\mathrm{d}\mathcal{N}_j/\mathrm{d}t$, in the present case, can be written in the following form:

$$\begin{aligned}\frac{\mathrm{d}\mathcal{N}_j}{\mathrm{d}t} &= \mathcal{N}_0\langle\hat{\sigma}_{0j}(w)w\rangle n_e - \mathcal{N}_j\langle\hat{\sigma}_{j0}(w)w\rangle n_e \\ &\quad - \mathcal{N}_j\langle\hat{\sigma}_{ji}(w)w\rangle n_e - \frac{D_j}{\Lambda^2}\mathcal{N}_j\ .\end{aligned} \tag{VI.5}$$

The first term on the RHS represents the way the intermediate state is populated by electron collisions on the atom in the ground state (reaction (VI.1)). The other terms correspond to the depopulation of the intermediate states, successively, as a result of their de-excitation by electron collision to the ground state (the inverse process to reaction (VI.1), by ionisation described by (VI.2), and by diffusion of the atoms to the walls. In equation (VI.5), N_0 is the density of the atoms in the ground state; $k_{0j} \equiv \langle\hat{\sigma}_{0j}(w)w\rangle$ is the electron excitation coefficient from ground state to the state j; $k_{j0} \equiv \langle\hat{\sigma}_{j0}(w)w\rangle$ is the electron de-excitation coefficient from the state j to the ground state and $k_{ji} \equiv \langle\hat{\sigma}_{ji}(w)w\rangle$ is the ionisation coefficient from the excited state j. D_j

is the diffusion coefficient for excited atoms in the plasma and Λ is the characteristic diffusion length ($\Lambda = R/2.405$ in cylindrical geometry, where R is the tube radius; this is described in more detail in Sect. 3.8). The processes included in (VI.5) are represented in Fig. VI.1, where the notation j defines the intermediate state. The diffusion of the atoms in the intermediate stage is represented in the same diagram by their characteristic diffusion time τ_j where (Sect. 3.8):

$$\frac{D_j}{\Lambda^2} = \frac{1}{\tau_j} \, . \tag{VI.6}$$

Two-step ionisation generally occurs from atoms in metastable states as the intermediate stage, because their radiative de-excitation time is very long, such that the depopulation is governed by electron collisions and by diffusion. We obtain the density of the intermediate states in a stationary plasma from equation (VI.5):

$$\mathcal{N}_j = \frac{\mathcal{N}_0 \langle \hat{\sigma}_{0j}(w) w \rangle n_e}{\tau_j^{-1} + \left(\langle \hat{\sigma}_{j0}(w) w \rangle + \langle \hat{\sigma}_{ji}(w) w \rangle \right) n_e} \, . \tag{VI.7}$$

The rate of depopulation of the intermediate stages by diffusion is principally determined by the gas pressure: as gas density increases, the diffusion time τ_j for these atoms also increases, reducing the loss by this mechanism (the two metastable states in Fig. VI.1 are considered as forming one single intermediate stage). When the depopulation of these states by diffusion is much smaller than that due to collisions (i.e. $\tau_j^{-1} \ll \left(\langle \hat{\sigma}_{j0}(w) w \rangle + \langle \hat{\sigma}_{ji}(w) w \rangle \right) n_e$), Eq. (VI.7) shows that the value of $\mathcal{N}_j$ is independent of the electron density: this effect also manifests itself when the electron density is extremely high.

Ionisation frequency

Equation (VI.3), defining the two-step ionisation frequency, with the substitution of $\mathcal{N}_j$ from (VI.7), leads to:

$$\nu_{ie} = \frac{\mathcal{N}_0 \langle \hat{\sigma}_{0j}(w) w \rangle \langle \hat{\sigma}_{ji}(w) w \rangle \tau_j n_e}{1 + \left(\langle \hat{\sigma}_{j0}(w) w \rangle + \langle \hat{\sigma}_{ji}(w) w \rangle \right) \tau_j n_e} \, . \tag{VI.8}$$

Setting:

$$\rho_{ie} = \mathcal{N}_0 \langle \hat{\sigma}_{0j}(w) w \rangle \langle \hat{\sigma}_{ji}(w) w \rangle \tau_j \, , \qquad \text{units: cm}^3\,\text{s}^{-1} \tag{VI.9}$$

which we call the *two-step ionisation coefficient*, and:

$$\eta = \left(\langle \hat{\sigma}_{j0}(w) w \rangle + \langle \hat{\sigma}_{ji}(w) w \rangle \right) \tau_j \, , \qquad \text{units: cm}^3 \tag{VI.10}$$

which we refer to as the *saturation coefficient of intermediate states*. The two-step ionisation frequency can then be written in the form (1.159):

$$\nu_{ie} = \frac{\rho_{ie} n_e}{1 + \eta n_e} . \tag{VI.11}$$

The case where the value of ν_{ie} remains constant when n_e increases is referred to as *saturation*; two-step ionisation then clearly exceeds direct ionisation. This situation occurs when the diffusion time τ_j is very large (large value of η) or when the value of n_e is very large. In these conditions, Eq. (VI.11) no longer depends on n_e, since:

$$\nu_{ie} \simeq \frac{\rho_{ie}}{\eta} . \tag{1.160}$$

Appendix VII
Basic Notions of Tensors

A tensor is characterised by the transformation required to express it in another frame.

- A scalar s is an invariant quantity with respect to a change of frame. It is a tensor of rank (or order) zero.
- A vector $\boldsymbol{w}$ can be written as:

$$\boldsymbol{w} = w_x \hat{\mathbf{e}}_x + w_y \hat{\mathbf{e}}_y + w_z \hat{\mathbf{e}}_z \ , \qquad \text{(VII.1)}$$

where $\hat{\mathbf{e}}_x$, $\hat{\mathbf{e}}_y$ and $\hat{\mathbf{e}}_z$ are the basis vectors in a given frame, which we will refer to as the old frame. Following a change of frame, the components $(w_x,\ w_y,\ w_z)$ of $\boldsymbol{w}$ are related to the components in a new frame by a transformation matrix A with elements α^i_j. Writing W^I for the components of $\boldsymbol{w}$ in the new coordinate system, we then have[184]:

$$W^I = \alpha^I_1 w_1 + \alpha^I_2 w_2 + \alpha^I_3 w_3 = \alpha^I_i w^i \qquad \text{(VII.2)}$$

where, in order to apply the *Einstein summation convention* (to avoid using the summation sign), we have written the indices for the components of $\boldsymbol{w}$ on the RHS of the equation as superscripts. This rule is such that the same index, repeated as subscript and superscript, implies a summation over all its values: the *index* is said to be *dummy* because its name can be arbitrarily changed. Following this notation, the vector $\boldsymbol{w}$ can be represented in compact fashion by $\boldsymbol{w} = w^i \hat{\mathbf{e}}_i$.
The inverse transformation (from the new frame into the old) is made using the matrix B with elements β^i_j, which is the inverse of matrix A of elements (α^i_j); this implies that $(A)(B) = (I)$, where (I) is the identity matrix. Hence, the components of $\boldsymbol{w}$ in the old coordinate system can be written:

---

[184] If we designate the basis vectors in the new frame by $\hat{\boldsymbol{E}}_x$, $\hat{\boldsymbol{E}}_y$, $\hat{\boldsymbol{E}}_z$, these can be obtained from the basis vectors in the old frame through the relation $\hat{\boldsymbol{E}}_I = \alpha^k_I \hat{\mathbf{e}}_k$.

M. Moisan, J. Pelletier, *Physics of Collisional Plasmas*,
DOI 10.1007/978-94-007-4558-2,

$$w^i = \beta_I^i W^I \ . \qquad \text{(VII.3)}$$

The vector $\boldsymbol{w}$ is a first-rank tensor.

- A second-rank tensor $\underline{\boldsymbol{T}}$ can be expressed in a given frame as:

$$\underline{\boldsymbol{T}} = t_{xx}\hat{\mathbf{e}}_x\hat{\mathbf{e}}_x + t_{xy}\hat{\mathbf{e}}_x\hat{\mathbf{e}}_y + t_{xz}\hat{\mathbf{e}}_x\hat{\mathbf{e}}_z + \cdots \ , \qquad \text{(VII.4)}$$

i.e. it requires two basis vectors for each component (written as such, $\underline{\boldsymbol{T}}$ is also termed a dyadic tensor): there are 9 components in total. In consequence, during a change of frame, it is necessary to use the transformation matrix twice. Thus the components of $\underline{\boldsymbol{T}}$ in the new frame are given by:

$$T^{IJ} = \alpha_i^I \alpha_j^J t^{ij} \ . \qquad \text{(VII.5)}$$

We can generalise the definition of $\underline{\boldsymbol{T}}$ to any particular rank by noting that the number of indices in a tensor, i.e. the number of transformation matrices required during a change of frame, defines the *rank of the tensor.*

Tensor products

- The *tensor product* of two vectors $\boldsymbol{A}$ and $\boldsymbol{B}$, written as $\boldsymbol{A} \otimes \boldsymbol{B}$, is defined by its elements:

$$T_{ij} = A_i B_j \ , \qquad \text{(VII.6)}$$

where the indices i and j can take the values 1, 2 or 3. This creates a second-rank tensor. The tensor product can be generally applied to a product of two tensors of any order: the rank of the resulting tensor is the sum of the ranks of the tensors which form the product.

- The *scalar product* or internal product is, in tensor formalism, a *contraction*, reducing the order of the initial tensors by two units[185]. Thus, for two vectors $\boldsymbol{A}$ and $\boldsymbol{B}$, this gives (using the implicit summation rule):

$$\boldsymbol{A} \cdot \boldsymbol{B} = \left(A^i \hat{\mathbf{e}}_i\right) \cdot \left(B^j \hat{\mathbf{e}}_j\right) = A^i B^j \hat{\mathbf{e}}_i \cdot \hat{\mathbf{e}}_j = A^i B^j \delta_{ij} = \sum_i A^i B^i \qquad \text{(VII.7)}$$

where δ_{ij} is the Kronecker delta ($\delta_{ij} = 1$ if $i = j$ and $\delta_{ij} = 0$ if $i \neq j$). To continue to use the implicit summation rule, we must write (VII.7) in the form:

$$\boldsymbol{A} \cdot \boldsymbol{B} = A^i B_i \ . \qquad \text{(VII.8)}$$

The result is a scalar, i.e. a zero order tensor.

- The *vector product* of two vectors $\boldsymbol{A}$ and $\boldsymbol{B}$ is considered as a vector, but in fact it is a pseudovector. When we move from a right-handed triad to

---

[185] The scalar product of two vectors can be seen as a tensor product followed by a contraction.

a left-handed one, the pseudovector changes its direction in space, which is contrary to the notion of a true vector (also called a polar vector). In reality, the vector product should be represented by an antisymmetric, second-order tensor, i.e $T_{ij} = -T_{ji}$, which implies that the diagonal elements on the matrix $\underline{\boldsymbol{T}}$ should be zero. This tensor comprises only three independent elements, allowing the vector product to be represented by a vector in 3-dimension space.

Operators

- The *gradient* is an operator that produces a tensor one order higher than that on which it operates. Thus, starting with a scalar s:

$$\boldsymbol{\nabla} s = \frac{\partial s}{\partial x}\hat{\mathbf{e}}_x + \frac{\partial s}{\partial y}\hat{\mathbf{e}}_y + \frac{\partial s}{\partial z}\hat{\mathbf{e}}_z = \sum_i \partial_i s \; \hat{\mathbf{e}}_i \; , \tag{VII.9}$$

  we obtain a vector.
  The convention of dummy indices cannot be applied here, since the operator "spatial derivative" is, strictly speaking, covariant[186] (yielding a subscript index), hence the necessity of writing explicitly the summation sign.
- The *divergence* is the result of the action of a gradient operator followed by a contraction. Thus, the divergence of a vector $\boldsymbol{w}$, $\boldsymbol{\nabla} \cdot \boldsymbol{w}$, is a scalar (see note 74, p. 138): the result is a tensor of one rank lower than the one that is operated on.
- The *curl* acts on a covariant vector with components a_i and produces a covariant, antisymmetric, second-rank tensor with elements b_{ij}:

$$b_{ij} = \frac{\partial a_j}{\partial x^i} - \frac{\partial a_i}{\partial x^j} \; . \tag{VII.10}$$

  This creates a tensor of one order higher than the initial tensor, due to the action of the spatial operator $\partial/\partial x^i$.

Example of the proof of a tensor identity

We wish to show that:

$$\boldsymbol{\nabla}_r \cdot (\boldsymbol{w}\boldsymbol{w} f) = \boldsymbol{w}(\boldsymbol{w} \cdot \boldsymbol{\nabla}_r f) \; . \tag{VII.11}$$

Developing the LHS:

[186] The concept of a covariant derivative is beyond the scope of the present plasma treatise.

$$\boldsymbol{\nabla}_r \cdot \boldsymbol{w}\boldsymbol{w}f = \sum_i \partial_i \left(\hat{\mathbf{e}}_i \cdot \left(w^p w^q \hat{\mathbf{e}}_p \hat{\mathbf{e}}_q\right) f\right) , \tag{VII.12}$$

where f is a scalar and ∂_i a derivative operator in position space (which therefore does not act on the microscopic velocities). Note that there is no implicit summation over index i because the two elements carrying this index are covariant. Expanding the product $\hat{\mathbf{e}}_i \cdot \hat{\mathbf{e}}_p = \delta_{ip}$ in (VII.12), which imposes $i = p$ (VII.7), we have:

$$\partial_i \left(\hat{\mathbf{e}}_i \cdot \left(w^p w^q \hat{\mathbf{e}}_p \hat{\mathbf{e}}_q\right) f\right) = \partial_i \left(w^i w^q \hat{\mathbf{e}}_q f\right) = \partial_i \left(f w^i\right) w^q \hat{\mathbf{e}}_q , \tag{VII.13}$$

where there is a summation (contraction) over the index i, such that:

$$\partial_i \left(f w^i\right) \boldsymbol{w} = \left(\boldsymbol{\nabla}_r f \cdot \boldsymbol{w}\right) \boldsymbol{w} . \tag{VII.14}$$

Finally, because $\boldsymbol{\nabla}_r f \cdot \boldsymbol{w}$ is a scalar, we can write:

$$\left(\boldsymbol{\nabla}_r f \cdot \boldsymbol{w}\right) \boldsymbol{w} = \boldsymbol{w}(\boldsymbol{\nabla}_r f \cdot \boldsymbol{w}) , \tag{VII.15}$$

which is the RHS of (VII.11), as required: QED.

Appendix VIII
Operations on Tensors

The fundamental properties of tensors were presented in Appendix VII. We will now give the rules for tensor operations, without recourse to the implicit summation (dummy indices) defined in Appendix VII.

Product of two vectors

Consider two vectors $\boldsymbol{A}$ and $\boldsymbol{B}$, with components A_i and B_i, with $i = x$, y, z or $i = 1$, 2, 3.

Scalar product of two vectors: $\boldsymbol{A} \cdot \boldsymbol{B}$

The result is a scalar C:

$$\boldsymbol{A} \cdot \boldsymbol{B} = \sum_{i=1}^{3} A_i B_i = \boldsymbol{B} \cdot \boldsymbol{A} = C \, . \tag{VIII.1}$$

The scalar product of two vectors is commutative.

Vector product of two vectors: $\boldsymbol{A} \wedge \boldsymbol{B}$

The result is a vector $\boldsymbol{C}$ (in reality, a pseudovector, Appendix VII) with components:

$$C_i = A_{i+1} B_{i-1} - A_{i-1} B_{i+1} \tag{VIII.2}$$

where, if $i = x$, then $x + 1 = y$ and $x - 1 = z$. It follows that:

$$\boldsymbol{A} \wedge \boldsymbol{B} = -\boldsymbol{B} \wedge \boldsymbol{A} = \boldsymbol{C} \, , \tag{VIII.3}$$

M. Moisan, J. Pelletier, *Physics of Collisional Plasmas*,
DOI 10.1007/978-94-007-4558-2,

The vector product is not commutative.

Note the important rule for the *double vector product*:

$$\boldsymbol{A} \wedge (\boldsymbol{B} \wedge \boldsymbol{C}) = (\boldsymbol{A} \cdot \boldsymbol{C})\boldsymbol{B} - (\boldsymbol{A} \cdot \boldsymbol{B})\boldsymbol{C} \; . \tag{VIII.4}$$

Tensor product of two vectors: $\boldsymbol{A} \otimes \boldsymbol{B}$

The result is a second-order tensor $\underline{\boldsymbol{T}}$, with its components T_{ij} being the algebraic product of the components A_i and B_j:

$$T_{ij} = A_i B_j \; . \tag{VIII.5}$$

It follows that:

$$\boldsymbol{A} \otimes \boldsymbol{B} = (\boldsymbol{B} \otimes \boldsymbol{A})^T \; . \tag{VIII.6}$$

The product is not commutative, unless the vectors are parallel. The superscript symbol T indicates that the tensor is transposed (T_{ij} becomes T_{ji}).

Remark: In the main text, for simplicity, we have represented the tensor product of two vectors $\boldsymbol{A}$ and $\boldsymbol{B}$ in the form $\boldsymbol{AB}$ rather than $\boldsymbol{A} \otimes \boldsymbol{B}$.

Product of two tensors

Consider two second-order tensors $\underline{\boldsymbol{S}}$ and $\underline{\boldsymbol{T}}$.

Tensor product: $\underline{\boldsymbol{S}} \otimes \underline{\boldsymbol{T}}$

The result is a 4th order tensor:

$$\underline{\boldsymbol{S}} \otimes \underline{\boldsymbol{T}} = \underline{\underline{\underline{\boldsymbol{U}}}} \tag{VIII.7}$$

whose components are:

$$U_{ijkl} = S_{ij} T_{kl} \; . \tag{VIII.8}$$

Singly-contracted product: $\underline{\boldsymbol{S}} \cdot \underline{\boldsymbol{T}}$

The result is a 2nd order tensor:

$$\underline{\boldsymbol{S}} \cdot \underline{\boldsymbol{T}} = \underline{\boldsymbol{U}} \tag{VIII.9}$$

whose components are:

$$U_{ij} = \sum_k S_{ik} T_{kj} \tag{VIII.10}$$

and if both tensors are symmetric:

$$\underline{\boldsymbol{S}} \cdot \underline{\boldsymbol{T}} = (\underline{\boldsymbol{T}} \cdot \underline{\boldsymbol{S}})^T = (\underline{\boldsymbol{U}})^T = \underline{\boldsymbol{U}} \,. \tag{VIII.11}$$

Doubly-contracted product: $\underline{\boldsymbol{S}} : \underline{\boldsymbol{T}}$

The result is a scalar:

$$\underline{\boldsymbol{S}} : \underline{\boldsymbol{T}} = \sum_i \sum_j S_{ij} T_{ji} = U \tag{VIII.12}$$

with:

$$\underline{\boldsymbol{S}} : \underline{\boldsymbol{T}} = \underline{\boldsymbol{T}} : \underline{\boldsymbol{S}} \,. \tag{VIII.13}$$

Product between a vector and a tensor

Consider $\boldsymbol{A}$, a vector, and $\underline{\boldsymbol{T}}$, a second-order tensor.

Tensor product of a vector with a 2nd order tensor

The result is a 3rd order tensor:

$$\boldsymbol{A} \otimes \underline{\boldsymbol{T}} = \underline{\underline{\boldsymbol{Q}}} \tag{VIII.14}$$

whose components are:

$$Q_{ijk} = A_i T_{jk} \,. \tag{VIII.15}$$

Contracted product of a vector with a 2nd order tensor

The result is a vector:

$$\underline{\boldsymbol{T}} \cdot \boldsymbol{A} = \boldsymbol{D} \tag{VIII.16}$$

whose components are:

$$D_i = \sum_j T_{ij} A_j \,. \tag{VIII.17}$$

Similarly, the product $\boldsymbol{A} \cdot \underline{\boldsymbol{T}}$ is a vector:

$$\boldsymbol{A} \cdot \underline{\boldsymbol{T}} = \boldsymbol{D}' \tag{VIII.18}$$

whose components are:

$$D'_i = \sum_j A_j T_{ji} \,. \tag{VIII.19}$$

Thus, if $\underline{\boldsymbol{T}}$ is a symmetric tensor ($T_{ij} = T_{ji}$), the product is commutative:

$$\underline{\boldsymbol{T}} \cdot \boldsymbol{A} = \boldsymbol{A} \cdot \underline{\boldsymbol{T}} \, . \tag{VIII.20}$$

If $\underline{\boldsymbol{T}}$ is obtained from the tensor product of two vectors:

$$\underline{\boldsymbol{T}} = \boldsymbol{B} \otimes \boldsymbol{C} \tag{VIII.21}$$

then:

$$\boldsymbol{A} \cdot (\boldsymbol{B} \otimes \boldsymbol{C}) = \boldsymbol{D} \, , \tag{VIII.22}$$

with components:

$$D_i = \sum_j A_j B_j C_i = (\boldsymbol{A} \cdot \boldsymbol{B}) C_i \, , \tag{VIII.23}$$

from which:

$$\boldsymbol{D} = \boldsymbol{A} \cdot (\boldsymbol{B} \otimes \boldsymbol{C}) = (\boldsymbol{A} \cdot \boldsymbol{B}) \boldsymbol{C} \tag{VIII.24}$$

and since the product is commutative:

$$\boldsymbol{D} = (\boldsymbol{A} \otimes \boldsymbol{B}) \cdot \boldsymbol{C} = \boldsymbol{A} (\boldsymbol{B} \cdot \boldsymbol{C}) \, . \tag{VIII.25}$$

Contracted product of a vector with a 3rd order tensor

Consider $\underline{\underline{\boldsymbol{Q}}}$, a 3rd order tensor. The result is a 2nd order tensor $\underline{\boldsymbol{T}}$:

$$\boldsymbol{A} \cdot \underline{\underline{\boldsymbol{Q}}} = \underline{\boldsymbol{T}} \tag{VIII.26}$$

whose components are:

$$T_{ij} = \sum_k A_k Q_{kij} \, . \tag{VIII.27}$$

Similarly:

$$\underline{\underline{\boldsymbol{Q}}} \cdot \boldsymbol{A} = \underline{\boldsymbol{T}}' \tag{VIII.28}$$

is a 2nd order tensor whose components are:

$$T'_{ij} = \sum_k Q_{ijk} A_k \, . \tag{VIII.29}$$

Vector product of a vector with a 2nd order tensor

The result is a 2nd order tensor:

$$\boldsymbol{A} \wedge \underline{\boldsymbol{T}} = \underline{\boldsymbol{U}} \tag{VIII.30}$$

whose components are:

$$U_{ij} = A_{i+1}T_{i-1,j} - A_{i-1}T_{i+1,j} \ . \qquad \text{(VIII.31)}$$

The vector column j of the tensor $\underline{\boldsymbol{U}}$ is the vector product of the vector $\boldsymbol{A}$ by the column vector j of the tensor $\underline{\boldsymbol{T}}$.

Operations involving the differential operator (Cartesian coordinates)

The differential operator $\boldsymbol{\nabla}$ (or $\partial/\partial\boldsymbol{r}$) can be considered as a vector, whose components in Cartesian coordinates are:

$$\boldsymbol{\nabla}_i = \frac{\partial}{\partial x_i} \ . \qquad \text{(VIII.32)}$$

Divergence of a vector

The result is a scalar:

$$\boldsymbol{\nabla} \cdot \boldsymbol{A} = \sum_{i=1}^{3} \frac{\partial A_i}{\partial x_i} = C \ . \qquad \text{(VIII.33)}$$

Divergence of a 2nd order tensor

The result is a vector:

$$\boldsymbol{\nabla} \cdot \underline{\boldsymbol{T}} = \boldsymbol{A} \qquad \text{(VIII.34)}$$

whose components are:

$$A_i = \sum_j \frac{\partial T_{ji}}{\partial x_j} \ . \qquad \text{(VIII.35)}$$

The divergence is the contracted product of the differential operator $\boldsymbol{\nabla}$ with a vector or a tensor.

Gradient of a scalar

The result is a vector:

$$\boldsymbol{\nabla} C = \boldsymbol{A} \qquad \text{(VIII.36)}$$

whose components are:

$$A_i = \frac{\partial C}{\partial x_i} \ . \qquad \text{(VIII.37)}$$

Gradient of a vector or a tensor

In general, the gradient is the tensor product of the differential operator $\boldsymbol{\nabla}$ with a scalar, a vector or a tensor.
Thus, operating on a vector $\boldsymbol{A}$, the result is a 2nd order tensor:

$$\boldsymbol{\nabla A} \equiv \boldsymbol{\nabla} \otimes \boldsymbol{A} = \underline{\boldsymbol{T}} \tag{VIII.38}$$

whose components are:

$$T_{ij} = \frac{\partial A_j}{\partial x_i} \,. \tag{VIII.39}$$

Note that $\boldsymbol{\nabla} B$ is a vector while $\boldsymbol{\nabla B}$ is a 2nd order tensor.

Curl of a vector

The curl is the vector product of the differential operator $\boldsymbol{\nabla}$ with a vector. We obtain a (pseudo) vector:

$$\boldsymbol{\nabla} \wedge \boldsymbol{A} = \boldsymbol{C} \tag{VIII.40}$$

with the components:

$$C_i = \frac{\partial A_{i-1}}{\partial x_{i+1}} - \frac{\partial A_{i+1}}{\partial x_{i-1}} \,. \tag{VIII.41}$$

Laplacian of a scalar

This is a scalar:

$$\Delta C = \sum_i \frac{\partial^2 C}{\partial x_i^2} \,. \tag{VIII.42}$$

Since:

$$\sum_i \frac{\partial^2 C}{\partial x_i^2} = \sum_i \frac{\partial}{\partial x_i}\left(\frac{\partial C}{\partial x_i}\right) = \boldsymbol{\nabla} \cdot (\boldsymbol{\nabla} C) \,, \tag{VIII.43}$$

thus:

$$\Delta C = \boldsymbol{\nabla} \cdot (\boldsymbol{\nabla} C) = \nabla^2 C \,. \tag{VIII.44}$$

Laplacian of a vector

This is a vector:

$$\Delta \boldsymbol{A} = \boldsymbol{C} \tag{VIII.45}$$

with the components:

$$C_i = \sum_j \frac{\partial^2 A_i}{\partial x_j^2} \; . \tag{VIII.46}$$

Since:

$$\Delta \boldsymbol{A} = \sum_j \frac{\partial^2 A_i}{\partial x_j^2} = \sum_j \frac{\partial}{\partial x_j}\left(\frac{\partial A_i}{\partial x_j}\right) = \boldsymbol{\nabla} \cdot (\boldsymbol{\nabla} \otimes \boldsymbol{A}) \; , \tag{VIII.47}$$

thus:

$$\Delta \boldsymbol{A} = \boldsymbol{\nabla} \cdot (\boldsymbol{\nabla} \otimes \boldsymbol{A}) = \nabla^2 \boldsymbol{A} \; . \tag{VIII.48}$$

Remark: One should not forget that $\boldsymbol{\nabla}$ is a differential operator, and that, as a consequence, when it is applied to a product, it is applied on each of the terms. For instance: $\boldsymbol{\nabla} \cdot (\boldsymbol{B} \otimes \boldsymbol{C}) = (\boldsymbol{\nabla} \cdot \boldsymbol{B})\boldsymbol{C} + (\boldsymbol{B} \cdot \boldsymbol{\nabla})\boldsymbol{C}$.

Appendix IX
Orientation of $\boldsymbol{w}_{2\perp}$ in the Reference Triad with Cartesian Axes $(\boldsymbol{E}_{0\perp} \wedge \boldsymbol{B},\ \boldsymbol{E}_{0\perp},\ \boldsymbol{B})$

We will make use of the reference triad shown in Fig. 2.8 and of the representation of the velocity $\boldsymbol{w}_{2\perp}$ in Fig. 2.9. From (2.137), for an electron ($q = -e$ and $\omega_c = \omega_{ce}$), we then have:

$$\boldsymbol{w}_{2\perp} = -\frac{e}{m_e(\omega_{ce}^2 - \omega^2)} \left\{ i\omega \boldsymbol{E}_{0\perp} - \frac{\omega_{ce}}{B}(\boldsymbol{E}_{0\perp} \wedge \boldsymbol{B}) \right\} \mathrm{e}^{\mathrm{i}\omega t} \tag{IX.1}$$

1. The case $\omega > \omega_{ce}$: major axis along $\boldsymbol{E}_{0\perp}$
   From (IX.1), we have:

$$\begin{aligned} \Re(\boldsymbol{w}_{2\perp}) = {} & \frac{e}{m_e(\omega^2 - \omega_{ce}^2)} \Re\Big\{ \mathrm{i}\omega(\cos\omega t + \mathrm{i}\sin\omega t)\boldsymbol{E}_{0\perp} \\ & \qquad - \frac{\omega_{ce}}{B}(\cos\omega t + \mathrm{i}\sin\omega t)(\boldsymbol{E}_{0\perp} \wedge \boldsymbol{B}) \Big\} \\ = {} & -A_1\omega(\sin\omega t)\boldsymbol{E}_{0\perp} - A_2\omega_{ce}(\cos\omega t)(\boldsymbol{E}_{0\perp} \wedge \boldsymbol{B}) \end{aligned} \tag{IX.2}$$

   where A_1 and A_2 are constants.
   At $t = 0$, we have $\boldsymbol{E}_\perp = \boldsymbol{E}_{0\perp}$, $\boldsymbol{w}_{2\perp} = -A_2\omega_{ce}(\boldsymbol{E}_{0\perp} \wedge \boldsymbol{B})$ and at $t = \mathcal{T}/2$, $\boldsymbol{E}_\perp = -\boldsymbol{E}_{0\perp}$, $\boldsymbol{w}_{2\perp} = A_2\omega_{ce}(\boldsymbol{E}_{0\perp} \wedge \boldsymbol{B})$, as shown in Fig. 2.9.

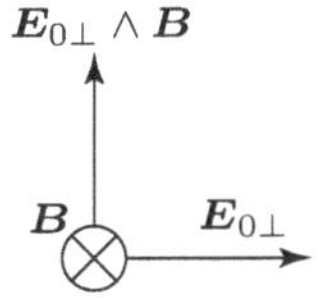

2. The case $\omega < \omega_{ce}$: major axis along $\boldsymbol{E}_{0\perp} \wedge \boldsymbol{B}$
   From (IX.1), we have:

$$\Re(\boldsymbol{w}_{2\perp}) = A_1\omega(\sin\omega t)\boldsymbol{E}_{0\perp} + A_2\omega_{ce}(\cos\omega t)(\boldsymbol{E}_{0\perp} \wedge \boldsymbol{B}) \,. \tag{IX.3}$$

   At $t = 0$, we have $\boldsymbol{E}_\perp = \boldsymbol{E}_{0\perp}$, $\boldsymbol{w}_{2\perp} = A_2\omega_{ce}(\boldsymbol{E}_{0\perp} \wedge \boldsymbol{B})$ and at $t = \mathcal{T}/2$, $\boldsymbol{E}_\perp = -\boldsymbol{E}_{0\perp}$, $\boldsymbol{w}_{2\perp} = -A_2\omega_{ce}(\boldsymbol{E}_{0\perp} \wedge \boldsymbol{B})$ (Fig. 2.9). Note that, in the present case, the velocity is out of phase by a factor π with respect to the case $\omega > \omega_{ce}$.

M. Moisan, J. Pelletier, *Physics of Collisional Plasmas*,
DOI 10.1007/978-94-007-4558-2,

Appendix X
Force Acting on a Charged Particle in the Direction of a Magnetic Field $\boldsymbol{B}$ Weakly Non-uniform Axially: Variant of (2.177)

The Lorentz force due to $\boldsymbol{B}$ along the z axis can be written $F_z = q(\boldsymbol{w} \wedge \boldsymbol{B}) \cdot \hat{\mathbf{e}}_z$. In cylindrical coordinates, due to the helical motion (components along $\hat{\mathbf{e}}_z$ and $\hat{\mathbf{e}}_\varphi$), we have:

$$\boldsymbol{w} \wedge \boldsymbol{B} = \begin{vmatrix} \hat{\mathbf{e}}_r & \hat{\mathbf{e}}_\varphi & \hat{\mathbf{e}}_z \\ 0 & w_\perp & w_\parallel \\ B_r & 0 & B_z \end{vmatrix} . \tag{X.1}$$

This leads to:

$$F_z = -q w_\perp B_r . \tag{X.2}$$

However, we also know that close enough to the axis of symmetry of the field $\boldsymbol{B}$, we can write:

$$r B_r \approx - \int_0^r r' \left(\frac{\partial B_z}{\partial z} \right)_{r'=0} \mathrm{d}r' = -\frac{1}{2} r^2 \left(\frac{\partial B_z}{\partial z} \right)_{r=0} . \tag{2.161}$$

Assuming that the particle moves about the guiding centre, defined by the axis of symmetry of the field $\boldsymbol{B}$, with a Larmor radius r_B, we can then set $r = r_B$ in the expression for B_r, from which:

$$F_z = q w_\perp \frac{r_B}{2} \frac{\partial B_z}{\partial z} \tag{X.3}$$

and

$$\begin{aligned} F_z &= q w_\perp \frac{w_\perp}{2\omega_c} \frac{\partial B_z}{\partial z} = q \frac{w_\perp^2}{2} \left(\frac{m_\alpha}{-q B_z} \right) \frac{\partial B_z}{\partial z} \\ &= -\frac{1}{2} \frac{m_\alpha w_\perp^2}{B_z} \frac{\partial B_z}{\partial z} = -\mu \left(\frac{\partial B_z}{\partial z} \right) \end{aligned} \tag{2.177}$$

since $\mu = \mathcal{E}_{\mathrm{kin}\perp}/B_z$ (2.148).

The axial non uniformity of the field $\boldsymbol{B}$ gives rise to a force proportional to the gradient of the field.

M. Moisan, J. Pelletier, *Physics of Collisional Plasmas*,
DOI 10.1007/978-94-007-4558-2,

Appendix XI
The Magnetic Moment, an Invariant in the Guiding Centre Approximation

We will consider the case where there is no applied electric field $\boldsymbol{E}$. We will also neglect the field $\boldsymbol{E}$ induced by the inhomogeneity of $\boldsymbol{B}$ in the frame of the particle: this is consistent with the zero-order guiding centre approximation. Under these conditions, the total particle kinetic energy, $W_T = W_\perp + W_\parallel$ is constant (2.179). It follows that:

$$\frac{\mathrm{d}}{\mathrm{d}t}(W_\parallel) \equiv \frac{\mathrm{d}}{\mathrm{d}t}\left(\frac{1}{2}m_\alpha w_z^2\right) = -\frac{\mathrm{d}}{\mathrm{d}t}(W_\perp) \tag{XI.1}$$

and, furthermore:

$$\frac{\mathrm{d}}{\mathrm{d}t}(W_\perp) \equiv \frac{\mathrm{d}}{\mathrm{d}t}\left(\frac{W_\perp B}{B}\right) = \frac{W_\perp}{B}\frac{\mathrm{d}B}{\mathrm{d}t} + B\frac{\mathrm{d}}{\mathrm{d}t}\left(\frac{W_\perp}{B}\right) . \tag{XI.2}$$

Knowing that:

$$F_z = -\mu\frac{\partial B_z}{\partial z} , \tag{2.177}$$

by multiplying each side of (2.177) by w_z, and since $\mu = \dfrac{W_\perp}{B}$:

$$w_z m_\alpha \frac{\mathrm{d}w_z}{\mathrm{d}t} \equiv \frac{\mathrm{d}}{\mathrm{d}t}\left(\frac{1}{2}m_\alpha w_\parallel^2\right) = -\frac{W_\perp}{B}\frac{\partial B_z}{\partial z}\frac{\mathrm{d}z}{\mathrm{d}t} = -\frac{W_\perp}{B}\frac{\mathrm{d}B_z}{\mathrm{d}t} , \tag{XI.3}$$

then, applying (XI.1), the LHS of (XI.3) can be written:

$$-\frac{\mathrm{d}}{\mathrm{d}t}\left(\frac{1}{2}m_\alpha w_\perp^2\right) = -\frac{W_\perp}{B}\frac{\mathrm{d}B_z}{\mathrm{d}t} . \tag{XI.4}$$

Taking account of (XI.2) and replacing the LHS of equation (XI.4) leads to:

$$\frac{W_\perp}{B}\frac{\mathrm{d}B_z}{\mathrm{d}t} + B\frac{\mathrm{d}}{\mathrm{d}t}\left(\frac{W_\perp}{B}\right) = \frac{W_\perp}{B}\frac{\mathrm{d}B_z}{\mathrm{d}t} , \tag{XI.5}$$

M. Moisan, J. Pelletier, *Physics of Collisional Plasmas*,
DOI 10.1007/978-94-007-4558-2,

which clearly imposes that:

$$\frac{\mathrm{d}}{\mathrm{d}t}\left(\frac{W_\perp}{B}\right) \equiv \frac{\mathrm{d}\mu}{\mathrm{d}t} = 0\,, \tag{XI.6}$$

i.e. the magnetic moment μ is independent of t. We obtained the same result in the introduction of Sect. 2.2.3 (p. 137).

Appendix XII
Drift Velocity $\boldsymbol{w}_d$ of a Charged Particle Subjected to an Arbitrary Force $\boldsymbol{F}_d$ in a Field $\boldsymbol{B}$: the Magnetic Field Drift

Generalisation of the expression for the drift velocity in a field $\boldsymbol{B}$ from the expression for the electric field drift

For the electric field drift, we found (2.114):

$$\boldsymbol{w}_{de} = \frac{\boldsymbol{E}_\perp \wedge \boldsymbol{B}}{B^2} \,, \tag{XII.1}$$

which is an expression that can be generalised by setting $q\boldsymbol{E}_\perp = \boldsymbol{F}_{de}$, where the meaning of $\boldsymbol{F}_{de}$ can be extended to include an arbitrary force $\boldsymbol{F}_d$, which leads us to the general expression:

$$\boldsymbol{w}_d = \frac{\boldsymbol{F}_d \wedge \boldsymbol{B}}{qB^2} \,. \tag{XII.2}$$

Application to the case of the magnetic field drift (rectilinear field lines) in a weakly inhomogeneous field $\boldsymbol{B}$

Since the field $\boldsymbol{B}$ is weakly inhomogeneous, to zeroth order μ is a constant of the motion and, following (2.177), which we generalise in three dimensions by writing:

$$\boldsymbol{F}_{dm} = \boldsymbol{\mu} \cdot \boldsymbol{\nabla} \boldsymbol{B} \,. \tag{XII.3}$$

In the case of rectilinear magnetic field lines, let $\boldsymbol{B}$ be directed along $\hat{\mathbf{e}}_z$ and the inhomogeneity along y; then $\boldsymbol{\mu}$, which is connected to the diamagnetic field, is directed along $-\hat{\mathbf{e}}_z$:

$$\boldsymbol{\mu} = -\mu_z \hat{\mathbf{e}}_z \quad \text{and} \quad \boldsymbol{B} = B_z(y)\hat{\mathbf{e}}_z \,, \tag{XII.4}$$

M. Moisan, J. Pelletier, *Physics of Collisional Plasmas*,
DOI 10.1007/978-94-007-4558-2,

we then have:

$$\boldsymbol{F}_{dm} \equiv -\mu_z \hat{\mathbf{e}}_z \cdot \left(\frac{\partial B_z}{\partial y} \hat{\mathbf{e}}_y \hat{\mathbf{e}}_z \right) = -\mu_z \frac{\partial B_z}{\partial y} \hat{\mathbf{e}}_y \ . \tag{XII.5}$$

Substituting the expression for the force generating the magnetic-field drift (XII.5) in (XII.2), we obtain:

$$\boldsymbol{w}_{dm} \equiv -\mu_z \frac{\partial B_z}{\partial y} \frac{\hat{\mathbf{e}}_y \wedge B\hat{\mathbf{e}}_z}{qB^2} = \frac{\mu}{q} \frac{(\boldsymbol{B} \wedge \boldsymbol{\nabla} B)}{B^2} \ , \tag{XII.6}$$

which is exactly the same equation we obtained in (2.217): this result supports our hypothesis that (XII.2) is valid for an arbitrary force $\boldsymbol{F}_d$.

Appendix XIII
Magnetic-Field Drift Velocity w_{dm} in the Frenet Frame Associated with the Lines of Force of a Magnetic Field with Weak Curvature

To first order, the particle follows a cyclotron motion around a line of force, which constitutes the axis of its helical motion.

Frenet-(Serret) frame

At each point on a magnetic field line of force (Fig. XIII.1), we can construct a Cartesian frame such that:

1. The unit vector $\hat{\mathbf{e}}_z$ is directed along the **tangent** to the magnetic field line of $\boldsymbol{B}$ at each point,
2. $\hat{\mathbf{e}}_y$ is **normal** to this tangent, and directed along the *radius of curvature* $\boldsymbol{\rho}$[187], this second vector pointing towards the field line, i.e. in the opposite direction to $\hat{\mathbf{e}}_y$, and
3. $\hat{\mathbf{e}}_x$ is along the **binormal**, i.e. in the direction perpendicular to the two other unit vectors, such as to form a right-handed triad.

The Frenet frame is also, to a first approximation, the *natural frame of the particle* in the present case.

Frenet relations

Classical mechanics teaches us that on a trajectory s connected to a Frenet frame:

$$\frac{\mathrm{d}\hat{\mathbf{e}}_z}{\mathrm{d}s} = \frac{\hat{\mathbf{e}}_y}{\rho} = -\frac{\boldsymbol{\rho}}{\rho^2}\,, \tag{XIII.1}$$

[187] The radius of curvature at a point A on a curve is the distance between that point and the intersection of two normals to the curve, situated immediately on either side of the point A.

M. Moisan, J. Pelletier, *Physics of Collisional Plasmas*,
DOI 10.1007/978-94-007-4558-2,

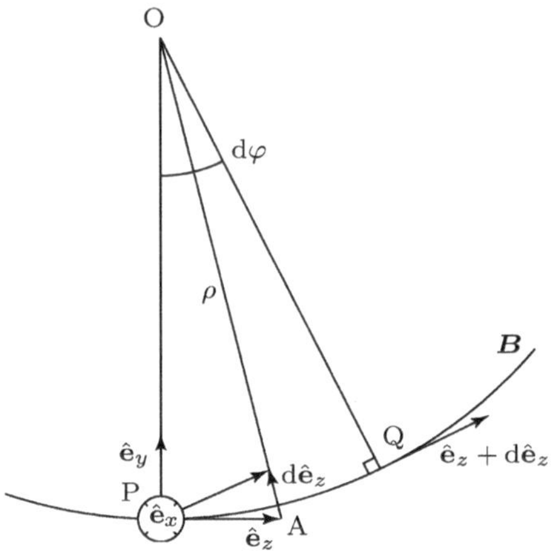

Fig. XIII.1 A Frenet frame, constructed on a magnetic field line with radius of curvature $\boldsymbol{\rho}$. The unit vector $\hat{\mathbf{e}}_x$ is directed into the page. The unit vector $\hat{\mathbf{e}}_z + \mathrm{d}\hat{\mathbf{e}}_z$, at point Q is transported, in parallel, to the point P in order to show the direction of $\mathrm{d}\hat{\mathbf{e}}_z$.

where the radius of curvature ρ is obtained from the derivative of the local tangents, as suggested in Fig. XIII.1. In the case where $y(z)$ describes the line of force, one can show that, provided $\mathrm{d}y/\mathrm{d}z$ is not too large:

$$\frac{\mathrm{d}^2 y}{\mathrm{d}z^2} = \frac{1}{\rho} \,. \tag{XIII.2}$$

Further, the components of the vector $\boldsymbol{\rho}$ can be written (Jancel and Kahan):

$$\frac{\rho_x}{\rho^2} = -\frac{1}{B}\frac{\partial B_x}{\partial z} \,, \tag{XIII.3}$$

$$\frac{\rho_y}{\rho^2} = -\frac{1}{B}\frac{\partial B_y}{\partial z} \,, \tag{XIII.4}$$

$$\frac{\rho_z}{\rho^2} = 0 \,. \tag{XIII.5}$$

Components of ∇B

The Maxwell equation $\boldsymbol{\nabla} \wedge \boldsymbol{B} = 0$ (without the RHS, in the framework of individual particle trajectories[85]) leads to:

$$\hat{\mathbf{e}}_x \left(\frac{\partial B_z}{\partial y} - \frac{\partial B_y}{\partial z} \right) + \hat{\mathbf{e}}_y \left(\frac{\partial B_x}{\partial z} - \frac{\partial B_z}{\partial x} \right) + \hat{\mathbf{e}}_z \left(\frac{\partial B_y}{\partial x} - \frac{\partial B_x}{\partial y} \right) = 0 \,. \tag{XIII.6}$$

Further, from the assumption that the inhomogeneity in $\boldsymbol{B}$ is independent of y, we have $\dfrac{\partial B_z}{\partial y} \neq 0$, such that (XIII.6) requires:

$$\frac{\partial B_z}{\partial y} = \frac{\partial B_y}{\partial z} \; . \tag{XIII.7}$$

The other terms in (XIII.6) are zero. Equations (XIII.3) and (XIII.4) then become:

$$\rho_x = 0 \; , \tag{XIII.8}$$

$$\frac{\rho_y}{\rho^2} = -\frac{1}{B_z}\frac{\partial B_z}{\partial y} \; , \tag{XIII.9}$$

such that $\rho_y = \rho$.

Parameterisation of a curved field line

In our case, $\boldsymbol{B} = B_z\hat{\mathbf{e}}_z + B_y\hat{\mathbf{e}}_y$, where $|B_y| \ll |B_z|$ is a first order correction to B_z, provided the curvature of the field is not too large. We are seeking a relation $y(z)$ to characterise this line of force.

We can perform a limited Taylor series development of the component B_y, with $B_y(0) = 0$, since this quantity is of order one, and:

$$B_y \approx \frac{\partial B_y}{\partial x}\,\mathrm{d}x + \frac{\partial B_y}{\partial y}\,\mathrm{d}y + \frac{\partial B_y}{\partial z}\,\mathrm{d}z \; , \tag{XIII.10}$$

where, according to (XIII.6) and (XIII.7), only the component $\frac{\partial B_y}{\partial z}$ is non zero, so that:

$$B_y \approx \frac{\partial B_y}{\partial z}\,\mathrm{d}z \; , \tag{XIII.11}$$

which, from (XIII.7), becomes:

$$B_y \simeq \frac{\partial B_z}{\partial y}\mathrm{d}z \; , \tag{XIII.12}$$

hence, for z small, from (2.221):

$$B_y \simeq B_0\beta z \; . \tag{XIII.13}$$

Further, by definition, locally (see Fig. 2.17):

$$\frac{B_y}{B_z} = \frac{\mathrm{d}y}{\mathrm{d}z} \; , \tag{XIII.14}$$

where:

$$\frac{B_y}{B_z} = \frac{B_0\beta z}{B_0(1+\beta y)} \simeq \beta z \; , \tag{XIII.15}$$

hence:

$$\frac{\mathrm{d}y}{\mathrm{d}z} = \beta z \,. \tag{XIII.16}$$

Integrating (XIII.2), we have (since $y = 0$ at $z = 0$):

$$\frac{\mathrm{d}y}{\mathrm{d}z} = \frac{z}{\rho} \,, \tag{XIII.17}$$

so we can deduce that:

$$\beta \approx 1/\rho \,. \tag{XIII.18}$$

The equation for $\boldsymbol{w}_{dm}$ in the Frenet frame

We have already shown that:

$$\boldsymbol{w}_{dm} = m_\alpha \frac{w_\perp^2}{2} \frac{1}{q_\alpha B^3} (\boldsymbol{B} \wedge \boldsymbol{\nabla} B) \,. \tag{2.216}$$

If we now set $\omega_c = -q_\alpha B/m_\alpha$, this can be written explicitly:

$$\boldsymbol{w}_{dm} = -\frac{1}{\omega_c B^2} \frac{w_\perp^2}{2} \left(\boldsymbol{B} \wedge \frac{\partial B_z}{\partial y} \hat{\mathbf{e}}_y \right) \,. \tag{XIII.19}$$

From (2.221) and (XIII.18), we obtain:

$$\frac{\partial B_z}{\partial y} = \frac{B_z}{\rho} \,, \tag{XIII.20}$$

and from (XIII.15):

$$\boldsymbol{w}_{dm} = \frac{1}{\omega_c B^2} \frac{w_\perp^2}{2} \left(\frac{B_z}{\rho} \hat{\mathbf{e}}_y \wedge \boldsymbol{B} \right) \,. \tag{XIII.21}$$

By introducing $\boldsymbol{\rho}$, directed opposite to $\hat{\mathbf{e}}_y$, we finally arrive at:

$$\boldsymbol{w}_{dm} = -\frac{1}{\omega_c B} \frac{w_\perp^2}{2} \left(\frac{\boldsymbol{\rho}}{\rho^2} \wedge \boldsymbol{B} \right) \,. \tag{XIII.22}$$

This expression, in contrast to (2.216), includes the (weak) curvature of the field lines.

Appendix XIV
Spherical Harmonics

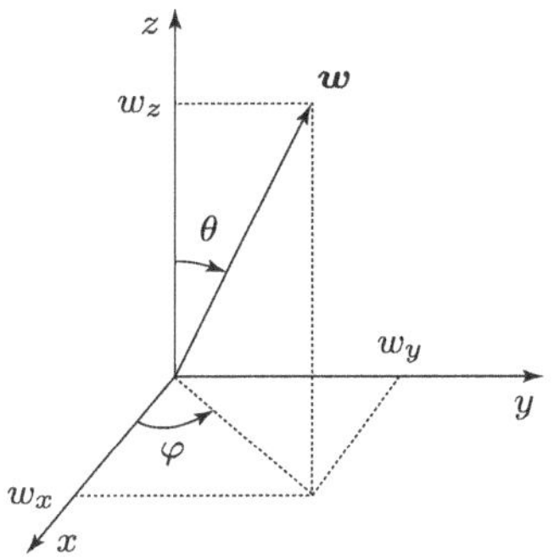

Fig. XIV.1 Spherical coordinate system in velocity space.

If the velocity $\boldsymbol{w}$ is expressed in spherical coordinates, the electron distribution function can be developed in *spherical harmonics*:

$$C_{lm} = w^l P_{lm}(\cos\theta)\cos m\varphi \,, \tag{XIV.1}$$

$$S_{lm} = w^l P_{lm}(\cos\theta)\sin m\varphi \,, \tag{XIV.2}$$

where $P_{lm}(\cos\theta)$ is the mth order *Legendre function*, defined for $l \geq 1$ and $0 \leq m \leq l$ by:

$$P_{lm}(\mu) = (1-\mu^2)^{\frac{m}{2}} \frac{\mathrm{d}^m}{\mathrm{d}\mu^m} P_l(\mu) \,, \tag{XIV.3}$$

and P_l is the Legendre *polynomial* of degree l. Note that for $m = 0$,

$$P_{lm}(\mu) = P_l(\mu) \,. \tag{XIV.4}$$

The first Legendre polynomials $P_{lm}(\mu)$ are:

M. Moisan, J. Pelletier, *Physics of Collisional Plasmas*,
DOI 10.1007/978-94-007-4558-2,

$$P_0(\mu) = 1\,, \quad P_1(\mu) = \mu\,, \quad P_{11}(\mu) = (1-\mu^2)^{\frac{1}{2}}\,,$$
$$P_2(\mu) = \frac{1}{2}\,(3\mu^2 - 1)\,, \quad P_{21}(\mu) = 3\mu(1-\mu^2)^{\frac{1}{2}}\,, \quad P_{22}(\mu) = 3(1-\mu^2)\,,$$
$$P_3(\mu) = \frac{1}{2}\,(5\mu^3 - 3\mu)\,\ldots \tag{XIV.5}$$

The first spherical functions, for $\mu = \cos\theta$, are therefore:

$$C_{00} = 1\,, \quad C_{10} = w\cos\theta = w_z\,, \quad C_{11} = w\sin\theta\cos\varphi = w_x\,,$$
$$S_{11} = w\sin\theta\sin\varphi = w_y\,, \quad C_{20} = w^2\left(\frac{3\cos^2\theta - 1}{2}\right)\,. \tag{XIV.6}$$

Assuming that the system is symmetrical in φ, the distribution function $f(\boldsymbol{r},\, \boldsymbol{w},\, t)$ can be expanded in terms of C_{i0} yielding:

$$f(\boldsymbol{r}, \boldsymbol{w}, t) = f_0(\boldsymbol{r}, w, t) + f_1(\boldsymbol{r}, w, t)\cos\theta + f_2(\boldsymbol{r}, w, t)\left(\frac{3\cos^2\theta - 1}{2}\right) + \ldots \tag{3.18}$$

Appendix XV
Expressions for the Terms $\underline{\underline{M}}$ and $\underline{\underline{\mathcal{R}}}_\alpha$ in the Kinetic Pressure Transport Equation (3.155)

Relationship between $\underline{\underline{M}}$ and the magnetic force

To calculate $\underline{\underline{M}}$ from (3.155) when the particles α are subjected to a magnetic field $\boldsymbol{B}$ (Laplace force[90]), it is sufficient to write (VIII.31):

$$M_{ij} = n_\alpha q_\alpha \langle (w_{i+1}B_{i-1} - w_{i-1}B_{i+1})u_j + (w_{j+1}B_{j-1} - w_{j-1}B_{j+1})u_i \rangle \ . \tag{XV.1}$$

Reordering the terms, we obtain:

$$\begin{aligned} M_{ij} &= n_\alpha q_\alpha \langle (u_{i+1} + v_{i+1})B_{i-1}u_j - (u_{i-1} + v_{i-1})B_{i+1}u_j \rangle \\ &\quad + n_\alpha q_\alpha \langle (u_{j+1} + v_{j+1})B_{j-1}u_i - (u_{j-1} + v_{j-1})B_{j+1}u_i \rangle \ , \end{aligned} \tag{XV.2}$$

hence:

$$\begin{aligned} M_{ij} &= \frac{q_\alpha}{m_\alpha} \left[B_{i-1}\Psi_{i+1,\ j} - B_{i+1}\Psi_{i-1,\ j} \right] \\ &\quad + \frac{q_\alpha}{m_\alpha} \left[B_{j-1}\Psi_{j+1,\ i} - B_{j+1}\Psi_{j-1,\ i} \right] \ , \end{aligned} \tag{XV.3}$$

which can be written in tensor form (VIII.6):

$$\underline{\underline{M}} = -\frac{q_\alpha}{m_\alpha} \left[\boldsymbol{B} \wedge \underline{\underline{\boldsymbol{\Psi}}} + (\boldsymbol{B} \wedge \underline{\underline{\boldsymbol{\Psi}}})^T \right] \ . \tag{XV.4}$$

Expression for the collision tensor $\underline{\underline{\mathcal{R}}}_\alpha$

Recall that:

$$\underline{\underline{\mathcal{R}}}_\alpha = \sum_{\beta \neq \alpha} \underline{\underline{\mathcal{R}}}_{\alpha\beta} \tag{3.141}$$

where:

M. Moisan, J. Pelletier, *Physics of Collisional Plasmas*,
DOI 10.1007/978-94-007-4558-2,

$$\underline{\boldsymbol{\mathcal{R}}}_{\alpha\beta} = \int m_\alpha(\boldsymbol{w}_\alpha - \boldsymbol{v}_\alpha)(\boldsymbol{w}_\alpha - \boldsymbol{v}_\alpha)S(f_\alpha)_\beta \, \mathrm{d}\boldsymbol{w}_\alpha \, . \tag{XV.5}$$

The tensor $\underline{\boldsymbol{\mathcal{R}}}_{\alpha\beta}$ can be written, after expansion:

$$\underline{\boldsymbol{\mathcal{R}}}_{\alpha\beta} = \int m_\alpha(\boldsymbol{w}_\alpha\boldsymbol{w}_\alpha - \boldsymbol{v}_\alpha\boldsymbol{w}_\alpha - \boldsymbol{w}_\alpha\boldsymbol{v}_\alpha + \boldsymbol{v}_\alpha\boldsymbol{v}_\alpha)S(f_\alpha)_\beta \, \mathrm{d}\boldsymbol{w}_\alpha \, , \tag{XV.6}$$

hence:

$$\begin{aligned}\underline{\boldsymbol{\mathcal{R}}}_{\alpha\beta} = {} & \int m_\alpha(\boldsymbol{w}_\alpha\boldsymbol{w}_\alpha)S(f_\alpha)_\beta \, \mathrm{d}\boldsymbol{w}_\alpha - \boldsymbol{v}_\alpha \int m_\alpha\boldsymbol{w}_\alpha S(f_\alpha)_\beta \, \mathrm{d}\boldsymbol{w}_\alpha \\ & - \left[\boldsymbol{v}_\alpha \int m_\alpha\boldsymbol{w}_\alpha S(f_\alpha)_\beta \, \mathrm{d}\boldsymbol{w}_\alpha\right]^T + \boldsymbol{v}_\alpha\boldsymbol{v}_\alpha \int m_\alpha S(f_\alpha)_\beta \, \mathrm{d}\boldsymbol{w}_\alpha \, . \end{aligned} \tag{XV.7}$$

If the number of particles is conserved during collisions, the last term in (XV.7) is zero. Including (3.120), which defines $\boldsymbol{\mathcal{P}}_{\alpha\beta}$ (XV.7), we obtain the final form:

$$\underline{\boldsymbol{\mathcal{R}}}_{\alpha\beta} = \int m_\alpha\boldsymbol{w}_\alpha\boldsymbol{w}_\alpha S(f_\alpha)_\beta \, \mathrm{d}\boldsymbol{w}_\alpha - \boldsymbol{v}_\alpha\boldsymbol{\mathcal{P}}_{\alpha\beta} - [\boldsymbol{v}_\alpha\boldsymbol{\mathcal{P}}_{\alpha\beta}]^T \, . \tag{XV.8}$$

Appendix XVI
Closure of the Hydrodynamic Transport Equation for Kinetic Pressure in the Case of Adiabatic Compression

We will consider the transport equation for kinetic pressure $\underline{\underline{\boldsymbol{\Psi}}}$ (3.160), with the assumption of adiabatic compression (see main text): $\boldsymbol{\nabla} \cdot \underline{\underline{\boldsymbol{Q}}} = \underline{\mathbf{0}}$ and $\underline{\boldsymbol{\mathcal{R}}} = \underline{\mathbf{0}}$. The equation then simplifies to give:

$$n\frac{\mathrm{d}}{\mathrm{d}t}\left(\frac{\underline{\boldsymbol{\Psi}}}{n}\right) + (\underline{\boldsymbol{\Psi}} \cdot \boldsymbol{\nabla})\, \boldsymbol{v} + [(\underline{\boldsymbol{\Psi}} \cdot \boldsymbol{\nabla})\, \boldsymbol{v}]^T - \underline{\boldsymbol{M}} = \underline{\mathbf{0}}\,. \tag{XVI.1}$$

The structure of the tensor $\underline{\boldsymbol{M}}$ (XV.2) is such that the off-diagonal terms are zero.

The equation (XVI.1), comprising second-order tensors, can lead to a scalar solution if we apply a contraction (Appendix VII) on the two indices of the various 2nd order tensors in the equation. Since a tensor $\underline{\boldsymbol{A}}$ can be expressed as $\underline{\boldsymbol{A}} = \hat{\mathbf{e}}_i\hat{\mathbf{e}}_j A_{ij}$, contracting the two indices ($i = j$) is equivalent to calculating the trace of $\underline{\boldsymbol{A}}$. The result of such a contraction on indices of the same variance (subscripts here) leads to a scalar, provided the coordinate systems considered are Cartesian.

Taking account of (3.115), the value of the trace Tr of the first term of (XVI.1) is:

$$\mathrm{Tr}\left[n\frac{\mathrm{d}}{\mathrm{d}t}\frac{\underline{\boldsymbol{\Psi}}}{n}\right] = n\frac{\mathrm{d}}{\mathrm{d}t}\frac{3p}{n}\,, \tag{XVI.2}$$

while the traces of the second and third terms give:

$$\mathrm{Tr}\,[(\underline{\boldsymbol{\Psi}} \cdot \boldsymbol{\nabla})\boldsymbol{v}] + \mathrm{Tr}\,[(\underline{\boldsymbol{\Psi}} \cdot \boldsymbol{\nabla})\boldsymbol{v}]^T = 2p\boldsymbol{\nabla} \cdot \boldsymbol{v}\,, \tag{XVI.3}$$

where we have set $\underline{\boldsymbol{\Psi}} = nk_BT(\underline{\boldsymbol{I}})$ (as in the warm plasma approximation). Finally, the complete trace of (XVI.1) is given by:

$$n\frac{\mathrm{d}}{\mathrm{d}t}\frac{3}{2}\frac{p}{n} + p\boldsymbol{\nabla} \cdot \boldsymbol{v} = 0\,. \tag{XVI.4}$$

This scalar equation thus replaces a 2nd order tensor equation, for which closure has been achieved by setting $\boldsymbol{\nabla} \cdot \underline{\underline{\boldsymbol{Q}}} = \underline{\mathbf{0}}$.

M. Moisan, J. Pelletier, *Physics of Collisional Plasmas*,
DOI 10.1007/978-94-007-4558-2,

Appendix XVII
Complementary Calculations to the Expression for $T_e(pR)$ (Sect. 3.13)

Maxwellian velocity distribution function expressed in terms of energy (eV)[188]

The following expression for the distribution function:

$$f(w) = \left(\frac{m_e}{2\pi k_B T_e}\right)^{\frac{3}{2}} \exp\left(-\frac{w^2}{v_{th}^2}\right) , \tag{3.293}$$

by its scalar velocity dependence, emphasises that we have neglected the anisotropy induced by the external field $\boldsymbol{E}$.
Introducing U_{eV}, the microscopic electron energy, expressed in eV:

$$U_{eV} = \frac{m_e w^2}{2e} , \tag{3.294}$$

we obtain:

$$w = \sqrt{\frac{2eU_{eV}}{m_e}} \tag{3.295}$$

[188] Substitution of the energy U_{eV} for w in $f(w)$ (isotropic) leads to the function $f(U_{eV})$, which is referred to as the *velocity distribution function expressed in terms of energy*. By the same token, we define an *energy distribution function* $F(U_{eV})$, by setting:

$$F(U_{eV})U_{eV}^{\frac{1}{2}} \, \mathrm{d}U_{eV} = f(w)4\pi w^2 \, \mathrm{d}w .$$

Thus, we have $F(U_{eV}) = 4\sqrt{2}\pi(e/m_e)^{3/2} f(w)$, hence:

$$F(U_{eV}) = \frac{2}{\pi^{\frac{1}{2}}(k_B T_e)^{\frac{3}{2}}} \exp\left(-\frac{eU_{eV}}{k_B T_e}\right) ,$$

the normalisation condition being:

$$\int_0^\infty F(U_{eV})U_{eV}^{\frac{1}{2}} \, \mathrm{d}U_{eV} = 1 .$$

M. Moisan, J. Pelletier, *Physics of Collisional Plasmas*,
DOI 10.1007/978-94-007-4558-2,

and since:

$$\frac{1}{2}m_e v_{th}^2 = k_B T_e = \frac{2}{3}\left(\frac{3}{2}k_B T_e\right) = \frac{2}{3}e\bar{U}_{eV} \ , \tag{XVII.1}$$

where $\bar{U}_{eV} = 3k_B T_e/2e$ is the electron mean energy, we finally have:

$$v_{th} = \sqrt{\frac{4}{3}\frac{e}{m_e}\bar{U}_{eV}} \ . \tag{3.297}$$

After substitution in (3.293), we obtain:

$$f\left(\sqrt{\frac{2eU_{eV}}{m_e}}\right) = \frac{1}{(2\pi)^{\frac{3}{2}}}\left(\frac{3m_e}{2e\bar{U}_{eV}}\right)^{\frac{3}{2}} \exp\left(-\frac{2eU_{eV}}{m_e} \Big/ \frac{4e\bar{U}_{eV}}{3m_e}\right) \ , \tag{XVII.2}$$

where:

$$f\left(\sqrt{\frac{2eU_{eV}}{m_e}}\right) = \left(\frac{3}{4\pi}\frac{m_e}{e}\right)^{\frac{3}{2}} \frac{1}{\bar{U}_{eV}^{\frac{3}{2}}} \exp\left(-\frac{3}{2}\frac{U_{eV}}{\bar{U}_{eV}}\right) \ . \tag{3.298}$$

The ionisation frequency in terms of the reduced energies $\mathcal{U}$ and $\mathcal{U}_i$

As a starting point, we have equation (3.301)

$$\langle \nu_i \rangle = 3\sqrt{\frac{3e}{m_e\pi}}\,\frac{p_0}{\bar{U}_{eV}^{\frac{3}{2}}}\,a_{i0}\int_{\mathcal{E}_i}^{\infty}(U_{eV} - \mathcal{E}_i)\,U_{eV}\,\exp\left(-\frac{3}{2}\frac{U_{eV}}{\bar{U}_{eV}}\right)\mathrm{d}U_{eV} \ . \tag{3.301}$$

Introducing the change of variable required for the energies to be normalised to the average energy (so-called *reduced energies*):

$$\frac{3}{2}\frac{U_{eV}}{\bar{U}_{eV}} = \mathcal{U} \ , \qquad \frac{3}{2}\frac{\mathcal{E}_i}{\bar{U}_{eV}} = \mathcal{U}_i \ , \tag{3.302}$$

it follows that:

$$\langle \nu_i \rangle = 3\sqrt{\frac{3e}{m_e\pi}}\,\frac{p_0 a_{i0}}{\bar{U}_{eV}^{\frac{3}{2}}}\int_{\mathcal{E}_i}^{\infty}\frac{2}{3}\bar{U}_{eV}(\mathcal{U}-\mathcal{U}_i)\frac{2}{3}\bar{U}_{eV}\mathcal{U}\exp(-\mathcal{U})\,\frac{2}{3}\bar{U}_{eV}\,\mathrm{d}\mathcal{U} \ , \tag{XVII.3}$$

from which:

$$\langle \nu_i \rangle = \frac{8}{9}\sqrt{\frac{3e}{m_e\pi}}\,p_0 a_{i0}\,\bar{U}_{eV}^{\frac{3}{2}}\int_{\mathcal{U}_i}^{\infty}(\mathcal{U}-\mathcal{U}_i)\mathcal{U}\exp(-\mathcal{U})\,\mathrm{d}\mathcal{U} \ . \tag{3.303}$$

Equation (3.309) expressed as a function of the reduced energies $\mathcal{U}$ and $\mathcal{U}_i$

This equation reads:

$$\frac{2}{3}\bar{U}_{eV}\mu_i\left(\frac{2.405}{R}\right)^2 = 2\left(\frac{4}{3}\right)^{\frac{3}{2}}\sqrt{\frac{e}{m_e\pi}}\,a_{i0}p_0\,\bar{U}_{eV}^{\frac{3}{2}}\left(\frac{3}{4}\frac{\mathcal{E}_i}{\bar{U}_{eV}}\right)\exp\left(-\frac{3}{2}\frac{\mathcal{E}_i}{\bar{U}_{eV}}\right). \tag{3.309}$$

Since $3\mathcal{E}_i/2\bar{U}_{eV} \equiv \mathcal{U}_i$, we can write:

$$\bar{U}_{eV}\mu_i\left(\frac{2.405}{R}\right)^2 = 2\sqrt{\frac{2e}{m_e\pi}}\,\frac{a_{i0}}{p_0}\,p_0^2\,\mathcal{E}_i^{\frac{1}{2}}\mathcal{U}_i^{\frac{1}{2}}\exp -\mathcal{U}_i \tag{XVII.4}$$

and then:

$$\mathcal{U}_i^{-\frac{1}{2}}\,(\exp\mathcal{U}_i) = \frac{2}{(2.405)^2}\sqrt{\frac{2e}{m_e\pi}}\underbrace{\left(\frac{a_{i0}\mathcal{E}_i^{\frac{1}{2}}}{\mu_i p_0}\right)}_{c_0^2}p_0^2R^2\,, \tag{XVII.5}$$

where $\mu_i p_0$ is the reduced ion mobility at 0°C, 1 torr (note the reference pressure here is 1 torr, rather than 760 torr).

Appendix XVIII
Propagation of an Electromagnetic Plane Wave in a Plasma and the Skin Depth

The propagation conditions of an electromagnetic (EM) wave in a plasma are governed by the four Maxwell equations, namely:

1. The Maxwell-Faraday equation:

$$\boldsymbol{\nabla} \wedge \boldsymbol{E} = -\frac{\partial \boldsymbol{B}}{\partial t} , \tag{2.2}$$

2. The Maxwell-Ampère equation:

$$\boldsymbol{\nabla} \wedge \boldsymbol{B} = \mu_0 \boldsymbol{J} + \mu_0 \epsilon_0 \frac{\partial \boldsymbol{E}}{\partial t} , \tag{2.3}$$

3. The Poisson (or Maxwell-Gauss) equation:

$$\boldsymbol{\nabla} \cdot \boldsymbol{E} = \frac{\rho}{\epsilon_0} , \tag{1.1}$$

4. The Maxwell-Thomson equation:

$$\boldsymbol{\nabla} \cdot \boldsymbol{B} = 0 . \tag{2.144}$$

Assuming that the plasma is neutral on the macroscopic scale ($\rho = 0$), and for the case of a dielectric description of the plasma (Sect. 2.2.1), (2.3) and (1.1) can be written, respectively:

$$\boldsymbol{\nabla} \cdot \boldsymbol{E} = 0 \tag{XVIII.1}$$

and:

$$\boldsymbol{\nabla} \wedge \boldsymbol{B} = \mu_0 \epsilon_0 \epsilon_p \frac{\partial \boldsymbol{E}}{\partial t} , \tag{2.45}$$

where ϵ_p represents the complex permittivity of the plasma relative to vacuum. The equation for the propagation of the EM wave is obtained by considering the rotational of (2.2):

$$\boldsymbol{\nabla} \wedge \boldsymbol{\nabla} \wedge \boldsymbol{E} = \boldsymbol{\nabla}(\boldsymbol{\nabla} \cdot \boldsymbol{E}) - \Delta \boldsymbol{E} = -\frac{\partial \boldsymbol{\nabla} \wedge \boldsymbol{B}}{\partial t} , \tag{XVIII.2}$$

M. Moisan, J. Pelletier, *Physics of Collisional Plasmas*,
DOI 10.1007/978-94-007-4558-2,

and then by taking account of (XVIII.1) and (2.45):

$$\Delta \boldsymbol{E} = \mu_0 \epsilon_0 \epsilon_p \frac{\partial^2 \boldsymbol{E}}{\partial t^2} \, . \tag{XVIII.3}$$

In the simple case of a plane EM wave (in the x, y plane) with angular frequency ω and propagating along the Oz axis, the electric field $\boldsymbol{E}$ can be written in the form:

$$\boldsymbol{E} = \boldsymbol{E}_0 \mathrm{e}^{\mathrm{i}(\omega t - k_z z)} \, , \tag{XVIII.4}$$

where k_z, the complex component along Oz of $\boldsymbol{k}$, the propagation vector (wavenumber[23]), can be written:

$$k_z = \beta + \mathrm{i}\alpha \, . \tag{XVIII.5}$$

Then from (XVIII.3) in Cartesian coordinates, we obtain the dispersion equation of a plane wave in an infinite homogeneous medium:

$$k_z^2 = \mu_0 \epsilon_0 \omega^2 \epsilon_p = \epsilon_p \left(\frac{\omega}{c} \right)^2 \, , \tag{XVIII.6}$$

where c is the speed of light in vacuum. In the general case, the permittivity ϵ_p of the plasma relative to vacuum can be written:

$$\epsilon_p = 1 + \frac{\sigma}{\mathrm{i}\omega\epsilon_0} \, , \tag{2.40}$$

or, taking account of the electric conductivity σ of electrons (2.39):

$$\epsilon_p = 1 - \frac{\omega_{pe}^2}{\omega(\omega - \mathrm{i}\nu)} \, . \tag{2.41}$$

From (XVIII.4) and (XVIII.5)), the electric field can be expressed as:

$$\boldsymbol{E} = \boldsymbol{E}_0 \mathrm{e}^{\mathrm{i}(\omega t - \beta z)} \mathrm{e}^{\alpha z} \, , \tag{XVIII.7}$$

which shows that the propagation of the wave is governed by β, the real part of k_z, while its attenuation is governed by α, the imaginary part of k_z. If k_z is strictly real ($\alpha = 0$), the wave propagates without any attenuation; if k_z includes a negative imaginary part ($\alpha < 0$), the wave is attenuated (a positive α value ($\alpha > 0$) would correspond to an amplification of the wave, which cannot be considered here as an acceptable physical solution). In the case of wave attenuation ($\alpha < 0$), the characteristic penetration depth δ_c of the HF field in the plasma is defined as the distance over which the field intensity of the plane EM wave reduces to 1/e (e is used exceptionally here for the base of the natural logarithm) of its initial value, i.e.:

$$\delta_c = \frac{1}{\Im(-k_z)} = -\frac{1}{\alpha} = \frac{(c/\omega)}{\Im\left(-\epsilon_p^{1/2}\right)} \, . \tag{XVIII.8}$$

The calculation of the skin depth can then be performed in a simple way for two particular cases:

Non-collisional plasmas: $\nu \ll \omega$

In this case ($\nu \ll \omega$), the value of the plasma permittivity (2.41), to a first approximation, is purely real:

$$\epsilon_p = 1 - \frac{\omega_{pe}^2}{\omega^2} \,. \tag{XVIII.9}$$

If the value of the permittivity is positive ($\omega_{pe} < \omega$), the wavenumber k_z is purely real ($k_z = \beta$) and the wave propagates without attenuation. If the value of the permittivity is negative ($\omega_{pe} > \omega$), the wavenumber k_z is totally imaginary ($k_z = \mathrm{i}\alpha$) and the wave cannot propagate in the plasma. In this latter case, the EM wave is reflected by the plasma (the plasma can be considered as a highly conductive medium) at the same time that there is attenuation of the HF field in the plasma along the z axis, related to the skin depth obtained from (XVIII.8):

$$\delta_c = \frac{c}{(\omega_{pe}^2 - \omega^2)^{1/2}} \simeq \frac{c}{\omega_{pe}} \,. \tag{XVIII.10}$$

Remark: The transition from the condition of propagation to that of no propagation of the EM plane wave (wave propagation cut-off) in a non-collisional plasma is obtained for the singular value $\epsilon_p = 0$ (or $k_z = 0$) resulting from the equality $\omega_{pe} = \omega$. The corresponding electron density n_c, the so-called *critical density* above which there is no propagation of a plane wave in a non-collisional plasma, is given by:

$$n_c = \frac{\epsilon_0 m_e \omega^2}{e^2} \,. \tag{XVIII.11}$$

Collisional plasmas: $\nu \gg \omega$

In this opposite case ($\nu \gg \omega$), the plasma permittivity of a high density plasma ($\omega_{pe} \gg \omega$) can be obtained from (2.41) as:

$$\epsilon_p = 1 - \frac{\mathrm{i}\omega_{pe}^2}{\nu\omega} \simeq -\frac{\mathrm{i}\omega_{pe}^2}{\nu\omega} \,. \tag{XVIII.12}$$

The complex square root of ϵ_p that corresponds to a physical solution for k_z can be written as:

$$\epsilon_p^{1/2} \simeq (1 - \mathrm{i}) \left(\frac{\omega_{pe}^2}{2\nu\omega} \right)^{1/2} . \tag{XVIII.13}$$

From (XVIII.8), the skin depth then takes the form:

$$\delta_c = \frac{c}{\omega_{pe}} \left(\frac{2\nu}{\omega} \right)^{1/2} , \tag{XVIII.14}$$

which, this time, depends on the collision frequency ν of the electrons and the angular frequency ω of the HF field. The wavenumber, derived from (XVIII.6) and (XVIII.14), can be expressed in the form:

$$k_z = \frac{1 - \mathrm{i}}{\delta_c} ,$$

which implies that, in a high density collisional plasma (ω_{pe} and $\nu \gg \omega$), the EM wave can propagate over a distance equal to the skin depth even though $\omega_{pe} > \omega$.

Remark: The electric conductivity of electrons:

$$\sigma = \frac{n_e^2}{m_e(\nu + \mathrm{i}\omega)} \tag{2.39}$$

in a collisional plasma ($\nu \gg \omega$), takes the purely real value:

$$\sigma = \frac{n_e^2}{m_e \nu} . \tag{XVIII.15}$$

In this case, the skin depth (XVIII.14) can be written in the form:

$$\delta_c = \left(\frac{2}{\sigma \omega \mu_0} \right)^{1/2} , \tag{XVIII.16}$$

which is an expression that exactly corresponds to the well-known penetration depth formula of a plane wave in metallic conductors.

Appendix XIX
Surface-Wave Plasmas (SWP)

This class of HF plasma has played a determining role in the understanding and modelling of the plasmas generated by RF and microwave fields, and even for the positive column of DC discharges. This is due, on one hand, to the great flexibility of the operating conditions for SWP and, on the other hand, to their intrinsic properties. The following brief overview of SWP will aid the reader to better comprehend how some of the results presented in Chap. 4 have been obtained.

Figure XIX.1 shows schematically the way in which an SWP is generated in a cylindrical dielectric tube (planar or flat SWPs can also be created). Note that the HF field applicator, in this case a wave launcher, covers only a small length of the plasma column which is produced. This is because the discharge is sustained, at each point along the column, by the propagation of an electromagnetic (EM) wave, which is excited from the launching gap (typically a few mm wide) of the field applicator. The propagating medium of the wave consists of the plasma, the dielectric tube containing it, the air surrounding the tube and, in some cases, a hollow cylindrical conductor enclosing the whole system and coaxial to it. The EM wave is referred to as a surface wave, because the intensity of its field $\boldsymbol{E}$ is a maximum, radially, at the plasma-tube interface of the discharge, such that the wave seems to cling to the discharge tube and, in fact should the case arise, follows the variations in its diameter and curvature, if they are not too abrupt.

A surface wave is excited in both the forward and backward directions from the launching gap, as shown in Fig. XIX.1. With some types of launchers (for example, a surfaguide), the forward column is symmetric to the backward column, with respect to the gap, while in contrast, with some other launchers (for example, a surfatron), the plasma is almost exclusively that of the forward column. The power flow $P(z)$ emerging from the launching gap attenuates along the discharge vessel as the wave transfers its energy to the gas in the discharge that it creates.

One particular property of SWP is that the power lost by the wave $\mathrm{d}P(z)/\mathrm{d}z$ between z and $z + \mathrm{d}z$ is absorbed by the discharge over the same

M. Moisan, J. Pelletier, *Physics of Collisional Plasmas*,
DOI 10.1007/978-94-007-4558-2,

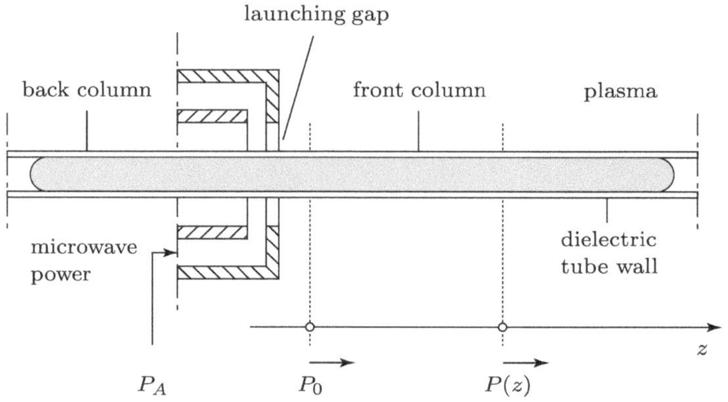

Fig. XIX.1 Schematic of the principle of formation of a plasma column generated by an electromagnetic surface wave in a dielectric tube, from the launching gap of a field applicator.

axial range z and $z + \mathrm{d}z$ (this is not true of HF plasmas in general), which simplifies the modelling.

The result of the progressive reduction in power flow $P(z)$ from the launching gap is a decrease, usually linear, in the electron density, as shown in Fig. XIX.2. In this figure, we further note that the slope of the curves becomes steeper as the wave frequency increases. In the low pressure case ($\nu/\omega \ll 1$), the wave ceases to propagate, and hence does not maintain a discharge, when n_e is below a certain value[189], while in the high pressure case, the wave ceases to propagate when the power flow is no longer sufficient to maintain the discharge, which determines, in both cases, the end of the plasma column.

Another remarkable property of SWPs, at least at pressures much below atmospheric pressure, is that an increase in the HF power delivered to the field applicator produces an increase in the length of the plasma column, without modifying the pre-existing segment of plasma with respect to the end of the column, this plasma segment being simply pushed away as a whole from the launching gap. Figure XIX.2 at 100 MHz is a good illustration of this behaviour of SWPs. The arrow indicating 36 W represents the axial position occupied by the applicator with respect to the end of the column ($z \simeq 0$) and we can see, as noted above, that the segment of the plasma column is not modified when the HF power is increased to 58 W. The average electron density $\bar{n}_e$ across a radial section of the plasma:

$$\overline{n_e}(z) = \frac{1}{\pi R^2} \int_0^R n_e(r,z)\, 2\pi r \mathrm{d}r \tag{XIX.1}$$

[189] The minimum value of $\bar{n}_e$ in this case is $\bar{n}_e \simeq 1.2 \times 10^4 (1 + \epsilon_v) f^2$ (cm^{-3}) where $\epsilon_v$is the relative permittivity of the discharge tube (for example, 3.78 for fused silica) and f is the wave frequency, expressed in MHz.

of the additional segment is higher, but its gradient $d\bar{n}_e/dz$ remains the same. Note that, for a frequency of 27 MHz and a pressure of 30 mtorr (4 Pa), the plasma column in argon extends to 4.5 m with less than 40 W transmitted to the surfatron.

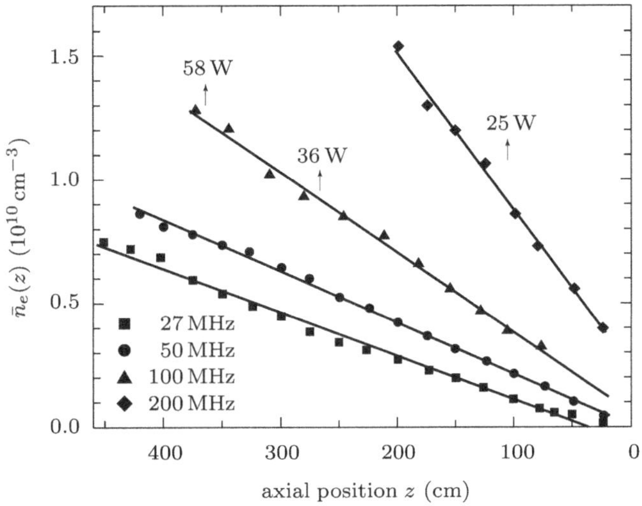

Fig. XIX.2 Axial distribution of the electron density observed along a plasma column, produced by a surface wave, at different excitation frequencies (tube radius $R = 6.4$ cm in free air, surfatron, argon 30 mtorr).

Figure XIX.3 shows that the value of $\bar{n}_e$ and its gradient $d\bar{n}_e/dz$ increase as the tube diameter is increased. This property is true for plasmas from atmospheric pressure (Fig. XIX.3) down to a few mtorr, and is related to the axial variation of the attenuation coefficient α of the surface wave [25]. This emphasises the fact that, given a plasma column along which $\bar{n}_e$ and $P(z)$ vary, we are able to perform a self-consistent study of the wave and plasma properties at each axial position z, without the need to modify the operating conditions. Having access to such a tremendous amount of data (from experiment and model) is another important and unique feature of SWP.

The range of possible operating conditions is the largest of all HF plasmas, and this makes it an instrument of choice for modelling, allowing ready comparison between experiment and theory, during which only one operating parameter can be modified at a time. As a matter of fact, it is possible to create these plasmas at frequencies from as low as 150 kHz to at least 2.45 GHz[190], producing a surface wave with the same azimuthal symmetry, i.e. the same EM configuration; we believe that this is not possible with any other

[190] SWPs have been achieved at 40 GHz. However, it is not clear whether the discharge was sustained on the $m = 0$ (azimuthally symmetric) surface wave mode [40].

type of HF discharges, since it would require changing the tube diameter or the type of field applicator, i.e. the EM field configuration, to cover such a large frequency domain. The range of pressures can extend from a few mtorr (much less even, in the presence of a confining magnetic field at ECR) up to at least 7 times atmospheric pressure, which we have been able to achieve, with the same EM field configuration. The diameter of the discharge tubes can range from less than 1 mm, up to 300 mm, for frequencies that are not too high (some restrictions apply, in effect, to the maximum discharge diameter, to avoid higher EM modes of the surface wave when the wave frequency is increased [18]. Due to this extreme adaptability of the operating conditions, we can say that the main application of SWPs is the modelling of HF plasmas, although there are now numerous industrial applications of SWPs.

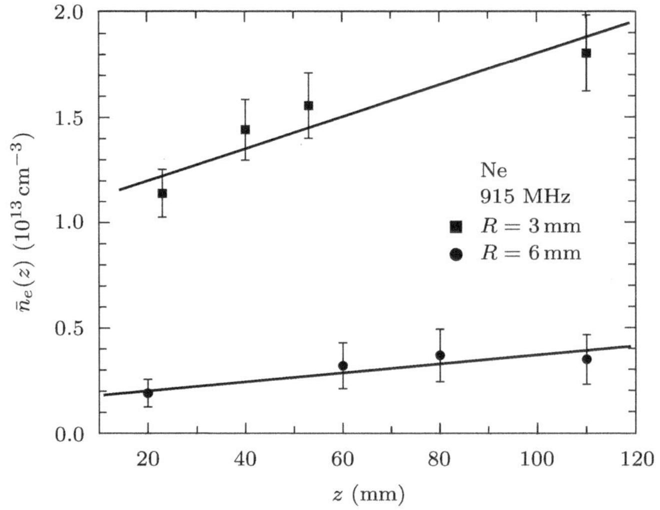

Fig. XIX.3 Axial distribution of the average electron density across a section of the discharge tube, for two values of internal radius (3 and 6 mm) in a neon plasma at 915 MHz, at atmospheric pressure.

Appendix XX
Useful Integrals and Expressions for the Differential Operators in Various Coordinate Systems

Useful integrals

Γ Function

$$\Gamma(x) = \int_0^\infty t^{z-1} \mathrm{e}^{-t} \, \mathrm{d}t \, , \quad \Re(z) > 0 \tag{XX.1}$$

For $z = n$, where n is an integer:

$$\Gamma(n) = (n-1)! \, , \quad \Gamma(n+1) = n\Gamma(n) \, .$$

Noteworthy values for the Γ function:

$$\Gamma\left(\frac{1}{2}\right) = \sqrt{\pi} \, , \quad \Gamma\left(\frac{3}{2}\right) = \frac{1}{2}\Gamma\left(\frac{1}{2}\right) = \frac{\sqrt{\pi}}{2} \, , \quad \Gamma(1) = 1 \, .$$

Other integrals

$$\int_{-\infty}^{\infty} \mathrm{e}^{-y^2} \, \mathrm{d}y = \sqrt{\pi} \tag{XX.2}$$

$$E(n) = \int_{-\infty}^{\infty} \mathrm{e}^{-ax^2} x^n \, \mathrm{d}x \qquad a > 0 \, , \tag{XX.3}$$

$$E(n) = 0 \qquad \text{for } n \text{ odd,}$$

$$E(n) = 2\int_0^\infty \mathrm{e}^{-ax^2} x^n \, \mathrm{d}x = \frac{\Gamma\left(\frac{n+1}{2}\right)}{a^{\left(\frac{n+1}{2}\right)}} \qquad \text{for } n \text{ even.}$$

M. Moisan, J. Pelletier, *Physics of Collisional Plasmas*,
DOI 10.1007/978-94-007-4558-2,

Noteworthy values

$$E(n) = \frac{1}{2}\Gamma\left(\frac{n+1}{2}\right) a^{-\frac{n+1}{2}} ,$$

$$E(0) = \frac{1}{2}\Gamma\left(\frac{1}{2}\right) a^{-\frac{1}{2}} = \frac{\sqrt{\pi}}{2} a^{-\frac{1}{2}} , \qquad E(1) = \frac{1}{2}\Gamma(1)\, a^{-1} = \frac{1}{2} a^{-1} ,$$

$$E(2) = \frac{1}{2}\Gamma\left(\frac{3}{2}\right) a^{-\frac{3}{2}} = \frac{\sqrt{\pi}}{4} a^{-\frac{3}{2}} , \qquad E(3) = \frac{1}{2}\Gamma(2)\, a^{-2} = \frac{1}{2} a^{-2},$$

$$E(4) = \frac{1}{2}\Gamma\left(\frac{5}{2}\right) a^{-\frac{5}{2}} = \frac{3\sqrt{\pi}}{8} a^{-\frac{5}{2}} , \qquad E(5) = \frac{1}{2}\Gamma(3)\, a^{-3} = a^{-3} .$$

Expression for the differential operators in an arbitrary coordinate system (orthogonal, rectilinear and curvilinear coordinates)

If x^1, x^2 and x^3 are the system coordinates and e_1, e_2, and e_3 are the (local) scale factors, we can express the differential operators in the following manner:

- The gradient

$$\boldsymbol{\nabla} = \left(\frac{1}{e_1}\partial_1,\ \frac{1}{e_2}\partial_2,\ \frac{1}{e_3}\partial_3\right) , \tag{XX.4}$$

  where $\partial_i \equiv \partial/\partial x^i$.

- The curl of a vector

$$\boldsymbol{\nabla} \wedge \boldsymbol{A} = \left(\frac{1}{e_2 e_3}(\partial_2 e_3 A_3 - \partial_3 e_2 A_2),\ \frac{1}{e_3 e_1}(\partial_3 e_1 A_1 - \partial_1 e_3 A_3),\ \frac{1}{e_1 e_2}(\partial_1 e_2 A_2 - \partial_2 e_1 A_1)\right) . \tag{XX.5}$$

- The divergence of a vector

$$\boldsymbol{\nabla} \cdot \boldsymbol{A} = \frac{1}{e_1 e_2 e_3}(\partial_1 e_2 e_3 A_1 + \partial_2 e_3 e_1 A_2 + \partial_3 e_1 e_2 A_3) . \tag{XX.6}$$

- The Laplacian of a scalar

$$\Delta\phi = \frac{1}{e_1 e_2 e_3}\left(\partial_1 \frac{e_2 e_3}{e_1}\partial_1\phi + \partial_2 \frac{e_3 e_1}{e_2}\partial_2\phi + \partial_3 \frac{e_1 e_2}{e_3}\partial_3\phi\right) . \tag{XX.7}$$

- The Laplacian of a vector in Cartesian coordinates

$$\Delta \boldsymbol{A} = \Delta A_1 \hat{\mathbf{e}}_1 + \Delta A_2 \hat{\mathbf{e}}_2 + \Delta A_3 \hat{\mathbf{e}}_3 \ . \tag{XX.8}$$

Quite generally:

$$\Delta \boldsymbol{A} = \boldsymbol{\nabla}(\boldsymbol{\nabla} \cdot \boldsymbol{A}) - \boldsymbol{\nabla} \wedge \boldsymbol{\nabla} \wedge \boldsymbol{A} \ . \tag{XX.9}$$

These operators have the following properties:

- The curl of a gradient is zero

$$(\boldsymbol{\nabla} \wedge \boldsymbol{\nabla}\phi)_1 = \frac{1}{e_2 e_3}\left(\partial_2 e_3 \frac{1}{e_3}\partial_3\phi - \partial_3 e_2 \frac{1}{e_2}\partial_2\phi\right) = \frac{1}{e_2 e_3}(\partial_2\partial_3 - \partial_3\partial_2) = 0 \ , \tag{XX.10}$$

and similarly for the two other components.

- The divergence of the curl of a vector is zero

$$(\boldsymbol{\nabla} \cdot \boldsymbol{\nabla} \wedge \boldsymbol{A}) = \frac{1}{e_1 e_2 e_3}\left(\partial_1 e_2 e_2 \frac{1}{e_2 e_3}(\partial_2 e_3 A_3 - \partial_3 e_2 A_2)\right.$$
$$\left. + \partial_2 e_3 e_1 \frac{1}{e_3 e_1}(\partial_3 e_1 A_1 - \partial_1 e_3 A_3) + \partial_3 e_1 e_2 \frac{1}{e_1 e_2}(\partial_1 e_2 A_2 - \partial_2 e_1 A_1)\right) = 0 \ . \tag{XX.11}$$

- The divergence of the gradient of a scalar is equal to the Laplacian

$$\boldsymbol{\nabla} \cdot \boldsymbol{\nabla}\phi = \frac{1}{e_1 e_2 e_3}\left(\partial_1 \frac{e_2 e_3}{e_1}\partial_1\phi + \partial_2 \frac{e_3 e_1}{e_2}\partial_2\phi + \partial_3 \frac{e_1 e_2}{e_3}\partial_3\right)\phi = \Delta\phi \ . \tag{XX.12}$$

Curvilinear, orthogonal, coordinate systems and scale factors

Cartesian coordinates

The variables are $x^1 = x$, $x^2 = y$ and $x^3 = z$ and the scale factors are $e_1 = e_2 = e_3 = 1$.

$$\boldsymbol{\nabla}\phi = \frac{\partial\phi}{\partial x}\hat{\mathbf{e}}_x + \frac{\partial\phi}{\partial y}\hat{\mathbf{e}}_y + \frac{\partial\phi}{\partial z}\hat{\mathbf{e}}_z \tag{XX.13}$$

$$\boldsymbol{\nabla} \cdot \boldsymbol{A} = \frac{\partial A_x}{\partial x} + \frac{\partial A_y}{\partial y} + \frac{\partial A_z}{\partial z} \tag{XX.14}$$

$$\boldsymbol{\nabla} \wedge \boldsymbol{A} = \left(\frac{\partial A_z}{\partial y} - \frac{\partial A_y}{\partial z}\right)\hat{\mathbf{e}}_x + \left(\frac{\partial A_x}{\partial z} - \frac{\partial A_z}{\partial x}\right)\hat{\mathbf{e}}_y + \left(\frac{\partial A_y}{\partial x} - \frac{\partial A_x}{\partial y}\right)\hat{\mathbf{e}}_z \tag{XX.15}$$

$$\Delta \phi = \frac{\partial^2 \phi}{\partial x^2} + \frac{\partial^2 \phi}{\partial y^2} + \frac{\partial^2 \phi}{\partial z^2} \tag{XX.16}$$

$$\begin{aligned}\Delta \boldsymbol{A} = &\left(\frac{\partial^2 A_x}{\partial x^2} + \frac{\partial^2 A_x}{\partial y^2} + \frac{\partial^2 A_x}{\partial z^2}\right) \hat{\mathbf{e}}_x + \left(\frac{\partial^2 A_y}{\partial x^2} + \frac{\partial^2 A_y}{\partial y^2} + \frac{\partial^2 A_y}{\partial z^2}\right) \hat{\mathbf{e}}_y \\ &+ \left(\frac{\partial^2 A_z}{\partial x^2} + \frac{\partial^2 A_z}{\partial y^2} + \frac{\partial^2 A_z}{\partial z^2}\right) \hat{\mathbf{e}}_z\end{aligned} \tag{XX.17}$$

$$\begin{aligned}(\boldsymbol{A} \cdot \boldsymbol{\nabla})\boldsymbol{B} = &\left(A_x \frac{\partial B_x}{\partial x} + A_y \frac{\partial B_x}{\partial y} + A_z \frac{\partial B_x}{\partial z}\right) \hat{\mathbf{e}}_x \\ &+ \left(A_x \frac{\partial B_y}{\partial x} + A_y \frac{\partial B_y}{\partial y} + A_z \frac{\partial B_y}{\partial z}\right) \hat{\mathbf{e}}_y \\ &+ \left(A_x \frac{\partial B_z}{\partial x} + A_y \frac{\partial B_z}{\partial y} + A_z \frac{\partial B_z}{\partial z}\right) \hat{\mathbf{e}}_z\end{aligned} \tag{XX.18}$$

Cylindrical coordinates

The variables are $x^1 = r$, $x^2 = \theta$ and $x^3 = z$ and the scale factors are $e_1 = 1$, $e_2 = r$, $e_3 = 1$.

$$\boldsymbol{\nabla}\phi = \frac{\partial \phi}{\partial r}\hat{\mathbf{e}}_r + \frac{1}{r}\frac{\partial \phi}{\partial \theta}\hat{\mathbf{e}}_\theta + \frac{\partial \phi}{\partial z}\hat{\mathbf{e}}_z \tag{XX.19}$$

$$\boldsymbol{\nabla} \cdot \boldsymbol{A} = \frac{1}{r}\frac{\partial (r A_r)}{\partial r} + \frac{1}{r}\frac{\partial A_\theta}{\partial \theta} + \frac{\partial A_z}{\partial z} \tag{XX.20}$$

$$\begin{aligned}\boldsymbol{\nabla} \wedge \boldsymbol{A} = &\left(\frac{1}{r}\frac{\partial A_z}{\partial \theta} - \frac{\partial A_\theta}{\partial z}\right) \hat{\mathbf{e}}_r + \left(\frac{\partial A_r}{\partial z} - \frac{\partial A_z}{\partial r}\right) \hat{\mathbf{e}}_\theta \\ &+ \frac{1}{r}\left(\frac{\partial (r A_\theta)}{\partial r} - \frac{\partial A_r}{\partial \theta}\right) \hat{\mathbf{e}}_z\end{aligned} \tag{XX.21}$$

$$\Delta \phi = \frac{1}{r}\frac{\partial}{\partial r}\left(r \frac{\partial \phi}{\partial r}\right) + \frac{1}{r^2}\frac{\partial^2 \phi}{\partial \theta^2} + \frac{\partial^2 \phi}{\partial z^2} \tag{XX.22}$$

$$\begin{aligned}\Delta \boldsymbol{A} = &\left(\Delta A_r - \frac{2}{r^2}\frac{\partial A_\theta}{\partial \theta} - \frac{A_r}{r^2}\right) \hat{\mathbf{e}}_r \\ &+ \left(\Delta A_\theta - \frac{2}{r^2}\frac{\partial A_r}{\partial \theta} - \frac{A_\theta}{r^2}\right) \hat{\mathbf{e}}_\theta + \Delta A_z \hat{\mathbf{e}}_z\end{aligned} \tag{XX.23}$$

$$(\boldsymbol{A}\cdot\boldsymbol{\nabla})\boldsymbol{B} = \left(A_r\frac{\partial B_r}{\partial r} + \frac{A_\theta}{r}\frac{\partial B_r}{\partial \theta} + A_z\frac{\partial B_r}{\partial z} - \frac{A_\theta B_\theta}{r}\right)\hat{\mathbf{e}}_r$$
$$+\left(A_r\frac{\partial B_\theta}{\partial r} + \frac{A_\theta}{r}\frac{\partial B_\theta}{\partial \theta} + A_z\frac{\partial B_\theta}{\partial z} + \frac{A_\theta B_r}{r}\right)\hat{\mathbf{e}}_\theta$$
$$+\left(A_r\frac{\partial B_z}{\partial r} + \frac{A_\theta}{r}\frac{\partial B_z}{\partial \theta} + A_z\frac{\partial B_z}{\partial z} +\right)\hat{\mathbf{e}}_z \qquad \text{(XX.24)}$$

Spherical coordinates

The variables are $x^1 = r$, $x^2 = \theta$ and $x^3 = \varphi$ and scale factors are $e_1 = 1$, $e_2 = r$, $e_3 = r\sin\theta$.

$$\boldsymbol{\nabla}\phi = \frac{\partial \phi}{\partial r}\hat{\mathbf{e}}_r + \frac{1}{r}\frac{\partial \phi}{\partial \theta}\hat{\mathbf{e}}_\theta + \frac{1}{r\sin\theta}\frac{\partial \phi}{\partial \varphi}\hat{\mathbf{e}}_\varphi \qquad \text{(XX.25)}$$

$$\boldsymbol{\nabla}\cdot\boldsymbol{A} = \frac{1}{r^2}\frac{\partial(r^2 A_r)}{\partial r} + \frac{1}{r\sin\theta}\frac{\partial(A_\theta\sin\theta)}{\partial\theta} + \frac{1}{r\sin\theta}\frac{\partial A_\varphi}{\partial\varphi} \qquad \text{(XX.26)}$$

$$\boldsymbol{\nabla}\wedge\boldsymbol{A} = \frac{1}{r\sin\theta}\left(\frac{\partial(A_\varphi\sin\theta)}{\partial\theta} - \frac{\partial A_\theta}{\partial\varphi}\right)\hat{\mathbf{e}}_r + \frac{1}{r}\left(\frac{1}{\sin\theta}\frac{\partial A_r}{\partial\varphi} - \frac{\partial(rA_\varphi)}{\partial r}\right)\hat{\mathbf{e}}_\theta$$
$$+\frac{1}{r}\left(\frac{\partial(rA_\theta)}{\partial r} - \frac{\partial A_r}{\partial\theta}\right)\hat{\mathbf{e}}_\varphi \qquad \text{(XX.27)}$$

$$\Delta\phi = \frac{1}{r^2}\frac{\partial}{\partial r}\left(r^2\frac{\partial\phi}{\partial r}\right) + \frac{1}{r^2\sin\theta}\frac{\partial}{\partial\theta}\left(\sin\theta\frac{\partial\phi}{\partial\theta}\right) + \frac{1}{r^2\sin^2\theta}\frac{\partial^2\phi}{\partial\varphi^2} \qquad \text{(XX.28)}$$

$$\Delta\boldsymbol{A} = \left(\Delta A_r - \frac{2A_r}{r^2} - \frac{2}{r^2}\frac{\partial A_\theta}{\partial\theta} - \frac{2A_\theta\cot\theta}{r^2} - \frac{2}{r^2\sin\theta}\frac{\partial A_\varphi}{\partial\varphi}\right)\hat{\mathbf{e}}_r$$
$$+\left(\Delta A_\theta + \frac{2}{r^2}\frac{\partial A_r}{\partial\theta} - \frac{A_\theta}{r^2\sin^2\theta} - \frac{2\cos\theta}{r^2\sin^2\theta}\frac{\partial A_\varphi}{\partial\varphi}\right)\hat{\mathbf{e}}_\theta$$
$$+\left(\Delta A_\varphi - \frac{A_\varphi}{r^2\sin^2\theta} + \frac{2}{r^2\sin\theta}\frac{\partial A_r}{\partial\varphi} + \frac{2\cos\theta}{r^2\sin^2\theta}\frac{\partial A_\theta}{\partial\varphi}\right)\hat{\mathbf{e}}_\varphi \qquad \text{(XX.29)}$$

$$
\begin{aligned}
(\boldsymbol{A}\cdot\boldsymbol{\nabla})\boldsymbol{B} &= \left(A_r\frac{\partial B_r}{\partial r} + \frac{A_\theta}{r}\frac{\partial B_r}{\partial\theta} + \frac{A_\varphi}{r\sin\theta}\frac{\partial B_r}{\partial\varphi} - \frac{A_\theta B_\theta + A_\varphi B_\varphi}{r}\right)\hat{\mathbf{e}}_r \\
&+ \left(A_r\frac{\partial B_\theta}{\partial r} + \frac{A_\theta}{r}\frac{\partial B_\theta}{\partial\theta} + \frac{A_\varphi}{r\sin\theta}\frac{\partial B_\theta}{\partial\varphi} + \frac{A_\theta B_r}{r} - \frac{A_\varphi B_\varphi\cot\theta}{r}\right)\hat{\mathbf{e}}_\theta \\
&+ \left(A_r\frac{\partial B_\varphi}{\partial r} + \frac{A_\theta}{r}\frac{\partial B_\varphi}{\partial\theta} + \frac{A_\varphi}{r\sin\theta}\frac{\partial B_\varphi}{\partial\varphi} + \frac{A_\varphi B_r}{r} + \frac{A_\varphi B_\theta\cot\theta}{r}\right)\hat{\mathbf{e}}_\varphi
\end{aligned}
\tag{XX.30}
$$

References

[1] Allis, W.P., Rose, D.J., *Phys. Rev.* **93**, 84 (1954)
[2] Allis, W.P., *Handbuch der Physik* **21**, 383 (1956) (Springer Verlag, Berlin)
[3] Calzada, M.D., Moisan, M., Gamero, A., Sola, A., *J. Appl. Phys.* **80**, 46 (1996)
[4] Castaños Martìnez, E., Kabouzi, Y., Makasheva, K., Moisan, M., *Phys. Rev. E* **70**, 066405 (2004)
[5] Castaños Martìnez, E., Moisan, M., *52nd Annual Technical Conference Proceedings of the Society of Vacuum Coaters*, 333 (2009)
[6] Castaños Martìnez, E., Moisan, M., Kabouzi, Y., *J. Phys. D: Appl. Phys.* **42**, 012003 (2009)
[7] Castaños Martìnez, E., Moisan, M., *IEEE Trans. Plasma Sci.* **39**, 2192 (2011)
[8] Claude, R., Moisan, M., Wertheimer, M.R., Zakrzewski, Z., *Plasma Chem. Plasma Proc.* **7**, 451 (1987)
[9] Cramer, W.H., Simons, J.H., *J. Chem. Phys.* **26**, 1272 (1957)
[10] Druyvesteyn, M.J., Penning, F.M., *Rev. Mod. Phys.* **12**, 87 (1940)
[11] Ferreira, C.M., Loureiro, J., *J. Phys. D: Appl. Phys.* **16**, 2471 (1983)
[12] Fozza, A.C., Moisan, M., Wertheimer, M.R., *J. Appl. Phys.* **88**, 20 (2000)
[13] Golubovskii, Yu.B., Sonneburg, R., *Zh. Tekh. Fiz.* **49**, 295 (1979) [*Sov. Phys.Tech. Phys.* **24**, 179 (1979)]
[14] Hansen, C., Reimann, A.B., Fajans, J., *Phys. Plasmas* **3**, 1820 (1996)
[15] Kabouzi, Y., Calzada, M.D., Moisan, M., Tran, K.C., Trassy, C., *J. Appl. Phys.* **91**, 1008 (2002)
[16] Koleva, I., Makasheva, K., Paunska, Ts., Schlueter, H., Shivarova, A., Tarnev, Kh., *Contrib. Plasma Phys.* **44** 552 (2004)
[17] Makasheva, K., Shivarova, A., *Phys. Plasmas* **8**, 836 (2001)
[18] Margot-Chaker, J., Moisan, M., Chaker, M., Glaude, V.M.M., Lauque, P., Paraszczak, J. and Sauvé, G., *J. Appl. Phys.* **66** 4134 (1989)
[19] Margot, J., Moisan, M., *J. Phys. D: Appl. Phys.* **24**, 1765 (1991)
[20] Margot, J., Moisan, M., chapter 8 of [26]
[21] McDaniel, E.W., *Collision phenomena in ionized gases* (Wiley, 1964)

M. Moisan, J. Pelletier, *Physics of Collisional Plasmas*,
DOI 10.1007/978-94-007-4558-2,

[22] Microwave discharges: fundamentals and applications, edited by C.M. Ferreira, M. Moisan, *NATO ASI series B: Physics* **302** (Plenum, New York, 1993)

[23] Moisan, M., Zakrzewski, Z., in *Radiative Processes in Discharge Plasmas*, edited by J.M. Proud, L.H. Luessen (Plenum, New York, 1986) p. 381

[24] Moisan, M., Barbeau, C., Claude, R., Ferreira, C.M., Margot, J., Paraszczak, J., Sá, A.B., Sauvé, G., Wertheimer, M.R., *J. Vac. Sci. Technol.* **B9**, 8 (1991)

[25] Moisan, M., Zakrzewski, Z., *J. Phys. D: Appl. Phys.* **24**, 1025 (1991)

[26] Moisan, M., Pelletier, J., *Microwave excited plasmas* (Elsevier, Amsterdam, 1992)

[27] Nachtrieb, R., Khan, F., Waymouth, J.F., *J. Phys. D: Appl. Phys.* **38**, 3226–3236 (2005)

[28] Nickel, J.C., Parker, J.V., Gould, R.W., *Phys. Rev. Lett.* **11**, 183 (1963)

[29] Royal, J., Orel, A.E., *Phys. Rev. A* **73**, 042706 (2006)

[30] Pollak, J., Moisan, M., Zakrzewski, Z., *Plasma Sources Sci. Technol.* **16**, 310 (2007)

[31] Ramsauer, C., Kollath, R., *Ann. Phys.* **12**, 527 (1932)

[32] Rose, D.J., Brown, S.C., *Phys. Rev.* **98**, 310 (1955)

[33] Sá, P.A., Loureiro, J., Ferreira, C.M., *J. Phys. D: Appl. Phys.* **25**, 960 (1992). Also see C.M. Ferreira, M. Moisan, chapter 3 of [26]

[34] Sá, P.A., Loureiro, J., Ferreira, C.M., *J. Phys. D: Appl. Phys.* **27**, 1171 (1994)

[35] Sauvé, G., Moisan, M., Paraszczak, J., Heidenreich, J., *Appl. Phys. Lett.* **53**, 470 (1988)

[36] Schlüter, H., Shivarova, A., *Physics Reports* **443**, 121 (2007)

[37] Simons, J.H., Fontana, C.M., Muschlitz Jr., E.E., Jackson, S.R., *J. Chem. Phys.* **11**, 307 (1943)

[38] Simons, J.H., Fontana, C.M., Francis, H.T., Unger, L.G., *J. Chem. Phys.* **11**, 312 (1943)

[39] Simon, A., *Phys. Rev.* **98**, 317 (1955)

[40] Vikharev, A.L., Ivanov, O.A., Kolysko, A.L., *Technical Physics Letters*, **22**, 832 (1996)

[41] Winkler, R., Deutsch, H., Wilhelm, J., Wilke, Ch., *Beitr. Plasmaphys.* **24**, 285 (1984)

[42] Winkler, R., Wilhelm, J., Hess, A., *Annalen der Physik* **42**, 537 (1985)

[43] Zakrzewski, Z., Moisan, M., Sauvé, G., chapter 4 of [26]

Recommended Reading

General treatises on plasma physics

- Badareu, E., Popescu, E., *Gaz ionisés* (Dunod, 1965)
- Delcroix, J.L., *Introduction à la théorie des gaz ionisés* (Dunod, 1959)
- Delcroix, J.L., *Physique des plasmas*, T. I et II (Monographie Dunod, 1966)
- Delcroix, J.L., Bers, A., *Physique des plasmas*, T. 1 et T. 2 (Interéditions-CNRS, 1994)
- Held, B., *Physique des plasmas froids* (Masson, 1994)
- Moisan, M., Pelletier, J., *Physique des plasmas collisionnels* (EDP Sciences, 2006)
- Papoular, R., *Phénomènes électriques dans les gaz* (Dunod, 1963)
- Quémada, D., *Gaz et plasmas*, chapters 4 and 5 in *Traité d'électricité : l'électricité et la matière*, tome III, edited by G. Goudet (Masson, 1975)
- Rax, J.M., *Physique des plasmas* (Dunod, 2005)

- Chen, F.F., *Introduction to plasma physics and controlled fusion*, volume 1 (Plenum, 1984)
- Golant, V.E., Zhilinsky, A.P., Sakharov, I.E., *Fundamentals of plasma physics*, edited by S.C. Brown (Wiley, 1977)
- Gurnett, D.A., Bhattacharjee, A., *Introduction to plasma physics* (Cambridge University Press, 2005)
- Jancel, R., Kahan, Th., *Electrodynamics of Plasmas* (John Wiley and Sons, 1966)
- Lieberman, M.A., Lichtenberg, A.J., *Principles of plasma discharges and materials processing*, Second Edition (Wiley, 2005)
- Nishikawa, K., Wakatani, M., *Plasma physics* (Springer, 2000)
- Raizer, Yu.P., *Gas discharge physics* (Springer, 1997)
- Seshadri, S.R., *Fundamentals of plasma physics* (Elsevier, 1973)
- von Engel, A., *Ionized gases* (Clarendon Press, 1957)
- von Engel, A., *Electric plasmas: Their nature and uses* (Taylor and Francis, 1983)

M. Moisan, J. Pelletier, *Physics of Collisional Plasmas*,
DOI 10.1007/978-94-007-4558-2,

Treatises on specific topics of plasma physics

- Aliev, Yu.M., Schlüter, H., Shivarova, A., *Guided-wave-produced plasmas* (Springer, 2000)
- Chabert, P., Braithwaite, N., *Physics of Radio-Frequency Plasmas* (Cambridge University Press, 2011)
- Ferreira, C.M., Moisan, M., *Microwave Discharges* (Plenum Press, 1993)
- Geller, R., *Electron Cyclotron Resonance Ion Sources and ECR Plasmas* (IOP Publishing, 1996)
- Kando, M., Nagatsu, M. (eds.) *Proc. of the VIIth Int. Workshop on Microwave discharges: fundamentals and applications* (Shizuoka University, 2009)
- Lebedev, Yu.A. (ed.), *Proc. of the VIth Int. Workshop on Microwave discharges: fundamentals and applications* (Yanus-K, 2006)
- Moisan, M., Pelletier, J., *Microwave excited plasmas* (Elsevier, 1992)
- Popov, O.A., *High Density Plasma Sources* (Noyes Publications, 1995)
- Raizer, Yu.P., Shneider, M.N., Yatsenko, N.A., *Radio-frequency capacitive discharges* (CRC Press, 1995)
- Schlüter, H., Shivrova, A., *Advanced technologies Based on Wave and Beam Generated Plasmas* (Kluwer Academic Publishers, 1999)

- Artsimovitch, L., Loukianov, S., *Mouvement des particules chargées dans des champs électriques et magnétiques* (Ed. de Moscou, 1975)
- Brown, S.C., *Basic data of plasma physics* (MIT, 1959)
- Makabe, T. (ed.), *Advances in low temperature RF plasmas* (North-Holland, 2002)
- Massey, H.S., Burhop, E.H.S., *Electronic and ionic impact phenomena* (Oxford, 1952)
- McDaniel, E.W., *Collision phenomena in ionized gases* (Wiley, 1964)

- Allis, W.P., Buchsbaum, S.J., Bers, A., *Waves in anisotropic plasmas* (MIT, 1963)
- Quémada, D., *Ondes dans les plasmas* (Hermann, 1968)
- Vandenplas, P.E., *Electron waves and resonances in bounded plasmas* (Wiley, 1968)

- Auciello, O., Flamm, D.L. (eds.), *Plasma diagnostics* (Academic Press, 1989)
- Heald, M.A., Wharton, C.B., *Plasma diagnostics with microwaves* (Wiley, 1965)
- Ricard, A., *Reactive plasmas* (Société Française du Vide, 1996)

- Yvon, J., *Les corrélations et l'entropie* (Monographie Dunod, 1966)

- Collection du Commissariat à l'énergie atomique (CEA, France) publiée sous la direction de R. Dautray, *La fusion thermonucléaire contrôlée par confinement magnétique* (Masson, 1987)

Popular treatise on plasmas

- Bradu, P., *L'univers des plasmas* (Flammarion, 2002)

Further related reading

- Brillouin, L., *Les tenseurs en mécanique et en élasticité* (Masson, 1960)
- Landau, L., Lifchitz, E., *Mécanique des fluides* (Mir, 1989)
- Lorrain, P., Corson, D.P., Lorrain, F., *Electromagnetic fields and waves* (Freeman, 1987)

Index

M. Moisan, J. Pelletier, *Physics of Collisional Plasmas*,
DOI 10.1007/978-94-007-4558-2,